An Introduction to Proof Theory

'This book deftly frames the philosophical context of proof theory while developing methods no less technical than other branches of logic, finally giving an answer to the question of where to start with proof theory. It cuts like an arrow through the most essential introductory topics: everything here is necessary and sufficient for the beginner to understand the chief aims of proof theory and to be ready to move into advanced work in the subject.'

Andrew Arana, Université de Lorraine

'Mancosu, Galvan, and Zach have done a remarkable thing: they present seminal results in logic in a way that does justice to both their historical origins and their contemporary relevance. The result is an engaging introduction to proof theory that is at the same time a definitive exposition of Gentzen's towering contributions.'

Jeremy Avigad, Carnegie Mellon University

This is an excellent and much-needed textbook. It is historically sensitive and technically first-rate, and covers exactly what's needed to bring students (and many of their teachers) up to speed in proof theory. Everyone who cares about logic should have it on their shelf.

Patricia Blanchette, University of Notre Dame

An Introduction to Proof Theory

Normalization, Cut-Elimination, and Consistency Proofs

Paolo Mancosu
Sergio Galvan
Richard Zach

OXFORD
UNIVERSITY PRESS

OXFORD

UNIVERSITY PRESS

Great Clarendon Street, Oxford, OX2 6DP,
United Kingdom

Oxford University Press is a department of the University of Oxford.
It furthers the University's objective of excellence in research, scholarship,
and education by publishing worldwide. Oxford is a registered trade mark of
Oxford University Press in the UK and in certain other countries

First Edition published in 2021

Impression: 1

Published in the United States of America by Oxford University Press
198 Madison Avenue, New York, NY 10016, United States of America

British Library Cataloguing in Publication Data
Data available

Library of Congress Control Number: 2021910782

ISBN 978–0–19–289593–6 (hbk.)
ISBN 978–0–19–289594–3 (pbk.)

DOI: 10.1093/oso/9780192895936.001.0001

Printed and bound by
CPI Group (UK) Ltd, Croydon, CR0 4YY

Links to third party websites are provided by Oxford in good faith and
for information only. Oxford disclaims any responsibility for the materials
contained in any third party website referenced in this work.

Contents

Preface

About this book

This book arose out of the desire to enable students, especially those in philosophy, who have only a minimal background in mathematics and logic, to appreciate the most important results of classical proof theory. Most proof theory textbooks begin at a rather demanding level. The more accessible among them only cover structural proof theory, and those that also cover ordinal proof theory assume a strong mathematical background. Unfortunately, there is no good, elementary introduction to proof theory in English that does both. Our book is meant to fill this gap.

We assume that the reader is familiar with both the propositional and predicate calculus, as is covered in most introductory courses on formal logic taught in philosophy departments. We do not assume familiarity with any specific deductive system. Indeed, the first few chapters are for the most part introductions to three of them: axiomatic derivations, natural deduction, and sequent calculus. We also do not assume familiarity with the metatheory of logic, not even the most elementary principle of reasoning about our calculi, namely the principle of induction on the natural numbers ($\mathbb{N}$). We try to ease the reader into the mechanics of inductive proofs, first presenting simple induction on $\mathbb{N}$, then double induction, and finally induction along more complicated well-orderings such as ε_0.

For those who desire to plunge immediately into structural proof theory, Chapter 2 on axiomatic derivations can be skipped. Keep in mind, however, that the principle of induction on $\mathbb{N}$ is explained there (in Section 2.7). Moreover, the chapter introduces the crucial distinctions between minimal, intuitionistic, and classical systems of logic. As is usual in proof theory, our syntax assumes that free and bound variables are distinct syntactic categories. An explanation of this and the difference to the usual approach that uses only a single kind of variable (but distinguishes between free and bound *occurrences*) can also be found there (in Section 2.13).

One of the main goals we set for ourselves is that of providing an introduction to proof theory which might serve as a companion to reading the original articles by Gerhard Gentzen. For this reason, while not slavishly following Gentzen, we never deviate much from his choice of systems and style of proof. This, as we shall see, already determines the choice of systems for our presentation of axiomatic calculi.

In structural proof theory, we cover, among other things, the Gödel-Gentzen translation of classical into intuitionistic logic (and arithmetic), natural deduction and the normalization theorems (for both **NJ** and **NK**), the sequent calculus, including cut-elimination and mid-sequent theorems, and various applications of these results. The second half of the book covers ordinal proof theory, specifically, Gentzen's consistency proof for first-order Peano arithmetic using ordinal induction up to ε_0. The theory of ordinal notations and other elements of ordinal theory are developed from scratch; no knowledge of set theory is presumed.

In order to make the content accessible to readers without much mathematical background, we carry out the details of proofs in much more detail than is usually done. For instance, although we follow Prawitz (1965) in our proof of the sub-formula property for natural deduction (for the system **NJ**), we verify all of the details left out of Prawitz's treatment. In the proof of Gentzen's consistency result, likewise, we prove all the required properties about ordinal notations, carry through all of the required lemmas, and verify that ordinal notations decrease in the reduction procedure in detail. We also work through many examples to illustrate definitions and to show how the proofs work in specific examples.

We've also diverged from the usual way of presenting results so that we can avoid assuming some background. For instance, ordinal notations $< \varepsilon_0$ are defined purely combinatorially, without assuming familiarity with ordinals (although we will also discuss set-theoretic definitions of ordinals and the relation to the combinatorial definition of ordinal notations). This has a philosophical payoff: from the start we emphasize that ordinal notations are not transfinite ordinals, and that the fact that they are well-ordered follows from elementary combinatorial principles about orderings of sequences. In other words, our presentation of ordinal notations is almost finitary.

Our proof of consistency of **PA** also has a significant feature not shared with usual presentations or Gentzen's original. Rather than show that a putative proof of the empty sequent can be transformed into one without induction and cuts on complex formulas, we show this for proofs of any sequent consisting of only atomic formulas. The philosophical payoff is that the consistency proof is formulated from the start as a conservativity result: induction and complex CUTS can be eliminated from proofs of elementary arithmetical facts in **PA**. Since the reduction procedure now applies to actually existing proofs (and not just to putative, and in fact non-existent, proofs of contradictions) we can actually present examples of how the procedure works. So this way of doing things also has a pedagogical payoff.

The book will be of interest to philosophers, logicians, mathematicians, computer scientists, and linguists. Through this material, philosophy students will acquire the tools required for tackling further topics in the philosophy of mathematics (such as the prospects for Hilbert's program and its relativized versions), and in the philosophy of logic and language (the meaning of the logical constants; proof-theoretic semantics; realism/anti-realism, Dummett's program, proof-theoretic harmony etc.).

For further reading

1. Gentzen's articles are translated in:

   Gentzen, Gerhard (1969), *The Collected Papers of Gerhard Gentzen*, ed. by Manfred E. Szabo, Amsterdam: North-Holland.

2. For a detailed overview of the history of mathematical logic from Russell to Gentzen, see:

   Mancosu, Paolo, Richard Zach, and Calixto Badesa (2009), "The development of mathematical logic from Russell to Tarski: 1900–1935," in *The Development of Modern Logic*, ed. by Leila Haaparanta, New York and Oxford: Oxford University Press, pp. 324–478, DOI: 10.1093/acprof:oso/9780195137316.003.0029.

3. For the philosophical background of intuitionism and Hilbert's program:

   Iemhoff, Rosalie (2020), "Intuitionism in the philosophy of mathematics," in *The Stanford Encyclopedia of Philosophy*, ed. by Edward N. Zalta, Fall 2020, https://plato.stanford.edu/entries/intuitionism/.
   Mancosu, Paolo, ed. (1998), *From Brouwer to Hilbert: The Debate on the Foundations of Mathematics in the 1920s*, New York and Oxford: Oxford University Press.
   Zach, Richard (2019), "Hilbert's program," in *Stanford Encyclopedia of Philosophy*, ed. by Edward N. Zalta, Fall 2019, https://plato.stanford.edu/archives/fall2019/entries/hilbert-program/.

4. For an overview of the history of proof theory up to Gentzen:

   Hendricks, Vincent F., Stig Andur Pedersen, and Klaus Frovin Jørgensen, eds. (2000), *Proof Theory: History and Philosophical Significance*, Dordrecht: Springer, DOI: 10.1007/978-94-017-2796-9.
   Von Plato, Jan (2018), "The development of proof theory," in *The Stanford Encyclopedia of Philosophy*, ed. by Edward N. Zalta, https://plato.stanford.edu/archives/win2018/entries/proof-theory-development/.

5. On Gentzen's life and career:

   Menzler-Trott, Eckart (2016), *Logic's Lost Genius: The Life of Gerhard Gentzen*, History of Mathematics, 33, American Mathematical Society.

6. For a more technical overview of Gentzen's contributions to logic:

Von Plato, Jan (2009), "Gentzen's logic," in *Logic from Russell to Church*, ed. by Dov M. Gabbay and John Woods, Handbook of the History of Logic, 5, Amsterdam: North-Holland, pp. 667–721, DOI: 10.1016/S1874-5857(09)70017-2.

7. Some introductory surveys on proof theory:

Prawitz, Dag (1971), "Ideas and results in proof theory," in *Proceedings of the Second Scandinavian Logic Symposium*, ed. by Jens E. Fenstad, Studies in Logic and the Foundations of Mathematics, 63, Amsterdam: North-Holland, pp. 235–307, DOI: 10.1016/S0049-237X(08)70849-8.

Rathjen, Michael and Wilfried Sieg (2020), "Proof theory," in *The Stanford Encyclopedia of Philosophy*, ed. by Edward N. Zalta, Fall 2020, https://plato.stanford.edu/archives/fall2020/entries/proof-theory/.

8. More advanced textbooks in proof theory:

Bimbó, Katalin (2014), *Proof Theory: Sequent Calculi and Related Formalisms*, Boca Raton: CRC Press.

Buss, Samuel R., ed. (1998), *Handbook of Proof Theory*, Amsterdam: Elsevier.

Diller, Justus (2019), *Functional Interpretations: From the Dialectica Interpretation to Functional Interpretations of Analysis and Set Theory*, World Scientific.

Girard, Jean-Yves (1987), *Proof Theory and Logical Complexity*, Studies in Proof Theory, 1, Naples: Bibliopolis.

Negri, Sara and Jan von Plato (2001), *Structural Proof Theory*, Cambridge: Cambridge University Press.

Pohlers, Wolfram (2009), *Proof Theory: The First Step into Impredicativity*, Berlin, Heidelberg: Springer, DOI: 10.1007/978-3-540-69319-2.

Prawitz, Dag (1965), *Natural Deduction: A Proof-Theoretical Study*, Stockholm Studies in Philosophy, 3, Stockholm: Almqvist & Wiksell.

Schütte, Kurt (1977), *Proof Theory*, Berlin and New York: Springer.

Takeuti, Gaisi (1987), *Proof Theory*, 2nd ed., Studies in Logic, 81, Amsterdam: North-Holland.

Troelstra, Anne Sjerp and Helmut Schwichtenberg (2000), *Basic Proof Theory*, 2nd ed., Cambridge: Cambridge University Press.

Acknowledgments

This book arose out of courses and seminars the authors have held over the years at Berkeley, Calgary, and Milan. We would like to thank our students for their valuable comments. We are especially grateful to James Walsh and Logan Heath who have commented on the entire text, often catching important mistakes, and for checking all the exercises. We would like to thank Luca Bellotti, Enrico Moriconi, Dario Palladino, Francesca Poggiolesi, and David Schrittesser for their comments on different sections of the book, and Yuuki Andou, Dag Prawitz, and Peter Schröder-Heister for answering our questions. Special thanks goes to Jeremy Avigad, who as reviewer for Oxford University Press provided thorough and constructive comments.

The order of the co-authors on the cover of the book only reflects the order in which they joined the project.

1

Introduction

1.1 Hilbert's consistency program

While reflection on mathematical proof goes as far back as the time of the ancient Greeks, it reached the late nineteenth century with the detailed formalization of mathematical proofs given first by Frege and Peano and later by Russell and Whitehead, and others. A mathematical analysis of proofs considered as mathematical objects only truly began with David Hilbert and his school.

David Hilbert's work in logic originated from his interest in foundational questions and in particular in the foundations of geometry (Hilbert, 1899). The work on the foundations of geometry led Hilbert to a conception of mathematical theories as formal systems that could be given varying interpretations, by choosing the domain of objects at will and assigning to the basic predicates arbitrary meanings constrained only by the requirement that the axioms of the theory be satisfied in the given interpretation. Thus, a basic predicate such as $L(x)$ that in the standard interpretation could be read as "x is a line", and thus having its intended meaning as the set of lines in the Euclidean plane, could be re-interpreted as a set of pairs of real numbers in an arithmetical interpretation of the theory. The role of predicates in this context is formal in the sense that they are independent of any specific interpretation, i.e., the axioms allow for different interpretations. However, at this early stage Hilbert did not work with a formalized logic. Rather, logic is assumed informally in the background of his axiomatic studies. In particular, the logical constants (connectives, quantifiers, etc.) are not treated formally, i.e., as mere symbols devoid of an interpretation, but rather have the standard informal meaning. Only when he became acquainted with Whitehead and Russell's *Principia Mathematica* (Whitehead and Russell, 1910–1913) did Hilbert use formalized logic. The influence of *Principia Mathematica* and Hilbert's thorough engagement with logic is very much in evidence in the set of lecture notes for the winter semester 1917/1918 titled "The principles of mathematics" (Hilbert, 1918). This set of lectures forms the basis of the later textbook *Principles of Theoretical Logic* (Hilbert and Ackermann, 1928). They contain a wealth of material on logic including

An Introduction to Proof Theory: Normalization, Cut-Elimination, and Consistency Proofs.
Paolo Mancosu, Sergio Galvan, and Richard Zach, Oxford University Press. © Paolo Mancosu,
Sergio Galvan and Richard Zach 2021. DOI: 10.1093/oso/9780192895936.003.0001

propositional, predicate and higher-order logic (namely, the type theory of Russell and Whitehead). Along with the beginning of what we call metatheory (studies of consistency and completeness for propositional logic, independence of the axioms etc.), the lectures are important because Hilbert was now able to formalize mathematical theories completely, i.e., not only in their non-logical parts (say, axioms for arithmetic, geometry or set theory) but in their logical parts as well. Hilbert's assistant Paul Bernays further developed these investigations in his 1918 dissertation (Bernays, 1918), where a clear distinction between syntax and semantics is formulated. Here, we find all the key elements of our contemporary understanding of propositional logic: a formal system, a semantic interpretation in terms of truth-values and a proof of soundness and completeness relative to that semantics. In addition, Bernays also showed that propositional logic was decidable by using normal forms.

The decidability of mathematics (or fragments thereof) was an important question for Hilbert. The roots of Hilbert's engagement with decidability can be traced as far back as 1900, although at the time he had no idea of what technical tools would have to be enlisted to address such a task. In his 1900 address to the International Congress of Mathematicians held in Paris, Hilbert claimed that every mathematical problem has a solution:

This conviction of the solvability of every mathematical problem is a powerful incentive to the worker. We hear within us the perpetual call: There is the problem. Seek its solution. You can find it by pure reason, for in mathematics there is no *ignorabimus*. (Hilbert, 1900, p. 1102)[1]

In "Axiomatic Thought" (Hilbert, 1917, p. 1113), the problem of "the *decidability* of a mathematical question in a finite number of operations" is listed as one of the fundamental problems for the axiomatic method. By now, the technical tools for addressing such issues were beginning to be developed, as witnessed by the 1917/18 lecture notes and other contemporary work by Löwenheim, Behmann, and others.[2]

Another problem that Hilbert had raised at the Paris Congress was that of the consistency of mathematics. The consistency of various systems of geometry (Euclidean and non-Euclidean) could be shown by reduction to arithmetical theories (broadly construed as to include the real numbers and some set theory) but the problem was pressing for the arithmetical theories themselves. This points to the second element, in addition to Hilbert's discovery of *Principia Mathematica*, that led to his engagement with fully formalized theories and a renewed interest in foundations. That element is the foundational debate that was forcefully brought to a crisis point by the radical proposals of Hermann Weyl and L. E. J. Brouwer. Weyl's and Brouwer's proposals can be seen as a reaction to the way classical

---

[1] Page references are to the translation(s) listed in the bibliography.
[2] On the origin and development of the decision problem see Mancosu and Zach (2015).

mathematics was developing, in particular to the non-constructive and infinitary turn it had taken since the late nineteenth century.

In *Das Kontinuum* (1918), Weyl criticized set theory and classical analysis as "a house built on sand" and proposed a remedy to the uncertainty that in his opinion affected a large part of classical mathematics. In the first part of the book, Weyl proposed a number of criticisms to contemporary analysis and set theory, but also to some of the accepted reconstructions of classical mathematics, such as Zermelo's set theory and Russell's theory of types. In the positive part, Weyl developed his own arithmetical continuum, a number system in which large portions of analysis can be carried out. This foundational position is called "predicativity."[3]

In 1921, just three years after the publication of *Das Kontinuum*, Weyl discovered the new foundational proposal that had meanwhile been championed by Brouwer: intuitionism. This was the most radical proposal in the foundations of mathematics. It was radical in that it urged an abandonment of the logical principle of the excluded middle for infinite totalities, the abandonment of non-constructive mathematics and thus most of infinitary mathematics. The costs to be paid were high: the intuitionistic reconstruction of mathematics had to sacrifice a great deal of classical mathematics. However, this did not dissuade Weyl, who spoke of Brouwer as *"die Revolution"*. Weyl joined Brouwer's critique of classical mathematics, and abandoned his previous "predicativist" position, mainly on epistemological grounds.

As a reaction to the threat that the Brouwerian and Weylean positions posed to classical mathematics, Hilbert initiated a far-reaching program in the foundations of mathematics. Developed with the help of Paul Bernays, who had joined Hilbert in 1917 and worked with him throughout the twenties and thirties, Hilbert's approach rejected the revisionist approaches to mathematics defended by Brouwer and Weyl and aimed at a foundation of mathematics that would guarantee the certainty of mathematics without giving up any part of classical mathematics. The branch of mathematical logic known as proof theory, of which this book presents the basic elements, originated with this foundational program.

Already in 1905, Hilbert had the remarkable insight that we can "consider the proof itself to be a mathematical object." However, at the time he did not try to spell out the logic and the logical language underlying the mathematical theories—he only referred to "the familiar modes of logical inference"—and even a clear

---

[3] The key aspect of Weyl's proposal is the restriction of quantification to individuals in the domain. In particular, Weyl excluded quantification over subsets of individuals of the domain, thereby blocking what are called "impredicative" definitions. An impredicative definition is one that defines a set X by quantifying over a collection of entities to which the set X belongs. A typical example would be the definition of the natural numbers as the set obtained as the intersection of all the sets containing a distinguished element (namely, 0) and closed under a one-to-one operation (successor). Since the natural numbers have the specified property, we are defining the natural numbers by quantifying over a totality of sets that contain the set of natural numbers itself. Russell, Poincaré, and Weyl considered this to be a vicious circle that had to be eliminated from the foundations of classical analysis.

distinction between theory and metatheory was missing. Nevertheless, Hilbert already in this early article gave metatheoretical arguments for the consistency of certain theories. In doing so he appealed to mathematical induction (on the length of proofs) to show that no proof in a certain system has a certain property.[4] Poincaré was quick to seize on this point and to object to Hilbert's approach to the foundations on the bais that it was circular, since attempting to prove the consistency of arithmetic (which includes the principle of induction) would seem to require arguing by induction! In 1922, Hilbert found a way out of this dilemma by distinguishing *mathematics proper* and *metamathematics*. In any case, we find in 1905 two essential features of Hilbert's later approach to the foundations of mathematics. The first is the idea that logic and arithmetic have to be developed simultaneously. This can be seen as excluding a development of mathematics out of pure logic, as in Frege's logicist program. The second is the idea that a consistency proof has to investigate the nature of proof directly, i.e., it must consider mathematical proofs as objects of mathematical investigation, an insight which will bear immense fruit.

As already mentioned, by the early twenties the situation in the foundations of mathematics had become dire. Brouwer had begun an alternative development of set theory along intuitionistic lines and Weyl had first presented an alternative foundation of analysis, in *Das Kontinuum*, and in 1921 he had joined the intuitionistic camp. In "New grounding of mathematics" (1922), Hilbert addresses the challenge posed by Brouwer and Weyl. After criticizing Brouwer and Weyl, Hilbert described his own new approach as requiring two stages.

In the first stage, an axiomatization of one specific area of mathematics, or of all of mathematics, is developed. This includes also a complete formalization of proofs by means of an axiomatic presentation of the logical calculus. However, according to views already present in Hilbert's earlier foundational work, the axiomatic development of a theory requires a proof for the consistency of the axioms. In order to reply to Poincaré's objection concerning the possible circularity involved in such a strategy, Hilbert draws a distinction between three mathematical levels: ordinary mathematics, mathematics in the strict sense (or "proper" mathematics), and metamathematics. Ordinary mathematics is mathematics as practiced by the community of mathematicians; proper mathematics is the fully formalized version of ordinary mathematics; and metamathematics consists of epistemologically reliable forms of inference used in proving results about proper mathematics. Before we say something more about metamathematics, also defined by Hilbert as proof theory, let us add that already in 1923 Hilbert recognized the partial validity of the intuitionistic charge while maintaining that he could outdo the intuitionists. Hilbert speaks in this context of "finitary" and "infinitary" inferential procedures:

---

[4] We will explain in the course of the book how mathematical induction on objects such as proofs work. For the reader who has not encountered such inductive arguments before let us say, as a preliminary approximation, that this is a form of inference that allows one to show that if a certain property holds for proofs of length 1 and that, in addition, if it holds for proofs of length n, it also holds for proofs of length $n + 1$, then the property holds for all proofs, no matter what their length might be.

We therefore see that, if we wish to give a rigorous grounding of mathematics, we are not entitled to adopt as logically unproblematic the usual modes of inference that we find in analysis. Rather, our task is precisely to discover why and to what extent we always obtain correct results from the application of transfinite modes of inference of the sort that occur in analysis and set theory. The free use and the full mastery of the transfinite is to be achieved on the territory of the finite! (Hilbert, 1923, p. 1140)

In other words, classical mathematics can be justified with merely finitary reasoning that must also be acceptable to the intuitionists.[5] A full description of Hilbert's program would take us too far afield (see the reading list in the Preface for further references). What we have said so far suffices to explain what the goal of Hilbert's proof theory was. The formalization of mathematical systems, including the formal rules of inference, is only the first part of Hilbert's program. In addition to formalized mathematics, which proceeds purely formally, Hilbert requires, as part of his proof theory, a metamathematics which can make use of contentual reasoning:[6]

In addition to this proper mathematics, there appears a mathematics that is to some extent new, a *metamathematics* which serves to safeguard [proper mathematics] by protecting it from the terror of unnecessary prohibitions as well as from the difficulty of paradoxes. In this metamathematics—in contrast to the purely formal modes of inference in mathematics proper—we apply contentual inference; in particular, to the proof of the consistency of the axioms. (Hilbert, 1922, p. 212/1132)

In his early approaches to the consistency problem (for instance, in 1905), Hilbert had not distinguished between the formalized theories and the metamathematical level of investigation. The present distinction allowed Hilbert to reply to Poincaré's objections that a vicious circle is involved in the Hilbertian attempt to prove the consistency of arithmetic. In 1922, Hilbert asserted that what is involved at the metamathematical level is only a small part of arithmetical reasoning which does not appeal to the full strength of the induction axiom. Moreover, the part of arithmetical reasoning used for metamathematical purposes is completely safe. Hilbert's foundational program thus proceeds by requiring a formalization of arithmetic, analysis, and set theory (in the early twenties, the central concern seems to be analysis, later number theory will take center stage) and associated with it a proof of consistency of such formalizations by means of proof theory. A proof of consistency will have to show, by appealing to contentual considerations which are completely unproblematic, that in the formalism in question it is never possible to derive the formula $a \neq a$, or alternatively it is not possible to prove both $a = b$ and $a \neq b$. These metamathematical considerations are, in contrast with the formal

---

[5] Hilbert's position is often called "finitism." The terms "finitary reasoning" and "finitistic reasoning" are both used in the literature; we will adopt the former.

[6] We translate Hilbert's *inhaltlich* as "contentual," for lack of a better word. In German, *Inhalt* means content, so the "contentual" considerations are those that are not formal but rather have a meaning associated to their use.

procedures contained in the formalized mathematics, provided with content. What has been gained by doing so? The formulas and proofs in a formalized system are, unlike the infinitary objects invoked by mathematicians, finite combinations of symbols on which it is possible to operate with the certainty of reasoning required by Hilbert's strict finitary demands. The consistency proof thus acts as a finitary guarantee of the admissibility of the infinitary objects considered in ordinary mathematics.

What exactly should have been accepted as part of "finitary methods" is a delicate historical and theoretical question that we need not address here (see Mancosu, 1998b and Zach, 2019a). Rather, we mention one more development before concluding this section. However broadly one might construct the finitary point of view, it seemed to most logicians that the totality of finitary modes of reasoning could be captured within some well-known systems (such as Peano arithmetic or perhaps an even stronger theory). But Kurt Gödel proved in 1931 that no specifiable consistent system in which a minimal amount of arithmetic can be developed could prove its own consistency.[7] And this means that if a system contains all the finitary modes of reasoning, no finitary proof of its consistency could be given. But this was not the end of the story (although it is a big part of it, whose consequences mathematical logicians are still exploring). This leads us to Gentzen's contributions.

1.2   Gentzen's proof theory

Hilbert's ideas in foundations of mathematics and especially consistency proofs were developed by a number of his students and collaborators, especially Bernays and Wilhelm Ackermann.[8] Other mathematicians not strictly part of Hilbert's school contributed as well, e.g., von Neumann (1927). Herbrand (1930) independently pursued the consistency problem from a different perspective. But it was Gerhard Gentzen who, in the 1930s, developed the basis for what we now understand as proof theory.

One of Gentzen's earliest contributions was a proof that any derivation in classical arithmetic could be translated into a derivation in a system of intuitionistic arithmetic (1933).[9] This was not a direct consistency proof of classical arithmetic in the sense required by Hilbert's program. Instead of showing directly using finitary means that a proof of a contradiction was impossible, it showed that if a contradiction could be derived in classical arithmetic, a contradiction would also be derivable in intuitionistic arithmetic. In the dialectical situation to which Hilbert

---

[7] Gödel's two main results in his 1931 are referred to as Gödel's incompleteness theorems.

[8] Early work on proof theory was carried out in a system called the ε-calculus, see Avigad and Zach (2020) for general background and Zach (2003; 2004) for the history of consistency proofs in this tradition.

[9] Gentzen withdrew the paper from publication since Gödel (1933) had already independently obtained the same result.

had responded by formulating the consistency program, this could be considered an advance. Because, as we saw above, Hilbert's program was a reaction to the challenge brought by intuitionistic critics of classical mathematics (Brouwer, Weyl). They charged that classical mathematics is threatened by inconsistency, and so should be replaced by the safer intuitionistic approach. Gentzen's result already constitutes a defense of classical arithmetic against this proposal. It shows that any threat of inconsistency faced by classical arithmetic is also faced by intuitionistic arithmetic. It also showed that Hilbert's finitary standpoint is more restrictive than intuitionistic mathematics, a fact that was not appreciated until Gentzen's proof. Until then, the participants in the debate assumed that finitary reasoning and intuitionistic reasoning are more or less the same, at least as far as elementary arithmetic is concerned. Thus, Gentzen's result was not yet a solution to Hilbert's problem of providing a finitary consistency proof. We discuss Gentzen's 1933 result in Chapter 2.

Proof theory, so far, had concentrated on studying derivations in an axiomatic calculus. Such derivations are sequences of formulas of a formal language in which every formula either is one of a number of axioms or follows from formulas preceding it by one of a small number of inference rules. In his dissertation "Investigations into logical deduction" (1935b,c), Gentzen introduced two systems which work rather differently. These systems are known as natural deduction and the sequent calculus, and form the basis for almost all subsequent work in proof theory.

In axiomatic systems, derivations proceed from a number of axioms using one of a number of inference rules. Each step in such a derivation establishes that the formula proved in the respective step is a valid principle of the corresponding logic, or follows from the axioms of the corresponding theory in the logic. Natural deduction systems formalize reasoning from hypothetical assumptions. An assumption is a formula assumed to be true; the rules of natural deduction allow us to infer consequences of these assumptions. Some of the rules allow us to "discharge" assumptions: the conclusion of such a rule then no longer depends on the formula that was assumed to be true at the outset. For instance, if we can deduce B from the assumption A, then we can infer $A \supset B$, but no longer assuming A.[10] A formula is proved outright if all assumptions are discharged.

In Gentzen's natural deduction, a proof is a tree of formulas (as opposed to a linear sequence). Each formula in it is either an assumption, or follows from the one, two, or three formulas immediately above it in the tree by a rule of inference. In contrast to axiomatic derivation systems, which have many axioms and only a few inference rules, Gentzen's natural deduction system has many inference

---

[10] Proof theorists often use Russell's symbol $\supset$ for the material conditional to avoid confusion with the sequent arrow to be introduced shortly.

rules, but no axioms.[11] The inference rules were designed to model the patterns of ordinary reasoning that mathematicians carry out in proving results, such as hypothetical proofs or reasoning by cases. Natural deduction will be the subject of Chapter 3.

Gentzen's other logical system is the sequent calculus. In the sequent calculus, a proof is a tree of what Gentzen called sequents: pairs of lists of formulas separated by a $\Rightarrow$ symbol,[12] e.g., $A, B \Rightarrow A \wedge B, C$. One way of thinking about sequents is as a shorthand for the statement that the conjunction of the formulas on the left implies the disjunction of the formulas on the right. The sequent calculus only has axioms of the form $A \Rightarrow A$, and, like natural deduction, a larger number of inference rules. One might also think of the sequent calculus as a formalization of the relation of deducibility from assumptions: the formulas on the left are the assumptions on which the formulas on the right depend. In contrast to natural deduction, the rules of the sequent calculus work on the left and right side of a sequent. For instance, the $\wedge$R rule allows us to infer $C \Rightarrow A \wedge B$ from the sequents $C \Rightarrow A$ and $C \Rightarrow B$. The $\wedge$L rule works on the left side of a sequent and allows us to infer $A \wedge B \Rightarrow C$ from $A \Rightarrow C$. While the "right" rules of the sequent calculus correspond directly to inference rules in natural deduction, the "left" rules do not. They do, however, correspond to facts about what can be deduced: e.g., if we can deduce C from the assumption A, then we can also deduce C from the assumption $A \wedge B$. We introduce the sequent calculus in Chapter 5.

The CUT rule allows us to infer from sequents such as $A \Rightarrow B$ and $B \Rightarrow C$ the sequent $A \Rightarrow C$. It is the only rule of the sequent calculus in which a formula which appears in the premises of a rule no longer appears in the conclusion. In the sequent calculus, a contradiction is represented by a sequent in which the two lists of formulas on either side of the $\Rightarrow$ symbol are empty. Since only the CUT rule can make formulas disappear from a proof, and the axioms contain formulas, only proofs using this rule can possibly be proofs of a contradiction, i.e., the sequent $\Rightarrow$.

Gentzen proved a major theorem about the sequent calculus, which is now known as the cut-elimination theorem. It shows that from a proof in the sequent calculus, any use of the CUT rule can be removed. We will prove this result in Chapter 6. This establishes that the sequent calculus itself is consistent, and is the basis of much of subsequent proof theory. Just as axiomatic systems for logic can be extended to provide systems for mathematical theories (such as arithmetic or set theory) by adding axioms, the sequent calculus can also be used to formalize mathematical theories by adding axiom sequents or additional rules of inference.

---

[11] There are variants of Gentzen's natural deduction system in which proofs are also linear sequences. The larger number of axioms in axiomatic proof systems is not a necessary feature; there are also axiomatic systems for propositional logic with just one axiom.

[12] Gentzen's original papers used a single arrow $\rightarrow$ as the separator, and nowadays the turnstile symbol $\vdash$ is also common.

If a cut-elimination theorem holds for such a system, we have a consistency proof for it. The cut-elimination theorem for the sequent calculus proceeds by a merely combinatorial modification of proofs, and so itself is finitary. When axioms are added to the calculus, as they are in arithmetic, full cut-elimination no longer holds. If axioms contain only atomic formulas, one can extend cut-elimination to remove CUTS on complex formulas; but CUTS on atomic formulas may remain. For Peano arithmetic, every axiom except for the induction axiom can be formulated as a sequent consisting of atomic formulas. The induction axiom itself can be replaced by an inference rule. To show the consistency of arithmetic, we show that from proofs of sequents with only atomic formulas (including the empty sequent), all complex CUTS and applications of the induction rule can be eliminated. But since proofs from the axioms of Peano arithmetic using only CUTS on atomic formulas cannot be proofs of contradictions, this shows that Peano arithmetic is consistent.

In 1936, Gentzen obtained the first consistency proof for arithmetic; two years later he offered a different one (Gentzen, 1938). Although the manipulation of proofs was still purely syntactic and combinatorial, proving that the process of removing applications of the CUT and induction rules always comes to an end relies on a new method. Gentzen's previous results relied on measures of complexity, namely, numbers measuring the complexity and number of occurrences of cut-formulas in the proof. For the proof of consistency of arithmetic, a much more sophisticated measure of complexity is necessary. This measure of complexity is known as an "ordinal notation." Although an ordinal notation is also a finite, syntactic object, the ordering Gentzen defined for them is very complex. The consistency proof of arithmetic relies on the fact that ordinal notations are *well-ordered*, i.e., that there cannot be a sequence of ordinal notations in which each one is "smaller" than the previous one (in Gentzen's complicated ordering) and which has no end. Gentzen established the consistency of arithmetic by giving a procedure in which proofs are successively replaced by new proofs with "smaller" ordinal notations. Since any decreasing sequence of ordinal notations must eventually stop, the replacement procedure must eventually stop as well—and when it stops, all applications of the induction rule and all complex CUT inferences have been eliminated from the proof. We will study Gentzen's proof and his ordinal notations in detail in Chapters 7 to 9.

In his approach to proof theory, Hilbert was interested mainly in derivability (i.e., whether something or other was derivable in a certain theory by certain means). With Gentzen, we reach a new level where proof theory is not restricted by the goal of reducing proofs that use infinitary considerations to finitary proofs. Some logicians speak of *reductive proof theory* for this latter approach, characterized by Hilbert's take on proof theory. Gentzen managed to expand the Hilbertian approach by considering mathematical proofs in themselves, their transformations, etc., without being restricted to the considerations of proofs as means of an epistemic reduction. Gentzen's approach was pivotal in the move from reductive proof theory to general proof theory (see Prawitz, 1971).

Gentzen's results emerged in the wake of Gödel's incompleteness theorems, when it had become clear that Hilbert's original program of proving the consistency of arithmetic by finitary means could not be achieved. But not being able to prove the consistency of arithmetic by finitary means did not exclude the possibility of finding other principles that, while not strictly finitist, were still sufficiently constructive. Gentzen's consistency proof of arithmetic was given by using a principle of induction restricted to quantifier-free formulas. It was extended, as we already mentioned, to an order type more complex than that of the natural numbers, namely the order type of the ordinal notations he introduced. This order type is characterized by an ordinal number called ε_0. In order to obtain his results, Gentzen introduced natural deduction and the sequent calculus and was led to study certain canonical forms for proofs that avoided any needless detours. In the sequent calculus, such proofs are called cut-free: the use of the CUT rule that we have described above has been eliminated. This exemplifies how results of *general proof theory* (to be contrasted with reductive proof theory) were to be found embedded in a development emerging from reductive proof theory. By general proof theory is meant an approach to the study of proofs that is not constrained by the epistemological aims that dominate reductive proof theory or even the extended program for reductive proof theory exemplified in Gentzen's work. In particular, one does not restrict the mathematical tools that can be enlisted in the study of proofs (thus, no restriction to finitary or constructive reasoning in the metamathematics). Proofs and their transformations are studied as objects of interest in themselves and not only as tools in the service of a foundational program.

1.3   Proof theory after Gentzen

Gentzen's work contains the beginnings of what we call structural proof theory (to which the first part of this book is devoted) as well as ordinal proof theory (to which the second part of the book is devoted). After Gentzen, both strands of proof theory were developed further. Natural deduction and the sequent calculus were introduced for systems other than intuitionistic and classical logic. New results in both structural and ordinal proof theory were established. One such result was an analog of the cut-elimination theorem for natural deduction, called the normalization theorem. This result was proved independently by Raggio (1965) and Prawitz (1965); we present it in Chapter 4.[13]

The contrast between reductive and general proof theory has been especially emphasized by Prawitz and Kreisel in the 1970s (Prawitz, 1971). As already mentioned, Prawitz proved in 1965 the normalization theorem for first-order intuitionistic logic and extended it to classical logic and stronger logics. The

---

[13] Indeed, it was recently discovered that Gentzen had already obtained a proof of normalization for the intuitionistic predicate calculus, but never published it; see Gentzen (2008).

normalization theorem for classical first-order logic not only extends the normalization theorem for intuitionistic logic but is also closely related to Gentzen's own result showing cut-elimination of the sequent calculus for first-order logic.

We conclude by adding a few more details on the extensions of the finitary point of view. First of all, we should mention that the 1950s saw the first important results concerning the consistency of systems going beyond arithmetic. In particular, proof theorists studied systems of mathematical analysis (logicians call this "second-order arithmetic") with set existence axioms, and tried to extend Gentzen's consistency proof. Set existence axioms ("comprehension") state that for every formula $A(x)$ of a certain kind, there is a set of objects containing all and only the objects satisfying $A(x)$. Full second-order arithmetic is the system in which no restrictions are placed on $A(x)$—in particular, $A(x)$ may contain quantifiers over sets itself. If $A(x)$ is restricted to formulas containing only quantifiers over numbers but not over sets of objects, we get a system called **ACA$_0$**. It can prove the same arithmetical statements (i.e., statements not involving quantifiers over sets) as **PA** does, and its consistency can be proved also by induction along ε_0. But systems based on stronger set existence axioms are much stronger. To prove that they are consistent, we need principles of induction along well-orders much more complicated than the one characterized by ε_0. One such order type is known as Γ_0. It was used by Feferman (1964) and Schütte (1965) to give consistency proofs of predicative systems inspired by the foundational position introduced by Weyl (1918), mentioned above. The first consistency proof of an impredicative system is due to Takeuti (1967), for a system in which the formula $A(x)$ in the set existence axiom is allowed to contain initial universal quantifiers over sets.

The extensions of the finitary point of view made necessary to carry out the consistency proofs for stronger systems does not only concern the principle of induction which is used to prove the consistency of the system in question. Schütte, and others following him, found it expedient to work with systems which themselves contain infinitary rules, namely rules that have an infinite number of premises and one conclusion, e.g., rules allowing the move from $A(0)$, $A(1)$, $A(2)$ ... to $\forall x\, A(x)$. This "ω-rule" has now become standard in proof theory but is definitely quite far from what Hilbert envisaged in his finitary presentation of theories.

Another important extension of the finitary point of view in the development of general proof theory is Gödel's consistency proof, published in 1958, for arithmetic by means of a system, called T, of functionals of finite type. A numerical function is simply a function that takes natural numbers as arguments and produces natural numbers as values, or outputs. A functional is a function which may in addition take functions as arguments, and produce functions as values. Gödel's system T is produced by starting with some specific numerical functions, producing new functions and functionals according to certain principles. If such operations are repeated an arbitrary finite number of times, we obtain the class of functionals of finite type. Gödel singled out a class called recursive functionals of finite type that had the property of being computable if the arguments to which they were applied were also computable. He then associated with every statement P of arithmetic an assertion concerning the existence of a specific recursive functional in such a way that the proof of P in arithmetic implies the existence assertion for the corresponding functional. One then shows that the statement of existence concerning the recursive functional of finite type corresponding to $0 = 1$ is false, thereby showing the unprovability of $0 = 1$ in arithmetic and hence the consistency of the latter. The assumption of existence of the recursive functionals of finite type is demonstrably equivalent to an induction along the well-ordering ε_0, thereby showing a deep connection between Gentzen's original proof of consistency and this new one proposed by Gödel.

While Gödel's consistency proof can be seen as being part of reductive proof theory, it also had embedded in it the germs for a possible extension to general proof theory. This was exploited already in the 1970s by establishing an association between recursive functionals and proofs by means of the λ-calculus. By defining the notion of a functional in normal form appropriately, one can prove that to every proof in normal form (i.e., without detours) there corresponds a functional in normal form and vice versa.

The developments described in the preceding three paragraphs go well beyond what we will manage to cover. The reader will find in this book a solid foundation comprising the more elementary parts of proof theory and, having mastered those, the more advanced textbooks we have listed in the preface will be the next step in exploring these fascinating topics.

2

Axiomatic calculi

Before Gentzen's introduction of natural deduction and sequent calculi in 1935b,c, the axiomatic presentation of logical systems was typical of the algebra of logic tradition, the postulate theorists, Whitehead and Russell's *Principia Mathematica*, and Hilbert's school.[1] The conception of logic embodied in such systems is one according to which logic is characterized by some logical laws, which are taken as starting points—called axioms or postulates—and by a number of inferential rules that allow one to move from the starting points to new logical validities. One can present such a calculus by developing the propositional and quantificational parts together, or by developing each one separately. To ease you into it, we will begin first with the propositional calculus and then move to the predicate calculus. Finally, we will define the system of classical and intuitionistic arithmetic and prove one of Gentzen's first important results. This result, also proven independently by Gödel, consists in showing that there is a way to translate classical arithmetic into intuitionistic arithmetic, thereby showing that if intuitionistic arithmetic is consistent so is classical arithmetic.

2.1 Propositional logic

Although Gentzen worked only with intuitionistic and classical logic, Johansson (1937) introduced minimal logic, a weaker system than intuitionistic logic, that has remained central to proof theory.[2] Thus we will axiomatize the three systems by building intuitionistic and classical logic on top of minimal logic. The axiomatization will follow those of Heyting (1930) and Johansson.

Definition 2.1. The language of propositional logic consists of:

1. A denumerable set of propositional variables $p_1, p_2, p_3, \ldots$

---

[1] For more historical details on these traditions see Mancosu, Zach, and Badesa (2009).
[2] A system of logic T is said to be "weaker" ("stronger") than another system S if the set of theorems of T (respectively, S) is properly included in the set of theorems of S (respectively, T).

An Introduction to Proof Theory: Normalization, Cut-Elimination, and Consistency Proofs.
Paolo Mancosu, Sergio Galvan, and Richard Zach, Oxford University Press. © Paolo Mancosu,
Sergio Galvan and Richard Zach 2021. DOI: 10.1093/oso/9780192895936.003.0002

2. Connectives: ¬, ∨, ∧, ⊃

3. Parentheses: (,)

The notion of formula is defined inductively. Inductive definitions are a way to characterize a class of objects by first stating which "elementary" objects belong to the class (basis clause) and then providing a rule (or rules) that allow(s) us to construct "more complex" objects from the ones already obtained (inductive clause). A final clause, the extremal clause, specifies that the class of objects we intend to characterize is the smallest of the possible classes satisfying the basis clause and the inductive clause(s).

For instance, we could define the class of Frege's (direct) ancestors by specifying a basis clause to the effect that his mother and father are his ancestors. Then the inductive clause would specify that if x is his ancestor, then the mother of x and the father of x are also Frege's ancestors. One way to think of this is that we have two functions, mother(x) and father(x), which when applied to any of Frege's ancestors, say a, output two new ancestors, mother(a) and father(a). The extremal clause would ensure that we only take the smallest class containing Frege's father and mother and that is closed under the functions mother(x) and father(x).[3]

One can also look at this process as a construction by stages. At stage 1 we put in the set of Frege's ancestors his father and his mother. At stage 2 we also add his paternal grandfather, paternal grandmother, maternal grandfather, and maternal grandmother. It should now be obvious that at stage 3 we obtain eight new elements of the set of Frege's ancestors. And so it goes with sixteen new elements added at stage 4, etc. Let us see how this works with the definition of formulas of our language.

Definition 2.2. The *formulas* are defined as follows:

1. *Basis clause:* Each propositional variable is a formula (called an *atomic formula*).

2. *Inductive clause:* If A and B are formulas so are $\neg A$, $(A \wedge B)$, $(A \vee B)$, and $(A \supset B)$.

3. *Extremal clause:* Nothing else is a formula.

Remark. The letters "A" and "B" (and later "C", "D", ...) are not symbols of our propositional calculus. Rather, we use them in our discussion to stand in for arbitrary formulas. Such schematic letters are called *metavariables*, since they are variables we use in the metalanguage in which we talk about symbols, formulas, derivations, etc. Later on we will also use metavariables for other symbols and expressions.

---

[3] A class or set is "closed under" a function f if it contains the value of f applied to any element of the class. So in this case it means that if anyone is in the class, so are their father and mother.

One can interpret the above definition as providing a rule for constructing, by stages, more and more complex formulas. For instance, by the basis clause, we know that p_3 is a formula. Using the inductive clause we now can build $\neg p_3$. But by the basis clause, p_5 is also a formula. And by applying the inductive clause again we obtain that $(\neg p_3 \wedge p_5)$, $(p_5 \supset \neg p_3)$, $(\neg p_3 \vee p_3)$—to give only a few examples—are also formulas. From now on, we will drop the outermost pair of parentheses around formulas, so e.g., $\neg p_3 \wedge p_5$ is short for $(\neg p_3 \wedge p_5)$. When the amount of parentheses becomes too large, we sometimes will also use square brackets and curly braces instead of parentheses to make it easier to parse formulas. So, for instance,

$$[p_1 \vee (p_2 \wedge p_3)] \supset \{(p_1 \wedge p_2) \vee (p_1 \wedge p_3)\}$$

is just a different way of writing what's officially

$$((p_1 \vee (p_2 \wedge p_3)) \supset ((p_1 \wedge p_2) \vee (p_1 \wedge p_3))).$$

Students often fail to see the relevance of the extremal clause. The extremal clause is necessary to block the possibility that other "extraneous" or "unintended" elements turn out to be in the class of formulas. In fact, without stating clause 3, a collection containing in addition to $p_1, p_2, \ldots$ also a primitive symbol, say, $\star$, to which we apply the inductive clause would satisfy both the basis and inductive clauses. It would satisfy the basis clause on account of the fact that we are assuming $p_1, p_2, p_3, \ldots$ are already present (together with $\star$). And it would satisfy the inductive clause since we are allowing the application of the inductive clause to all the generated formulas. But then the class defined by this definition would contain also $\star$, $\neg \star$, $(\neg \star \wedge p_3)$, $\ldots$ In other words, the basis and inductive clauses on their own fail to characterize the "intended" class of formulas without the explicit requirement—formulated in the extremal clause—that the class of formulas must be the smallest containing $p_1, p_2, \ldots$, and closed under the successive finite application of $\neg, \wedge, \vee, \supset$ to previously obtained formulas.[4] Having clarified this, we will henceforth often avoid explicitly stating the extremal clause in our inductive definitions.

2.2 Reading formulas as trees

If one thinks of a formula as the product of a construction from atomic formulas by the repeated application of connectives to the already available formulas, then it is easy to visualize the construction of a formula by means of a tree. For the moment, we will not give a formal definition of "tree" but rather introduce the concept informally.

---

[4] One could even think of ")" as exemplifying the issue. How do we know that ")" and "$\neg$" are not formulas? Because they do not belong to the smallest class satisfying conditions (1) and (2).

Consider the formula $(\neg p_3 \supset p_2)$. The construction by means of the clauses of the inductive definition has already been explained. It consists in accepting that p_3 is a formula on account of the basis clause (1). That starting point will correspond to a leaf (i.e., a terminal node) of our tree:

$$p_3$$

The next step consists in showing how $\neg p_3$ was built from p_3. In the case with the application of a unary connective like $\neg$, we show a line segment one of whose endpoints is the node corresponding to p_3 and the new endpoint is a node labeled $\neg p_3$.

Now, p_2 is accepted as a formula independently of the above construction. We then add a new leaf labeled p_2 (since we will connect p_2 with $\neg p_3$ by means of a conditional we already position p_2 on the right of $\neg p_3$):

$$p_3$$
$$|$$
$$\neg p_3 \quad p_2$$

Finally, we finish the construction of the formula by adding a new node labelled $(\neg p_3 \supset p_2)$ and joining the nodes for $\neg p_3$ and p_2 to the new node:

Note that this way of representing the construction of the formula displays very clearly which are the starting points of the process. If you were to compute the truth-table for such a formula, the tree illustrates that you first have to assign truth-values to the atomic formulas and then proceed down the tree to calculate the truth-value of the formula.

Problem 2.3.

1. Give a construction of the following formulas, justifying each step of the construction by appealing to the appropriate clause of the inductive definition:

    (a) $((\neg p_1 \vee p_2) \supset p_3)$
    (b) $\neg((p_1 \wedge p_2) \vee p_4)$

(c) $(\neg(\neg p_5 \wedge \neg p_6) \vee \neg p_5)$

2. Give tree representations for the construction of these formulas.

2.3 Sub-formulas and main connectives

It will often be important to know what the last step in the construction of a formula was. If the last step, say, invoked the clause "If A and B are formulas, so is $(A \supset B)$" then the formula constructed is called a *conditional*; the corresponding formula A is called its *antecedent* and B its *consequent*. When we say that a formula is "of the form $(A \supset B)$" we mean it is a formula which in its construction invoked the mentioned clause as the last step. Likewise, a formula "of the form $(A \wedge B)$" is one where we invoked the clause "If A and B are formulas, so is $(A \wedge B)$" as the last step. Such formulas are called *conjunctions*, and A and B its *conjuncts*. A formula "of the form $(A \vee B)$" is called a *disjunction* and A and B its *disjuncts*. And a formula "of the form $\neg A$" is called a *negated formula*.

The formula $\neg p_3 \supset p_2$ of the last section, for instance, was generated using the clause "If A and B are formulas, so is $(A \supset B)$." (Note that by convention, we leave out the outermost parentheses, so really the formula in question is $(\neg p_3 \supset p_2)$.) So we're dealing with a conditional; the formula is "of the form $A \supset B$." In this case, the antecedent A is $\neg p_3$ and the consequent B is p_2.

It will also often be useful to have a name for the connective introduced in the last step of the formation of a formula: it is called the *main connective*. So the main connective of $A \supset B$ is $\supset$; of $A \wedge B$ is $\wedge$; of $A \vee B$ is $\vee$; and of $\neg A$ is $\neg$.

For every formula, the main connective and hence what "form" it has is uniquely determined. That means: there is no formula which is, say, both of the form $A \supset B$ and $\neg A$, has both $\supset$ and $\neg$ as a main connective. For instance, you might think $\neg p_3 \supset p_2$ is both of the form $A \supset B$ and $\neg A$. But remember that we've silently dropped the outermost parentheses! So the formula really is $(\neg p_3 \supset p_2)$. Now it is clear that this formula could not have been produced by taking some formula A and prefixing it with a single $\neg$. The only way to construct $(\neg p_3 \supset p_2)$ is to take $\neg p_3$ and p_2 as A and B, and then form $(A \supset B)$ from them.

Lastly, when we speak of an *immediate sub-formula* of a formula, we mean the A (and B) used in the last step of its construction. So the immediate sub-formulas of $A \wedge B$ are A and B, similarly of $A \vee B$ and $A \supset B$, and the immediate sub-formula of $\neg A$ is A. A *sub-formula* of a formula is any formula involved in its construction (including the formula itself).

We can also give an inductive definition of the sub-formulas of a formula A:

Definition 2.4. *Sub-formulas* are inductively defined as follows:

1. *Basis clause:* If A is atomic, then A is a sub-formula of A.

2. *Inductive clauses:*

(a) If A is a formula of the form $\neg B$, then the sub-formulas of A are A itself and the sub-formulas of B.

(b) If A is a formula of the form $(B \wedge C)$, $(B \vee C)$, or $(B \supset C)$, then the sub-formulas of A are A itself and the sub-formulas of B and C.

3. *Extremal clause:* Nothing else is a sub-formula of A.

If we have the tree representation of the construction of a formula A, then the connective introduced in the root is the main connective, and the immediate sub-formulas are the formulas (or the single formula) standing immediately above it. The sub-formulas of A are all the formulas occurring in the tree.

Let the *degree* of a formula be the number of logical symbols in it. Then the degree of an atomic formula is 0, and the degree of any immediate sub-formula of a formula is smaller than the degree of the formula itself. E.g., the degree of $(p_1 \wedge (p_2 \vee p_3))$ is 2. Its immediate sub-formulas are p_1 (degree 0) and $p_2 \vee p_3$ (degree 1). In fact, the degree of any formula of the form $(A \wedge B)$, $(A \vee B)$, $(A \supset B)$ is 1 plus the sum of the degrees of A and B, and the degree of $\neg A$ is 1 plus the degree of A.

Following the inductive definition of formulas, one can now define a function $d(A)$, by first defining the value of the function when A is atomic and then specifying the values of the function when A is not atomic.

Definition 2.5. The *degree* $d(A)$ of a formula A is inductively defined[5] as follows:

1. *Basis clause:* If A is atomic, then $d(A) = 0$.

2. *Inductive clauses:*

    (a) If A is a formula of the form $\neg B$, then $d(A) = d(B) + 1$.

    (b) If A is a formula of the form $(B \wedge C)$, $(B \vee C)$, or $(B \supset C)$, then $d(A) = d(B) + d(C) + 1$.

3. *Extremal clause:* $d(A)$ is undefined otherwise.

2.4 Logical calculi

We can now state the axiomatic propositional calculi we will be concerned with: minimal ($\mathbf{M}_0$), intuitionistic ($\mathbf{J}_0$), and classical ($\mathbf{K}_0$).[6] Each calculus is stronger than

---

[5] Definitions like this in which d is defined in terms of itself (as in, e.g., $d(\neg B) = d(B) + 1$) are also called *recursive*.

[6] The subscript 0 indicates that we are considering the propositional fragment; we will later also consider full predicate logics in their minimal, intuitionistic, and classical versions. Gentzen's names for our $\mathbf{J}_1$ and $\mathbf{K}_1$ are **LHJ** and **LHK**. 'L' stands for logistic (i.e., axiomatic) calculus; 'H' for Hilbert, as Hilbert made extensive use of such calculi; and 'J' and 'K' for Intuitionistic and Classical ('J' and 'I' were considered typographical variants of the same uppercase letter in German). We will call the minimal, intuitionistic, and classical predicate calculi $\mathbf{M}_1$, $\mathbf{J}_1$, and $\mathbf{K}_1$.

the previous one(s), and it is obtained by adding more powerful assumptions to its predecessor.

2.4.1   Minimal logic

In order to avoid needing to include a rule of substitution (as Gentzen does following Heyting), our axiomatization consists of axiom schemes. What that means is that each instance of the schema is an axiom of our system. For example, axiom PL1 is given as $A \supset (A \wedge A)$. It is understood that every formula *of this form* counts as an axiom, such as $p_1 \supset (p_1 \wedge p_1)$, $\neg(p_1 \vee p_2) \supset (\neg(p_1 \vee p_2) \wedge \neg(p_1 \vee p_2))$, etc.

We will first present the systems in a purely formal way. In Section 2.11, we will discuss their informal interpretation.

PL1.	$A \supset (A \wedge A)$	[2.1]
PL2.	$(A \wedge B) \supset (B \wedge A)$	[2.11]
PL3.	$(A \supset B) \supset [(A \wedge C) \supset (B \wedge C)]$	[2.12]
PL4.	$[(A \supset B) \wedge (B \supset C)] \supset (A \supset C)$	[2.13]
PL5.	$B \supset (A \supset B)$	[2.14]
PL6.	$(A \wedge (A \supset B)) \supset B$	[2.15]
PL7.	$A \supset (A \vee B)$	[3.1]
PL8.	$(A \vee B) \supset (B \vee A)$	[3.11]
PL9.	$[(A \supset C) \wedge (B \supset C)] \supset [(A \vee B) \supset C]$	[3.12]
PL10.	$[(A \supset B) \wedge (A \supset \neg B)] \supset \neg A$	[4.11]

The numbers on the right give the numbering in Heyting (1930). Gentzen (1933) also refers to the same axiomatization.

2.4.2   Intuitionistic logic

The axioms of $\mathbf{J}_0$ are obtained by adding to the axioms of $\mathbf{M}_0$ the following axiom schema:

PL11.	$\neg A \supset (A \supset B)$	[4.1]

2.4.3   Classical logic

The axioms of $\mathbf{K}_0$ are obtained by adding to the axioms of $\mathbf{J}_0$ the following axiom schema:

PL12. $\neg\neg A \supset A$

2.5   Inference rules

All three calculi have one single rule of inference, namely *modus ponens* (MP): from A and $A \supset B$, infer B.[7]

More formally, we define a derivation in any of the above systems as follows: Let S_0 stand for either M_0, J_0, or K_0.

Definition 2.6. A *derivation* in S_0 is a finite sequence of formulas $A_1, \ldots, A_n$ each one of which is either an axiom of S_0 or is obtained from previous formulas by modus ponens. The last formula in the derivation, A_n, is called the *end-formula*.

B is a *theorem* of S_0 if there is a derivation with end-formula B in S_0.[8] When this is the case we write $\vdash_{S_0} B$. We will also say that B is provable in S_0.

Our first example of a derivation shows how to derive a sentence of M_0. But in general we will prove schematic theorems that go proxy for an infinite number of proofs of their instances.[9] We derive $p_1 \supset (p_2 \vee p_1)$ in M_0. (Recall the bracketing conventions of Section 2.1: we use square brackets and braces in addition to parentheses for readability, and leave out the outermost parentheses.)

1. $\vdash p_1 \supset (p_1 \vee p_2)$ (axiom PL7)

2. $\vdash [p_1 \supset (p_1 \vee p_2)] \supset [((p_1 \vee p_2) \supset (p_2 \vee p_1)) \supset (p_1 \supset (p_1 \vee p_2))]$ (axiom PL5)

3. $\vdash ((p_1 \vee p_2) \supset (p_2 \vee p_1)) \supset (p_1 \supset (p_1 \vee p_2))$ (MP 1, 2)

4. $\vdash [((p_1 \vee p_2) \supset (p_2 \vee p_1)) \supset (p_1 \supset (p_1 \vee p_2))]$
$\supset [\{((p_1 \vee p_2) \supset (p_2 \vee p_1)) \wedge ((p_1 \vee p_2) \supset (p_2 \vee p_1))\} \supset$
$\{(p_1 \supset (p_1 \vee p_2)) \wedge ((p_1 \vee p_2) \supset (p_2 \vee p_1))\}]$ (axiom PL3)

5. $\vdash \{((p_1 \vee p_2) \supset (p_2 \vee p_1)) \wedge ((p_1 \vee p_2) \supset (p_2 \vee p_1))\} \supset$
$\{(p_1 \supset (p_1 \vee p_2)) \wedge ((p_1 \vee p_2) \supset (p_2 \vee p_1))\}$ (MP 3, 4)

6. $\vdash [(p_1 \vee p_2) \supset (p_2 \vee p_1)] \supset$
$[((p_1 \vee p_2) \supset (p_2 \vee p_1)) \wedge ((p_1 \vee p_2) \supset (p_2 \vee p_1))]$ (axiom PL1)

7. $\vdash (p_1 \vee p_2) \supset (p_2 \vee p_1)$ (axiom PL8)

8. $\vdash ((p_1 \vee p_2) \supset (p_2 \vee p_1)) \wedge ((p_1 \vee p_2) \supset (p_2 \vee p_1))$ (MP 6, 7)

9. $\vdash (p_1 \supset (p_1 \vee p_2)) \wedge ((p_1 \vee p_2) \supset (p_2 \vee p_1))$ (MP 5, 8)

10. $\vdash [((p_1 \supset (p_1 \vee p_2)) \wedge ((p_1 \vee p_2) \supset (p_2 \vee p_1))] \supset (p_1 \supset (p_2 \vee p_1))$ (axiom PL4)

11. $\vdash p_1 \supset (p_2 \vee p_1)$ (MP 9, 10)

---

[7] In an application of modus ponens the order in which the premises occur does not matter. When justifying the inference we will always cite the lines in the order in which they occur in the derivation.

[8] We of course extend the notion of a derivation to include lines justified by appeal to previously derived theorems. The extension is inessential since every appeal to a theorem can in principle be eliminated by reproving the theorem in question from axioms and modus ponens.

[9] By "go proxy" we mean that the schematic theorem in its general structure already subsumes all the specific instances, which are simply obtainable from the general theorem by substitution of specific sentences for the variables.

The first thing that will strike you about this derivation is that it is incredibly hard to read. The idea of *finding* such a derivation from scratch is daunting. We will consider some strategies for doing that below. But first, let's use this example to get clearer about what a derivation is. First of all, only the formulas in the middle make up the actual derivation. The numbers on the left are there for convenience, so we can refer to the individual formulas. The justifications on the right help us ascertain which axioms or prior formulas in the derivation the current furmula follows from, and according to which rule.[10] The formula on line 1 is the first formula A_1 in the derivation, the second is A_2, etc. The formula on line 11, $p_1 \supset (p_2 \lor p_1)$ is the end-formula. By giving this sequence of formulas we have produced a derivation of $p_1 \supset (p_2 \lor p_1)$.

As required by the definition, each formula in the sequence is either an instance of an axiom, or follows from previous formulas by modus ponens. For instance, the formula on line 1, $p_1 \supset (p_1 \lor p_2)$, is an instance of axiom PL7, i.e., of $A \supset (A \lor B)$: it is obtained by replacing A by p_1 and B by p_2. We note which axiom a formula is an instance of on the right. The formula on line 2 is an instance of axiom PL5, i.e., $B \supset (A \supset B)$. Here, the instance is obtained by replacing A by $((p_1 \lor p_2) \supset (p_2 \lor p_1))$ and B by $p_1 \supset (p_1 \lor p_2)$. Line 3 follows from lines 1 and 2 by modus ponens. For line 2 is *also* an instance of $A \supset B$, if A is replaced by $p_1 \supset (p_1 \lor p_2)$ and B is replaced by $((p_1 \lor p_2) \supset (p_2 \lor p_1)) \supset (p_1 \supset (p_1 \lor p_2))$. And *these* formulas A and B also appear on lines 1 and 3. So line 1 is of the form A, line 2 of the form $A \supset B$, and line 3 of the form B. This is what it means to say that 3 follows from 1 and 2 by modus ponens. We indicate this on the right side of line 3 by "MP 1, 2."

The first thing to observe is that although the formulas in our derivation involve the propositional variables p_1 and p_2, the derivation actually shows something more general: Not only is $p_1 \supset (p_2 \lor p_1)$ a theorem, but any formula of the form $C \supset (D \lor C)$. For if we uniformly replace p_1 and p_2 in our derivation by formulas C and D, respectively, we still have a correct derivation. This is because whatever formulas C and D are, $C \supset (C \lor D)$ is an instance of axiom PL7, so line 1 will still be an axiom. The formula

$$[C \supset (C \lor D)] \supset [((C \lor D) \supset (D \lor C)) \supset (C \supset (C \lor D))]$$

is an instance of $B \supset (A \supset B)$, so line 2 will still be an axiom. In general: an instance of an instance of an axiom is an axiom, and an instance of a correct application of MP is a correct application of MP.

It will be instructive to analyze the structure of the derivation. One way to understand the proof we just gave *schematically* is to see its first nine lines as an instance of the following (meta)derivation, where we take C to abbreviate $p_1 \supset (p_1 \lor p_2)$ and D to abbreviate $(p_1 \lor p_2) \supset (p_2 \lor p_1)$.

---

[10] Of course, when we ask you in a problem to produce a derivation, you should also provide such line numbers and justifications.

1. $\vdash C$ (hypothesis)
2. $\vdash C \supset (D \supset C)$ (axiom PL5)
3. $\vdash D \supset C$ (MP 1, 2)
4. $\vdash (D \supset C) \supset [(D \wedge D) \supset (C \wedge D)]$ (axiom PL3)
5. $\vdash (D \wedge D) \supset (C \wedge D)$ (MP 3, 4)
6. $\vdash D \supset (D \wedge D)$ (axiom PL1)
7. $\vdash D$ (hypothesis)
8. $\vdash D \wedge D$ (MP 6, 7)
9. $\vdash C \wedge D$ (MP 5, 8)

In our example, C and D are specific instances of axioms. But we can see from the above derivation schema that whenever C and D are derivable, so is $C \wedge D$. Any derivation that contains C and D can be extended by the above inferences (with the justifications on the left suitably adjusted), and turned into a derivation of $C \wedge D$.

In our derivations we will often need to make use repeatedly of the following strategy. Suppose we have that $\vdash_{M_0} A \supset B$ and $\vdash_{M_0} B \supset C$. Intuitively, it should be possible to prove that $\vdash_{M_0} A \supset C$ as well. But modus ponens is our only rule, and does not allow us to make this inference directly. But there is a somewhat more complicated proof:

1. $\vdash A \supset B$ (hypothesis)
2. $\vdash B \supset C$ (hypothesis)
3. $\vdash (A \supset B) \wedge (B \supset C)$ (by the above proof)
4. $\vdash [(A \supset B) \wedge (B \supset C)] \supset (A \supset C)$ (axiom PL4)
5. $\vdash A \supset C$ (MP 3, 4)

We can think of the derivation as establishing the property of transitivity for $\supset$. We can abbreviate the above derivation in the form of a *derived rule*. We will be able to appeal to such a rule to show in M_0 (and in the other calculi as well), that if we have derived the formulas $A \supset B$ and $B \supset C$ then we can also derive $A \supset C$.

Thus, we have shown that in addition to modus ponens we have at our disposal two derived rules:

∧INTRO: If $\vdash_{M_0} A$ and $\vdash_{M_0} B$, then $\vdash_{M_0} A \wedge B$.

⊃TRANS: If $\vdash_{M_0} A \supset B$ and $\vdash_{M_0} B \supset C$, then $\vdash_{M_0} A \supset C$.

Note that since the derivations these rules abbreviate only use axioms of M_0 and modus ponens, the rules also work in J_0 and K_0. The first nine lines in our original derivation, formulated schematically, can now be abbreviated as follows:

1. $\vdash A \supset (A \vee B)$ (axiom PL7)
2. $\vdash (A \vee B) \supset (B \vee A)$ (axiom PL8)
3. $\vdash (A \supset (A \vee B)) \wedge ((A \vee B) \supset (B \vee A))$ (∧INTRO 1, 2)

Using ⊃TRANS, we can abbreviate the entire derivation as:

1. $\vdash A \supset (A \vee B)$ (axiom PL7)
2. $\vdash (A \vee B) \supset (B \vee A)$ (axiom PL8)
3. $\vdash A \supset (B \vee A)$ ($\supset$TRANS 1, 2)

The above example shows that working in an axiomatic calculus is not an easy task. The problem is that we have only one rule of inference, modus ponens, and it is not possible to develop simple heuristic strategies that would allow us, starting from a sentence B which we would like to prove, to find sentences A and $A \supset B$ that are provable and would thus yield B. In the absence of any effective heuristics, one develops (a) a knack for what might work, and (b) a set of schematic routine derivations. While developing (a) is a matter of intuition and extended familiarity with deriving sentences, one can say something more precise about (b). In general, the establishment of routine schematic derivations such as those we have presented for $\wedge$INTRO and $\supset$TRANS provide very effective tools for proving theorems in an axiomatic context. Even more useful will be the deduction theorem to be presented in Section 2.8. For this reason, the exercises in Problem 2.7, where the use of the deduction theorem is still not available, will be accompanied by very detailed hints given in Appendix D. You may appeal at this stage to $\wedge$INTRO and $\supset$TRANS. Moreover, you may appeal to the result of all the previous exercises when proving the statement of any exercise. One final word about the rationale behind the choice of the exercises. The exercises are set up in such a way as to provide all the theorems needed for the results to be proved in Section 2.15. Thus, they are not meant to provide an exhaustive set of theorems about the various interactions between the connectives, or later, among the connectives and the quantifiers.

Problem 2.7. Prove the following:[11]

E1. $\vdash_{M_0} (A \wedge B) \supset A$ [2.2]

E2. $\vdash_{M_0} A \supset A$ [2.21]

E3. $\vdash_{M_0} (A \wedge B) \supset B$ [2.22]

E4. If $\vdash_{M_0} B$ then $\vdash_{M_0} A \supset (A \wedge B)$. We will call this derived rule E4.

E5. If $\vdash_{M_0} B$ then $\vdash_{M_0} A \supset B$. We will call this derived rule E5.

E6. $\vdash_{M_0} [A \supset (B \wedge C)] \supset [A \supset (C \wedge B)]$

E7. If $\vdash_{M_0} A \supset (B \supset C)$ and $\vdash_{M_0} A \supset (C \supset D)$, then $\vdash_{M_0} A \supset (B \supset D)$. We will call this derived rule E7.

E8. $\vdash_{M_0} [(A \supset B) \wedge (C \supset D)] \supset [(A \wedge C) \supset (B \wedge D)]$ [2.23]

E9. $\vdash_{M_0} [(A \supset B) \wedge (A \supset C)] \supset [A \supset (B \wedge C)]$ [2.24a]

E10. $\vdash_{M_0} [A \supset (B \wedge C)] \supset [(A \supset B) \wedge (A \supset C)]$ [2.24b]

---

[11] The numbers in square brackets again refer to the numbering of the same theorems in Heyting (1930).

E11. $\vdash_{M_0} [B \wedge (A \supset C)] \supset [A \supset (B \wedge C)]$ [2.25]

E12. $\vdash_{M_0} B \supset [A \supset (A \wedge B)]$ [2.26]

E13. $\vdash_{M_0} [A \supset (B \supset C)] \supset [(A \wedge B) \supset C]$ [2.27a]

E14. $\vdash_{M_0} [(A \wedge B) \supset C] \supset [A \supset (B \supset C)]$ [2.27b]

With the above exercises in place we can now prove a theorem which will be central for establishing the deduction theorem (in Section 2.8).

Theorem 2.8. $\vdash_{M_0} [A \supset (B \supset C)] \supset [(A \supset B) \supset (A \supset C)]$

Proof.

1.	$\vdash \{[A \supset (B \supset C)] \wedge (A \supset B)\} \supset \{A \supset [(B \supset C) \wedge B]\}$	(E9 of Problem 2.7)
2.	$\vdash \{A \supset [(B \supset C) \wedge B]\} \supset \{A \supset [B \wedge (B \supset C)]\}$	(E6 of Problem 2.7)
3.	$\vdash \{[A \supset (B \supset C)] \wedge (A \supset B)\} \supset \{A \supset [B \wedge (B \supset C)]\}$	($\supset$TRANS 1, 2)
4.	$\vdash [B \wedge (B \supset C)] \supset C$	(axiom PL6)
5.	$\vdash \{[A \supset (B \supset C)] \wedge [A \supset B]\} \supset \{[B \wedge (B \supset C)] \supset C\}$	(E5 of Problem 2.7, 4)
6.	$\vdash \{[A \supset (B \supset C)] \wedge (A \supset B)\} \supset (A \supset C)$	(E7 of Problem 2.7, 3, 5)
7.	$\vdash (6) \supset \{[A \supset (B \supset C)] \supset [(A \supset B) \supset (A \supset C)]\}$	(E14 of Problem 2.7)
8.	$\vdash [A \supset (B \supset C)] \supset [(A \supset B) \supset (A \supset C)]$	(MP 6, 7)

In line (7), the antecedent of the formula is the entire preceding formula; we've abbreviated it by its line number (6).   □

2.6 Derivations from assumptions and provability

If you have been diligent and at least attempted the previous derivations, you will have noticed how frustrating proving theorems in an axiomatic system for logic can be. We wanted you to experience first hand this complexity of axiomatic calculi, however, because only then you can develop an appreciation for our first proof-theoretic result. The theorem is due to Herbrand and it is called *the deduction theorem*. Before turning to this theorem (see Section 2.8) we need to describe proofs by induction (see Section 2.7). We first need to extend the notion of derivation. To that effect, we expand our meta-variables to also include Greek letters Γ, Δ, etc., for sets of formulas.[12]

Definition 2.9. A formula C is *derivable* in any of our calculi M_0, J_0, or K_0 from a set Γ of assumptions if there is a finite sequence of formulas $A_1, \ldots, A_n$ such that

1. for each $1 \leq i \leq n$, A_i is either an axiom of the relevant calculus or a formula in Γ, or is obtained by modus ponens from formulas A_k, A_ℓ (with $1 \leq k, \ell < i$) and

---

[12] The Greek alphabet is provided in Appendix A.

2. $A_n = C$

When this is the case, we say that $A_1, \ldots, A_n$ is a *derivation of C from a set Γ of assumptions* in the relevant calculus $\mathbf{S}_0$ and we write $\Gamma \vdash_{\mathbf{S}_0} C$. When the system we are working in is understood, we simply write $\Gamma \vdash C$.

If $\Gamma = \{B_1, \ldots, B_m\}$ we often write $B_1, \ldots, B_m \vdash C$ for $\{B_1, \ldots, B_m\} \vdash C$. If Γ contains just one formula, e.g., $\Gamma = \{B\}$, we write $B \vdash C$ for $\{B\} \vdash C$.[13]

Notice that Γ can be empty, in which case we have a derivation from the empty set of assumptions and we write $\vdash C$. If a formula C is derivable from the empty set of assumptions then we say that C is *provable* (in the appropriate system). Thus, provability is the special case of derivability which occurs when the set of assumptions is empty.

It is occasionally useful to display one of the assumptions. For instance, we can write $\Gamma \cup \{A\}$ to display the assumption A, whether or not $A \in \Gamma$. $\Gamma \cup \{A\}$ will also be written as Γ, A.[14]

Here are some obvious consequences of the definition of derivability and provability for all our systems:

1. If $A \in \Gamma$, then $\Gamma \vdash A$.

2. If $\Gamma \vdash A$ and $\Gamma \subseteq \Gamma^\star$, then $\Gamma^\star \vdash A$.[15]

   In particular, if Γ is empty, then A can also be proved using any number of additional assumptions.

3. If A is an axiom then $\vdash A$.

One slightly less obvious consequence is the following:

Proposition 2.10. *If $\Gamma \vdash B$ and $\Delta \vdash B \supset C$, then $\Gamma \cup \Delta \vdash C$.*

Proof. By assumption, there are sequences of formulas

$$B_1, \ldots, B_n = B \text{ and}$$
$$C_1, \ldots, C_m = B \supset C$$

satisfying the conditions of Definition 2.9. Then

$$B_1, \ldots, B_n, C_1, \ldots, C_m, C$$

---

[13] The curly braces $\{\ldots\}$ are here used to collect $B_1, \ldots, B_m$ into a set (see Appendix B). Do not confuse them with the { and } we have used before instead of parentheses in formulas to increase readability.

[14] If they are unfamiliar to you, the symbols $\cup, \in, \subseteq$, and other notations of set theory are defined in Appendix B.

[15] This property is called *monotonicity*. It holds for all the logics we will investigate in this book. There are logics in which this property does not hold (non-monotonic logics), however.

is a derivation of C from $\Gamma \cup \Delta$, since C follows from the formulas $B_n = B$ and $C_m = B \supset C$ by modus ponens, and these occur before C in the sequence. □

Proposition 2.11.   1. *If $\Gamma \vdash A$ and $\Delta \vdash B$, then $\Gamma \cup \Delta \vdash A \wedge B$.*

   2. *If $\Gamma \vdash A \supset B$ and $\Delta \vdash B \supset C$, then $\Gamma \cup \Delta \vdash A \supset C$.*

Proof. As in the previous proposition, using derived rules ∧INTRO and ⊃TRANS. □

This proposition and Proposition 2.10 justify the use of MP, ∧INTRO, and ⊃TRANS also when we give derivations from assumptions.

2.7   Proofs by induction

In order to establish proof-theoretic results, i.e., results about all derivations, we need a proof method that we can apply to derivations. This is the method of *proof by induction*. We introduce it in this section; in the next section we will apply it to prove the deduction theorem.

Induction is a method of proof that's ubiquitous in mathematics. In its most basic form, it is a method of proof that applies to (properties of) the natural numbers, $0, 1, 2, \ldots$. P may be a property such as "is even or odd," which is short for "can be expressed as $2k$ or $2k + 1$ for some natural number k." We'll write $P(n)$ for "P applies to the number n." So for instance, if P is "is either even or odd," then $P(n)$ means "there is some number k such that $n = 2k$ or $n = 2k + 1$." To prove $P(n)$ for all n, we use "induction on n." The most basic form of induction is what's called "successor induction."

Definition 2.12 (Successor induction). A property P applies to all natural numbers provided:

(0) P holds for 0, and

     for arbitrary $n \geq 0$,

(n#)   if P holds for n it also holds for $n + 1$.

Suppose that (0) is true, i.e., we have $P(0)$. Since (n#) is true for every n, it is true in particular for $n = 0$. Let's call that special case (0#): if $P(0)$ then $P(0 + 1)$, i.e., $P(1)$. But we know, from (0), that $P(0)$. So, $P(1)$. But we also have (1#): if $P(1)$ then $P(2)$. Since we've just shown that $P(1)$, we can infer $P(2)$, etc. We see that if both (0) is true, and (n#) true for every n, then $P(n)$ holds for all n.

To give a proof by induction that, for all n, $P(n)$ holds, we have to prove both (0) and prove (n#) for arbitrary n (for the P in question). Let's prove that every natural number n can be written as $2k$ or $2k + 1$, for some number k. Then $P(n)$ is "there is a k such that $n = 2k$ or $n = 2k + 1$." We have to prove both (0) and (n#), i.e., we have to prove:

(0) There is some k such that $0 = 2k$ or $0 = 2k + 1$, and,

for arbitrary $n \geq 0$,

(n#) if there is a k such that $n = 2k$ or $n = 2k + 1$, then there is some k', such that $n + 1 = 2k'$ or $n + 1 = 2k' + 1$.

(0) is called the *induction basis*. It's clearly true: take $k = 0$. Then $0 = 2k$, and so also either $0 = 2k$ or $0 = 2k + 1$. $P(0)$ holds.

(n#) is called the *inductive step*. We have to prove $P(n + 1)$ for arbitrary n, but in our proof we are allowed to assume that $P(n)$ is true. So suppose $P(n)$ is true: there is some k such that $n = 2k$ or $n = 2k + 1$. This is called the *inductive hypothesis*. From it, we have to show that $P(n + 1)$ is also true: there is some k' such that $n + 1 = 2k'$ or $n + 1 = 2k' + 1$. We'll prove this by "proof by cases": if it follows separately from $n = 2k$ and also from $n = 2k + 1$, then we know it follows from "either $n = 2k$ or $n = 2k + 1$."

1. $n = 2k$. Then $n + 1 = 2k + 1$. So there is a k' (namely k itself) such that $n + 1 = 2k' + 1$, and so also either $n + 1 = 2k'$ or $n + 1 = 2k' + 1$.

2. $n = 2k + 1$. Then $n + 1 = (2k + 1) + 1$. But $(2k + 1) + 1 = 2k + 2 = 2(k + 1)$. So there is a k' (namely $k + 1$) such that $n + 1 = 2k'$, hence also either $n + 1 = 2k'$ or $n + 1 = 2k' + 1$.

In a proof by successor induction, the crucial inductive step is from $P(n)$ to $P(n + 1)$ for all $n \geq 0$. This is of course equivalent to the step from $P(n - 1)$ to $P(n)$ for all $n \geq 1$. So we can think of the inductive step as a proof of $P(n)$, for arbitrary $n \geq 1$, in which we're allowed to assume $P(n - 1)$, i.e., allowed to assume that P applies to a number $< n$. It is often useful to allow not just the specific case for $n - 1$, but *any* number $< n$, in the proof of $P(n)$. This is called *strong induction*.

Definition 2.13 (Strong induction). A property P applies to all natural numbers provided:

(0) P holds for 0, and

for arbitrary $n > 0$,

(n^*) if $P(m)$ holds for all $0 \leq m < n$, then $P(n)$.

It is not hard to see how (0) together with (n^*) for arbitrary n, guarantees that $P(n)$ itself is true for all n. For given that $P(0)$ is true—by (0)—$P(m)$ is then true for all $m < 1$. That's the left-hand side of (1^*), so we get $P(1)$. Now we know $P(0)$ and $P(1)$ are both true, i.e., $P(m)$ is true for all $m < 2$. By (2^*), $P(2)$ is true. And so on.

A proof by strong induction for any specific property P will generally proceed by showing two things. First we prove the induction basis (0), i.e., that $P(0)$ holds. Then we prove the inductive step: show that for an arbitrary $n > 0$, $P(n)$ holds on the assumption that $P(m)$ holds for all $m < n$. The assumption that $P(m)$ already holds for all $m < n$ is again called the *inductive hypothesis*. The proof of (n^*)

typically does not use the full inductive hypothesis; often it just involves showing that $P(n)$ holds on the assumption that $P(m_i)$ holds for some $m_1, \ldots, m_k$, all $< n$. How many and which $m_1, \ldots, m_k$ we "reduce" the truth of $P(n)$ to depends on what we're proving and how we prove it, of course. Since often it won't be known which m_i we need, it is convenient to assume the truth of all $P(m)$ for all $m < n$, i.e., (n^*).

Sometimes the property P does not, in fact, hold of all numbers, but, e.g., only of all numbers ≥ 1 or $\geq k$ for some k. The same principles of induction apply, except instead of proving $P(0)$ as the induction basis, we prove $P(k)$, and instead of proving $(n\#)$ for arbitrary $n \geq 0$ we prove it for arbitrary $n \geq k$. In the case of strong induction, we prove $P(k)$ and that, for arbitrary $n \geq k$,

(n^*_k) if $P(m)$ for all m where $k \leq m < n$, then $P(n)$.

We'll leave it as an exercise for you to convince yourself that this works, i.e., it guarantees that $P(n)$ is true for all $n \geq k$.

Let's again consider an example: let P now be the property "is a product of one or more primes." A prime is any number $p > 1$ that is only evenly divisible by 1 and p itself. (By convention, we let a single prime count as a product of primes.) Neither 0 nor 1 count as primes, so $P(n)$ is not true for 0 or 1. But, every number ≥ 2 is a product of primes. Let's prove this by strong induction.

The induction basis in this case is $P(2)$. But 2 is prime, so a product of primes.

For the inductive step, we have to show $P(n)$ for arbitrary $n > 2$, but in our proof we are allowed to assume that $P(m)$ holds whenever $2 \leq m < n$. Now either n is a prime, or it is not. If it is a prime, it is a product of a single prime (itself) and we have nothing more to show. Otherwise, it is not a prime: then it is divisible by some number m_1 that is neither 1 nor n itself. In other words, $n = m_1 \cdot m_2$ for some m_1 and m_2. Clearly, both m_1 and m_2 are $< n$ and ≥ 2 (since neither is $= n$, $= 0$, or $= 1$). We're allowed to assume that P holds for any number m such that $2 \leq m < n$, so in particular that P holds for m_1 and m_2. So, m_1 and m_2 are both products of primes by inductive hypothesis. But then $n = m_1 \cdot m_2$, i.e., n is also a product of primes. So in the second case as well, $P(n)$ holds.

We've seen above why, in the general case, (0) and (n^*) together suffice to show that $P(n)$ is true for all n. There's another way to see it, based on the fact that there are no infinite decreasing sequences of natural numbers. Suppose the proof of the inductive step involves just one $m_1 < n$. Now consider any n. The proof of (n^*) shows that $P(n)$ is true provided $P(m_1)$ is true. Now apply the proof of (n^*) to $P(m_1)$ instead of $P(n)$: $P(m_1)$ is true if $P(m_2)$ is true for some $m_2 < m_1$. By repeating this, we get a decreasing sequence of numbers, $n > m_1 > m_2 > \ldots$. Now in the natural numbers—and this is the important part—any such sequence must eventually terminate at 0, say, $m_k = 0$. But (0) establishes that $P(0)$, i.e., $P(m_k)$. And since we have a chain of conditionals, "if $P(m_k)$ then $P(m_{k-1})$," $\ldots$, "if $P(m_2)$ then $P(m_1)$," "if $P(m_1)$ then $P(n)$," we get the truth of $P(n)$.

If the proof of (n^*) doesn't depend on just one $m_1 < n$ but several, then instead of generating one decreasing sequence of numbers we get many. But each of them eventually must reach 0, and the induction basis and the corresponding chain of conditionals eventually shows that $P(n)$ must be true.

Let's consider another application of induction, this time not to purely arithmetical properties of numbers but to properties of other kinds of objects. Recall that a formula is an expression that can be generated from propositional variables using the inductive clause of Definition 2.2. Let's say that the propositional variables themselves are generated at stage 1, anything that we get from them by the inductive clause is generated in stage 2, formulas of stage 3 are those, not previously generated, that can be generated from formulas generated in the previous two stages by the inductive clause, etc. Clearly, something is a formula if, and only if, it is generated at some stage n. Now, if we wanted to prove that all *formulas* have some property Q, how could we use induction? We would prove that the following property P is true for all $n \geq 1$: "Q holds of all formulas generated at stage n." Since every formula is generated at some stage n, this would show that Q holds of all formulas.

For instance, here's a property that all formulas have: the number of left and right parentheses is always the same. So let that be our property Q. Then $P(n)$ is: "all formulas generated in stage n have property Q." This is now a property of natural numbers, so we can use induction.

Induction basis: $P(1)$ is "all formulas generated in stage 1 have the same number of left and right parentheses." This is clearly true, since the formulas generated in stage 1 are all (and only) the propositional variables: they have 0 left and 0 right parentheses.

Inductive hypothesis: All formulas generated in any stage m where $1 \leq m < n$ have the same number of left and right parentheses.

Inductive step: We now have to show that all formulas generated in stage n have the same number of left and right parentheses. A formula generated in stage n is either $\neg A$, $(A \vee B)$, $(A \wedge B)$, or $(A \supset B)$, where A and B are formulas generated in some stage stage $< n$. By inductive hypothesis, the number of left and right parentheses in A are equal, and so are the number of left and right parentheses in B. But then the number of left parentheses in, say, $(A \wedge B)$, is equal to the number of right parentheses in it: The number of left parentheses is 1 plus the sum of the number of left parentheses in A and of those in B, which equals 1 plus the sum of right parentheses in A and B by inductive hypothesis, which is the number of right parentheses in $(A \wedge B)$.

Problem 2.14.

E1. Prove by induction that for all $n \geq 1$, $1 + 3 + 5 + \cdots + (2n - 1) = n^2$.

E2. Prove by induction that for all $n \geq 2$, $(1 - 1/2)(1 - 1/3)\ldots(1 - 1/n) = 1/n$.

Problem 2.15. We define by induction the class of terms T given as follows. Let a, b be individual constants and $+$ and $\times$ binary functions.

1. *Basis: a, b are terms in T.*

2. *Inductive clause:* if t_1 and t_2 are in T so are $(t_1 + t_2)$ and $(t_1 \times t_2)$.

We now define the notion of denotation of a term in T.

1. *Basis:* The denotation of a and b is either 0 or 1.

2. *Inductive clause:* If the denotation of t_1 is k_1 and the denotation of t_2 is k_2 then the denotation of $(t_1 + t_2)$ is 1 in all cases except when k_1 and k_2 are both 0; and the denotation of $(t_1 \times t_2)$ is 1 just in case both k_1 and k_2 are 1, and 0 otherwise.

Define now for any term $t \in T$ the following function $\deg(t)$:

1. *Basis:* If t is a or b, then $\deg(t) = 0$.

2. *Inductive clause:* If t is $(t_1 + t_2)$ or $(t_1 \times t_2)$ then $\deg(t) = \deg(t_1) + \deg(t_2) + 1$.

E1. Prove by induction that every term in T has an odd number of symbols.

E2. Prove by induction the following statement: For all $t \in T$, for all $n \in \mathbb{N}$, if $\deg(t) = n$ then whenever a and b denote the value 0, t also denotes the value 0.

2.8   The deduction theorem

The deduction theorem justifies why one can prove a conditional in the calculus by making use of the antecedent of the conditional as a hypothetical assumption to derive the consequent of the conditional. It is exactly the kind of move we would have liked to use when proving the results of Problem 2.7. However, before employing such a derivational strategy we must first justify it.

Theorem 2.16. *If $\Gamma, A \vdash B$, then $\Gamma \vdash A \supset B$.*

Proof. The proof of the result will be by induction on the length of the sequence $B_1, \ldots, B_n$, which is the derivation of B (i.e., B_n) from $\Gamma \cup \{A\}$.[16] Although we are proving something about all derivations, our proof proceeds by induction

---

[16] Here $\{A\}$ indicates the set containing the formula A and this use of curly braces should not be confused with that we use to make formulas more readable.

on a *measure* of derivations, namely its length—the number n of formulas in the sequence $B_1, \ldots, B_n$. If you want to think of this proof instead as a proof that all $n \geq 1$ have a property $P(n)$, you can think of $P(n)$ as:

For every derivation of B from $\Gamma \cup \{A\}$ of length n, there is a derivation of $A \supset B$ from Γ.

In order to apply the principle of induction, we have to prove $P(1)$ (the induction basis), and for arbitrary $n > 1$, if $P(m)$ for all $m < n$, then $P(n)$ (the inductive step). In other words, we show that if B has a derivation from $\Gamma \cup \{A\}$ of length 1 (in which case $B_1 = B$), the result holds. Then we assume that the result holds for all derivations of length $m < n$ and show that it holds for any derivation of length n.

   Induction basis: In the basis case the length of the derivation is 1, i.e., $n = 1$. There are two sub-cases:

Case 1. B is an axiom. We have $\Gamma \cup \{A\} \vdash B$. We need to prove $\Gamma \vdash A \supset B$.

1.	$\vdash B$	(axiom)
2.	$\vdash A \supset B$	(E5 of Problem 2.7)
3.	$\Gamma \vdash A \supset B$	(monotonicity)

Case 2. $B \in \Gamma \cup \{A\}$. Here we have two further sub-cases:

Case 2a. $A = B$

1.	$\vdash A \supset A$	(E2 of Problem 2.7)
2.	$\Gamma \vdash A \supset A$	(monotonicity)

   This is our desired result, since by assumption $A = B$.

Case 2b. $A \neq B$

1.	$\Gamma \vdash B$	($B \in \Gamma$)
2.	$\vdash B \supset (A \supset B)$	(axiom PL5)
3.	$\Gamma \vdash A \supset B$	(MP 1, 2)

   Again, this is our desired result.

Notice that while we gave the argument for a derivation of B of length 1, the result established is more general. Namely, if we have a derivation of B of length n, and B_i for $i < n$ is either an axiom or belongs to Γ, then we automatically know that if $\Gamma \cup \{A\} \vdash B_i$ then $\Gamma \vdash A \supset B_i$. We will exploit this fact in the inductive step.

   Inductive hypothesis: For every derivation of B from $\Gamma \cup \{A\}$ of length $j < n$, there is a derivation of $A \supset B$ from Γ.

   Inductive step: Assume $\Gamma \cup \{A\} \vdash B$, where $B_1, \ldots, B_n$ is the derivation of B ($= B_n$). That is, each B_j with $j \leq n$ is either an axiom, or a sentence in $\Gamma \cup \{A\}$, or there are i and $k < j$ such that $B_k = B_i \supset B_j$ (i.e., B_j follows from some previous formulas B_i and B_k by modus ponens).

Now consider B_n: If B_n is an axiom or a formula in $\Gamma \cup \{A\}$ then we simply apply the considerations used in the basis case and reach the desired result. Thus, the only interesting case is when B_n is obtained by modus ponens from previous formulas B_i, B_k (where $i, k < n$). Without loss of generality assume $i < k$. Then B_k has the form $B_i \supset B_n$. Now clearly

$$B_1, \ldots, B_i \text{ and } \qquad\qquad B_1, \ldots, B_k = (B_i \supset B_n)$$

are derivations of B_i and $B_k = B_i \supset B_n$, respectively, from $\Gamma \cup \{A\}$. Their lengths are i and k, and both are $< n$. So the inductive hypothesis applies to them. We have:

$$\Gamma \vdash A \supset B_i \text{ and } \qquad\qquad \Gamma \vdash A \supset (B_i \supset B_n)$$

We have seen in Theorem 2.8 that in all our calculi we have:

$$\vdash [A \supset (B_i \supset B_n)] \supset [(A \supset B_i) \supset (A \supset B_n)].$$

Since we have $\Gamma \vdash A \supset (B_i \supset B_n)$, by MP we obtain:

$$\Gamma \vdash (A \supset B_i) \supset (A \supset B_n)$$

But we also have $\Gamma \vdash A \supset B_i$. One more application of modus ponens gives us $\Gamma \vdash A \supset B_n$. □

Of course, the converse of the deduction theorem also holds: it merely requires an application of modus ponens. Suppose $\Gamma \vdash A \supset B$. Add to the derivation the formulas A and B. B is justified by MP from the preceding two lines. The resulting sequence of formulas, therefore, is a derivation of B from $\Gamma \cup \{A\}$. (Alternatively, we can apply Proposition 2.10, observing that $A \vdash A$.)

One example will be sufficient to show how much easier it is now to prove conditional theorems. Suppose we want to prove

$$[A \wedge ((A \wedge B) \supset C)] \supset (B \supset C)$$

Assume we have reached

$$\{A \wedge ((A \wedge B) \supset C)\} \cup \{B\} \vdash C$$

by means of the following derivation:

1.	$A \wedge ((A \wedge B) \supset C)$	(assumption)
2.	B	(assumption)
3.	$\vdash [A \wedge ((A \wedge B) \supset C)] \supset A$	(E1 of Problem 2.7)
4.	$(1) \vdash A$	(MP 1, 3)
5.	$\vdash [A \wedge ((A \wedge B) \supset C)] \supset [(A \wedge B) \supset C]$	(E3 of Problem 2.7)
6.	$(1) \vdash (A \wedge B) \supset C$	(MP 1, 5)
7.	$(1), (2) \vdash A \wedge B$	($\wedge$INTRO 2, 4)
8.	$(1), (2) \vdash C$	(MP 6, 7)

Note that lines 1 and 2 do not have a ⊢ in front of them. They are assumptions, not theorems. We indicate on the left of the turnstile which assumptions a line makes use of. For instance, line 4 is proved from lines 1 and 3, but line 3 is a theorem—it does not use any assumptions.

We have thus established that $\{A \wedge ((A \wedge B) \supset C)\} \cup \{B\} \vdash C$. Now by one application of the deduction theorem we know that there is a derivation from the assumption $A \wedge ((A \wedge B) \supset C)$ to $B \supset C$. By a second application we also know that from our assumptions we can derive

$$[A \wedge ((A \wedge B) \supset C)] \supset (B \supset C)$$

Instead of describing these processes of applying the deduction theorem, we can also just continue the derivation as follows:

8. $(1), (2) \vdash C$ (MP 6, 7)
9. $\quad (1) \vdash B \supset C$ (deduction theorem)
10. $\quad\quad \vdash [A \wedge ((A \wedge B) \supset C))] \supset (B \supset C)$ (deduction theorem)

Problem 2.17. Using the deduction theorem, prove the following in $\mathbf{M}_0$:

E1. $\vdash [A \supset (B \supset C)] \supset [B \supset (A \supset C)]$ [2.271]

       Hint: Use assumptions $A \supset (B \supset C)$, B, and A.

E2. $\vdash A \supset \{C \supset [(A \wedge B) \supset C]\}$ [2.28]

E3. $\vdash A \supset \{B \supset [A \supset (C \supset B)]\}$ [2.281]

E4. $\vdash (A \supset B) \supset [(B \supset C) \supset (A \supset C)]$ [2.29]

E5. $\vdash (B \supset C) \supset [(A \supset B) \supset (A \supset C)]$ [2.291]

2.9 Derivations as trees

Earlier on we said that one can consider the construction of a formula A as a tree whose nodes are labeled by the formulas that go into constructing the formula A, which will itself be the root of the tree. Trees can also be useful to represent derivations. In that case the root of the tree will be the formula to be derived and each node is either a terminal node (or leaf) or is the node joining two branches. Every leaf is labelled with an axiom or an assumption. Every non-leaf node is labelled by the conclusion, by modus ponens, of the formulas labelling the two nodes above it. We will soon rewrite the proof that $(C \wedge D)$ is a theorem if both C and D are theorems in this "tree" representation.

How do we represent a derivation in tree form? A tree is made up of nodes connected by lines called edges. Each node has a formula associated with it. Trees are defined formally in graph theory as simple connected graphs with no cycles, where a graph is an ordered pair $\langle V, E \rangle$ consisting of a set V of vertices (or nodes) and E a set of "edges" or connecting lines between nodes (technically, a set of ordered pairs of vertices). We do not need to explain the mathematical subtleties of this definition for our purposes.

It is enough to point out that a derivation can be represented as a tree with the end-formula of the derivation at the bottom of the tree and where the initial nodes are the axioms (or assumptions) of the derivations. An application of modus ponens connects two nodes of the tree to a single formula which is the consequence of the inference. For instance:

$$\frac{A \qquad A \supset (B \supset A)}{B \supset A}$$

Unlike the axiomatic derivations we have studied, in a tree representation of the derivation each use of a formula requires a different derivation of that formula. Consider for instance the following axiomatic derivation.

1.		$A \supset B$	(assumption)
2.		$(A \supset B) \supset [C \supset ((A \supset B) \supset D)]$	(assumption)
3.		C	(assumption)
4.	$(1), (2) \vdash$	$C \supset ((A \supset B) \supset D)$	(MP 1, 2)
5.	$(1), (2), (3) \vdash$	$(A \supset B) \supset D$	(MP 3, 4)
6.	$(1), (2), (3) \vdash$	D	(MP 1, 5)

In the tree representation the two appeals to line (1) require two different leaves in the tree.

$$\frac{\dfrac{\boxed{A \supset B} \qquad (A \supset B) \supset [C \supset ((A \supset B) \supset D)]}{\dfrac{C \supset ((A \supset B) \supset D) \qquad\qquad C}{(A \supset B) \supset D} \qquad \boxed{A \supset B}}}{D}$$

We boxed $A \supset B$ to emphasize the need to single out by a different node the repeated use of $A \supset B$ (which in the axiomatic derivation corresponds to appealing twice to line 1). However, the set of premises or assumptions required for proving D remains the same. More formally we could define inductively the notion of "δ is a derivation of E from a set Δ of assumptions" as follows:

1. *Basis:* If E is an axiom then E is a derivation of E from an empty set Δ of assumptions. If E is an assumption then E is a derivation of E from $\{E\}$.

2. *Inductive clause:* Let now δ_1 be a derivation of A from assumptions Δ_1 and δ_2 be a derivation of $A \supset B$ from assumptions Δ_2. Then

$$\delta = \left\{ \begin{array}{cc} \Delta_1 & \Delta_2 \\ \vdots\, \delta_1 & \vdots\, \delta_2 \\ \vdots & \vdots \\ \dfrac{A \qquad A \supset B}{B} & \text{MP} \end{array} \right.$$

is a derivation of B from $\Delta_1 \cup \Delta_2$.

If we revisit our previous example (starting at the top), we have that $A \supset B$ is a derivation from $\{A \supset B\}$. $(A \supset B) \supset [C \supset ((A \supset B) \supset D)]$ is a derivation from $\{(A \supset B) \supset [C \supset ((A \supset B) \supset D)]\}$. Now

$$\frac{A \supset B \qquad (A \supset B) \supset [C \supset ((A \supset B) \supset D)]}{C \supset ((A \supset B) \supset D)} \text{ MP}$$

is a derivation of $C \supset ((A \supset B) \supset D)$ from $\Delta_1 \cup \Delta_2$ where

$$\Delta_1 = \{A \supset B\} \text{ and}$$
$$\Delta_2 = \{(A \supset B) \supset [C \supset ((A \supset B) \supset D)]\}$$

Proceeding inductively in this way we see that the final tree derivation is a derivation of D from $\Delta_1 \cup \Delta_2 \cup \Delta_3 \cup \Delta_4$, when $\Delta_3 = \{C\}$ and $\Delta_4 = \{A \supset B\}$. Since $\Delta_1 = \Delta_4$, the derivation is a derivation of D from the set of premises $\{A \supset B, (A \supset B) \supset [C \supset ((A \supset B) \supset D)], C\}$ which is of course what one would expect. When we have a derivation of E from the empty set of premises, then we say that E is a theorem.

You will notice that axioms and assumptions always appear as leaves (terminal nodes) of the tree (with the possible exception of the root).

2.10   Negation

Recall that the intuitionistic calculus J_0 has axiom PL11,

$$\neg A \supset (A \supset B),$$

which the minimal calculus M_0 does not have. Moreover, K_0 is obtained by adding to J_0 axiom PL12:

$$\neg \neg A \supset A.$$

In this section we will try to get a better grasp of what holds in these different systems.

2.10.1   The informal interpretation of M_0, J_0, and K_0

The calculi J_0 and M_0 were first presented by Heyting (1930) and Johansson (1937), respectively. J_0 can be seen as incorporating only those principles of classical logic that are intuitionistically valid. Classical logic was codified in the work of Frege (1879) and Whitehead and Russell (1910–1913). The propositional part of classical logic is partly characterized by the axiom $\neg \neg A \supset A$ or, equivalently (over J_0), by the axiom $A \vee \neg A$. These principles correspond to the intuitive interpretation underlying classical logic, namely that the truth of a proposition is independent of the knowing subject and that each proposition is either true or false (bivalence). These assumptions are then codified in the truth-tables for the connectives which decree that if a proposition A is true (respectively, false) then

¬*A* is false (respectively, true). Thus, it is always the case that *A* ∨ ¬*A* is true since (under a classical interpretation of ∨) it must be the case that either *A* is true or ¬*A* is true and thus the disjunction must hold.

Heyting wrote to Oskar Becker in 1933 (see Troelstra, 1990) that he found his axiomatization for intuitionistic logic by going through the axioms of *Principia Mathematica* and checking which axioms would remain valid according to the intuitionistic interpretation of the logical connectives and quantifiers. We will not attempt here to give a precise exposition of the intuitionistic interpretation of the connectives (a matter which is to a certain extent still being debated).[17] However, the basic intuition is that the logical connectives express the successful carrying out of certain constructions by a knowing subject. For instance, to assert the correctness of *A* ∧ *B* is to assert that we have a successful construction (a proof) of *A* and a successful construction (a proof) of *B*. Specifying what such constructions (or proofs) are is vital to the foundational understanding of intuitionism, but we do not need to settle the matter here.

More important for us is to point out that by connecting the notion of correctness (or "truth") to that of a successful construction by a knowing subject, the meaning of the connectives is profoundly altered compared to the classical case. For instance, intuitionistic negation ¬*A* will be interpreted as meaning that we have (or can effectively find) a method that will transform every construction (proof) of *A* into a construction (proof) of a contradictory statement (such as, say, 0 = 1). Intuitionistic disjunction, *A* ∨ *B*, will mean that we have (or can effectively find) either a construction (proof) of *A* or a construction (proof) of *B*.

But now consider *A* ∨ ¬*A*. Under the interpretation just sketched it is in general not true that we either have a construction (proof) of *A* or a construction (proof) that the assumption *A* leads to a contradiction, nor is there any guarantee that one of the two can effectively be found. For this reason the intuitionist abandons the excluded middle.

By narrowly connecting the meaning of the connectives to the carrying out of certain constructions or proofs, the intuitionist rejects the classical interpretation of truth, as independent of the knowing subject (which, by contrast, grounds the interpretation of the excluded middle in classical logic) and ties "truth" to the knowing subject. But then we lose bivalence for propositions since given any proposition *A* there is no guarantee that we have a proof of *A* or a proof that the assumption of *A* leads to a contradiction.[18]

---

[17] Heyting (1956) and Dummett (2000) provide a more extended treatment. See van Atten (2017) for a discussion of the historical development of intuitionistic logic.

[18] For simplicity of exposition we avoid a discussion of the relation between bivalence and the excluded middle. Recent work on supervaluationist semantics shows that the law of the excluded middle (*A* ∨ ¬*A*) can hold even if bivalence fails. For our purposes it is enough to observe that in classical logic the law of the excluded middle is often grounded by reference to bivalence (namely, the semantic claim that every proposition is either true or false).

One way to capture intuitionistic negation, then, is to construe it as follows: We introduce a new symbol, $\bot$, to play the role of an arbitrary contradiction.[19] Syntactically, $\bot$ is an atomic formula. We can now take $\neg A$ is equivalent to $A \supset \bot$. Thus, $A \vee \neg A$ is to be read intuitionistically as $A \vee (A \supset \bot)$. But this presupposes an understanding of $\supset$, which cannot be the standard one given by truth-tables. $A \supset B$ means intuitionistically that we have a method that transforms every proof of A into an effective proof of B. You should convince yourself that with the interpretations just given the axioms of J_0 are indeed correct, whereas $A \vee \neg A$ cannot be intuitionistically accepted as correct.

The abandonment of the principle of excluded middle has some serious consequences for classical mathematics as ordinarily understood. For instance, implementing the intuitionistic strictures on excluded middle leads to the abandonment of standard patterns of reasoning which are used in mathematics and in ordinary reasoning. One such is the pattern of "reductio ad absurdum." Suppose we want to prove A. Classically, we can assume $\neg A$ and show that the assumption leads us to a contradiction, that is, $\neg A \supset \bot$. According to the intuitionistic reading that means $\neg\neg A$. But the classical mathematician goes on to infer A from $\neg\neg A$, thereby using the principle of double negation (which, when we are working within J_0 as background theory, is equivalent to $A \vee \neg A$). Thus reductio ad absurdum does not go through according to intuitionistic strictures. The intuitionist accepts the principle that if from A we obtain a contradiction then $\neg A$ (this simply captures the definition of negation). But *reductio* in the form leading from $\neg A \supset \bot$ to A is not intuitionistically valid.

Minimal logic was formulated by Johannson as a restriction of intuitionistic logic. In minimal logic one accepts all intuitionistic principles except $\neg A \supset (A \supset B)$. Intuitionistically the principle says that we can transform every proof that A leads to a contradiction into a proof that transforms the proof of A into a proof of B. Classically the principle is equivalent to the principle "ex falso quodlibet," i.e., the fact that from any contradiction we can infer anything. Upon being introduced to classical logic, students often find this difficult to accept but do so, perhaps reluctantly, upon seeing that it does follow from the classical definition of logical consequence. Even among intuitionists the principle led to disagreements as to its acceptability (see van Atten, 2017) and it is not surprising that Johansson was led to explore a calculus without presupposing it.

The name minimal logic derives from the fact that it is a logic that respects the intuitionistic meaning of $\supset, \wedge, \vee$ and adds only those axioms for negation that should be accepted by any minimal notion of negation compatible with the intuitionistic interpretation of those connectives. Such a minimal notion of negation shares with intuitionism the definition of $\neg A$ as $A \supset \bot$. However, in minimal logic we think of $\bot$ as a propositional variable (see Johansson, 1937,

---

[19] Gentzen used $\wedge$, but $\bot$ is now standard.

p. 130). When negation is given such an etiolated meaning, axiom PL10 of the minimal calculus becomes:

$$[(A \supset B) \wedge (A \supset (B \supset \perp))] \supset (A \supset \perp)$$

But notice that this formula can be derived from axioms PL1–PL9. To regain all of intuitionistic logic one would need a principle $\perp \supset B$ as an extra assumption. In intuitionistic logic, this principle is acceptable. But the minimalist logician does not interpret $\perp$ as intuitionistic "contradiction," instead it is treated like a propositional variable. Under this interpretation, there is no ground for accepting the principle, just as the principle $p_5 \supset B$ would not be acceptable. In other words, the intuitionistic principle

$$\neg A \supset (A \supset B)$$

cannot be obtained without interpreting $\perp$ as a "full-blown" contradiction. For, the formula

$$(A \supset \perp) \supset (A \supset B)$$

can be obtained using the logic of the conditional only under the assumption $\perp \supset B$. But the latter principle is not compelling when $\perp$ goes proxy for a propositional variable.[20]

2.10.2 Minimal logic

$\mathbf{M}_0$ is weaker than $\mathbf{J}_0$. Axiom PL11, i.e., $\neg A \supset (A \supset B)$, is an axiom of $\mathbf{J}_0$ but not of $\mathbf{M}_0$. It can also not be proved from the other axioms of $\mathbf{M}_0$. Another formula which is provable in $\mathbf{J}_0$ but not in $\mathbf{M}_0$ is $(A \wedge \neg A) \supset B$.. We will prove this using the method of interpretations in Section 2.11, and proof-theoretically in Corollary 4.49.

Despite being weaker than $\mathbf{J}_0$, $\mathbf{M}_0$ proves a good deal of important properties about negation (which then are also automatically theorems of $\mathbf{J}_0$ and $\mathbf{K}_0$).

Problem 2.18. Prove in $\mathbf{M}_0$:

E1. $\vdash \neg(A \wedge \neg A)$ [4.13]

E2. $\vdash (A \supset B) \supset (\neg B \supset \neg A)$ [4.2]

E3. $\vdash (A \supset B) \supset (\neg\neg A \supset \neg\neg B)$ [4.22]

E4. $\vdash (A \supset \neg B) \supset (B \supset \neg A)$ [4.21]

E5. $\vdash A \supset \neg\neg A$ [4.3]

E6. $\vdash \neg A \supset \neg\neg\neg A$ [4.31]

E7. $\vdash \neg\neg\neg A \supset \neg A$ [4.32]

---

[20] An alternative way to characterize minimal negation is by defining $\perp$ in terms of $\neg$ as an abbreviation for $\neg p \wedge \neg\neg p$. The definition is independent of the propositional variable p (as $\neg p \wedge \neg\neg p$ is equivalent to $\neg q \wedge \neg\neg q$ in minimal logic; see Johansson, 1937, p. 129).

E8. $\vdash \neg A \supset (A \supset \neg B)$ [4.4]

E9. $\vdash \neg\neg(A \wedge B) \supset (\neg\neg A \wedge \neg\neg B)$ [4.61]

E10. $\vdash \neg\neg(A \vee \neg A)$ [4.8]

E11. $\vdash (A \supset B) \supset \neg(A \wedge \neg B)$ [4.9]

E12. $\vdash (A \wedge B) \supset \neg(\neg A \vee \neg B)$ [4.92]

Hints are given in Appendix D.2. You may use the deduction theorem.

The following key properties of double negation will be useful when giving the translation of classical arithmetic into intuitionistic arithmetic.

Lemma 2.19. *Suppose that* $\vdash_{M_0} \neg\neg A \supset A$ *and* $\vdash_{M_0} \neg\neg B \supset B$. *Then:*

1. $\vdash_{M_0} \neg\neg(A \wedge B) \supset (A \wedge B)$
2. $\vdash_{M_0} \neg\neg(A \supset B) \supset (A \supset B)$

Proof. We prove (1) and leave (2) as an exercise. Of course, we will make use of the deduction theorem.

1.	$\vdash \neg\neg A \supset A$	(hypothesis)
2.	$\vdash \neg\neg B \supset B$	(hypothesis)
3.	$\neg\neg(A \wedge B)$	(assumption)
4.	$\vdash \neg\neg(A \wedge B) \supset (\neg\neg A \wedge \neg\neg B)$	(E9 of Problem 2.18)
5.	$(3) \vdash \neg\neg A \wedge \neg\neg B$	(MP $3, 4$)
6.	$\vdash (\neg\neg A \wedge \neg\neg B) \supset \neg\neg A$	(E1 of Problem 2.7)
7.	$(3) \vdash \neg\neg A$	(MP $5, 6$)
8.	$\vdash (\neg\neg A \wedge \neg\neg B) \supset (\neg\neg B \wedge \neg\neg A)$	(axiom PL2)
9.	$(3) \vdash \neg\neg B \wedge \neg\neg A$	(MP $5, 8$)
10.	$\vdash (\neg\neg B \wedge \neg\neg A) \supset \neg\neg B$	(E1 of Problem 2.7)
11.	$(3) \vdash \neg\neg B$	(MP $9, 10$)
12.	$(3) \vdash A$	(MP $1, 7$)
13.	$(3) \vdash B$	(MP $2, 11$)
14.	$(3) \vdash A \wedge B$	($\wedge$INTRO $12, 13$)

In the above, lines 1 and 2 should be thought of as the results of two derivations which, by hypothesis, exist. Line 3 is a genuine assumption, so the complete derivation is one of $A \wedge B$ from $\neg\neg(A \wedge B)$. We can now conclude that

$$\vdash_{M_0} \neg\neg(A \wedge B) \supset (A \wedge B)$$

by applying the deduction theorem. □

Problem 2.20. Prove Lemma 2.19(2).

2.10.3   *From intuitionistic to classical logic*

The difference between the systems J_0 and K_0 consists in the fact that K_0 includes axiom PL12, i.e., $\neg\neg A \supset A$. When this schema is added to J_0 it enables us to prove the law of excluded middle, which characterizes the classical calculus.

We now show: $\vdash_{K_0} A \vee \neg A$. For any A, $\neg\neg A \supset A$ is a theorem of K_0. In particular, by substitution of $A \vee \neg A$ for A, we have:

1. $\vdash \neg\neg(A \vee \neg A) \supset (A \vee \neg A)$ (axiom PL12)
2. $\vdash \neg\neg(A \vee \neg A)$ (E10 of Problem 2.18)
3. $\vdash A \vee \neg A$ (MP 1, 2)

The fact that J_0 is weaker than K_0 can also be appreciated by mentioning that some of the standard equivalences among connectives (which hold in K_0) fail for J_0.

For instance, in J_0 we have

$$\vdash_{J_0} (A \vee B) \supset \neg(\neg A \wedge \neg B)$$
$$\vdash_{J_0} (A \wedge B) \qquad\qquad\qquad\qquad \supset \neg(\neg A \vee \neg B)$$

but not the converses, which hold only in K_0:

$$\vdash_{K_0} \neg(\neg A \wedge \neg B) \supset (A \vee B)$$
$$\vdash_{K_0} \neg(\neg A \vee \neg B) \supset (A \wedge B)$$

2.11   Independence

We have confidently asserted that $\neg A \supset (A \supset B)$ cannot be proved in M_0 and that $\neg\neg A \supset A$ cannot be proved in J_0. But how do we know that? In this section we want to explain one of the techniques for proving such results. The technique is model-theoretic (as opposed to proof-theoretic) and consists in showing that we can concoct interpretations (or models) of our axiom systems that verify the axioms and preserve the validity of the inferential rules but falsify the statement we want to show to be independent. The basic idea is that we can consider connectives as function symbols operating on "truth-values." These function symbols output certain values depending on the input value of their arguments according to a specific table of values.

These systems look very similar to the truth-tables familiar from a first logic course, except that in many cases, instead of having only two truth-values (0, 1 or T, F) we might have more than two, and occasionally more than one truth-value can play the role that T (or F) played for truth-tables. Just as we show that A does not follow from $A \vee B$ in propositional calculus by showing that there is an assignment of truth-values to A and B (namely 0, 1) which makes $A \vee B$ true (= 0) but A false (= 1), we will in our case use more complex truth-tables to show that one axiom schema is independent of the other axioms by giving a system of "truth"-tables that makes the latter axioms "true" (according to our designated values for truth) and the former proposition "false" (according to our designated values for falsity). The assignment of the numbers 0 as "true" and 1 (and later also 2) as "false" is purely conventional; often 1 is used as "true" and 0 as "false." Because we need to show independence from an axiom schema (recall that in an axiom such as $A \supset (A \wedge A)$, A is schematic) we will have to check that the truth-tables make the axiom schema true under all possible values of its propositional letters according to the values of the table and that MP preserves "truth" in the table.

Let us first prove that $\neg A \supset (A \supset B)$ and $\neg\neg A \supset A$ are not derivable in $\mathbf{M}_0$. Consider the following truth-tables for $\supset, \wedge, \vee, \neg$ where we have only two truth-values $(0, 1)$ and 0 plays the role of truth. (When using the table for a binary connective, the first argument is given by the leftmost vertical column and the second argument by the horizontal row at the top.)

$\supset$	0	1		$\wedge$	0	1		$\vee$	0	1		$\neg$	
0	0	1		0	0	1		0	0	0		0	0
1	0	0		1	1	1		1	0	1		1	0

Notice that the interpretation is purely algebraic.[21] For instance, it certainly does not capture our intuition about negation to see that whether the input is 0 or 1, our table always outputs 0. But that does not matter, as we are only interested in showing the possibility of an interpretation of the connectives as functions over truth-values which will show that the value of the two propositions is "falsity" while the other axiom and their consequences can only be "true."

We can now show, by induction on the length of proofs, that every formula provable in $\mathbf{M}_0$ has the property that it is a tautology in the sense of the non-standard truth-tables (regardless of what combination of 0 or 1 is assigned to the propositional variables, the entire formula receives value 0). If we can show that a formula doesn't have this property (it takes value 1 for some assignment), it cannot be derivable in $\mathbf{M}_0$. You should verify that all axioms of $\mathbf{M}_0$ always have value 0 according to the above tables, and further, that the rules of inference always lead from formulas with values 0 to a formula with value 0. Our two formulas, however, obtain value 1 when we assign suitable values to A and B. For the first formula, assign 0 to A and 1 to B. With these values assigned to A and B, we have that $A \supset B$ has value 1, $\neg A$ has value 0, and $\neg A \supset (A \supset B)$ has value 1. To show the possible "falsity" of the second formula, instead assign 1 to A. Then $\neg\neg A = 0$ and so $\neg\neg A \supset A$ has value 1. So, neither $\neg A \supset (A \supset B)$ nor $\neg\neg A \supset A$ is provable in $\mathbf{M}_0$.

As a consequence we see that $(A \wedge \neg A) \supset B$ cannot be proved in $\mathbf{M}_0$. Indeed suppose it were the case that $\vdash_{\mathbf{M}_0} (A \wedge \neg A) \supset B$. Then:

---

[21] In this case, all the connectives except negation are interpreted as in classical logic.

1. $\vdash (A \wedge \neg A) \supset B$ $\qquad$ (hypothesis)
2. $\vdash (\neg A \wedge A) \supset (A \wedge \neg A)$ $\qquad$ (axiom PL2)
3. $\vdash (\neg A \wedge A) \supset B$ $\qquad$ ($\supset$TRANS 1, 2)
4. $\vdash [(\neg A \wedge A) \supset B] \supset [\neg A \supset (A \supset B)]$ $\qquad$ (E14 of Problem 2.7)
5. $\vdash \neg A \supset (A \supset B)$ $\qquad$ (MP 3, 4)

But we have shown that $\nvdash_{M_0} \neg A \supset (A \supset B)$. Thus, $\nvdash_{M_0} (A \wedge \neg A) \supset B$.

To show that $A \vee \neg A$ is independent of J_0 requires a more complex truth-table with three values 0, 1, 2:

$\supset$	0	1	2
0	0	1	2
1	0	0	0
2	0	1	0

$\wedge$	0	1	2
0	0	1	2
1	1	1	1
2	2	1	2

$\vee$	0	1	2
0	0	0	0
1	0	1	2
2	0	2	2

$\neg$	
0	1
1	0
2	1

You can check that all the axioms of J_0 are verified in the above tables (with 0 as the value "true") and that modus ponens preserves "truth" (i.e., 0). However, the formula $\neg A \vee A$ receives the value 2 when we let $A = 2$, as $\neg 2 = 1$ and $1 \vee 2 = 2$. The same table shows that $\neg \neg A \supset A$ is falsified by also letting $A = 2$: $\neg \neg 2 = 0$ and $0 \supset 2 = 2$.

An intuitive interpretation of the above tables can be given by taking 0 to mean that the statement is correct (or true), 1 that the statement is false, and 2 that the statement cannot be false but it also cannot be proved correct. Regardless of the intuitive interpretation, independence is guaranteed once we show that for appropriate values of A (namely, 2), $A \vee \neg A$ and $\neg \neg A \supset A$ do not yield 0 according to the above tables, while all the axioms do and the tables preserve value 0 under modus ponens. Later in the book we will show how such independence results can also be obtained proof-theoretically (see Corollaries 4.47 and 4.49, and Proposition 6.33 and Corollary 6.34).

Problem 2.21. Using the first set of tables above, show that $A \vee \neg A \nvdash_{M_0} \neg \neg A \supset A$.

Since $\neg \neg A \supset A \vdash_{M_0} A \vee \neg A$, this shows that the equivalence between $\neg \neg A \supset A$ and $A \vee \neg A$ requires a background logical theory which is stronger than M_0. J_0 will suffice.

Problem 2.22. Show that $\neg \neg (A \vee \neg A)$ always has value 0 in the second set of tables.

Problem 2.23. Check that axioms PL1, PL2, PL5, and PL7 always yield 0 in the second set of tables for any values of A and B. Check also that modus ponens preserves value 0.

2.12 An alternative axiomatization of J_0

The following is the axiomatization of intuitionistic logic given in Gentzen's 1935c paper.[22] Let us denote it by $J_0^\star$.

G1. $A \supset A$

G2. $A \supset (B \supset A)$

G3. $(A \supset (A \supset B)) \supset (A \supset B)$

G4. $(A \supset (B \supset C)) \supset (B \supset (A \supset C))$

G5. $(A \supset B) \supset ((B \supset C) \supset (A \supset C))$

G6. $(A \wedge B) \supset A$

G7. $(A \wedge B) \supset B$

G8. $(A \supset B) \supset ((A \supset C) \supset (A \supset (B \wedge C)))$

G9. $A \supset (A \vee B)$

G10. $B \supset (A \vee B)$

G11. $(A \supset C) \supset ((B \supset C) \supset ((A \vee B) \supset C))$

G12. $(A \supset B) \supset ((A \supset \neg B) \supset \neg A)$

G13. $\neg A \supset (A \supset B)$

Inference rule:

$$\frac{A \qquad A \supset B}{B} \text{ MP}$$

Gentzen claims that the system $J_0^\star$ is equivalent to J_0.

Problem 2.24. Prove that if $\vdash_{J_0^\star} C$ then $\vdash_{J_0} C$. You may use the deduction theorem for J_0.

Problem 2.25. The deduction theorem holds for $J_0^\star$. To establish this, prove the following:

E1. If $\vdash_{J_0^\star} A \supset B$ and $\vdash_{J_0^\star} B \supset C$ then $\vdash_{J_0^\star} A \supset C$.

E2. If $\vdash_{J_0^\star} A$ and $\vdash_{J_0^\star} B$ then $\vdash_{J_0^\star} A \wedge B$.

E3. $\vdash_{J_0^\star} [A \supset (B \supset C)] \supset [(A \supset B) \supset (A \supset C)]$.

Problem 2.26. Using the deduction theorem for $J_0^\star$, prove the following:

E1. $\vdash_{J_0^\star} ((C \supset A) \wedge (C \supset B)) \supset (C \supset (A \wedge B))$.

E2. $\vdash_{J_0^\star} ((A \supset C) \wedge (B \supset C)) \supset (A \vee B \supset C)$.

---

[22] The axiomatization is due to Glivenko (1929). The axioms are the same as used in Section V, §2, of Gentzen (1935c).

E3. If $\vdash_{J_0^\star} A \supset B$, then $\vdash_{J_0^\star} (A \wedge C) \supset (B \wedge C)$.

E4. If $\vdash_{J_0^\star} B \supset A$, then $\vdash_{J_0^\star} (B \vee C) \supset (A \vee C)$.

E5. If $\vdash_{J_0^\star} A \supset B$, then $\vdash_{J_0^\star} (C \wedge A) \supset (C \wedge B)$.

E6. If $\vdash_{J_0^\star} B \supset A$, then $\vdash_{J_0^\star} (C \vee B) \supset (C \vee A)$.

E7. $\vdash_{J_0^\star} (A \wedge (A \supset B)) \supset B$.

Problem 2.27. Assuming the results of all the previous exercises, show that if $\vdash_{J_0} C$, then $\vdash_{J_0^\star} C$.

2.13   Predicate logic

We now move from propositional logic to predicate logic. Recall that in predicate logic we increase our expressive power by being able to "look inside" the structure of propositions. Propositions, which are taken as basic in the propositional calculus, are seen now as constructed by means of individual variables, individual constants, predicates, functions, connectives, and quantifiers. In this way we are able to account for the validity of many patterns of reasoning that propositional logic cannot account for. First of all we specify the basic symbols of an arbitrary predicate language and we inductively define the class of terms and formulas. We then give the additional inference rules for quantifiers that are needed to extend propositional logic to predicate logic. Finally we state the deduction theorem for predicate logic.

The basic expressions for a first-order language are variables for individuals, individual constants, predicate symbols, function symbols, connectives, and quantifiers. We do not treat equality as a logical constant. This is in a sense a matter of convention but, following Gentzen, we prefer to add equality axioms as mathematical axioms to our mathematical theories. Nothing essential hinges on this.

Definition 2.28. The language of the predicate calculus consists of the language of propositional logic plus the following:

1. Free variables: $a_1, a_2, a_3, \ldots$

2. Bound variables: $x_1, x_2, x_3, \ldots$

3. Individual constants: $k_1, k_2, k_3, \ldots$

4. Predicate symbols: P_i^n where n denotes the number of arguments of the predicate. Thus: $P_1^1, P_2^1, \ldots, P_1^2, P_2^2, \ldots$, etc.

5. Function symbols: f_i^n where n denotes the number of arguments of the function. Thus: $f_1^1, f_2^1, \ldots, f_1^2, f_2^2, \ldots$, etc.

6. Quantifiers: $\forall, \exists$

We will use a, b, c as metavariables for the official free variables, x, y, z as metavariables for the official bound variables, and P, Q, R, etc., as metavariables ranging over the official predicate symbols.

Definition 2.29. The *terms* are inductively defined as follows:

1. *Basis clause:* Individual constants and free variables are terms.

2. *Inductive clause:* If $t_1, t_2, \ldots, t_n$ are terms and f is an n-ary function symbol, then $f(t_1, \ldots, t_n)$ is a term.

3. *Extremal clause:* Nothing else is a term.

A term in which no variables appear is called a *closed term*.

Note that bound variables do *not* count as terms, and terms, properly speaking, cannot contain bound variables.

Definition 2.30. The *formulas* are inductively defined as follows:

1. *Basis clause:* If $t_1, \ldots, t_n$ are terms and P is an n-ary predicate symbol, then $P(t_1, \ldots, t_n)$ is an atomic formula.

2. *Inductive clause:*

   (a) If A is a formula, then $\neg A$ is a formula.

   (b) If A and B are formulas, then so are $(A \wedge B)$, $(A \vee B)$, and $(A \supset B)$.

   (c) If A is a formula, a is a free variable occurring in A, and x is a bound variable not occurring in A, then $\forall x\, A[x/a]$ and $\exists x\, A[x/a]$ are formulas, where $A[x/a]$ is the result of replacing every occurrence of a in A by x.

3. *Extremal clause:* Nothing else is a formula.

A *sentence* is a formula with no free variables.

Before proceeding, we should explain a difference in our approach to dealing with variables that, although standard in the proof-theoretic literature, is perhaps unfamiliar. Most general logic textbooks do not make a categorical distinction between bound and free variables—there is just one kind of variable, usually written like our bound variables x_1, x_2, etc. These variables occur in terms and formulas, and quantified formulas $\forall x\, A$ and $\exists x\, A$ simply result from a formula A by prefixing it with $\forall$ or $\exists$ plus a variable x. Then one defines when an *occurrence* of a variable x in a formula is "free" or "bound" (namely when it occurs in the scope of a matching $\forall x$ or $\exists x$). In our approach, we instead start with two separate kinds of variables—the *free* variables a_1, a_2, $\ldots$, and the *bound* variables x_1, x_2, $\ldots$ Only free variables play a part in the formation of terms, and so an atomic formula cannot contain a bound variable. However, when we form formulas with quantifiers, we introduce the bound variable x into the formula following the quantifier, and we do this by replacing some free variable in it everywhere by x. Let's define a notation for this operation.

Definition 2.31. If A is a formula, t a term or a bound variable, and a is a free variable, we denote by $A[t/a]$ the result of replacing every occurrence of a in A by t.

It is not hard to see that $A[t/a]$ is a formula if t is a term. However, $A[x/a]$ is *not* a formula if a occurs in A. For instance, if A is $P(a, b)$ then $A[x/a]$ is $P(x, b)$, which is not a formula since x is not a term. However, in Definition 2.30(c) this $A[x/a]$ occurs only after a $\forall x$ or $\exists x$, and the entire expression is a formula. In other words, since $P(a, b)$ is a formula, and x does not occur in it but a does, $\forall x\, P(x, b)$ is a formula (and so is $\exists x\, P(x, b)$). Similarly, since $P(a, a)$ is a formula, so are $\forall x\, P(x, x)$ and $\exists x\, P(x, x)$.[23] The role of a can be played by some other free variable. So, e.g., $P(c, a)$ is a formula containing the free variable c and not containing the bound variable x. Applying clause (c), we have that $\forall x\, P(x, a)$ and $\exists x\, P(x, a)$ are formulas, since $P(x, a)$ is the result of replacing c in $P(c, a)$ by x. Finally, we should highlight that the formation rules require the variable a to actually occur in A. That is, we do not allow vacuous quantification; the variable x must occur in $A[x/a]$. $\forall x\, P(a, a)$ is not a formula.

In order to formulate the predicate calculus we add, following Gentzen, two axioms and two rules of inference to the previous propositional systems. In the following, let t be a term and x a bound variable not occurring in A. When we write $A(a)$ we mean that A contains the free variable a. In the same context, $A(t)$ then denotes $A[t/a]$, and $A(x)$ is $A[x/a]$.

Axioms For any formula $A(a)$, term t, and bound variable x not occurring in $A(a)$:

QL1. $\forall x\, A(x) \supset A(t)$

QL2. $A(t) \supset \exists x\, A(x)$

Note that, just as in the case of the propositional axioms, the above are schemas of axioms; that is, each instance of the above schemes is an axiom.

Rules of Inference

QR₁. If $A \supset B(a)$ is derivable, a does not occur in A, and x is a bound variable not occurring in $B(a)$, then $A \supset \forall x\, B(x)$ is derivable.

QR₂. If $B(a) \supset A$ is derivable, a does not occur in A, and x is a bound variable not occurring in $B(a)$, then $\exists x\, B(x) \supset A$ is also derivable.

---

[23] If you are used to the formation rules of predicate formulas in a language that does not distinguish between free and bound variables, then think of our formulas as the result of replacing every free occurrence of a variable y by a special variable b. Conversely, if you take one of our formulas A and replace every free variable b by a variable y not occurring in A, the result will be a formula in which every occurrence of y is a free occurrence.

The variable a is called the *eigenvariable* of the inference.[24]

As noted above, in textbooks that allow only one kind of variable, whether an occurrence of a variable in a formula is free or bound depends on whether it has a "matching" quantifier. In that case, it becomes necessary to define when a term can be substituted for a variable, and the corresponding axioms have to take this into account. For instance, axiom QL1, $\forall x\, A(x) \supset A(t)$ must be restricted to the case where the term t is "free for" the variable x in $A(x)$. This means that t does not contain some variable y which would fall in the scope of a quantifier $\forall y$ or $\exists y$ at a place where t is substituted for x. If it does, then the variable y in t would be "captured," and this results in incorrect inferences. For instance, suppose $A(x)$ is really $\exists y\, P(x, y)$. If we allow t to be y then the corresponding instance of QL1 would read,

$$\forall x \exists y\, P(x, y) \supset \exists y\, P(y, y).$$

This is bad: for instance, interpret $P(x, y)$ as "x is less than y." Then the antecedent is true (e.g., about the natural numbers) but the consequent is false! If free and bound variables are distinguished, as they are here, this cannot happen, simply because t is a term and so cannot contain a bound variable y. We do not have to worry about capture of variables and a term being free for a variable in a formula. When we distinguish free and bound variables, substituting a term for a free variable is simply replacement (since free variables are never bound) and never problematic (in the sense that accidental capture of variables cannot happen, since terms don't contain bound variables).

Example 2.32. The formulas

$\forall x\, P(x, x) \supset P(t, t)$	$P(t, t) \supset \exists x\, P(x, x)$
$\forall x\, P(x, t) \supset P(t, t)$	$P(t, t) \supset \exists x\, P(x, t)$
$\forall x\, P(x, t) \supset P(s, t)$	$P(s, t) \supset \exists x\, P(x, t)$
$\forall x\, P(s, x) \supset P(s, t)$	$P(s, t) \supset \exists x\, P(s, x)$

are all instances of QL1 and QL2, respectively, but

$\forall x\, P(x, x) \supset P(s, t)$	$P(s, t) \supset \exists x\, P(x, x)$

are not (unless s and t are the same term). For instance, if we take $A(a)$ to be $P(a, t)$ then $\forall x\, A(x)$ is $\forall x\, P(x, t)$ and $A(t)$ is $P(t, t)$, so $\forall x\, P(x, t) \supset P(t, t)$ is a correct instance of QL1. But $P(x, x)$ can only result from $P(a, a)$ by replacing all occurrences of a by x. (Recall that bound variables are not terms, so $P(a, x)$ and $P(x, a)$ are not formulas.)

---

[24] The word *eigenvariable* comes from the German *Eigenvariable* with the second half of the word changed into English. The prefix *Eigen-* is hard to translate. It means something like "characteristic" or "proper." Because it is hard to translate, "eigenvariable" is now a term of art in proof theory, just like "eigenvalue" and "eigenvector" are in linear algebra.

Intuitively, to get from a quantified formula $\forall x\, A(x)$ to $A(t)$, you remove the quantifier and replace *all* occurrences of x by t. To get from a formula $A(t)$ to $\forall x\, A(x)$ or $\exists x\, A(x)$, you replace *one or more*, and perhaps, but not necessarily, all occurrences of t by x. That is because both $A(x)$ and $A(t)$ are actually the result of replacing, in some formula $A(a)$, the free variable a by x and t, respectively. This formula $A(a)$ may contain occurrences of t as well, since t is a term (e.g., $P(a, t)$), and when it does, the corresponding $A(x)$ also contains occurrences of t (e.g., $P(x, t)$)—we have "replaced" only some of the terms t by x. Since $A(a)$ cannot contain x (with or without matching quantifier), every occurrence of x in $A(x)$ is one that results from substituting x for a in $A(a)$. So $A(t)$, the result of substituting t for every occurrence of a in $A(a)$, is the same as the result of "replacing" every occurrence of x in $A(x)$ by t.

In the rules QR$_1$ and QR$_2$, the requirement that a not occur in A is critical. For suppose we did not have it. Now take A and $B(a)$ to be the same. $B(a) \supset B(a)$ is derivable. So by QR$_1$ (ignoring the requirement that a not occur in the antecedent), we could derive $B(a) \supset \forall x\, B(x)$. Now we can apply QR$_2$ to derive $\exists x\, B(x) \supset \forall x\, B(x)$. (Similarly, we can first apply QR$_2$ while ignoring the requirement to get $\exists x\, B(x) \supset B(a)$ and then QR$_1$.) That is obviously not desirable.

When QL1, QL2, QR$_1$, and QR$_2$ are added to $\mathbf{M_0}$, $\mathbf{J_0}$, and $\mathbf{K_0}$ the resulting systems are denoted by $\mathbf{M_1}$, $\mathbf{J_1}$, and $\mathbf{K_1}$, respectively. Note that when applying the axioms of the propositional calculus in this context, we may now replace the schematic letters in the axioms by formulas of the predicate calculus (including variables and quantifiers). For instance, we have $\vdash_{\mathbf{M_1}} F(a) \supset (F(a) \wedge F(a))$ by PL1.

Derivations in the predicate calculus

Let us now look at an example of a theorem in our new calculi. $\mathbf{S_1}$ will denote any of the three systems $\mathbf{M_1}$, $\mathbf{J_1}$, $\mathbf{K_1}$.

Definition 2.33. A *derivation* in $\mathbf{S_1}$ is a finite sequence of formulas $A_1, \ldots, A_n$, each one of which is either an axiom of $\mathbf{S_1}$ or is obtained by previous formulas by modus ponens or by one of the two quantificational rules (QR$_1$ and QR$_2$). The last formula of the derivation, A_n, is called the *end-formula* of the derivation. The definitions of *theorem of* $\mathbf{S_1}$ and *provable in* $\mathbf{S_1}$ are the natural modifications of those given for the propositional calculus.

Example 2.34. $\vdash_{\mathbf{S_1}} F(t) \supset \neg\forall x\, \neg F(x)$

Proof.

1. $\vdash \forall x \neg F(x) \supset \neg F(t)$ (QL1)
2. $\vdash (\forall x \neg F(x) \supset \neg F(t)) \supset (\neg\neg F(t) \supset \neg\forall x\, \neg F(x))$ (E2 of Problem 2.18)
3. $\vdash \neg\neg F(t) \supset \neg\forall x\, \neg F(x)$ (MP 1, 2)
4. $\vdash F(t) \supset \neg\neg F(t)$ (E5 of Problem 2.18)

5. $\vdash (F(t) \supset \neg\neg F(t)) \land (\neg\neg F(t) \supset \neg\forall x \, \neg F(x))$ ($\land$INTRO 3, 4)

6. $\vdash [(F(t) \supset \neg\neg F(t)) \land (\neg\neg F(t) \supset \neg\forall x \neg\neg F(x))] \supset (F(t) \supset$
    $\neg\forall x \neg F(x))$ (axiom PL4)

7. $\vdash F(t) \supset \neg\forall x \neg F(x)$ (MP 5, 6)

The reason why the above is true in any system $\mathbf{S}_1$ is that the derivation is justified using only axioms and rules of inference of $\mathbf{M}_1$. For an example of a derivation requiring the application of the new quantifier rules (as opposed to axioms) consider the following theorem, which we will need later.

Example 2.35. If $F(a)$ is a formula, and if $\neg\neg F(a) \supset F(a)$ is provable in $\mathbf{S}_1$ then

$$\vdash_{\mathbf{S}_1} \neg\neg\forall x \, F(x) \supset \forall x \, F(x)$$

Proof.

1. $\vdash \neg\neg F(a) \supset F(a)$ (hypothesis)
2. $\vdash \forall x \, F(x) \supset F(a)$ (axiom QL1)
3. $\vdash (\forall x \, F(x) \supset F(a)) \supset (\neg\neg\forall x \, F(x) \supset \neg\neg F(a))$ (E3 of Problem 2.18)
4. $\vdash \neg\neg\forall x \, F(x) \supset \neg\neg F(a)$ (MP 2, 3)
5. $\vdash \neg\neg\forall x \, F(x) \supset F(a)$ ($\supset$TRANS 4, 1)
6. $\vdash \neg\neg\forall x \, F(x) \supset \forall x \, F(x)$ (QR$_1$)

Note that a does not occur in the antecedent $\forall x \, F(x)$, so the condition on QR$_1$ in line 6 is satisfied. $\square$

Now for the definition of derivation in $\mathbf{S}_1$ from a set of premises Γ:

Definition 2.36. A formula A is *derivable* (in any of our calculi $\mathbf{S}_1$) *from a set of assumptions* Γ if there is a finite sequence of formulas $A_1, \ldots, A_n$ such that

1. For each $1 \le i \le n$, A_i is either an axiom of the relevant calculus or a formula in Γ or it is obtained by modus ponens or QR$_1$ or QR$_2$ applied to previous formulas A_k, A_ℓ for $k, \ell < i$.

2. $A_n = A$.

When this is the case we say that $A_1, \ldots, A_n$ is a *derivation of A from Γ* in the relevant calculus.

For instance, from $\Gamma = \{F(a)\}$ we can derive $\forall x \, F(x)$. Let G be any instance of an axiom of $\mathbf{M}_1$ in which a does not occur.

1. $\vdash F(a) \supset (G \supset F(a))$ (axiom PL5)
2. $F(a)$ ($F(a) \in \Gamma$)
3. (2) $\vdash G \supset F(a)$ (MP 1, 2)
4. (2) $\vdash G \supset \forall x \, F(x)$ (QR$_1$)
5. $\vdash G$ (axiom)
6. (2) $\vdash \forall x \, F(x)$ (MP 4, 5)

Recall that a does not occur in G, hence the condition on QR$_1$ in line (4) is satisfied.

2.13.1   *Dependence*

Assume $B \in \Gamma$ and let $A_1 \ldots A_n$ be a derivation from Γ with the justifications for each step. A_i *depends on* B if, and only if, B is the formula A_i and the justification for B is that it belongs to Γ; or, there are A_k, A_ℓ with $k, \ell < i$ such that A_i is justified by an application of modus ponens, QR₁ or QR₂ applied to A_k, A_ℓ and at least one of A_k, A_ℓ depends on B.

Let us revisit the previous example in light of the definition of dependence. Line 1 does not depend on $F(a)$; line 2 depends on $F(a)$; line 3 depends on $F(a)$ because it is obtained by MP from lines 1 and 2, and line 2 depends on $F(a)$. Since line 4 is obtained by QR₁ from line 3, and the latter depends on $F(a)$, line 4 depends on it, too. Similar considerations show that line 5 does not depend on $F(a)$ while line 6 does.

Before we can formulate the deduction theorem we prove the following lemma.

Lemma 2.37. *Suppose A is the end-formula of a derivation from $\Gamma \cup \{B\}$. If A does not depend on B in $\Gamma, B \vdash A$, then $\Gamma \vdash A$.*

Proof. We prove the result by induction on the length of a derivation of A from Γ, B in which A does not depend on B.

Induction basis: For derivations of length 1, A can only be an axiom or an element of Γ which is not B. In this case obviously $\Gamma \vdash A$.

Inductive hypothesis: Assume the result holds for all $i < n$.

Inductive step: If $\Gamma, B \vdash A_i$ and A_i does not depend on B, then $\Gamma \vdash A_i$ by inductive hypothesis. Consider now how A_n is obtained in the derivation $\Gamma, B \vdash A_n$. There are two major cases:

1. A_n could be obtained either as an axiom or an element of Γ which is not B. Then $\Gamma \vdash A_n$ as in the induction basis.

2. A_n is obtained from previous formulas A_ℓ, A_k (where $\ell, k < n$) by modus ponens, QR₁, or QR₂. Since A_n does not depend on B in the derivation $\Gamma, B \vdash A_n$, then neither A_k nor A_ℓ depend on B. By inductive hypothesis,

$$\Gamma \vdash A_k \text{ and } \Gamma \vdash A_\ell$$

   and thus we can derive $\Gamma \vdash A_n$ using the corresponding rules (either MP, QR₁ or QR₂) from Γ alone without appealing to B. □

2.14   The deduction theorem for the predicate calculus

In Section 2.8, we proved the deduction theorem for propositional logic. It states that if we can derive B from formulas Γ and A, we can also derive $A \supset B$ from Γ alone. A similar result holds for the predicate calculus; however, we have to be more careful in our proof, taking into account the complications introduced by the eigenvariable condition on the QR₁ and QR₂ rules.

Theorem 2.38. *Suppose* $\Gamma, B \vdash A$, *and whenever* QR_1 *or* QR_2 *is applied in the derivation to a formula that depends on* B, *then the eigenvariable* a *of that inference does not occur in* B. *Then* $\Gamma \vdash B \supset A$.

Proof. Let $A_1, \ldots, A_n$ (with $A_n = A$) be the derivation from Γ, B satisfying the assumption of the theorem. We will prove the result by induction on n. In the course of the proof we will use certain steps whose justification is given by referring to exercises at the end of the proof.

Induction basis: If $n = 1$, then A_1 is either an axiom or it belongs to Γ, B. Here the proof resembles that already given for the propositional case.

If A_1 is an axiom then we have:

1.	$\vdash A_1$	(axiom)
2.	$\vdash B \supset A_1$	(E5 of Problem 2.7)
3.	$\Gamma \vdash B \supset A_1$	(monotonicity)

Suppose now $A_1 \in \Gamma$ and A_1 is different from B. Then

1.	$\Gamma \vdash A_1$	($A_1 \in \Gamma$)
2.	$\vdash A_1 \supset (B \supset A_1)$	(axiom PL5)
3.	$\Gamma \vdash B \supset A_1$	(MP 1, 2)

If A_1 is B then

1.	$\vdash B \supset B$	(E2 of Problem 2.7)
2.	$\Gamma \vdash B \supset B$	(monotonicity)
3.	$\Gamma \vdash B \supset A_1$	(by identity of B and A_1)

Inductive hypothesis: Assume now the result to hold for all $m < n$.

Inductive step: Consider $\Gamma, B \vdash A_n$. We have four cases. A_n could be B; A_n could be in Γ and different from B; A_n could be an axiom; A_n could be obtained from previous formulas A_k, A_ℓ (where $k, \ell < n$) by MP, QR_1, or QR_2. The first three cases are dealt with as in the case for $n = 1$. The last case contains three sub-cases depending on the rule of inference used to derive A_n. We will treat MP and QR_1, leaving QR_2 as an exercise.

Let us assume $\Gamma, B \vdash A_n$ is obtained from $\Gamma, B \vdash A_k$ and $\Gamma, B \vdash A_\ell$ where A_k has the form $A_\ell \supset A_n$ and where each derivation satisfies the hypothesis of the theorem. By inductive hypothesis,

$$\Gamma \vdash B \supset A_\ell \text{ and } \Gamma \vdash B \supset (A_\ell \supset A_n)$$

So:

1.	$\Gamma \vdash B \supset A_\ell$	(hypothesis)
2.	$\Gamma \vdash B \supset (A_\ell \supset A_n)$	(hypothesis)
3.	$\vdash (B \supset (A_\ell \supset A_n)) \supset ((B \supset A_\ell) \supset (B \supset A_n))$	(Theorem 2.8)
4.	$\Gamma \vdash (B \supset A_\ell) \supset (B \supset A_n)$	(MP 2, 3)
5.	$\Gamma \vdash B \supset A_n$	(MP 1, 4)

Assume now that $\Gamma, B \vdash A_n$ where A_n is obtained by QR$_1$ from $\Gamma, B \vdash A_\ell$, the conclusion A_ℓ has the form $D \supset C(a)$, and A_n has the form $D \supset \forall x\, C(x)$.

From the fact that $\Gamma, B \vdash A_n$ is obtained through a correct application of QR$_1$ we know that D does not contain the free variable a occurring in it, and $C(a)$ does not contain x. Thus starting with $\Gamma, B \vdash D \supset C(a)$ we distinguish two sub-cases. The first sub-case consists in assuming that $D \supset C(a)$ does not depend on B. Then we have:

1. $\Gamma \vdash D \supset C(a)$ (Lemma 2.37)
2. $\Gamma \vdash D \supset \forall x\, C(x)$ (QR$_1$ 1)
3. $\Gamma \vdash A_n$ (definition of A_n)
4. $ \vdash A_n \supset (B \supset A_n)$ (axiom PL5)
5. $\Gamma \vdash B \supset A_n$ (MP 3, 4)

For the second sub-case, assume $D \supset C(a)$ depends on B. By the assumption of the theorem, a does not occur in B. By inductive hypothesis, $\Gamma \vdash B \supset (D \supset C(a))$. Thus, $\Gamma \vdash B \supset \forall x(D \supset C(x))$ by QR$_1$. Now, $\vdash \forall x(D \supset C(x)) \supset (D \supset \forall x\, C(x))$ (exercise). So, $\Gamma \vdash B \supset (D \supset \forall x\, C(x))$, by $\supset$TRANS. So, $\Gamma \vdash B \supset A_n$, by definition of A_n. □

Problem 2.39. Carry out the case of QR$_2$ in the proof of Theorem 2.38.

From the above theorem it is also possible to immediately infer that if we have a derivation $\Gamma, B \vdash A$ and the derivation does not apply QR$_1$ or QR$_2$ to a free variable which occurs in B, then $\Gamma \vdash B \supset A_i$. Obviously, the same result holds if B contains no free variables at all (i.e., if it is a sentence).

Problem 2.40. Prove the following in $\mathbf{M}_1$ without using the deduction theorem:

E1. If $\vdash A(a)$ then $\vdash \forall x\, A(x)$.

E2. $\vdash \forall x(A(x) \supset A(x))$.

E3. $\vdash \forall x\, A(x) \supset \exists x\, A(x)$.

E4. $\vdash \forall x(A \supset B(x)) \supset (A \supset \forall x\, B(x))$, if x does not occur in A.

 Prove the following using the deduction theorem:

E5. $\vdash \forall x_1 \forall x_2\, C(x_1, x_2) \supset \forall x_2 \forall x_1 C(x_2, x_1)$.

E6. $\vdash \forall x(A(x) \supset B(x)) \supset (\forall x\, A(x) \supset \forall x\, B(x))$.

Assume in each case that $A(a)$, $B(a)$, and $C(a_1, a_2)$ do not contain x, or x_1 and x_2. Hints for E1, E2, and E3 are given in Appendix D.

2.15   Intuitionistic and classical arithmetic

2.15.1   Historical background

We now proceed to present a remarkable result due to Gentzen (1933) and independently to Gödel (1933). It concerns the relative consistency of classical arithmetic with respect to intuitionistic arithmetic. The result itself does not depend essentially on using the axiomatic presentation of logic and could be proved using systems of arithmetic based on natural deduction or sequent calculi. We present it here for two reasons. First, Gentzen himself obtained the result working within an axiomatic presentation of logic. So this entire section can be seen as a detailed working out of Gentzen (1933). Second, the proof provides further opportunities to develop your own understanding of the axiomatic presentation of logic. Even if you are not so interested in the specific details of the axiomatic derivations of the theorems required to prove the main result in this section, you will still benefit from perusing the statements of the key theorems and the general strategy for proving the result. Alternatively, you may later try to prove the same theorems within natural deduction or the sequent calculus, once you have become familiar with the latter systems. The result is another interesting application of induction on the length of proofs.

The background for the result presented here, an interpretation of classical arithmetic into intuitionistic arithmetic, is to be found in the debate that opposed Hilbert and Brouwer on the foundations of mathematics and in the specific approach to proof theory championed by Hilbert. It is of course impossible to recount in detail the Brouwer-Hilbert debate or the specifics of Hilbert's program (for an introduction and further references, see Mancosu, 1998b and Zach, 2019a) but a few comments are in order.

Hilbert's aim in developing proof theory, or "metamathematics" as he called it, was epistemological. Mathematicians with "constructivist" leanings such as Kronecker, Poincaré, Weyl, and Brouwer had questioned the meaningfulness and validity of certain parts and forms of reasoning of classical mathematics on account of more stringent criteria about what was to count as showing existence of mathematical entities or validity of patterns of reasoning.

Thus, for instance, Kronecker only wanted to allow sets in which the membership in the set could be effectively decided, Poincaré was rejecting certain forms of set construction given by quantification over a collection of sets to which the set being defined belongs (impredicative definitions) and Brouwer rejected any existence proof for which one could not effectively provide an instance of the type of object claimed to exist. Principles of classical mathematics such as excluded middle and double negation seemed unjustified or, worse, invalid under such conceptions. These mathematicians proposed to abandon the principles of mathematics which were objectionable from their respective standpoints, and to proceed with an epistemologically more well-grounded mathematics. The parts of classical mathematics that did not satisfy the more stringent "constructivist" criteria were to be excised once and for all.

Hilbert could not accept such a mutilation of mathematics. He realized that what the constructivist had done was to isolate a core of constructive mathematics which was more certain and epistemologically privileged in comparison to the rest of classical mathematics. His thought was that classical mathematics could be shown to be certain by using forms of reasoning that the constructivist would have found unobjectionable. He thus tried to articulate what in his opinion could count as such an unobjectionable starting point for a proof of the reliability of classical mathematics and used the word "finitism" to capture this standpoint. The second part of his program was to fully formalize areas of classical mathematics, such as number theory, analysis, geometry, etc., and show with "finitary" reasoning that the theories in question could never generate a contradiction. On the assumption that "finitary" reasoning is absolutely reliable, the outcome of the program would have been to show that the prima facie unreliable theories, such as full number theory or analysis, could not be inconsistent unless "finitary" forms of reasoning were.

The program had to face major challenges but we now know enough to put the main result of this chapter in perspective. For a large part of the 1920s, the exact extent of "finitary" reasoning was unclear. Many identified Hilbert's "finitism" with "intuitionism," i.e., the position defended by Brouwer.

At the core of the intuitionistic reconstruction of mathematics is intuitionistic arithmetic. This is a formal system for arithmetic which has the same non-logical axioms as the ones for classical arithmetic but it uses only intuitionistic logic; in other words, the law of double negation (or that of excluded middle) is not assumed in intuitionistic arithmetic. Semantically, this corresponds to the fact that the intuitionistic system is meant to capture a domain of numbers which are not given in their totality but are rather thought of as *in fieri*, namely in a never ending process of construction. Using **PA** for the axioms of Peano arithmetic let us add subscripts M, I, and K to denote the systems $\mathbf{PA}_M$, $\mathbf{PA}_I$, $\mathbf{PA}_K$ obtained by restricting the logic in the background to minimal, intuitionistic, and classical.[25]

Since no one doubted that $\mathbf{PA}_I$ was one of the systems that was finitistically and intuitionistically problematic (if such a distinction could even be made at the time), the general idea was that $\mathbf{PA}_I$ (known in the literature as **HA**, i.e., Heyting's arithmetic) was part of the safe core of mathematics. It then came as a surprise when in 1933 both Gödel and Gentzen independently proved that if $\mathbf{PA}_I$ is consistent then $\mathbf{PA}_K$ is consistent.

One can read a formal result of this sort in various ways. For instance, one might question whether the patterns of reasoning involved in showing that if $\mathbf{PA}_I$ is consistent then $\mathbf{PA}_K$ is consistent, are "safe" patterns of reasoning. An analysis of

---

[25] We will use I, not J, to indicate intuitionistic systems in this section.

the proof shows that that reasoning is "finitarily" ("intuitionistically") acceptable. With that objection out of the way one can then philosophically interpret the result as showing either that PA_I is just as problematic as PA_K and thus itself in need of justification; or, that PA_K's prima facie unreliability was only apparent. The lesson drawn from the result was actually the former. It revealed that "intuitionism" allows forms of reasoning that are epistemologically problematic. This finally led to a sharper definition of Hilbert's "finitism" as being more restrictive than intuitionism.

Hopefully, the above will convey to you a sense of the philosophical importance of the result we will now delve into. From a technical point of view the result is not difficult, but it displays with great effectiveness a standard proof-theoretic technique, namely the method of interpretability (see Tarski, Mostowski, and Robinson, 1953 and Feferman, 1988).

2.15.2 Systems of arithmetic

Formal systems of arithmetic can be treated very much the same way we handled logical calculi in Section 2.13. Terms and formulas of arithmetic are defined the same way we defined terms and formulas there. However, in arithmetic we introduce a special vocabulary of non-logical symbols rather than use the general vocabulary of logic. We also add to the axioms of logic a number of axioms that govern the non-logical vocabulary. The non-logical vocabulary consists only of constant, function, and predicate symbols. We of course still have all the variables, the logical connectives, and the quantifiers, and our derivations will allow all the logical axioms and inference rules we have discussed so far.

In arithmetic, we single out one constant symbol, 0, to stand for the number zero. We have three function symbols: $+$ and $\cdot$, to stand for addition and multiplication, respectively, and a one-place function symbol ' to stand for the successor function (i.e., the function that maps any number n to $n + 1$). We'll write $+$ and $\cdot$ between its arguments as usual, but write ' after its argument. We have only one predicate symbol, $=$, which we write between its arguments. In Definitions 2.29 and 2.30 we defined, in general, what is to count as a term or a formula in predicate logic. Now, we define terms and formulas of arithmetic. The definitions are essentially the same, except for the fact that we restrict the non-logical vocabulary to the symbols $0, ', +, \cdot$, and $=$, and we allow a more intuitive notation where symbols may appear between or after arguments, rather than in front of them (e.g., $0 = 0$ rather than $=(0, 0)$).[26]

---

[26] The axiom system for intuitionistic arithmetic given by Gentzen, which we use here, follows that presented by Herbrand (1931). It differs from more modern axiomatizations in that equality is considered a non-logical symbol and the axioms for equality as non-logical axioms. We have modified the Herbrand-Gentzen system so that the number sequence starts with 0 as opposed to 1.

Definition 2.41. *Terms of arithmetic* are inductively defined as follows:

1. *Basis clause:* The constant symbol 0 and the free variables $a_1, a_2, \ldots$, are terms.
2. *Inductive clause:* If t_1, t_2 are terms, so are t_1', $(t_1 + t_2)$, and $(t_1 \cdot t_2)$.

Definition 2.42. *Formulas of arithmetic* are inductively defined as follows:

1. *Basis clause:* If t_1, t_2 are terms then $t_1 = t_2$ is an (atomic) formula.
2. *Inductive clause:*

    (a) If A is a formula, then $\neg A$ is a formula.
    (b) If A and B are formulas, then $(A \wedge B)$, $(A \vee B)$, and $(A \supset B)$ are formulas.
    (c) If $A(a)$ is a formula containing the free variable a, and x is a bound variable not occurring in $A(a)$, then $\forall x\, A(x)$ and $\exists x\, A(x)$ are formulas.

Note that the only difference to Definition 2.30 is that we restrict atomic formulas to those of the form $t_1 = t_2$.

The axioms of **PA**$_I$ are obtained from **J**$_1$ by adding all instances of the following:

PA1. $a = a$

PA2. $a = b \supset b = a$

PA3. $(a = b \wedge b = c) \supset a = c$

PA4. $\neg a' = 0$

PA5. $a = b \supset a' = b'$

PA6. $a' = b' \supset a = b$

PA7. $[F(0) \wedge \forall x(F(x) \supset F(x'))] \supset \forall x\, F(x)$ for any formula $F(a)$

PA8. $\neg\neg a = b \supset a = b$

PA9. $(a + 0) = a$

PA10. $(a + b') = (a + b)'$

PA11. $(a \cdot 0) = 0$

PA12. $(a \cdot b') = ((a \cdot b) + a)$

We allow every instance of these formulas as axioms, i.e., any formula resulting from replacing a and b by terms. For instance, $0 = 0$ and $(a_1 + 0') = (a_1 + 0')$ count as instances of axiom PA1.

With this system we can develop intuitionistic arithmetic **PA**$_I$. If the background logic is classical logic then the resulting system **PA**$_K$ is equivalent to first-order Peano arithmetic.[27]

---

[27] First-order Peano arithmetic (**PA**) in the standard literature is axiomatized by means of axioms PA4, PA6, PA7 and PA9, PA10, PA11, PA12. Axioms PA1, PA2, PA3, and PA5, are provided by the

Obviously intuitionistic logic is contained in classical logic (that is, formally, i.e., if we do not raise philosophical issues about the meaning of intuitionistic connectives).

It would also seem clear that $\mathbf{PA}_K$ is more powerful than $\mathbf{PA}_I$. For instance, in $\mathbf{PA}_K$ we can show that for any $F(a)$, the instance of the induction principle with $F(a)$ is equivalent to an instance of the least number principle, namely,[28]

$$\exists x\, F(x) \supset \exists x \forall y (y < x \supset \neg F(y)).$$

Let us use IP(F) to denote the induction principle for F, and LNP(F) for the least number principle for F. Let $\mathbf{PA}_K^-$ be the system obtained from $\mathbf{PA}_K$ by dropping the induction schema PA7. Then both

$$\vdash_{\mathbf{PA}_K^-} \text{IP}(F) \supset \text{LNP}(F) \text{ and}$$

$$\vdash_{\mathbf{PA}_K^-} \text{LNP}(F) \supset \text{IP}(F).$$

In the corresponding system $\mathbf{PA}_I^-$, however, only the second conditional LNP(F) $\supset$ IP(F) is provable; the first is not provable for all F in $\mathbf{PA}_I^-$. As a consequence, $\nvdash_{\mathbf{PA}_I}$ LNP(F), for at least some F.

2.15.3   *The Gödel-Gentzen translation*

Despite the fact that prima facie $\mathbf{PA}_K$ seems stronger than $\mathbf{PA}_I$, Gentzen and Gödel proved that when interpreting the classical connectives according to an intuitionistically acceptable meaning, it is possible to translate $\mathbf{PA}_K$ into $\mathbf{PA}_I$. The translation is defined in such a way as to yield the result that if $\mathbf{PA}_I$ is consistent, also $\mathbf{PA}_K$ is consistent.

We have already established the following theorem.

Theorem 2.43. *If* $\neg\neg B \supset B$ *and* $\neg\neg C \supset C$ *are derivable in* J_1, *then so are:*

$\neg\neg\neg B \supset \neg B$	(E7 of Problem 2.18)
$\neg\neg(B \wedge C) \supset (B \wedge C)$	(Lemma 2.19(1))
$\neg\neg(B \supset C) \supset (B \supset C)$	(Lemma 2.19(2))

If B contains a, then also

$\neg\neg\forall x\, B[x/a] \supset \forall x\, B[x/a]$	(Example 2.35)

background logic, which treats = as a logical constant. In classical logic, axiom PA8 is a special case of axiom PL12.

[28] The least number principle involves a sign < for "less than," which is not officially part of our language. We can introduce it to **PA** with a suitable axiom, or take $y < x$ as an abbreviation for $\exists z\, (z' + y) = x$.

Theorem 2.44. *If A is a formula of arithmetic in which neither ∨ nor ∃ occurs, then* ¬¬A ⊃ A *is derivable in* **PA**$_I$.

The proof of Theorem 2.44 is by induction on the complexity of A, an idea we first outlined in Section 2.7. Formulas are inductively defined, and so we can prove something for all formulas by proceeding as follows:

(a) establishing the induction basis: proving the claim for atomic formulas;

(b) assuming as inductive hypothesis that the claim is true for B (and C);

(c) carrying out the inductive step: show, using the inductive hypothesis, that the claim is also true for ¬B, $B \wedge C$, $B \vee C$, $B \supset C$, $\forall x\, B[x/a]$, and $\exists x\, B[x/a]$.

Formally, this is a form of strong induction on the degree of A. The degree of A can be defined for arithmetical formulas by analogy with Definition 2.5:

Definition 2.45. The *degree* $d(A)$ of a formula A is inductively defined as follows:

1. *Basis clause:* If A is atomic, then $d(A) = 0$.

2. *Inductive clauses:*

   (a) If A is a formula of the form ¬B, $\forall x\, B[x/a]$, or $\exists x\, B[x/a]$, then $d(A) = d(B) + 1$.

   (b) If A is a formula of the form $(B \wedge C)$, $(B \vee C)$, or $(B \supset C)$, then $d(A) = d(B) + d(C) + 1$.

Proof of Theorem 2.44. By induction on the degree of A.

Induction basis: If $d(A) = 0$, then A is atomic and so has the form $t_1 = t_2$. But ⊢$_{\mathbf{PA}_I}$ ¬¬$t_1 = t_2 \supset t_1 = t_2$, for it is an axiom.

Inductive hypothesis: Assume that the claim is true for all formulas of degree $< d(A)$.

Inductive step: We prove that the claim is true for A on the basis of the inductive hypothesis. If $d(A) > 0$, and A contains neither ∨ nor ∃, it is of one of the following forms:

(a) A is ¬B;

(b) A is $B \wedge C$;

(c) A is $B \supset C$; or

(d) A is $\forall x\, B[x/a]$ if B contains a.

In each case, the degree of B and C is less than $d(A)$, so the inductive hypothesis applies. That means that ¬¬$B \supset B$, and ¬¬$C \supset C$ are derivable in J$_1$. By Theorem 2.43, the claim follows for A. ☐

Definition 2.46. We inductively define a translation G^* of formulas G as follows:

1. *Basis case:* If G is atomic, then $G^* = G$. Otherwise,

2. *Inductive clause:* If G is not atomic, G^* is defined by:

$$(\neg F)^* = \neg F^*$$
$$(F \wedge H)^* = F^* \wedge H^*$$
$$(F \vee H)^* = \neg(\neg F^* \wedge \neg H^*)$$
$$(F \supset H)^* = F^* \supset H^*$$
$$(\forall x\, F[x/a])^* = \forall x\, F^*[x/a]$$
$$(\exists x\, F[x/a])^* = \neg\forall x\, \neg F^*[x/a]$$

Recall that any quantified formula $\forall x\, F(x)$ is obtained from some formula $F(a)$ containing the free variable a by replacing a everywhere by x and then prefixing $\forall x$ (and similarly for $\exists x\, A(x)$). So, any such formula is of the form $\forall x\, F[x/a]$. The degree of F is lower than that of $\forall x\, F(x)$. Thus F is a bona fide formula for which F^* is defined. Although * is not defined for expressions like $F(x)$ which contain bound variables x without a matching quantifier, it is defined for $F(a)$. The translation of $\forall x\, F(x)$ is obtained by computing the translation of $F(a)$, and then replacing the free variable a in the result by x and prefixing $\forall x$ again.

As can easily be proved, if G does not contain $\vee$ or $\exists$, then $G^* = G$. It can also be easily proved that both

$$\vdash_{\mathbf{K}_1} G \supset G^* \text{ and}$$
$$\vdash_{\mathbf{K}_1} G^* \supset G.$$

Problem 2.47. Prove by induction on $d(G)$ that G^* does not contain $\vee$ or $\exists$.

Our main result is the following:

Theorem 2.48. *A derivation in $\mathbf{PA}_K$ with end-formula G can be transformed into a derivation in $\mathbf{PA}_I$ with the end-formula G^*.*

Proof. The proof proceeds in two steps. First we transform a derivation δ in $\mathbf{PA}_K$ of G into a derivation δ^* of G^* (also in $\mathbf{PA}_K$). The transformation yields a derivation δ^* without any occurrences of $\vee$ or $\exists$. Then we turn δ^* into an intuitionistic derivation δ^{**} of G^*.

Suppose the derivation δ in $\mathbf{PA}_K$ has the following form:

$$S_1, \qquad\qquad S_2, \qquad\qquad \ldots, \qquad\qquad S_n = G$$

where each S_i is either an axiom or is obtained from previous formulas by one of the rules, namely modus ponens, QR_1, and QR_2. We might hope that

$$S_1^*, \qquad\qquad S_2^*, \qquad\qquad \ldots, \qquad\qquad S_n^* = G^*$$

is also a derivation in $\mathbf{PA}_K$.

However, the above scheme is too simple. For suppose the original proof begins with an instance of axiom PL7, $A \supset (A \vee B)$. Say S_1 is

$$0 = 0 \supset (0 = 0 \vee \neg(0 = 0)).$$

The translation S_1^* reads

$$(0 = 0)^* \supset (0 = 0 \vee \neg(0 = 0))^*, \text{ i.e.,}$$
$$(0 = 0) \supset \neg(\neg(0 = 0) \wedge \neg\neg(0 = 0)).$$

But a derivation cannot begin with S_1^*, for S_1^* is not an axiom. However, we can show that S_1^* is derivable in $\mathbf{PA}_K$. Thus what is needed is a proof that whenever the translation $*$ might affect an axiom or an inference rule we can fill the gaps by providing the right derivation. We need to consider mathematical axioms, logical axioms, as well as the rules of inference.

We now prove that wherever the translation $*$ affects the form of axioms or inference rules, one can complete the derivation by adding the missing links so as to obtain a new derivation of G^* in $\mathbf{PA}_K$.

Let's begin by considering mathematical axioms. The axioms for arithmetic do not contain $\exists$ or $\vee$. So the translation $*$ does not affect them; for any mathematical axiom A, $A^* = A$. But what about the induction schema? In

$$[F(0) \wedge \forall x(F(x) \supset F(x'))] \supset \forall x \, F(x),$$

the formula F might contain $\vee$ and $\exists$, so, e.g., $F(0)^* \neq F(0)$. However, the translation of the instance of induction as a whole,

$$[F^*(0) \wedge \forall x(F^*(x) \supset F^*(x'))] \supset \forall x \, F^*(x)$$

is also an instance of the induction schema, and so counts as an axiom. By Problem 2.47, $F^*(a)$ does not contain $\vee$ or $\exists$. We will use the result of Problem 2.47 implicitly also in the remaining part of the proof.

Let us now look at the logical axioms. We then have to deal with propositional axioms (such as $(A \vee B) \supset (B \vee A)$) or quantifier axioms (such as $F(t) \supset \exists x \, F(x)$), in which $\vee$ or $\exists$ occur. If the axiom itself does not mention $\vee$ or $\exists$, its translation is still an axiom. For example, axiom PL1, i.e., $A \supset (A \wedge A)$ does not mention $\vee$ or $\exists$. Of course, in an instance of this schema, A may contain $\vee$ or $\exists$. But the translation of axiom PL1 is $A^* \supset (A^* \wedge A^*)$, which is also an instance of axiom PL1.

So we only have work to do in four cases, corresponding to axioms PL7, PL8, PL9, and QL2. Those are:

PL7. $A \supset (A \vee B)$

PL8. $(A \vee B) \supset (B \vee A)$

PL9. $((A \supset C) \wedge (B \supset C)) \supset ((A \vee B) \supset C)$

QL2. $F(t) \supset \exists x \, F(x)$

What is required is a proof of their translations, namely,

PL7*. $A^* \supset \neg(\neg A^* \wedge \neg B^*)$

PL8*. $\neg(\neg A^* \wedge \neg B^*) \supset \neg(\neg B^* \wedge \neg A^*)$

PL9*. $((A^* \supset C^*) \wedge (B^* \supset C^*)) \supset (\neg(\neg A^* \wedge \neg B^*) \supset C^*)$

QL2*. $F^*(t) \supset \neg \forall x \, \neg F^*(x)$

(QL2*) has already been proved in Example 2.34. Note that the derivation there did not make use of any axioms involving $\vee$ or $\exists$.
    Derivation of PL7*:

1.	$\vdash (\neg A^* \wedge \neg B^*) \supset \neg A^*$	(E1 of Problem 2.7)
2.	$\vdash [(\neg A^* \wedge \neg B^*) \supset \neg A^*] \supset [A^* \supset \neg(\neg A^* \wedge \neg B^*)]$	(E4 of Problem 2.18)
3.	$\vdash A^* \supset \neg(\neg A^* \wedge \neg B^*)$	(MP 1, 2)

    Derivation of PL8*:

1.	$\vdash (\neg B^* \wedge \neg A^*) \supset (\neg A^* \wedge \neg B^*)$	(axiom PL2)
2.	$\vdash ((\neg B^* \wedge \neg A^*) \supset (\neg A^* \wedge \neg B^*)) \supset$	
	$\qquad (\neg(\neg A^* \wedge \neg B^*) \supset \neg(\neg B^* \wedge \neg A^*))$	(E2 of Problem 2.18)
3.	$\vdash \neg(\neg A^* \wedge \neg B^*) \supset \neg(\neg B^* \wedge \neg A^*)$	(MP 1, 2)

    Derivation of PL9*:

1.	$(A^* \supset C^*) \wedge (B^* \supset C^*)$	(assumption)
2.	$\vdash (1) \supset (A^* \supset C^*)$	(E1 of Problem 2.7)
3.	$(1) \vdash A^* \supset C^*$	(MP 1, 2)
4.	$\vdash (A^* \supset C^*) \supset (\neg C^* \supset \neg A^*)$	(E2 of Problem 2.18)
5.	$(1) \vdash \neg C^* \supset \neg A^*$	(MP 3, 4)
6.	$\vdash (1) \supset (B^* \supset C^*)$	(E3 of Problem 2.7)
7.	$(1) \vdash B^* \supset C^*$	(MP 1, 6)
8.	$\vdash (B^* \supset C^*) \supset (\neg C^* \supset \neg B^*)$	(E2 of Problem 2.18)
9.	$(1) \vdash \neg C^* \supset \neg B^*$	(MP 7, 8)
10.	$(1) \vdash (5) \wedge (9)$	($\wedge$INTRO 5, 9)
11.	$\vdash ((5) \wedge (9)) \supset [\neg C^* \supset (\neg A^* \wedge \neg B^*)]$	(E9 of Problem 2.7)
12.	$(1) \vdash \neg C^* \supset (\neg A^* \wedge \neg B^*)$	(MP 10, 11)
13.	$\vdash (12) \supset [\neg(\neg A^* \wedge \neg B^*) \supset \neg\neg C^*]$	(E2 of Problem 2.18)
14.	$(1) \vdash \neg(\neg A^* \wedge \neg B^*) \supset \neg\neg C^*$	($\supset$TRANS 12, 13)
15.	$\vdash \neg\neg C^* \supset C^*$	(axiom PL12)
16.	$(1) \vdash \neg(\neg A^* \wedge \neg B^*) \supset C^*$	($\supset$TRANS 14, 15)
17.	$\vdash ((A^* \supset C^*) \wedge (B^* \supset C^*)) \supset$	
	$\qquad [\neg(\neg A^* \wedge \neg B^*) \supset C^*]$	(deduction theorem, 16)

    It is important that none of the derivations we are using add any formulas containing $\vee$ or $\exists$. In particular, the derivations of all the exercises used avoid $\vee$ and $\exists$.

We now need to check the rules of inference. If B follows from A and $A \supset B$ by modus ponens, then B^* also follows from A^* and $A^* \supset B^*$ by modus ponens. So, applications of modus ponens translate into applications of modus ponens. The same goes for QR_1:

$$\frac{A^* \supset B^*(a)}{A^* \supset \forall x\, B^*(x)}$$

The only form of inference we need to worry about is QR_2:

$$\frac{B(a) \supset A}{\exists x\, B(x) \supset A}\; QR_2$$

It is translated into

$$\frac{B^*(a) \supset A^*}{\neg \forall x \neg B^*(x) \supset A^*}\; QR_2^*$$

Here is a derivation:

1.	$\vdash B^*(a) \supset A^*$	(hypothesis)
2.	$\vdash (1) \supset (\neg A^* \supset \neg B^*(a))$	(E2 of Problem 2.18)
3.	$\vdash \neg A^* \supset \neg B^*(a)$	(MP 1, 2)
4.	$\vdash \neg A^* \supset \forall x \neg B^*(x)$	(QR$_1$ 3)
5.	$\vdash (4) \supset (\neg \forall x \neg B^*(x) \supset \neg\neg A^*)$	(E2 of Problem 2.18)
6.	$\vdash \neg \forall x\, B^*(x) \supset \neg\neg A^*$	(MP 4, 5)
7.	$\vdash \neg\neg A^* \supset A^*$	(axiom PL12)
8.	$\vdash \neg \forall x \neg B^*(x) \supset A^*$	($\supset$TRANS 6, 7)

Since a does not occur in A^*, the condition on QR_1 is satisfied on line 4. Furthermore, the cited exercises make no use of $\lor$ or $\exists$.

In the above derivation we have used an instance of PL12 of the form $\neg\neg A^*$. All instances of PL12 that were used in previous theorems concerning the translation (for instance, in the derivation of PL9*) were also of this form.

We can now justify each individual step in the sequence $S_1^*, \ldots, S_n^*$. In other words, if we have a derivation δ in $\mathbf{PA}_K$ of G of the form:

$$S_1, \qquad\qquad S_2, \qquad\qquad \ldots, \qquad\qquad S_n = G$$

we can now fill the gaps between $S_1^*, S_2^*, \ldots, S_n^* = G^*$ that might be created by the translation so as to obtain a derivation of the following form:

$$\ldots, S_1^*, \qquad\qquad \ldots, S_2^*, \qquad\qquad \ldots \qquad\qquad \ldots, S_n^* = G^*$$

where some or all of the $\ldots$ might be empty. This will be the derivation δ^* of G^* in $\mathbf{PA}_K$. Notice that in both δ and δ^* we might have occurrences of the classical axiom PL12, i.e., $\neg\neg A \supset A$. But in the transformed δ^* that becomes $\neg\neg A^* \supset A^*$. By Theorem 2.44, we know that for a formula such as A^*, which does not contain $\lor$ or $\exists$, $\neg\neg A^* \supset A^*$ is intuitionistically valid. So we can now go through the derivation of δ^* and eliminate each occurrence of $\neg\neg A^* \supset A^*$ by providing its explicit intuitionistic derivation. The resulting derivation δ^{**} is derivation of G^* in $\mathbf{PA}_I$. $\quad\square$

Corollary 2.49. *Let δ be a derivation in PA_K whose end-formula G does not contain $\vee$ or $\exists$. Then δ can be transformed into a derivation δ^{**} in PA_I of G.*

Proof. Since G does not contain $\vee$ or $\exists$, its translation G^* is identical to G. Thus using Theorem 2.48 we first obtain a proof δ^* of G in PA_K and then a proof δ^{**} of G in PA_I. □

Corollary 2.50. *For every formula A of arithmetic there exists a classically equivalent formula B such that, if $\vdash_{PA_K} A$, then $\vdash_{PA_I} B$.*

Proof. In classical predicate calculus we can prove

$$\vdash_{K_1} (A \supset A^*) \wedge (A^* \supset A)$$

(with A^* defined as in the translation above). For any choice of A we then choose B to be A^*. The result now follows since

$$\vdash_{PA_K} A \Rightarrow \vdash_{PA_K} A^* \Rightarrow \vdash_{PA_I} A^*$$

by Theorem 2.48.

Furthermore, notice that if $\vdash_{PA_I} A^*$, then $\vdash_{PA_K} A^*$, and hence $\vdash_{PA_K} A$ (since $PA_I \subseteq PA_K$ and A and A^* are logically equivalent in PA_K). □

Definition 2.51. A theory T (such as classical or intuitionistic arithmetic) is *consistent* if it does not derive both A and $\neg A$, and otherwise *inconsistent*.

In the debate between advocates of intuitionism such as Brouwer and those of classical mathematics such as Hilbert, intuitionists raised the worry that classical mathematics might in fact be inconsistent. The worry was not too far fetched, since some theories such as Cantor's set theory and Frege's axiomatization of mathematics had in fact turned out to be inconsistent. Intuitionists thought the problem lay in classical principles, such as the law of excluded middle, and thought contradiction could be avoided if we remove these principles. The next theorem, then, is a vindication of classical arithmetic: it shows that if intuitionistic arithmetic is consistent, then so is classical arithmetic.

Corollary 2.52. *If PA_I is consistent then PA_K is consistent.*

Proof. If $\vdash_{PA_K} \neg 0 = 0$, then by Corollary 2.49, $\vdash_{PA_I} \neg 0 = 0$. But under the assumption that PA_I is consistent, since $\vdash_{PA_I} 0 = 0$, we cannot also have $\vdash_{PA_I} \neg 0 = 0$. Hence by reductio $\nvdash_{PA_K} \neg 0 = 0$, i.e., PA_K is consistent. □

In the proof above, we used the following fact about consistency of intuitionistic and classical theories: since $\vdash_{PA_K} 0 = 0$, PA_K is consistent if, and only if, $\nvdash_{PA_K} \neg(0 = 0)$.

Proposition 2.53. *If $\vdash_T B$, then T is consistent if, and only if, $\nvdash_T \neg B$.*

Proof. If $\vdash_T \neg B$, since also $\vdash_T B$, it is inconsistent according to the definition. In the other direction, suppose T is inconsistent, i.e., $\vdash_T A$ and $\vdash_T \neg A$. Then $\vdash_T \neg B$ by axiom PL11.   □

Problem 2.54. The derivations of PL7*–PL9* given in the proof of Theorem 2.48 and of QL2* in Example 2.34 made use of the derived rules ∧INTRO, ⊃TRANS, a number of exercises, and the deduction theorem. Revisit the proofs of these to convince yourself that none require the use of an axiom involving ∨ or ∃.

Problem 2.55. In order to get some practice with the translation ∗ consider the following proof of $B \vee \neg B$:

1.	$\vdash B \supset (B \vee \neg B)$	(axiom PL7)
2.	$\vdash \neg B \supset (\neg B \vee B)$	(axiom PL7)
3.	$\vdash (\neg B \vee B) \supset (B \vee \neg B)$	(axiom PL8)
4.	$\vdash \neg B \supset (B \vee \neg B)$	(⊃TRANS 2, 3)
5.	$\vdash [\neg B \supset (B \vee \neg B)] \supset [\neg(B \vee \neg B) \supset \neg\neg B]$	(E2 of Problem 2.18)
6.	$\vdash \neg(B \vee \neg B) \supset \neg\neg B$	(MP 4, 5)
7.	$\vdash [B \supset (B \vee \neg B)] \supset [\neg(B \vee \neg B) \supset \neg B]$	(E2 of Problem 2.18)
8.	$\vdash \neg(B \vee \neg B) \supset \neg B$	(MP 1, 7)
9.	$\vdash [\neg(B \vee \neg B) \supset \neg B] \wedge [\neg(B \vee \neg B) \supset \neg\neg B]$	(∧INTRO 6, 8)
10.	$\vdash \{[\neg(B \vee \neg B) \supset \neg B] \wedge [\neg(B \vee \neg B) \supset \neg\neg B]\} \supset \neg\neg(B \vee \neg B)$	(axiom PL10)
11.	$\vdash \neg\neg(B \vee \neg B)$	(MP 9, 10)
12.	$\vdash \neg\neg(B \vee \neg B) \supset (B \vee \neg B)$	(axiom PL12)
13.	$\vdash B \vee \neg B$	(MP 11, 12)

Now suppose, given our previous translation ∗, that $B = B^*$ (this will be the case, for instance, when B is atomic).

E1. Write down $(B \vee \neg B)^*$.

E2. Write down the classical proof δ^* for $(B \vee \neg B)^*$.

E3. Write down the intuitionistic proof δ^{**} for $(B \vee \neg B)^*$.

3

Natural deduction

3.1   Introduction

The axiomatic method is incredibly successful at making precise the principles underlying a domain of science, as well as at clarifying the logical connections between the primitives of that science. When Frege, Peano, Russell, and Hilbert applied the axiomatic method to logic itself, the results were powerful and simple systems. These systems made clear which principles can form an axiomatic basis for logic, and how the different logical primitives—connectives, quantifiers, variables—interact. Different systems, nevertheless, are successful to differing degrees in this regard. Both Frege and Russell's systems reduced the number of primitives and hence the axioms only directly showed connections between items of these reduced vocabularies. In the case of Frege, the basic logical constants were the conditional, negation, and the universal quantifier; in the case of Russell, disjunction, negation, and the universal and existential quantifiers. Common to these approaches to formalizations of logic is also the focus on just a handful of inference rules. Modus ponens is often the only one dealt with explicitly, a generalization rule and a substitution rule frequently being used but not made precise.

Work on axiomatizations of logic in Hilbert's school addressed this to some extent. In his dissertation on the axiomatization of the propositional calculus, Bernays (1918) considered the question of replacing axioms by rules of inference, and proved the equivalence of Russell's axiomatization of classical propositional logic with a number of alternative systems using a variety of inference rules, but still only used disjunction and negation as primitives. In later work, he produced axiomatizations of propositional logic which make use of a larger number of axioms than needed, but these axiom systems divide into groups covering different connectives (e.g., axioms for the conditional, axioms for disjunction, etc., see Zach, 1999).

All of this, however, does not address a further issue anyone who's tried to use such logical calculi will be painfully aware of (see previous chapter): proofs

An Introduction to Proof Theory: Normalization, Cut-Elimination, and Consistency Proofs.
Paolo Mancosu, Sergio Galvan, and Richard Zach, Oxford University Press. © Paolo Mancosu,
Sergio Galvan and Richard Zach 2021. DOI: 10.1093/oso/9780192895936.003.0003

can be incredibly hard to find. There are a number of reasons for this. One is that although modus ponens is undoubtedly an inference rule often used in mathematical reasoning, it is by no means the only one. Other basic reasoning patterns, for instance, include indirect proof, proof by cases, general conditional proof, or existential instantiation. Whenever a mathematical proof contains the words, "Suppose that ..." or "Let x be such that ...," an *assumption* is being made. For instance, proof by cases: "X is either finite or infinite. Case 1: Assume X is finite ...," or general proof: "Let p be a prime number. ... Since p was any prime number, all primes are ..." This use of assumptions, introduced only hypothetically and *pro tem* in a proof, cannot be dealt with directly in an axiomatic derivation. There, every formula written down is already established, derived from the axioms using the rules of inference.

Another reason axiomatic deduction is difficult to use is that the axioms and rules don't help you to find a proof. They are set down as basic principles involving the primitives, but aren't selected in order to make discovery of proofs easy. In cases where the main (or only) inference rule is modus ponens, an added difficulty is this: Every proof, if it doesn't contain just axioms, must use modus ponens, that is, it must derive the conclusion B from premises A and $A \supset B$. But there is no prima facie connection between the theorem we're aiming to prove (B) and the premise (A) we're proving it from. A may have no connection, syntactic or otherwise, with B. We are left to guess.

Gentzen attempted to produce a proof system which doesn't suffer from these drawbacks. In his dissertation he introduced a system of *natural deduction* (Gentzen, 1935b).[1] Instead of deriving theorems from a selection of axioms using a few simple rules, natural deduction puts the inference rules at the forefront. Gentzen attempted to model the rules of his system on inference patterns mathematicians in fact use. For instance, in order to prove a conditional claim of the form "if P then Q," a mathematician typically begins by *assuming* the antecedent P, and proceeds to demonstrate the consequent Q on the basis of this assumption. Once the conclusion Q is reached, the conditional is established. Overall, the proof of the conditional then no longer depends on the assumption P made at the beginning. Although the proof of Q depends on the assumption P, the proof of "if P then Q" does not—the assumption P is "discharged."

A similarly common pattern of proof is arguing by cases. For instance, to prove that every prime number p has a certain property, it is often useful to treat the case of $p = 2$ separately, so the proof is split into the case where p is even ($p = 2$) and the case where p is odd. The two cases will require a different argument, but in the end each of these arguments reaches the conclusion that p has the sought for property. Generally speaking, the pattern is this: we know that one of two

---

[1] Stanisław Jaśkowski developed similar ideas around 1926 but didn't publish them until later (Jaśkowski, 1934).

cases obtains: either p is even or p is odd. We prove that p has the property first on the assumption that p is even, then on the assumption that p is odd. Since one of the assumptions must obtain, the conclusion (p has the property) follows independently of these assumptions. This scenario also highlights another proof pattern: to prove that *every* prime number has the property, we suppose that p is an arbitrary prime number and show that p has the property. In the proof, we make use of the assumption that p is a prime number. But once we have established that p has the property, we can conclude that every prime has the property, and forget about the variable p and discharge the assumption that p is prime.

These proof patterns can, to a degree, be accommodated in axiomatic calculi. Proof by cases, for instance, corresponds to axiom PL9 of minimal logic,

$$((A \supset C) \wedge (B \supset C)) \supset ((A \vee B) \supset C).$$

Conditional proof corresponds to the deduction theorem. And rule QR_1 can be used to justify inferring $\forall x(A(x) \supset B(x))$ from a proof of $A(a) \supset B(a)$ (read a as our p, $A(a)$ as the assumption that p is prime, and $B(a)$ as the claim that p has the sought-after property). In natural deduction, all three directly correspond to an inference rule.

Before we formally present the systems of natural deduction, we would like to alert the reader to a difference in the way we proceed with respect to Chapter 2. In Chapter 2, we first presented three systems of propositional calculus (minimal (M_0), intuitionistic (J_0), and classical (K_0)). We then developed the predicate calculus by adding quantifier axioms and rules and obtained the three systems M_1, J_1, and K_1. In this chapter we will assume all the definitions of formulas and terms given in Chapter 2 and present the predicate systems from the beginning without making the transition from propositional to predicate systems. For this reason, we will simply talk of **NM**, **NJ**, and **NK** for the three major systems of natural deduction, corresponding to minimal, intuitionistic, and classical predicate logic. In other words, the systems **NM**, **NJ**, and **NK** stand for NM_1, NJ_1, and NK_1. (If one needs to specify that one is working only with the propositional fragment of the systems then one can write NM_0, NJ_0, and NK_0.) In addition, most of our focus will be on **NJ** and **NK**, the two systems which have been the focus of attention since Gentzen introduced them in 1935. However, many of the things we will say about **NJ** will also apply to **NM** and we will occasionally ask the reader to prove things in **NM**.

3.2 Rules and deductions

Like axiomatic derivations, deductions in Gentzen's systems **NJ** and **NK** (or "natural deduction proofs") follow certain inference rules, and in fact the familiar modus ponens rule is one of them. But in contrast to derivations, a deduction in **NJ** or **NK** need not have (and usually doesn't have) *any* axioms. Instead, the starting points of deductions are *assumptions*, and a natural deduction proof *of a formula A* is a deduction in which all assumptions are "discharged." Thus, in natural deduction the inference rules become of central importance. In Gentzen's system **NJ**, the natural deduction system for intuitionistic logic, each logical operator (connective or quantifier) comes with two rules, one an "introduction" and the other an "elimination" rule. For instance, the conditional $\supset$ is governed by the following two rules:

$$\frac{\displaystyle \begin{array}{c} [A] \\ \vdots \\ B \end{array}}{A \supset B} \supset\!\text{I} \qquad \frac{A \supset B \quad A}{B} \supset\!\text{E}$$

The $\supset$E rule is the familiar modus ponens. The $\supset$I rule is the formal version of the informal "conditional proof:" to establish "if A then B," prove B from the (temporary) assumption A. In a deduction in **NJ**, when a temporary assumption has been used to, say, justify $A \supset B$ on the basis of a deduction of B from A, the assumption A may be discharged. In the rule above, this is indicated by putting the assumption in square brackets. The $\supset$I rule should be read as allowing the following: if a deduction ending in B is given, then we may extend this to a deduction of $A \supset B$ justified by the $\supset$I rule, and at the same time we may discharge assumptions of the form A used in the deduction of B. To keep track of discharges of assumptions, we label the inference at which the assumptions are discharged and the assumptions that are there discharged with the same (numerical) label.

Because deductions in **NJ** make use of assumptions, and the correct form of some inference rules depends on which premises "depend" on which assumptions, it becomes necessary to keep track of which premise of a rule depends on which assumptions. This is hard to do if, as in axiomatic derivations, we think of a deduction simply as a sequence of formulas. In an axiomatic derivation, say, a modus ponens inference of B is correct if both a formula of the form A and one of the form $A \supset B$ occur before the inference in the sequence of formulas. In a deduction in **NJ**, some inferences can also be justified by modus ponens—which in **NJ** is called the "$\supset$-elimination rule." But the $\supset$-introduction rule corresponds to conditional proof, i.e., it justifies a formula of the form $A \supset B$ on the basis of a premise of the form B which may depend on an assumption of the form A. So both A and B will appear before $A \supset B$ in the deduction, but they do so in different roles.

Moreover, and more importantly, we must ensure that the assumption A can't be appealed to (used as a premise) in another part of the proof that does not lead to the premise of the $\supset$-introduction rule. This is most easily accomplished by giving deductions a two-dimensional structure: instead of a simple sequence, we think of a deduction as a tree in which each conclusion of an inference is linked to the premises, these premises are linked to the premises of inference rules justifying them, etc., until a formula is not justified by an inference rule, i.e., it is an assumption. Every occurrence of a formula in a deduction can only serve as a premise of a single inference, and also only as the conclusion of a single inference. Thus, in contrast to axiomatic derivations, formulas may not be "reused," i.e., they may not be used as premises of two different inferences. This is not a limitation, since any formula that is required as a premise of a second inference can simply be deduced a second time. (Recall from Section 2.9 that every axiomatic derivation can also be converted into tree form.)

A deduction in **NJ** then is not a sequence, but a tree of formulas. The bottommost formula is the *end-formula*, and the topmost formulas (the formulas at the leaves of the tree) are all assumptions. Since the definition of a deduction requires a bottommost formula, deductions without inferences can only consist of a single assumption, which is also the end-formula. Some of the rules allow assumptions to be *discharged* in parts of the deduction tree ending at the premises of the rule. At any given point in the deduction, those assumptions that are not discharged by that point are called *open*. We'll make these notions precise later. At this point it is just important to introduce the notion of an open assumption, since some of the rules either allow open assumptions to be discharged, or have restrictions which depend on the open assumptions in the part of the deduction leading to the premises of the rule.

Definition 3.1. A *deduction of a formula A* is a tree of formulas in which every formula that is not an assumption is the conclusion of a correct application of one of the inference rules. The assumptions in the deduction that are not discharged by any rule in it are the *open assumptions* of the deduction. If every assumption is discharged, i.e., there are no open assumptions at all, we say the deduction is a *proof* of A, and that A is a *theorem*.

Every step in a deduction must be justified by an inference rule. The inference rules of **NJ** come in pairs, two for each connective and quantifier. They will be denoted by the connective (or quantifier) they deal with, followed by a small ɪ (for introduction) or ᴇ (for elimination). In some cases, namely for ∧-elimination and ∨-introduction, they come in two variants. This duality of rules will be important later when we discuss the normalization theorem.

An introduction rule is a rule the conclusion of which is a formula with the relevant connective or quantifier as the main operator. In other words, introduction rules justify conclusions of the form $A \wedge B$, $A \vee B$, $A \supset B$, $\neg A$, $\forall x\, A(x)$, or $\exists x\, A(x)$. The elimination rules all have the corresponding formula as one of the premises—so they allow us to pass, in a deduction, from a formula of the form—say, $A \wedge B$—to another formula. The conclusion is an immediate sub-formula of the major premise in the cases of ∧ᴇ, ⊃ᴇ, ∀ᴇ. In ∨ᴇ and ∃ᴇ, an immediate sub-formula may be discharged as an assumption in the sub-proof(s) leading to the minor premise(s). The premise that (in the schematic form of the rule) contains the connective or quantifier is called the *major premise*, the other are the *minor premises*.[2] In $\neg$E, the conclusion is $\bot$, which we may count as a sub-formula of $\neg A$—think of $\neg A$ as short for $A \supset \bot$.[3]

In Sections 3.2.1 to 3.2.6 we will present all the rules for connectives and quantifiers. These rules are common to **NM**, **NJ**, and **NK**. The difference between **NM** and **NJ** emerges when discussing the inference rule for $\bot$ (Section 3.2.4). The discussion of the relation between **NJ** and **NK** is postponed to Section 3.3. A summary of the rules of natural deduction given in Table 3.1.

3.2.1 *Rules for* $\wedge$

The rules, or "inference figures," as Gentzen calls them, for $\wedge$ are relatively simple: they don't make reference to assumptions or have restrictions. The $\wedge$E rule comes in two versions.

$$\frac{A \quad B}{A \wedge B} \wedge\text{I} \qquad \frac{A \wedge B}{A} \wedge\text{E} \qquad \frac{A \wedge B}{B} \wedge\text{E}$$

The upper formulas are customarily called the premise(s) of the respective rule, and the lower formula the conclusion. A correct application of these rules results in a deduction which consists of a deduction ending in the premise of $\wedge$E, or two deductions ending in the two premises of $\wedge$I, above a horizontal line, and below it the single formula which matches the conclusion of the rule. The open assumptions of the resulting deduction are the same as the open assumptions of the deduction(s) ending in the premise(s).

For instance, the following is a deduction of $A \wedge B$ from $B \wedge A$:

$$\frac{\dfrac{B \wedge A}{A} \wedge\text{E} \quad \dfrac{B \wedge A}{B} \wedge\text{E}}{A \wedge B} \wedge\text{I}$$

Note that the assumption $B \wedge A$ occurs twice, and neither occurrence is discharged. We use the first version of $\wedge$E on the right, and the second on the left.

Problem 3.2.

1. Show that from the assumption A one can infer $A \wedge A$.

2. Show that from the assumption $(A \wedge B) \wedge C$ one can infer $A \wedge (C \wedge B)$.

3.2.2 *Rules for* $\supset$

The rules for the conditional $\supset$ are the following:

---

[2] Immediate sub-formulas are defined in Section 2.3.
[3] Recall the interpretation of negation in Section 2.10.1.

$$\frac{\overset{\displaystyle [A]}{\overset{\vdots}{B}}}{A \supset B} \supset_{\text{I}} \qquad \frac{A \supset B \quad A}{B} \supset_{\text{E}}$$

The $\supset_{\text{E}}$ rule is again nothing out of the ordinary: it is simply modus ponens. But in contrast to the elimination rule $\wedge_{\text{E}}$, it requires a second premise. We can "eliminate" $\supset$ from $A \supset B$ only if we also have a deduction of A. The result of the elimination then is B. There is no analogous elimination that passes from $A \supset B$ and B to A—that would be a plainly invalid inference. In elimination rules with more than one premise, like $\supset_{\text{E}}$, the premise that contains the connective being eliminated (in this case $A \supset B$) is called the *major premise,* and the other premises are called *minor.* ($\wedge_{\text{E}}$ only has a major, but no minor premise.)[4] The open assumptions of the deduction resulting from applying $\supset_{\text{E}}$ are the same as the open assumptions of the deductions ending in the two premises.

The interesting rule is the $\supset_{\text{I}}$ rule. Intuitively, it corresponds to the principle of conditional proof: if we can prove B from some assumption A, then we're entitled to infer that the conditional "if A then B" is true (and true independently of the assumption A). In our three systems, a deduction of B from assumption A stands in for such a proof, and the convention of "discharging" assumptions marks the transition from a hypothetical proof which depends on an assumption to an outright proof of a claim.

The $[A]$ indicates that at the point in a deduction where we pass from the premise B to the conclusion $A \supset B$, we may discharge any open assumptions of the form A in the deduction ending in the premise. Note that B alone counts as the premise of the $\supset_{\text{I}}$ inference; the assumptions of the form A being discharged at the inference are not premises. We indicate which assumptions are discharged by a given inference by assigning a label to the inference and to its corresponding discharged assumptions. Of course, we'll require that no two inferences discharge assumptions labelled the same way. This can always be accomplished by switching labels.

Let's first look at a simple example of a deduction. The formula $(A \wedge B) \supset (B \wedge A)$ is of course valid, classically as well as intuitionistically. Now note that the conclusion of the $\supset_{\text{I}}$ rule is of the form $C \supset D$, and the formula we would like to deduce is also of that form. So when looking for a deduction, one way of proceeding would be to find a deduction the last inference of which is an application of $\supset_{\text{I}}$. A correct application of $\supset_{\text{I}}$ requires as a premise the consequent of the conditional, in this case, $B \wedge A$. In this application of $\supset_{\text{I}}$, we may discharge any open assumptions of

---

[4] Note that we have made the major premise the left premise. In Gentzen's original presentation, the major premise is on the right. This is of course only a notational difference, but our usage is now standard in the literature.

the same form as the antecedent of the conditional we're proving, i.e., $A \wedge B$.[5] In other words, a deduction of this form will look as follows:

$$\begin{array}{c} A \wedge B^{\,1} \\ \vdots \\ 1 \ \dfrac{B \wedge A}{(A \wedge B) \supset (B \wedge A)} \ \supset\!\text{I} \end{array}$$

The deduction is not complete yet, of course. We still actually have to derive $B \wedge A$ from $A \wedge B$. However, since our last $\supset$I inference allows us to discharge any assumptions of the form $A \wedge B$ we know that we can use assumptions of this form without worrying that they will remain undischarged at the end. Thus, our task is to find a deduction of $B \wedge A$ from (perhaps multiple copies of) the assumption $A \wedge B$. Again, it is helpful to try to imagine what a deduction would look like that ends with the introduction rule for the main connective of the conclusion, i.e., $B \wedge A$. The $\wedge$I rule requires two premises, of the form B and A, respectively. It does not allow discharging any premises. Each one of the two deductions ending in the two premises, however, may use the assumption $A \wedge B$. Hence the middle part of our deduction will be:

$$\begin{array}{cc} A \wedge B & A \wedge B \\ \vdots & \vdots \\ \dfrac{B \qquad A}{B \wedge A} & \wedge\text{I} \end{array}$$

We are still not quite done: we must deduce A from $A \wedge B$ (on the right), and B from $A \wedge B$ (on the left). These, however, require just one inference, namely $\wedge$E:

$$\dfrac{A \wedge B}{B} \ \wedge\text{E} \qquad \dfrac{A \wedge B}{A} \ \wedge\text{E}$$

We insert these into the deduction of $B \wedge A$ from $A \wedge B$ to obtain:

$$\dfrac{\dfrac{A \wedge B}{B} \ \wedge\text{E} \qquad \dfrac{A \wedge B}{A} \ \wedge\text{E}}{B \wedge A} \ \wedge\text{I}$$

And this, we now use as the deduction providing the premise of $\supset$I. Recalling that this application of $\supset$I discharges assumptions of the form $A \wedge B$, we label the two assumptions of that form with the label 1 corresponding to the $\supset$I inference:

$$1 \ \dfrac{\dfrac{\dfrac{A \wedge B^{\,1}}{B} \ \wedge\text{E} \qquad \dfrac{A \wedge B^{\,1}}{A} \ \wedge\text{E}}{B \wedge A} \ \wedge\text{I}}{(A \wedge B) \supset (B \wedge A)} \ \supset\!\text{I}$$

[5] This is a general strategy: when looking for a deduction of a non-atomic formula, find a deduction that ends with the I-rule for its main connective. Match the formula to be deduced to the inference rule: it will tell you what the premise(s) of the last inference have to be, and what formulas the inference may discharge. Then apply this strategy to the premises, etc. See Section 2.3 to remind yourself of what the main connective of a formula is.

This is now a deduction of $(A \wedge B) \supset (B \wedge A)$ in which every assumption is discharged, and every formula that is not an assumption is the conclusion of a correct application of an inference rule.

Rules like $\supset$I are *permissive* as far as discharging assumptions is concerned: any open assumptions of the form A may be discharged, but it is not *required* that we discharge any of them, nor is it required that any open assumptions of the form A be present in the first place. For instance, here's a simple deduction in which the first application of $\supset$I does not discharge any assumptions (but the second one does):

$$1 \,\cfrac{\cfrac{A^1}{B \supset A}\supset\text{I}}{A \supset (B \supset A)}\supset\text{I}$$

In a deduction, we often have a choice as to which assumptions are discharged by which inferences, without changing the end-formula or the set of open assumptions of the entire proof. We insist that the choice is made explicit: every inference discharging assumptions has a unique label and discharges all and only those assumptions with the same label. If two deductions are the same except for which assumption occurrence is discharged by which inference, we will consider them different deductions.[6] The following are all different deductions:

$$2 \,\cfrac{1\,\cfrac{\cfrac{A^1 \quad A^2}{A \wedge A}\wedge\text{I}}{A \supset (A \wedge A)}\supset\text{I}}{A \supset (A \supset (A \wedge A))}\supset\text{I} \qquad 2\,\cfrac{1\,\cfrac{\cfrac{A^2 \quad A^1}{A \wedge A}\wedge\text{I}}{A \supset (A \wedge A)}\supset\text{I}}{A \supset (A \supset (A \wedge A))}\supset\text{I} \qquad \cfrac{1\,\cfrac{\cfrac{A^1 \quad A^1}{A \wedge A}\wedge\text{I}}{A \supset (A \wedge A)}\supset\text{I}}{A \supset (A \supset (A \wedge A))}\supset\text{I} \qquad 2\,\cfrac{\cfrac{\cfrac{A^2 \quad A^2}{A \wedge A}\wedge\text{I}}{A \supset (A \wedge A)}\supset\text{I}}{A \supset (A \supset (A \wedge A))}\supset\text{I}$$

A formula in a deduction can serve both as a premise of an inference, and as an assumption discharged at the same inference. The simplest possible deduction can already serve to illustrate this:

$$1\,\cfrac{A^1}{A \supset A}\supset\text{I}$$

A, by itself, is a deduction of A from the assumption A (the very formula A itself). $\supset$I allows us to extend this deduction by adding $A \supset A$: A is the premise of the inference. But $\supset$I also allows us to discharge any assumptions of the form A on which the premise depends; in this case the premise just is an assumption, and so it depends on itself.

Problem 3.3.   1. Give a proof (i.e., a deduction from no open assumptions) of $A \supset (A \wedge A)$.   (PL1)

2. Prove $(A \supset B) \supset [(A \wedge C) \supset (B \wedge C)]$.   (PL3)

---

[6] It is possible to consider different conventions. For instance, the "complete discharge convention" requires that every inference discharges all the open assumptions it can. The result in terms of what can be proved is the same, but the results of the next chapter do not hold; see Leivant (1979).

3. Prove $[(A \supset B) \wedge (B \supset C)] \supset (A \supset C)$. (PL4)

4. Prove $(A \wedge (A \supset B)) \supset B$. (PL6)

5. Prove $[A \supset (B \supset C)] \supset [(A \supset B) \supset (A \supset C)]$. (Theorem 2.8)

6. Prove $(A \supset B) \supset ((A \supset C) \supset (A \supset (B \wedge C)))$. (G8)

3.2.3 Rules for ∨

The rules for disjunction ∨ are the following:

$$\frac{A}{A \vee B} \text{ ∨I} \qquad \frac{B}{A \vee B} \text{ ∨I} \qquad \frac{A \vee B \quad \overset{[A]}{\overset{\vdots}{C}} \quad \overset{[B]}{\overset{\vdots}{C}}}{C} \text{ ∨E}$$

The introduction rule for ∨ is again not problematic: it allows us to deduce $A \vee B$ from either disjunct. Like the ∧E rule, it comes in two alternative versions, one for the left and one for the right disjunct. The ∨E rule is more complicated. It is the equivalent in **NJ** of the informal principle of a case distinction: If we know that a disjunction $A \vee B$ follows from our assumptions, then we know that anything that follows from either alternative, follows from the same assumptions. In other words, if $A \vee B$ follows, C follows from A, and C also follows from B, then C follows.

Like the $\supset$E rule, the ∨E allows us to "eliminate" a disjunction. In ∨E rule, the left premise containing ∨ is again called the *major premise,* and the middle and right premises are called *minor.* As in the case of $\supset$I, the notation $[A]$ and $[B]$ indicates that at an application of ∨E, any open assumption of the form A in the deduction ending in the middle premise, and any open assumption of the form B in the deduction ending in the right premise may be discharged. The rule does not allow us to discharge assumptions in the deduction ending in the leftmost premise, or assumptions not of the correct form in the deductions leading to the other premises. When writing out deductions, we keep track of which assumptions are discharged at which inferences by labelling the rule and the corresponding discharged assumptions.

Let's see if we can apply these rules to deduce the following formula, which can be seen as a direct statement of proof by cases, or, "if A implies C and B implies C as well, then $A \vee B$ implies C."

$$((A \supset C) \wedge (B \supset C)) \supset ((A \vee B) \supset C)$$

We begin our search for a deduction again with a guess as to how the deduction will end: The main connective is the middle $\supset$, and the theorem to be deduced is a conditional with antecedent the conjunction $(A \supset C) \wedge (B \supset C)$ and consequent the conditional $(A \vee B) \supset C$. So our deduction should end with $\supset$I, inferring the theorem from the premise $(A \lor B) \supset C$. The deduction of the premise itself will use the assumption $(A \supset C) \land (B \supset C)$ (again, possibly more than one copy!), and these assumptions will be discharged by the $\supset$I rule. So our overall deduction will look like this:

$$(A \supset C) \land (B \supset C)^1$$
$$\vdots$$
$$1 \, \frac{(A \lor B) \supset C}{((A \supset C) \land (B \supset C)) \supset ((A \lor B) \supset C)} \supset I$$

Now we have to deduce the premise from the indicated assumption. The main connective of $(A \lor B) \supset C$ is again $\supset$, so a deduction of it will end with $\supset$I with premise C (the consequent), which in turn will be the result of a deduction which may use the assumption $A \lor B$. This assumption will be discharged by the new $\supset$I rule. Of course, this deduction of C will also use the assumptions discharged by the first (lowermost) $\supset$I rule, i.e., $(A \supset C) \land (B \supset C)$. So we are looking for a deduction of the following overall form:

$$(A \supset C) \land (B \supset C) \qquad A \lor B^2$$
$$\vdots$$
$$2 \, \frac{C}{(A \lor B) \supset C} \supset I$$

Now our task is to deduce the premise C from the two indicated assumptions. Our previous strategy of identifying the main connective and applying the corresponding introduction rule as the last inference does not work in this case, since C may not have a main connective. On the other hand, one of our assumptions is a disjunction, $A \lor B$. So one promising avenue is to attempt to deduce C by cases: once from the assumption A and once from the assumption B. We can then combine these two deductions with the assumption $A \lor B$ as the major premise of an application of $\lor$E. Again, the deductions of C may use the assumption $(A \supset C) \land (B \supset C)$ which will be discharged later on in the deduction. (It may also use the other assumption which will be discharged in the second $\supset$I rule, $A \lor B$. However, it will turn out it is not needed, so we will leave it out.) We are looking for a deduction of the form:

$$(A \supset C) \land (B \supset C) \quad A^3 \qquad\qquad (A \supset C) \land (B \supset C) \quad B^3$$
$$\vdots \qquad\qquad\qquad\qquad \vdots$$
$$3 \, \frac{A \lor B \qquad\qquad C \qquad\qquad\qquad\qquad C}{C} \lor E$$

We "zoom in" on the middle deduction of C from $(A \supset C) \land (B \supset C)$ and A. It is not difficult to guess how this deduction should go: from the first assumption we obtain $A \supset C$ by $\land$E, and then an application of $\supset$E from that formula and the assumption A gives C:

$$\dfrac{\dfrac{(A \supset C) \wedge (B \supset C)}{A \supset C} \; {}_{\wedge E} \qquad A}{C} \; {}_{\supset E}$$

The deduction of C from $(A \supset C) \wedge (B \supset C)$ and B is similar, but of course uses $B \supset C$ instead of $A \supset C$. If we plug these deductions into the $\vee$-elimination from $A \vee B$ we get:

$$\dfrac{{}_3 \; A \vee B \qquad \dfrac{\dfrac{(A \supset C) \wedge (B \supset C)}{A \supset C} \; {}_{\wedge E} \quad A \, {}^3_{\supset E}}{C} \qquad \dfrac{\dfrac{(A \supset C) \wedge (B \supset C)}{B \supset C} \; {}_{\wedge E} \quad B \, {}^3_{\supset E}}{C}}{C} \; {}_{\vee E}$$

We obtain the complete deduction by using this to yield the premise of the $\supset$I rule with label 2, and the result in turn as the deduction of the premise of the $\supset$I rule labelled ^{1}. In full, the deduction reads:

$$\dfrac{\dfrac{{}_3 \; A \vee B^2 \qquad \dfrac{\dfrac{(A \supset C) \wedge (B \supset C)^1}{A \supset C} \; {}_{\wedge E} \quad A \, {}^3_{\supset E}}{C} \qquad \dfrac{\dfrac{(A \supset C) \wedge (B \supset C)^1}{B \supset C} \; {}_{\wedge E} \quad B \, {}^3_{\supset E}}{C} \; {}_{\vee E}}{{}_2 \; \dfrac{C}{(A \vee B) \supset C} \; {}_{\supset I}}}{{}_1 \; \dfrac{}{((A \supset C) \wedge (B \supset C)) \supset ((A \vee B) \supset C)}} \; {}_{\supset I}$$

In the process of deducing these theorems we have constructed partial deductions in which some assumptions were left open—they were discharged only later when we put the pieces together. Sometimes we may consider deductions with open assumptions on their own: they show that the end-formula is a *consequence* of the open assumptions of the deduction. Here is an example deduction of the formula $A \vee (B \wedge C)$ from the assumption $(A \vee B) \wedge (A \vee C)$. As an exercise, attempt to construct this deduction as we did before, considering step-by-step which inferences and corresponding premises and assumptions would yield the desired conclusion.

$$\dfrac{{}_1 \; \dfrac{\dfrac{(A \vee B) \wedge (A \vee C)}{A \vee B} \; {}_{\wedge E} \quad \dfrac{A^1}{A \vee (B \wedge C)} \; {}_{\vee I} \quad {}_2 \dfrac{\dfrac{(A \vee B) \wedge (A \vee C)}{A \vee C} \; {}_{\wedge E} \quad \dfrac{A^2}{A \vee (B \wedge C)} \; {}_{\vee I} \quad \dfrac{\dfrac{B^1 \quad C^2}{B \wedge C} \; {}_{\wedge I}}{A \vee (B \wedge C)} \; {}_{\vee I}}{A \vee (B \wedge C)} \; {}_{\vee E}}{A \vee (B \wedge C)}}{}$$

In both applications of $\vee$E, the role of "C" in the statement of the rule, i.e., the minor premise, is played by $A \vee (B \wedge C)$. The two assumptions $(A \vee B) \wedge (A \vee C)$ are not discharged, and they are the only undischarged assumptions. We could add a $\supset$I inference at the end which discharges them. This inference would have as conclusion $((A \vee B) \wedge (A \vee C)) \supset (A \vee (B \wedge C))$.

Just as in the case of $\supset$I, discharging assumptions in the deductions leading to the minor premises of $\vee$E is *allowed* but not *required*. For instance, the following is a correct deduction of $(A \vee B) \supset A$ from A:

$$2 \frac{\dfrac{A \vee B^1 \quad A^2 \quad A}{A} \vee_E}{(A \vee B) \supset A} \supset_I 1$$

Here, the sub-proof (consisting just of the assumption A) leading to the left premise A of the $\vee_E$ inference discharges A (with label 2). The sub-proof leading to the right premise A does not, however, discharge B (it can't, since B is not an open assumption of that sub-proof). So the second assumption of the form A is undischarged, and A is an open assumption of the entire proof. Note in this case we could have let the $\vee_E$ inference not discharge anything. This would have resulted in a slightly different proof (different assumptions are discharged) but of the same end-formula from the same set of open assumptions.

Problem 3.4.   1. Prove $A \supset (A \vee B)$.   (PL7)

2. Prove $(A \vee B) \supset (B \vee A)$.   (PL8)

3. Give a deduction of $(A \vee B) \wedge (A \vee C)$ from the assumption $A \vee (B \wedge C)$.

4. Prove $[(A \supset B) \vee (A \supset C)] \supset [A \supset (B \vee C)]$.

3.2.4   Rules for $\neg$ and $\bot$

Recall that $\bot$ is an atomic formula that stands for an outright contradiction (see Section 2.10.1). **NJ** contains one rule for $\bot$, called $\bot_J$, the intuitionistic absurdity rule:

$$\frac{\bot}{D} \bot_J$$

for any formula D.

If we leave out $\bot_J$, we obtain a natural deduction system for minimal logic (see Section 2.4.1); i.e., we let **NM** consist of the I- and E-rules of **NJ**, but not the $\bot_J$ rule.

If $\neg A$ is introduced as a definitional abbreviation for $A \supset \bot$, the $\supset_I$ and $\supset_E$ rules for the special case of $B \equiv \bot$,

$$\begin{array}{c} [A] \\ \vdots \\ \frac{\bot}{A \supset \bot} \supset_I \end{array} \qquad \frac{A \supset \bot \quad A}{\bot} \supset_E$$

would serve to introduce and eliminate the negation sign. Gentzen originally introduced negation as a primitive and added the explicit rules $\neg_I$ and $\neg_E$:

$$\begin{array}{c} [A] \\ \vdots \\ \frac{\bot}{\neg A} \neg_I \end{array} \qquad \frac{\neg A \quad A}{\bot} \neg_E$$

As you can see, $\neg_I$ and $\neg_E$ would be special cases of the $\supset_I$ and $\supset_E$ rules, if we take $\neg A$ to be an abbreviation for $A \supset \bot$. In our proofs below we wouldn't have to treat ¬I and ¬E separately, if we adopted this abbreviation. For clarity, however, we will discuss these rules explicitly.

Let's give some examples of deductions using the ¬ rules: First, let's show that the explicit rules in fact define $\neg A$ as $A \supset \bot$, that is, that they allow us to deduce both $\neg A \supset (A \supset \bot)$ and $(A \supset \bot) \supset \neg A$. The deductions are not hard to find; as an exercise, try to develop them yourself without looking at the solution:

$$
\cfrac{\cfrac{\cfrac{\neg A^1 \quad A^2}{\bot}{\scriptstyle \neg E}}{2 \; \cfrac{}{A \supset \bot}{\scriptstyle \supset I}}{1 \; \cfrac{}{\neg A \supset (A \supset \bot)}{\scriptstyle \supset I}}
\qquad
\cfrac{\cfrac{\cfrac{A \supset \bot^1 \quad A^2}{\bot}{\scriptstyle \supset E}}{2 \; \cfrac{}{\neg A}{\scriptstyle \neg I}}{1 \; \cfrac{}{(A \supset \bot) \supset \neg A}{\scriptstyle \supset I}}
$$

For another simple example, consider $\neg(A \wedge \neg A)$. In order to use ¬I, we should attempt to deduce $\bot$ from $A \wedge \neg A$. Of course, we can use multiple copies of the assumption $A \wedge \neg A$. If we employ $\wedge E$, this allows us to deduce both A and $\neg A$, the premises required for an application of ¬E:

$$
\cfrac{\cfrac{A \wedge \neg A^1}{\neg A}{\scriptstyle \wedge E} \qquad \cfrac{A \wedge \neg A^1}{A}{\scriptstyle \wedge E}}{1 \; \cfrac{\bot}{\neg(A \wedge \neg A)}{\scriptstyle \neg I}}
$$

A simple example of a deduction that requires $\bot_J$, i.e., of a theorem of intuitionistic logic which isn't a theorem of minimal logic, consider this deduction of ex falso quodlibet, $(A \wedge \neg A) \supset B$:

$$
\cfrac{\cfrac{\cfrac{A \wedge \neg A^1}{\neg A}{\scriptstyle \wedge E} \qquad \cfrac{A \wedge \neg A^1}{A}{\scriptstyle \wedge E}}{\cfrac{\bot}{B}{\scriptstyle \bot_J}}{\scriptstyle \neg E}}{1 \; \cfrac{}{(A \wedge \neg A) \supset B}{\scriptstyle \supset I}}
$$

Now consider the intuitionistically acceptable direction of contraposition, $(A \supset B) \supset (\neg B \supset \neg A)$. (The other direction requires classical logic; see Section 3.3.) The last inference of our deduction will use ⊃I from premise $\neg B \supset \neg A$, discharging assumptions of the form $A \supset B$. To deduce the premise, we have to deduce $\neg A$ from the assumptions $\neg B$ and $A \supset B$ together. This, in turn, will require a deduction of $\bot$ from the assumption A (together with $\neg A$ and $A \supset B$), to which we can apply ¬I. This line of thought leads us to:

$$
\cfrac{\cfrac{\cfrac{\neg B^2 \quad \cfrac{A \supset B^1 \quad A^3}{B}{\scriptstyle \supset E}}{3 \; \cfrac{\bot}{\neg A}{\scriptstyle \neg I}}{\scriptstyle \neg E}}{2 \; \cfrac{}{\neg B \supset \neg A}{\scriptstyle \supset I}}{1 \; \cfrac{}{(A \supset B) \supset (\neg B \supset \neg A)}{\scriptstyle \supset I}}
$$

Problem 3.5. Find proofs in **NM** of the following formulas from Problem 2.18:

1. $(A \supset B) \supset (\neg\neg A \supset \neg\neg B)$.   (E3)

2. $(A \supset \neg B) \supset (B \supset \neg A)$.   (E4)

3. $A \supset \neg\neg A$.   (E5)

4. $\neg\neg\neg A \supset \neg A$.   (E7)

5. $\neg\neg(A \wedge B) \supset (\neg\neg A \wedge \neg\neg B)$.   (E9)

6. $\neg\neg(A \vee \neg A)$.   (E10)

Problem 3.6. Prove in **NJ**:

1. $((A \vee B) \wedge \neg A) \supset B$.

2. $\neg A \supset (A \supset B)$.   (PL11)

3.2.5   Rules for $\forall$

As in Chapter 2, we will use a, b, c, etc., as metavariables for free variables, and x, y, z, etc., as metavariables for bound variables. The rules for $\forall$ mirror the rule QR_1 and axiom QL1:

$$\frac{A(c)}{\forall x\, A(x)}\, \forall\text{I} \qquad\qquad \frac{\forall x\, A(x)}{A(t)}\, \forall\text{E}$$

Recall the conventions from Section 2.13. We write $A(c)$ to indicate that the formula A contains the free variable c. In this context, $A(x)$ stands for $A[x/c]$ and $A(t)$ stands for $A[t/c]$. This means that in $\forall$I, the conclusion $\forall x\, A(x)$ does not contain c (since every occurrence of c is replaced by x in $A[x/c]$). Of course, $A(c)$ may contain t, and so we can pass, e.g., from $\forall x\, P(x, t)$ to $P(t, t)$ using $\forall$E.

Whereas the QR_1 rule (see Section 2.13) allows us to introduce a universal quantifier in the consequent of a conditional, the $\forall$I rule operates on a single formula. This is possible because of the potential presence of open assumptions; the conditional can then be introduced separately and the assumptions corresponding to the antecedent formula discharged. However, just as the rule QR_1 must require that the variable that is being universally quantified not appear in the antecedent of the conditional, the corresponding natural deduction rule $\forall$I must also be restricted: it can only be applied if the variable indicated in the premise, c, does not occur in any assumptions which are open in the deduction ending in the premise. A free variable so restricted is called the *eigenvariable* of the inference.

The restriction is necessary to prevent moves like the following:

$$1\ \frac{\dfrac{A(c)^1}{\forall x\, A(x)}\, \forall\text{I}}{A(c) \supset \forall x\, A(x)}\, \supset\text{I}$$

The end-formula of this "deduction" is not valid, so it should not have a proof. This is not a *correct* deduction, however, since at the ∀I inference, the assumption $A(c)$ is still open (it is only discharged at the next inference), so the eigenvariable occurs in an open assumption and the restriction on ∀I is violated.

For instance, one might think that using the rules for ∀, together with those for ⊃, we can derive the QR₁ rule, i.e., deduce $A \supset \forall x\, B(x)$ from $A \supset B(c)$, provided c does not occur in A or in $\forall x\, B(x)$.

$$\cfrac{\cfrac{\cfrac{A \supset B(c) \quad A^1}{B(c)} \supset\text{E}}{\forall x\, B(x)} \forall\text{I}}{{}_1\; A \supset \forall x\, B(x)} \supset\text{I}$$

The restriction on ∀I, however, is not satisfied because even though we assumed that c does not occur in $\forall x\, B(x)$ (that is, $B(x)$ is the result of replacing every occurrence of c in $B(c)$ by x), and not in A, it does occur in the open assumption $A \supset B(c)$. So this is not a correct deduction on its own. (If it were, we would be able to use ⊃I to obtain a proof of

$$(A \supset B(c)) \supset (A \supset \forall x\, B(x)),$$

which is not valid.)

However, if we replace the open assumption $A \supset B(c)$ with the conclusion of a complete deduction of $A \supset B(c)$ in whose open assumptions c does not occur, we would obtain a correct deduction. For instance, if that hypothetical deduction of $A \supset B(c)$ does not have any open assumptions at all, it would show that $A \supset B(c)$ is a theorem, and hence valid. In that case, $A \supset \forall x\, B(x)$ would also be a theorem. So what we do have is the following: if $A \supset B(c)$ has a correct deduction, c does not occur in an open assumption or in A, then $A \supset \forall x\, B(x)$ has a correct deduction.

Finally, recall again that by convention $A(x)$ is the result of replacing *every* occurrence of c in $A(c)$ by x. Thus, c does not appear in the conclusion of ∀I. This is also important. For suppose we allowed the inference

$$* \cfrac{P(c, c)}{\forall y\, P(c, y)} \forall\text{I}$$

This would allow us to prove $\forall x\, P(x, x) \supset \forall x \forall y\, P(x, y)$. But this is not valid.

The ∀E rule has no restriction imposed on it, and together with ⊃I it allows us to prove QL1:

$$\cfrac{\cfrac{\forall x\, A(x)^1}{A(t)} \forall\text{E}}{{}_1\; \forall x\, A(x) \supset A(t)} \supset\text{I}$$

As a simple example, consider the formula $\forall x(A \wedge B(x)) \supset (A \wedge \forall x\, B(x))$. In order to deduce it, we proceed as before and attempt first to deduce the consequent $A \wedge \forall x\, B(x)$ from the assumption $\forall x(A \wedge B(x))$. In order to prove the conclusion—a conjunction—we want to prove both conjuncts. Suppose a is a free variable not occurring in A or $\forall x\, B(x)$. We can deduce $A \wedge B(a)$ from $\forall x(A \wedge B(x))$. By $\wedge$E, we obtain, on the one hand, A, and, on the other, $B(a)$. Since a was a variable assumed not to occur in either A or $\forall x\, B(x)$, it also does not occur in the assumption $\forall x(A \wedge B(x))$. Hence, we can use the formula $B(a)$ so obtained as the premise of $\forall$I and obtain $\forall x\, B(x)$. Putting things together using $\wedge$I followed by $\supset$I (and discharging the assumption), we get the deduction:

$$\cfrac{\cfrac{\cfrac{\forall x(A \wedge B(x))^{\,1}}{A \wedge B(a)}\,\forall\text{E}}{A}\,\wedge\text{E} \qquad \cfrac{\cfrac{\cfrac{\forall x(A \wedge B(x))^{\,1}}{A \wedge B(a)}\,\forall\text{E}}{\cfrac{B(a)}{\forall x\, B(x)}\,\forall\text{I}}\,\wedge\text{E}}{}}{\cfrac{\cfrac{A \wedge \forall x\, B(x)}{}\,\wedge\text{I}}{\forall x(A \wedge B(x)) \supset (A \wedge \forall x\, B(x))}\,\supset\text{I}}\;{}_{1}$$

Problem 3.7. Give proofs in **NM** of the following formulas (cf. Problem 2.40):

1. $\forall x(A(x) \supset A(x))$.   (E2)
2. $\forall x(A \supset B(x)) \supset (A \supset \forall x\, B(x))$.   (E4)
3. $\forall x_1 \forall x_2\, B(x_1, x_2) \supset \forall x_2 \forall x_1 B(x_2, x_1)$.   (E5)
4. $\forall x(A(x) \supset B(x)) \supset (\forall x\, A(x) \supset \forall x\, B(x))$.   (E6)

3.2.6   Rules for $\exists$

The rules for the existential quantifier $\exists$ are given by:

$$[A(c)]$$
$$\vdots$$

$$\cfrac{A(t)}{\exists x\, A(x)}\,\exists\text{I} \qquad\qquad \cfrac{\exists x\, A(x) \qquad C}{C}\,\exists\text{E}$$

The introduction rule for $\exists$ works the same way as the elimination rule for $\forall$, except in the opposite direction. Together with $\supset$I, we can use it to prove the characteristic axiom QL2 for $\exists$:

$$\cfrac{\cfrac{A(t)^{\,1}}{\exists x\, A(x)}\,\exists\text{I}}{A(t) \supset \exists x\, A(x)}\,\supset\text{I}\;\;{}_{1}$$

Recall again that $A(t)$ and $A(x)$ are short for $A[t/c]$ and $A[x/c]$ for some formula $A(c)$. Since $A(c)$ may contain t, e.g., we are allowed to infer $\exists x\, P(x, t)$ from $P(t, t)$ using $\exists$E—not every occurrence of t in the premise must be "replaced" by the bound variable x.

The elimination rule is much trickier. Like the $\forall$E rule, it has a major premise $\exists x\, A(x)$ and a minor premise C. Intuitively, the rule corresponds to the principle that if we know there are A's (at least one), we can pretend that the free variable c ranges over any of these, and so assume that $A(c)$. Once we assumed that c is one of the A's, we can use this assumption to prove C. When this is accomplished, the assumption $A(c)$ is no longer needed—provided that c doesn't appear in our conclusion C. Of course, not any free variable would be suitable to use for this purpose: only variables about which we haven't already assumed something, i.e., only variables that don't appear in the major premise or any open assumptions on which it or the subsidiary conclusion C depends. These rather complicated preconditions on the use of the variables c are expressed in a restriction on the use of $\exists$E: Only applications of $\exists$E are allowed where the variable c does not appear in C or in any assumptions that are open after the application of the inference. This means in particular that c may not be open in an assumption of the sub-deduction leading to the major premise $\exists x\, A(x)$, in any assumption not of the form $A(c)$ in the sub-deduction leading to the minor premise C, and that *all* open assumptions of the form $A(c)$ in that right sub-deduction must be discharged at this inference. As in the $\forall$I rule, we assume that the indicated free variable c is replaced by x to form $A(x)$. In other words, $\exists x\, A(x)$ is short for $\exists x\, A[x/c]$. Thus, c cannot occur in the major premise of the rule either. The variable c in $A(c)$ which has to satisfy these restrictions is called the *eigenvariable* of the $\exists$E inference. At an application of $\exists$E, any assumptions of the form $A(c)$ which are open in the deduction ending in the minor premise may be discharged.

To see what can go wrong if the eigenvariable condition is not satisfied, consider the obviously invalid formula $\exists x\, A(x) \supset \forall x\, A(x)$. One attempt to "prove" it would look like this:

$$
\cfrac{\cfrac{\cfrac{\exists x\, A(x)^1 \quad A(c)^2}{A(c)}\,{}_{\exists\mathrm{E}}}{\forall x\, A(x)}\,{}_{\forall\mathrm{I}}}{\exists x\, A(x) \supset \forall x\, A(x)}\,{}_{\supset\mathrm{I}}
$$

This violates the condition on $\exists$E, since the conclusion contains the eigenvariable c. Note that we also can't "prove" it by exchanging the order of $\forall$I and $\exists$E, i.e., the following way:

$$
\cfrac{\cfrac{\exists x\, A(x)^1 \quad \cfrac{A(c)^2}{\forall x\, A(x)}\,{}_{\forall\mathrm{I}}}{\forall x\, A(x)}\,{}_{\exists\mathrm{E}}}{\exists x\, A(x) \supset \forall x\, A(x)}\,{}_{\supset\mathrm{I}}
$$

Here, the condition on $\exists$E is satisfied: c does not occur in the conclusion $\forall x\, A(x)$ or any open assumption (other than $A(c)$, of course). However, we have here violated the condition on $\forall$I, since the free variable we universally generalized on there—i.e., c—occurs in an assumption that is still open where $\forall$I is applied, namely that formula itself: $A(c)$. It is only discharged by the $\exists$E rule below the $\forall$I inference.

Let's try to deduce the formula

$$\exists x(A(x) \supset B) \supset (\forall x\, A(x) \supset B).$$

Note that B cannot contain the bound variable x, as otherwise the occurrence of B on the right would contain x without a quantifier binding it, and so would not be well-formed.

The first step in the construction of our deduction, again, consists in setting up the last inference using $\supset$I. The deduction will have the form:

$$\exists x(A(x) \supset B)^1$$
$$\vdots$$
$$1 \,\frac{\forall x\, A(x) \supset B}{\exists x(A(x) \supset B) \supset (\forall x\, A(x) \supset B)} \supset\text{I}$$

To fill in the missing deduction of the premise, we repeat this, keeping in mind that the deduction of B from $\forall x\, A(x)$ may also use the assumption $\exists x(A(x) \supset B)$. In other words, we now want to find a deduction of the form:

$$\exists x(A(x) \supset B) \qquad \forall x\, A(x)^2$$
$$\vdots$$
$$2 \,\frac{B}{\forall x\, A(x) \supset B} \supset\text{I}$$

So how do we deduce B from $\exists x(A(x) \supset B)$ and $\forall x\, A(x)$? Here we will have to use the $\exists$E rule to make use of the first assumption. It allows us to deduce something—in this case, we're looking for a deduction of B—from the existentially quantified major premise $\exists x(A(x) \supset B)$ if we also manage to find a deduction of the conclusion from an instance of the major premise. In other words, we are looking for a deduction of the form:

$$(A(c) \supset B)^3 \qquad \forall x\, A(x)$$
$$\vdots$$
$$3 \,\frac{\exists x(A(x) \supset B) \qquad\qquad B}{B} \exists\text{E}$$

We have put $\forall x\, A(x)$ at the top to remind us that it is an assumption that we will later be able to discharge, so our deduction of B can use it, if needed. Again, we would also be able to use the assumption $\exists x(A(x) \supset B)$. Since we in fact won't need to use it, we have left it off. Of course we also have to be careful to pick a free variable c that does not occur in $\exists x(A(x) \supset B)$, B, or any other undischarged assumption such as $\forall x\, A(x)$—otherwise the eigenvariable condition on $\exists$E would be violated. The last piece of our deduction leading to the minor premise B now is simply:

$$\frac{A(c) \supset B \quad \dfrac{\dfrac{\forall x\, A(x)}{A(c)}\, \forall E}{B}\, \supset E}{B}$$

The deduction ending with $\exists E$, with the . . . filled in becomes:

$$\frac{\exists x (A(x) \supset B) \quad \dfrac{A(c) \supset B^{\,3} \quad \dfrac{\dfrac{\forall x\, A(x)}{A(c)}\, \forall E}{B}\, \supset E}{B}\, \exists E}{B}\; 3$$

The full deduction is:

$$\frac{\dfrac{\exists x (A(x) \supset B)^{\,1} \quad \dfrac{A(c) \supset B^{\,3} \quad \dfrac{\dfrac{\forall x\, A(x)^{\,2}}{A(c)}\, \forall E}{B}\, \supset E}{B}\, \exists E \;\; 3}{\dfrac{B}{\forall x\, A(x) \supset B}\, \supset I \;\; 2}}{\exists x (A(x) \supset B) \supset (\forall x\, A(x) \supset B)}\, \supset I \;\; 1$$

Problem 3.8. Give proofs in **NM** for the following:

1. $\exists x (A(x) \wedge B(x)) \supset (\exists x\, A(x) \wedge \exists x\, B(x))$.
2. $\exists x (A(x) \vee B(x)) \supset (\exists x\, A(x) \vee \exists x\, B(x))$.
3. $\exists x\, \forall y\, C(y, x) \supset \forall x\, \exists y\, C(x, y)$.
4. $\forall x (A(x) \supset B) \supset (\exists x\, A(x) \supset B)$.

As a final example, we'll construct a deduction of $\exists x\, \neg A(x) \supset \neg \forall x\, A(x)$. We'll use $\supset I$ as the last inference, and first deduce $\neg \forall x\, A(x)$ from $\exists x\, \neg A(x)$ as an assumption. Having $\exists x\, \neg A(x)$ available suggests we should use $\exists E$: deduce our goal $\neg \forall x\, A(x)$ from an assumption of the form $\neg A(c)$, which will be discharged by the $\exists E$ inference. To deduce $\neg \forall x\, A(x)$ from $\neg A(c)$ we should use $\neg I$, which requires a deduction of $\bot$ from $\forall x\, A(x)$ (together with the other assumption $\neg A(c)$). Finally, we can deduce $\bot$ by first applying $\forall E$ to the assumption $\forall x\, A(x)$ to obtain $A(c)$, and then $\neg E$ to the assumption $\neg A(c)$ and the conclusion of that inference. This results in the deduction below:

$$\frac{\dfrac{\exists x\, \neg A(x)^{\,1} \quad \dfrac{\dfrac{\neg A(c)^{\,2} \quad \dfrac{\dfrac{\forall x\, A(x)^{\,3}}{A(c)}\, \forall E}{}\, \neg E}{\bot}}{\neg \forall x\, A(x)}\, \neg I \;\; 3 \;\; \exists E \;\; 2}{\neg \forall x\, A(x)}}{\exists x\, \neg A(x) \supset \neg \forall x\, A(x)}\, \supset I \;\; 1$$

Problem 3.9. Prove in **NM**:

1. $\forall x\, \neg A(x) \supset \neg \exists x\, A(x)$.
2. $\neg \exists x\, A(x) \supset \forall x\, \neg A(x)$.

3.3 Natural deduction for classical logic

In this section we focus on the difference between **NJ** and **NK**. The system **NJ** is intuitionistic. To expand it to classical logic, a number of avenues are available. As we saw before (Section 2.4), classical logic can be obtained from intuitionistic logic by adding the axiom schema $\neg\neg A \supset A$ (the double negation rule). This suggests one way of expanding **NJ** to classical logic: allow formulas of this form as assumptions which do not have to be discharged. Alternatively, we could add the schema $A \vee \neg A$ (excluded middle), which also results in all classically valid formulas becoming provable when added to intuitionistic logic as an axiom. And in fact, Gentzen (1935b) defined the classical natural deduction calculus **NK** this way. He also mentioned that instead of adding $A \vee \neg A$ as an axiom one could add the additional rule

$$\frac{\neg\neg A}{A} \ \neg\neg$$

but mused that the nature of this rule would "fall outside the framework ... because it represents a new elimination of the negation whose admissibility does not follow at all from our method of introducing the $\neg$-symbol by the $\neg\text{I}$" rule.

Prawitz (1965) defined a classical natural deduction system by adding to **NJ** a generalized $\perp_J$ rule:

$$
\begin{array}{c}
[\neg A] \\
\vdots \\
\dfrac{\perp}{A} \ \perp_K
\end{array}
$$

It generalizes $\perp_J$ because open assumptions of the form $\neg A$ may, but do not have to be discharged by $\perp_K$. In other words, a correct application of $\perp_J$ is also a correct application of $\perp_K$, just one where we do not avail ourselves of the possibility of discharging an assumption.

Let's see how Prawitz's additional rule allows us to prove theorems which cannot be proved in **NJ** alone. The most famous example is, of course, the principle of excluded middle $A \vee \neg A$. If we want to use $\perp_K$ to prove it, we should attempt to deduce $\perp$ from $\neg(A \vee \neg A)$. We could get $\perp$ from $\neg\text{E}$, by first deducing two formulas $\neg B$ and B. While there is no clearly obvious candidate, we could pick A for B. In other words, let's deduce both $\neg A$ and A from $\neg(A \vee \neg A)$.

To deduce $\neg A$, we should use $\neg\text{I}$: for this, deduce $\perp$ from A. Of course, we can also use $\neg(A \vee \neg A)$. And this yields the solution almost immediately, since $\vee\text{I}$ gives us $A \vee \neg A$ from the assumption A:

$$
\cfrac{\neg(A \vee \neg A) \quad \cfrac{\cfrac{A^{\,2}}{A \vee \neg A} \ \vee\text{I}}{}}{2 \ \cfrac{\perp}{\neg A} \ \neg\text{I}} \ \neg\text{E}
$$

Almost the same deduction gives us A from $\neg(A \lor \neg A)$; we just have to assume $\neg A$ instead of A and apply $\bot_K$ instead of $\neg_I$:

$$\cfrac{\neg(A \lor \neg A) \quad \cfrac{\cfrac{\neg A^{\,3}}{A \lor \neg A}\,{}_{\lor_I}}{}}{3\ \cfrac{\cfrac{\bot}{}}{A}\,\bot_K}{}_{\neg_E}$$

Putting both deductions together using $\land_E$ and the final $\bot_K$ we can discharge $\neg(A \lor \neg A)$:

$$\cfrac{\cfrac{\neg(A \lor \neg A)^{\,1} \quad \cfrac{\cfrac{A^{\,2}}{A \lor \neg A}\,{}_{\lor_I}}{}}{2\ \cfrac{\bot}{\neg A}\,\neg_I}{}_{\neg_E} \qquad \cfrac{\neg(A \lor \neg A)^{\,1} \quad \cfrac{\cfrac{\neg A^{\,3}}{A \lor \neg A}\,{}_{\lor_I}}{}}{3\ \cfrac{\cfrac{\bot}{}}{A}\,\bot_K}{}_{\neg_E}}{1\ \cfrac{\bot}{A \lor \neg A}\,\bot_K}$$

As an exercise, construct a deduction where the last $\neg_E$ applies not to $\neg A$ and A, but to $\neg(A \lor \neg A)$ and $A \lor \neg A$.

Above we proved the law of contraposition in one direction and mentioned that the other direction requires classical logic. Here is that deduction:

$$\cfrac{}{1\ \cfrac{\cfrac{A^{\,2} \quad \cfrac{\neg B \supset \neg A^{\,1} \quad \neg B^{\,3}}{\neg A}\,\supset_E}{3\ \cfrac{\cfrac{\bot}{B}\,\bot_K}{2\ \cfrac{}{A \supset B}\,\supset_I}{}}\,\neg_E}{(\neg B \supset \neg A) \supset (A \supset B)}\,\supset_I}$$

We have here displayed the two premises of $\neg_E$, A and $\neg A$ in reverse order, to make the structural similarity of the deductions of the two directions of the law of contraposition more obvious.

As a final example we will deduce the theorem $\neg \forall x\, A(x) \supset \exists x\, \neg A(x)$, the converse of the theorem deduced above. It, likewise, requires classical logic and is not a theorem of **NJ** alone. We begin by observing that if we had a deduction of $\exists x\, \neg A(x)$ from the assumption $\neg \forall x\, A(x)$, we would be able to establish the theorem using an $\supset_I$ inference. In this case, a profitable strategy would be to deduce $\exists x\, \neg A(x)$ using $\bot_K$, i.e., we are looking for a deduction of the form:

$$\cfrac{\neg \forall x\, A(x)^{\,1} \qquad \neg \exists x\, \neg A(x)^{\,2}}{\vdots}$$

$$1\ \cfrac{2\ \cfrac{\bot}{\exists x\, \neg A(x)}\,\bot_K}{\neg \forall x\, A(x) \supset \exists x\, \neg A(x)}\,\supset_I$$

How do we get $\bot$? Usually we use the $\neg_E$ rule, but it requires premises $\neg B$ and B. One candidate for B is $\forall x\, A(x)$ since the corresponding premise $\neg B$, i.e., $\neg \forall x\, A(x)$, is already available as an assumption. So the challenge is now to deduce $\forall x\, A(x)$ from the assumption $\neg\exists x\, \neg A(x)$. We will want to use $\forall$I for this, i.e., attempt to deduce as follows:

$$\neg\exists x\, \neg A(x)$$
$$\vdots$$
$$\frac{A(c)}{\forall x\, A(x)}\ \forall\text{I}$$

In order to construct the missing deduction of $A(c)$ from $\neg\exists x\, \neg A(x)$ we will once more appeal to $\perp_K$: find a deduction of $\perp$ from $\neg A(c)$ together with $\neg\exists x\, \neg A(x)$. But this is now easy, for from $\neg A(c)$ we get $\exists x\, \neg A(x)$ by $\exists$I:

$$\cfrac{\neg\exists x\, \neg A(x) \qquad \cfrac{\cfrac{\neg A(c)^{\,3}}{\exists x\, \neg A(x)}\ \exists\text{I}}{}}{3\ \cfrac{\perp}{A(c)}\ \perp_K}\ \neg\text{E}$$

Fitting this deduction into the previous one, and that into the initial setup gives us:

$$\cfrac{\cfrac{\neg\forall x\, A(x)^{\,1} \qquad \cfrac{\cfrac{\neg\exists x\, \neg A(x)^{\,2} \quad \cfrac{\cfrac{\neg A(c)^{\,3}}{\exists x\, \neg A(x)}\ \exists\text{I}}{}}{3\ \cfrac{\perp}{A(c)}\ \perp_K}\ \neg\text{E}}{\cfrac{A(c)}{\forall x\, A(x)}\ \forall\text{I}}}{2\ \cfrac{\perp}{\exists x\, \neg A(x)}\ \perp_K}\ \neg\text{E}}{1\ \cfrac{}{\neg\forall x\, A(x) \supset \exists x\, \neg A(x)}}\ \supset\text{I}$$

If this seems cumbersome and opaque—it is. Finding deductions in **NK** is much less straightforward than in **NJ**, because often $\perp_K$ is the only possible way to deduce something. Even though both $\perp_K$ and $\neg$I are, in a sense, "indirect" proofs, while it is clear when to use $\neg$I (namely when the formula to be deduced is a negation) it is often not clear when to use $\perp_K$.

Problem 3.10. Prove in **NK**:

1. $\neg(A \land B) \supset (\neg A \lor \neg B)$.
2. $(A \supset B) \lor (B \supset A)$.
3. $\forall x\, \neg\neg A(x) \supset \neg\neg\forall x\, A(x)$.
4. $(B \supset \exists x\, A(x)) \supset \exists x(B \supset A(x))$.
5. $\forall x(A(x) \lor B) \supset (\forall x\, A(x) \lor B)$.

Problem 3.11. Show that Gentzen's and Prawitz's versions of **NK** are equivalent:

1. Derive A from $\neg\neg A$ using the $\perp_K$ rule (and other rules of **NJ**, if needed).
2. Show how you can turn a deduction of $\perp$ from open assumptions $\Gamma \cup \{\neg A\}$ into one of A from open assumptions Γ, using the double negation rule.

$$\frac{A \quad B}{A \wedge B} \text{ } \wedge\text{I} \qquad \frac{A \wedge B}{A} \text{ } \wedge\text{E} \qquad \frac{A \wedge B}{B} \text{ } \wedge\text{E}$$

$$\begin{array}{c} [A] \\ \vdots \\ \dfrac{B}{A \supset B} \text{ } \supset\text{I} \end{array} \qquad\qquad \frac{A \supset B \quad A}{B} \text{ } \supset\text{E}$$

$$\frac{A}{A \vee B} \text{ } \vee\text{I} \qquad \frac{B}{A \vee B} \text{ } \vee\text{I} \qquad \frac{A \vee B \quad \overset{\displaystyle [A]}{\underset{\displaystyle C}{\vdots}} \quad \overset{\displaystyle [B]}{\underset{\displaystyle C}{\vdots}}}{C} \text{ } \vee\text{E}$$

$$\begin{array}{c} [A] \\ \vdots \\ \dfrac{\bot}{\neg A} \text{ } \neg\text{I} \end{array} \qquad \frac{\neg A \quad A}{\bot} \text{ } \neg\text{E} \qquad \frac{\bot}{D} \text{ } \bot_J \qquad \begin{array}{c} [\neg A] \\ \vdots \\ \dfrac{\bot}{A} \text{ } \bot_K \end{array}$$

$$\frac{A(c)}{\forall x \, A(x)} \text{ } \forall\text{I} \qquad\qquad \frac{\forall x \, A(x)}{A(t)} \text{ } \forall\text{E}$$

$$\frac{A(t)}{\exists x \, A(x)} \text{ } \exists\text{I} \qquad\qquad \frac{\exists x \, A(x) \quad \overset{\displaystyle [A(c)]}{\underset{\displaystyle C}{\vdots}}}{C} \text{ } \exists\text{E}$$

Table 3.1: Rules of natural deduction

3.4   Alternative systems for classical logic

There are other ways to obtain classical systems for natural deduction, in addition to adding $A \vee \neg A$ as an axiom or adding $\bot_K$ as a rule. For instance, one can add one of the rules

$$\frac{\overset{\displaystyle [A]}{\underset{\displaystyle B}{\vdots}} \quad \overset{\displaystyle [\neg A]}{\underset{\displaystyle B}{\vdots}}}{B} \text{ } \text{GEM} \qquad \frac{\overset{\displaystyle [\neg A]}{\underset{\displaystyle A}{\vdots}}}{A} \text{ } \text{ND}$$

Rule GEM was introduced by Tennant (1978), the rule ND by Curry (1963) and studied by Seldin (1989) and Zimmermann (2002).

Problem 3.12. Show that the rules GEM and ND given above are equivalent to $\bot_K$.

Another way to obtain a classical system for natural deduction is to allow sequences (or sets) of formulas in deductions instead of only single formulas (see, e.g., Boričić, 1985; Cellucci, 1992; Parigot, 1992; Zach, 2021). Each node in a deduction tree is then not a single formula, but possibly multiple formulas, and an inference rule can apply to any one of them (but only one). Suppose we write Γ for a set of formulas, and A, Γ for the set $\Gamma \cup \{A\}$. The $\wedge\text{I}$ and $\supset\text{I}$ rules, for instance, then become

$$\frac{A, \Gamma \quad B, \Delta}{A \wedge B, \Gamma, \Delta} \wedge\text{I} \qquad \frac{\begin{array}{c}[A]\\ \vdots\\ B, \Gamma\end{array}}{A \supset B, \Gamma} \supset\text{I}$$

In such a system, no additional rule is required to derive $A \vee \neg A$. The $\neg\text{I}$ rule takes the form

$$\frac{\begin{array}{c}[A]\\ \vdots\\ \Sigma\end{array}}{\neg A, \Sigma} \neg\text{I}$$

An empty Σ is interpreted as $\bot$. If Σ is empty, the above rule just is the $\neg\text{I}$ rule. But in the multiple conclusion setting, Σ need not be empty in the above rule. Then the law of excluded middle can simply be derived as follows:

$$\frac{\dfrac{\dfrac{A^1}{\neg A, A} \neg\text{I} \; 1}{A \vee \neg A, A} \vee\text{I}}{\dfrac{A \vee \neg A, A \vee \neg A}{A \vee \neg A} =} \vee\text{I}$$

The last step is in fact not an inference: The set of formulas $\{A \vee \neg A, A \vee \neg A\}$ just is $\{A \vee \neg A\}$—if a rule yields as conclusion a formula already present, the two copies are automatically "contracted" into one.

3.5 Measuring deductions

Much of proof theory is concerned with the manipulations of deductions, and with proving results about the structure of deductions using such manipulations. The central method of proof used is induction. In our discussions of axiomatic derivations—e.g., in the proof of the deduction theorem—the induction proceeds according to the length of derivations. To show that something is true of *all* derivations, we prove that, for every n, every derivation of length n has the property in question. proof-theoretic questions thus re-formulated slightly as questions about natural numbers allow us to apply induction. So to prove that every derivation has a certain property, we show: derivations of length 1 have it, and when all derivations of length $< n$ have it, then so do derivations of length n.

Length, of course, is not the only number we can associate with a derivation. Since a deduction in **NK** is a tree of formulas, not a sequence, it is unclear what its length should be. We will define another measure of the complexity of deductions:

its "size." When proving results about **NK**-deductions, we sometimes also use the "height" of a deduction (the maximum number of inferences on a branch between the end-formula and an assumption).[7] These measures can themselves be defined inductively, by considering the case of deductions consisting only of a single assumption, and those that end in an inference and the sub-deductions ending in that inference's premises. For instance, the number of inferences in a deduction δ—the size of δ—can be defined by induction on the construction of δ.

Definition 3.13. The *size* of a deduction δ is defined inductively as follows:

1. *Basis clause:* If δ consists only of an assumption, the size of δ is 0.
2. *Inductive clause:* If δ ends in an inference, the size of δ is 1 plus the sum of the sizes of the sub-deductions ending in these premises.

Since we count inferences and not formula occurrences, the simplest deductions, i.e., those consisting of an assumption by itself, without any inferences, have size 0. A proof by induction on size would then start with the case for size 0 as the induction basis.

Definition 3.14. The *height* of δ is defined inductively as follows:

1. *Basis clause:* If δ consists only of an assumption, the height of δ is 1.
2. *Inductive clause:* If δ ends in an inference, the height of δ is 1 plus the maximum of the heights of the sub-deductions ending in the premises of the last inference.

The important aspect here is that the simplest deductions (single assumptions by themselves) have the lowest possible value (size, height), and combining deductions using inferences (or extending a single deduction using an inference) increases the value. Conversely, if a deduction contains at least one inference, the sub-deductions ending in the premises of the lowest inference have a lower value than the entire deduction.

Let's consider an example:

$$\dfrac{\dfrac{A \wedge B}{B} \wedge_{\mathrm{E}} \quad A}{B \wedge A} \wedge_{\mathrm{I}}$$

Since this ends in an inference and is not an assumption by itself, its size is 1 plus the sizes of the two sub-deductions,

$$\dfrac{A \wedge B}{B} \wedge_{\mathrm{E}} \qquad A$$

[7] Of course, size and height for deductions in **NM** and **NJ** are defined the same way, and the same proof methods apply there.

The sub-deduction on the right is just an assumption by itself, so its size is 0. The sub-deduction on the left ends in an inference, so its size is 1 plus the size of the sub-deduction ending in the left premise,

$$A \wedge B$$

This is again an assumption on its own, so its size is 0, and hence the size of the left sub-deduction is 1. The size of the entire deduction thus is $1 + (1 + 0)$, i.e., 2. It's clear that in general, the size of a deduction is just the number of inferences in it.

The height of the deduction, by contrast, is 1 plus the maximum of the heights of the two sub-deductions. The sub-deduction on the right, being an assumption alone, is 1. The height of the sub-deduction on the left is 2. The maximum of these is 2, so the height of the entire deduction is $1 + 2 = 3$. In general, the height of a deduction is the maximum number of formulas along a "branch," i.e., a sequence of formula occurrences starting from an assumption, passing from premises to conclusions, ending in the end-formula. Although in this case, the height is larger than the size, it is typically the other way around.

Problem 3.15. Rewrite the deduction of $A \vee \neg A$ given in Section 3.3. Annotate the deduction by computing the size and height of the (sub-)deduction(s) at each single formula and/or inference line.

3.6 Manipulating deductions, proofs about deductions

Let us begin with an example where we prove a property of deductions by induction on their size. If there is a deduction of A from assumptions Γ, and t is any term and c any free variable, then we can substitute t for c throughout the deduction. The result is not guaranteed to be a correct deduction itself: the term t may contain a free variable a which is used as an eigenvariable somewhere, and so this substitution may interfere with eigenvariable conditions in the deduction. For instance:

$$\cfrac{A(a) \quad \cfrac{\cfrac{\cfrac{A(a)^1 \quad A(b)}{A(a) \wedge A(b)} \wedge_I}{A(a) \supset (A(a) \wedge A(b))} \, 1 \; \supset_I}{\forall x (A(x) \supset (A(x) \wedge A(b)))} \forall_I}{A(a) \wedge \forall x (A(x) \supset (A(x) \wedge A(b)))} \wedge_I$$

Replacing a everywhere by t would turn the premise of $\forall_I$ into $(A(t) \supset (A(t) \wedge A(b)))$. If t is not a free variable, this would not be a correct inference at all. And if t happened to be b, then the uniform replacement of a by b would result in a violation of the eigenvariable condition. It is true, however, that if A has a deduction from Γ, then $A[t/c]$ has a deduction from $\Gamma[t/c]$—we just have to be much more careful. So let us prove this result (Theorem 3.20 below).

As a preparatory step, let us establish the following lemmas:

Lemma 3.16. *If δ is an NK-deduction of A from open assumptions Γ, and a and c are free variables not used as eigenvariables in δ, then replacing a everywhere by c results in a correct deduction $\delta[c/a]$ of $A[c/a]$ from $\Gamma[c/a]$.*[8]

Proof. What we have to show here is that for any deduction of size n, the result $\delta[c/a]$ of replacing a everywhere by c is a correct deduction, provided the conditions of the lemma are met. Since the smallest possible size of a deduction is 0, the induction basis is the case where $n = 0$.

Induction basis: If $n = 0$, δ is just the assumption A by itself. Then $\delta[c/a]$ is just $A[c/a]$ by itself. This is trivially a correct deduction of $A[c/a]$ from the open assumptions $\Gamma[c/a] = \{A[c/a]\}$.

Inductive step: If $n > 0$, then δ and $\delta[c/a]$ end in an inference. We inspect two cases as examples, the rest are left as exercises.

Suppose the last inference is $\wedge\textsc{i}$, i.e.,

$$
\begin{array}{cc}
\Gamma_1 & \Gamma_2 \\
\vdots\,\delta_1 & \vdots\,\delta_2 \\
B & C \\
\hline
\multicolumn{2}{c}{B \wedge C}
\end{array}\ \wedge\textsc{i}
$$

By inductive hypothesis, the lemma applies to δ_1 and δ_2, i.e., $\delta_1[c/a]$ is a correct deduction of $B[c/a]$ from $\Gamma_1[c/a]$, and $\delta_2[c/a]$ is a correct deduction of $C[c/a]$ from $\Gamma_2[c/a]$. If we combine these two deductions using $\wedge\textsc{i}$ we get:

$$
\begin{array}{cc}
\Gamma_1[c/a] & \Gamma_2[c/a] \\
\vdots\,\delta_1[c/a] & \vdots\,\delta_2[c/a] \\
B[c/a] & C[c/a] \\
\hline
\multicolumn{2}{c}{B[c/a] \wedge C[c/a]}
\end{array}\ \wedge\textsc{i}
$$

But this is just $\delta[c/a]$, since

$$(B \wedge C)[c/a] = (B[c/a] \wedge C[c/a]).$$

So $\delta[c/a]$ is a correct deduction of $A[c/a] = (B \wedge C)[c/a]$ from $\Gamma[c/a]$.

Suppose δ ends in $\forall\textsc{i}$:

$$
\begin{array}{c}
\Gamma \\
\vdots\,\delta_1 \\
A(b) \\
\hline
\forall x\, A(x)
\end{array}\ \forall\textsc{i}
$$

---

[8] We have stated the result for **NK**, since proofs for **NK** will cover all inference rules we consider. The results in this section hold for all three systems **NM**, **NJ**, and **NK** equally. This is not completely obvious: verify for yourself that none of the transformations we describe introduce $\perp_J$ or $\perp_K$ inferences into deductions in **NM** or **NJ**, respectively.

By the inductive hypothesis, $\delta_1[c/a]$ is a correct deduction of $A(b)[c/a]$ from $\Gamma[c/a]$. We have to verify that the last step, i.e., inferring $(\forall x\, A(x))[c/a]$, is a correct $\forall$I inference. In order for this inference to be correct, the expression following $\forall x$ in the conclusion must be identical to the premise, with the eigenvariable replaced by x. The expression following $\forall x$ is $A(x)[c/a]$, i.e., $(A(b)[x/b])[c/a]$. The premise is now $A(b)[c/a]$ and the eigenvariable is still b. So we must verify that

$$A(x)[c/a] = (A(b)[c/a])[x/b].$$

By the condition of the lemma, b is not a. Thus, replacing a by c in $A(b)$ does not disturb the occurrences of b. Since c is also not b, replacing a by c does not introduce any new occurrences of b. So, substituting x for b and substituting c for a do not interact.[9]

Furthermore, since b does not occur in Γ by the eigenvariable condition, and c is not b, replacing a by c in Γ will not introduce b to the open assumptions. In other words, b does not occur in $\Gamma[c/a]$. Thus we can correctly add an $\forall$I inference to $\delta_1[c/a]$:

$$\Gamma[c/a]$$
$$\vdots$$
$$\vdots\, \delta_1[c/a]$$
$$\vdots$$
$$\frac{A(b)[c/a]}{\forall x\, A(x)[c/a]}\ \forall\text{I}$$

The resulting correct deduction is just $\delta[c/a]$.

The remaining cases are left as exercises. $\qquad\square$

Problem 3.17. Complete the proof of Lemma 3.16 by verifying the remaining cases.

Lemma 3.18. *Let δ be an **NK**-deduction of A from Γ, and let V be some finite set of free variables. There is an **NK**-deduction δ' of A from Γ in which no two $\forall$I or $\exists$E inferences concern the same eigenvariable, and in which no eigenvariable of δ' is in V.*

Proof. Induction basis: If δ has size 0, it contains no inferences. Consequently, it contains no $\forall$I and $\exists$E inferences, and no eigenvariables. So, trivially, all eigenvariables are distinct, and not in V. Hence, we may take $\delta' = \delta$ in that case.

Inductive step: If δ has size $n > 0$, it ends in an inference. The only interesting cases are those where the last inference is $\forall$I or $\exists$E. Of the remaining cases we will

---

[9] A rigorous proof of this would define the substitution operation $A[t/a]$ inductively and then prove by induction on the complexity of A that under the conditions of the lemma, the order of substitution does not matter. Without the conditions, it does matter. Here are some counterexamples. If a is b, then $P(b)[c/a] = P(b)[c/b] = P(c)$. Then we'd have $(P(b)[c/a])[x/b] = P(c)$ but $(P(b)[x/b])[c/a] = P(x)$. On the other hand, if c is b, then $P(a,b)[c/a] = P(a,b)[b/a] = P(b,b)$. In this case we'd have $(P(a,b)[c/a])[x/b] = P(x,x)$ but $(P(a,b)[x/b])[c/a] = P(b,x)$.

deal with $\wedge\text{I}$ as an example; the others are all handled very similarly, and are left as exercises.

Suppose δ ends in $\wedge\text{I}$:

$$
\begin{array}{cc}
\Gamma_1 & \Gamma_2 \\
\vdots\,\delta_1 & \vdots\,\delta_2 \\
A & B \\
\hline
\multicolumn{2}{c}{A \wedge B}
\end{array}\ \wedge\text{I}
$$

The sizes of δ_1 and δ_2 are $< n$, so by induction hypothesis there is a deduction δ_1' of A with all eigenvariables distinct and also distinct from the free variables in V. The inductive hypothesis covers all deductions of size $< n$, for any end-formula, assumptions, and sets V. In particular, taking V' the union of V and of the eigenvariables in δ_1', the inductive hypothesis also gives us a deduction δ_2' of B from Γ_2, with eigenvariables distinct and not in V', so in particular no two eigenvariables in δ_1' and δ_2' are the same. Now we apply $\wedge\text{I}$ to obtain δ':

$$
\begin{array}{cc}
\Gamma_1 & \Gamma_2 \\
\vdots\,\delta_1' & \vdots\,\delta_2' \\
A & B \\
\hline
\multicolumn{2}{c}{A \wedge B}
\end{array}\ \wedge\text{I}
$$

In the resulting deduction, all eigenvariables are distinct, and no eigenvariable occurs in V.

Now for the interesting cases. Suppose δ ends in $\forall\text{I}$:

$$
\begin{array}{c}
\Gamma \\
\vdots\,\delta_1 \\
\dfrac{A(c)}{\forall x\,A(x)}
\end{array}\ \forall\text{I}
$$

The inductive hypothesis applies to δ_1, so there is a deduction δ_1' of $A(c)$ in which all eigenvariables are distinct and not in V. Let d be a free variable which is not used as an eigenvariable in δ_1', which is not in V, and which does not occur in Γ. (The variable c itself may satisfy this condition already, but if not, there always is another free variable satisfying it.) By the preceding lemma, $\delta_1'[d/c]$ is a correct deduction of $A(d)$ from $\Gamma[d/c]$. Because of the eigenvariable condition on $\forall\text{I}$, c does not occur in Γ, so $\Gamma[d/c]$ is just Γ, and d does not occur in Γ. Therefore, the eigenvariable condition for an application of $\forall\text{I}$ is satisfied. We can add an $\forall\text{I}$

inference to the deduction $\delta_1'[d/c]$ and we can take as δ' the deduction:

$$
\begin{array}{c}
\Gamma \\
\vdots \; \delta_1'[d/c] \\
\\
\dfrac{A(d)}{\forall x\, A(x)} \; \forall\mathrm{I}
\end{array}
$$

Since we chose d to be distinct from all the eigenvariables in δ_1' and not in V, the resulting deduction δ' has all eigenvariables distinct and not in V.

Suppose δ ends in $\exists\mathrm{E}$:

$$
\begin{array}{cc}
\Gamma_1 & \Gamma_2, A(c)^{\,1} \\
\vdots \; \delta_1 & \vdots \; \delta_2 \\
\exists x\, A(x) & C \\
\end{array}
\;\exists\mathrm{E}
$$
$$
\overline{C}
$$

The inductive hypothesis applies to δ_1, so there is a deduction δ_1' of $\exists x\, A(x)$ in which all eigenvariables are distinct and not in V. It also applies to δ_2, where we now consider not V but the union of V and the eigenvariables of δ_1'. Then the eigenvariables in δ_1' and δ_2' are also distinct.

Let d again be a free variable which is not used as an eigenvariable in δ_1' or δ_2', which is not in V, and which does not occur in Γ_1, Γ_2 or C. As before, $\delta_2'[d/c]$ is a correct deduction of C from $\Gamma_2 \cup \{A(d)\}$ (by the eigenvariable condition, c does not occur in C or Γ_2). d does not occur in Γ_1, Γ_2, or C. So the eigenvariable condition for an application of $\exists\mathrm{E}$ is satisfied, and we can take as δ' the deduction

$$
\begin{array}{cc}
\Gamma_1 & \Gamma_2, A(d)^{\,1} \\
\vdots \; \delta_1' & \vdots \; \delta_2'[d/c] \\
\exists x\, A(x) & C \\
\end{array}
\;\exists\mathrm{E}
$$
$$
\overline{C}
$$

Since we chose d to be distinct from all the eigenvariables in δ_1' and δ_2' and not in V, the resulting deduction δ' has all distinct eigenvariables which are not in V.□

Problem 3.19. Complete the proof of Lemma 3.18 by verifying the remaining cases.

Theorem 3.20. *If δ is an* **NK***-deduction of A from Γ, then there is an* **NK***-deduction of $A[t/c]$ from $\Gamma[t/c]$, for any free variable c and term t.*

Proof. Let V be the set of all free variables in t together with c. By the preceding Lemma 3.18, there is a deduction δ' in which no eigenvariable is in V, i.e., the eigenvariables are all distinct from c and the free variables in t. Replacing c everywhere by t in such a deduction is a correct deduction of $A[t/c]$ from $\Gamma[t/c]$. (We leave the verification of this last step as an exercise.)   □

Problem 3.21. Verify by induction on size that $\delta'[t/c]$ is in fact a correct deduction of $A[t/c]$ from $\Gamma[t/c]$.

3.7   Equivalence of natural and axiomatic deduction

In order to justify his new proof systems, Gentzen (1935c) had to show that they prove exactly the same formulas as the existing axiomatic systems did. This might be seen as a verification that, say, **NJ** is sound (proves no more than J_1) and complete (proves everything that J_1 proves), and similarly for **NK** and K_1. A modern presentation of **NJ** and **NK** might do this by showing directly that they are sound and complete for intuitionistic and classical semantics. However, at the time Gentzen was writing, no semantics for intuitionistic logic was available relative to which soundness and completeness could be proved, and Gödel's completeness proof for classical logic was just three years old. The remaining—and easier— option was to prove directly that J_1 and **NJ**, and K_1 and **NK**, prove the same formulas. Gentzen did this by showing that every K_1-derivation can be translated into a **NK**-deduction; every **NK**-deduction can be turned into a derivation in Gentzen's sequent calculus **LK** (see Chapter 5), and by showing that a derivation in **LK** can be translated into an **NK**-deduction, all of the same formula, of course.[10]

Theorem 3.22. *If A has a derivation in* K_1, *then there is an **NK** proof of A.*

Proof. We show this by induction on the number of inferences in the derivation of A.

The induction basis is the case where the number of inferences is 0, i.e., the derivation consists only of axioms. In that case, A must be an instance of one of the axiom schemas in K_0, of QL1, or of QL2.

To verify the claim, then, we have to show that all axioms of K_0 as well as QL1 and QL2 have proofs in **NK**. We have already given **NJ** proofs of the following axioms as examples:

PL2. $(A \wedge B) \supset (B \wedge A)$

PL5. $B \supset (A \supset B)$

PL9. $((A \supset C) \wedge (B \supset C)) \supset ((A \vee B) \supset C)$

PL11. $\neg A \supset (A \supset B)$

QL1. $\forall x\, A(x) \supset A(t)$

---

[10] Specifically, Gentzen proved this equivalence first for J_1, **NJ**, and **LJ**, and then explained how the proof can be extended to cover the classical calculi K_1, (his version of) **NK**, and **LK**. He did not treat the minimal calculi, as his paper predates the introduction of minimal logic by Johansson (1937). However, the same idea works in this case too: the natural deduction calculus **NM** for minimal logic is **NJ** without the $\perp_J$ rule, and a minimal sequent calculus **LM** would be **LJ** without the WR rule (see Section 5.2).

Here is an **NJ**-proof of axiom PL10, $((A \supset B) \wedge (A \supset \neg B)) \supset \neg A$:

$$\cfrac{\cfrac{\cfrac{(A \supset B) \wedge (A \supset \neg B)^2}{A \supset \neg B} \wedge_E \quad A^1}{\cfrac{\neg B}{}} \supset_E \qquad \cfrac{\cfrac{(A \supset B) \wedge (A \supset \neg B)^2}{A \supset B} \wedge_E \quad A^1}{B} \supset_E}{} \neg_E$$

$$_2 \cfrac{_1 \cfrac{\bot}{\neg A} \neg_I}{((A \supset B) \wedge (A \supset \neg B)) \supset \neg A} \supset_I$$

QL2 is trivial. The remaining cases have already been assigned as problems (PL1, PL3, PL4, and PL6 in Problem 3.3 and PL7, PL8 in Problem 3.4).

Note that only axiom PL12 ($\neg\neg A \supset A$) requires the $\bot_K$ rule, and only PL11 ($\neg A \supset (A \supset B)$) requires the $\bot_J$ rule. This means that a formula provable from the axioms of $\mathbf{M}_1$ has a deduction in **NM** (i.e., in the system without the rules for $\bot$); and a formula provable from the axioms of $\mathbf{J}_1$ has a deduction in **NJ**.

Now for the inductive step. Suppose the derivation of A contains n inferences, with $n > 0$. We may assume that the last formula is the conclusion of an inference. The inductive hypothesis is that the claim holds for any formula with a derivation consisting of at most $n - 1$ inferences. In particular, this means that every formula in the derivation before the last one also has a proof in **NK**. Specifically, each of the premises of the inference justifying A has a proof in **NK**.

We now distinguish cases according to which inference rule was applied to obtain the end-formula A.

If A was obtained by modus ponens, then the premises are of the form $B \supset A$ and B. Since these occur before the last formula in the derivation, they each have derivations with $< n$ inferences. By inductive hypothesis, they have proofs in **NK**: let's call those δ_1 and δ_2, respectively. We can combine these to obtain a proof of A using an application of $\supset_E$:

$$\cfrac{\begin{matrix} \vdots \, \delta_2 & \vdots \, \delta_1 \\ B \supset A & B \end{matrix}}{A} \supset_E$$

If A was obtained by QR_1, then A is of the form $B \supset \forall x \, C(x)$, and the premise of the inference is of the form $B \supset C(a)$. The condition on QR_1 guarantees that a does not occur in B. The premise has a derivation using $< n$ inferences, so by inductive hypothesis, there is an **NK**-proof δ of $B \supset C(a)$, and a does not occur in B.

We've already shown above how to expand this into a proof ending in A, but here it is again:

$$_1 \cfrac{\cfrac{\cfrac{\vdots \, \delta}{B \supset C(a)} \quad B^1}{\cfrac{C(a)}{\forall x \, C(x)} \forall_I} \supset_E}{B \supset \forall x \, C(x)} \supset_I$$

Since a does not occur in B, and the deduction δ of $B \supset C(a)$ has no open assumptions, the eigenvariable condition is satisfied.

We leave the other case (A is the conclusion of a QR$_2$ inference) as an exercise. $\square$

Problem 3.23. Complete the proof of Theorem 3.22 by dealing with the case where the last formula is the conclusion of a QR$_2$ inference.

The converse is a bit harder to prove, and was not proved directly by Gentzen. We have to show that the way **NK** formalizes reasoning with assumptions can be simulated in axiomatic deductions. We do this by showing that if A has an **NK**-deduction from assumptions Γ, then there is a $\mathbf{K_1}$-derivation of a conditional $G \supset A$, where G is the conjunction of all the formulas in Γ. Of course, Γ is a set, so in general there is no single such formula G. But, in $\mathbf{K_1}$, any two conjunctions G, G' of all the formulas in Γ are provably equivalent, i.e., the conditionals $G \supset G'$ and $G' \supset G$ are derivable.

Theorem 3.24. *If A has a deduction in **NK**, then there is a derivation of A in $\mathbf{K_1}$.*

Proof. We will show that if there is an **NK**-deduction δ of A from open assumptions $B_1, \ldots, B_k$, then there is a derivation in $\mathbf{K_1}$ of

$$(B_1 \wedge (B_2 \wedge \ldots (B_{k-1} \wedge B_k) \ldots)) \supset A$$

by induction on the size n of δ.

Induction basis: If $n = 0$, then δ consists simply in the assumption A by itself. The corresponding derivation in $\mathbf{K_1}$ shows $A \supset A$ (see Problem 2.7, E2).

Inductive step: If $n > 0$, we distinguish cases on the type of the last (bottommost) inference in δ. The conclusion of that inference is A. A sub-deduction ending in a premise C of that last inference is a correct deduction itself. The number of inferences in each of these sub-deductions is $< n$, so the inductive hypothesis applies: there is a derivation in $\mathbf{K_1}$ of $G \supset C$, where G is a conjunction of all open assumptions of the sub-deduction ending in the premise C.

The approach to the individual cases will be similar: The deduction of each premise of an inference yields a derivation of the conditional in $\mathbf{K_1}$ with those premises as consequent and the conjunction of the open assumptions in the sub-deductions as antecedent.

If a premise C depends on an assumption B which the **NK**-rule discharges, we have to give a derivation of the conditional corresponding to the entire deduction in which B no longer appears in the antecedent.

Suppose that A is inferred from $A \wedge B$ using $\wedge$E from open assumptions Γ, i.e., the deduction ends like this:

$$\Gamma$$
$$\vdots \; \delta_1$$
$$\frac{A \wedge B}{A} \; \wedge\text{E}$$

The deduction δ_1 ending in $A \wedge B$ has size $n - 1$; the entire deduction size n (the number of inferences in δ plus the last $\wedge$E). The open assumptions Γ of δ are the same as those of the entire deduction, since $\wedge$E does not discharge any assumptions. The induction hypothesis applies to the deduction δ_1 of $A \wedge B$. So, there is a derivation in $\mathbf{K}_1$ of $G \supset (A \wedge B)$, where G is a conjunction of all formulas in Γ. By E1 of Problem 2.7, there also is a derivation of $(A \wedge B) \supset A$. So we have $\vdash G \supset (A \wedge B)$ and $\vdash (A \wedge B) \supset A$. By the derived rule $\supset$TRANS of Section 2.5, $\vdash G \supset A$.

Now suppose that A is inferred from $B \supset A$ and B by $\supset$E. Suppose that the open assumptions of the sub-deductions ending in the premises $B \supset A$ and B are Γ_1 and Γ_2, respectively (and $\Gamma = \Gamma_1 \cup \Gamma_2$). So the deduction ends in

$$
\begin{array}{cc}
\Gamma_1 & \Gamma_2 \\
\vdots\, \delta_1 & \vdots\, \delta_2 \\
\dfrac{B \supset A \qquad B}{A} \,\supset\text{E}
\end{array}
$$

If the size of this deduction is n, then the size of δ_1 and of δ_2 are both $< n$. By inductive hypothesis, there are $\mathbf{K}_1$-derivations of $G_1 \supset (B \supset A)$ and of $G_2 \supset B$, where G_1 and G_2 are the conjunctions of the formulas in Γ_1 and Γ_2, respectively. By the derived rule $\wedge$INTRO of Section 2.5,

$$\vdash (G_1 \supset B) \wedge (G_2 \supset (B \supset A))$$

From Problem 2.7, E8, we know that

$$\vdash [(G_1 \supset B) \wedge (G_2 \supset (B \supset A))] \supset [(G_1 \wedge G_2) \supset (B \wedge (B \supset A))]$$

By modus ponens, $\vdash (G_1 \wedge G_2) \supset (B \wedge (B \supset A))$. An instance of PL6 is $(B \wedge (B \supset A)) \supset A$, So by $\supset$TRANS, $(G_1 \wedge G_2) \supset A$. $(G_1 \wedge G_2)$ is a conjunction of all the formulas in Γ.

Now suppose that A is of the form $B \supset C$, and that the deduction ends in $\supset$I:

$$
\begin{array}{c}
\Gamma, B^1 \\
\vdots\, \delta_1 \\
1\,\dfrac{C}{B \supset C} \,\supset\text{I}
\end{array}
$$

The deduction δ_1 of C from assumptions Γ, B has size $n - 1$. The inductive hypothesis applies; there is a derivation in $\mathbf{K}_1$ that shows $\vdash G_1 \supset C$, where G_1 is a conjunction of the formulas in $\Gamma \cup \{B\}$. Let G_2 be the conjunction $G \wedge B$ where G is the conjunction of all formulas in Γ alone. G_1 and G_2 are both conjunctions of all formulas in $\Gamma \cup \{B\}$, so $\vdash G_2 \supset G_1$. By $\supset$TRANS, $\vdash G_2 \supset C$, i.e., $\vdash (G \wedge B) \supset C$. By Problem 2.7, E14,

$$\vdash ((G \wedge B) \supset C) \supset (G \supset (B \supset C))$$

By modus ponens, finally, we have $\vdash G \supset (B \supset C)$, i.e., $\vdash G \supset A$.

Suppose δ ends in $\forall$E:

$$
\begin{array}{c}
\Gamma \\
\vdots \\
\vdots\,\delta \\
\vdots \\
\dfrac{\forall x\, A(x)}{A(t)}\ \forall\text{E}
\end{array}
$$

By inductive hypothesis applied to δ_1, $\vdash G \supset \forall x\, A(x)$. $\forall x\, A(x) \supset A(t)$ is an instance of QL1. By $\supset$TRANS, we have $\vdash G \supset A(t)$, as required.

Suppose δ ends in $\forall$I:

$$
\begin{array}{c}
\Gamma \\
\vdots \\
\vdots\,\delta \\
\vdots \\
\dfrac{A(a)}{\forall x\, A(x)}\ \forall\text{I}
\end{array}
$$

The eigenvariable condition requires that a not occur in any of the formulas in Γ. By inductive hypothesis, $\vdash G \supset A(a)$. Since a does not occur in Γ, it does not occur in G, the conjunction of those formulas. So the condition of QR1 is met, and we can infer in $\mathbf{K}_1$: $G \supset \forall x\, A(x)$.

We leave the other cases as exercises. $\qquad\qquad\qquad\qquad\qquad\qquad\qquad$ $\square$

Problem 3.25. Carry on the proof of Theorem 3.24 for deductions not involving $\bot$: Show that if the last inference is one of: $\wedge$I, $\vee$I, $\vee$E, $\exists$I, $\exists$E, then $G \supset A$ has a derivation in $\mathbf{K}_1$.

Problem 3.26. Complete the proof of Theorem 3.24 by dealing with $\neg$I, $\neg$E, $\bot_J$, and $\bot_K$. Assume, as we did in Section 2.10.1, that when dealing with $\mathbf{K}_1$, $\neg A$ is an abbreviation for $A \supset \bot$. Prove the inductive step for the cases of deductions ending in $\neg$I, $\neg$E, $\bot_J$, or $\bot_K$, by showing that in these cases $G \supset A$ has a derivation in $\mathbf{K}_1$ as well.

4
Normal deductions

4.1 Introduction

If proof theory is the study of structural properties of deductions and of methods for manipulating deductions, then a natural question to ask is: what makes a deduction simple? Or more to the point, how can deductions be simplified? This is of course not a precise question, since there is yet no clear notion of simplicity according to which deductions can be judged simple, or simpler than others. There are many approaches to the question as well: the simplest one, perhaps, is to count a deduction as simpler than another if it is shorter (contains fewer inferences). But, it turns out, short deductions can be more complex in other ways. Proof theorists from Gentzen onward have focused on a notion of simplicity that is perhaps better called *directness*.

Let us call a deduction *direct* if it contains no "detours." Here's a very simple example of a detour:

$$\cfrac{\cfrac{\vdots \qquad \vdots}{A \qquad B}}{\cfrac{A \wedge B}{A} \wedge_{\mathrm{E}}} \wedge_{\mathrm{I}}$$

The detour consists in deducing a formula $A \wedge B$ on the way to the conclusion A which could have been avoided: we already had a deduction ending in A before we introduced $A \wedge B$. Generalizing, we might say: any inference using an introduction rule followed by an elimination rule applied to the conclusion of the introduction rule is a detour. Here, we only count *major* premises of elimination rules as those to which the elimination rule applies. The minor premises may well be conclusions of introduction rules—without this constituting a detour.

Let's consider an example. Suppose we wanted to prove that $\neg A \supset \neg(A \wedge B)$, and suppose we've already proved $(A \wedge B) \supset A$ and $(C \supset D) \supset (\neg D \supset \neg C)$ for any C and D.

An Introduction to Proof Theory: Normalization, Cut-Elimination, and Consistency Proofs.
Paolo Mancosu, Sergio Galvan, and Richard Zach, Oxford University Press. © Paolo Mancosu, Sergio Galvan and Richard Zach 2021. DOI: 10.1093/oso/9780192895936.003.0004

$$\dfrac{\dfrac{A \wedge B\ ^1}{A}\ _{\wedge \text{E}}}{1\ \dfrac{}{(A \wedge B) \supset A}}\ _{\supset \text{I}}$$

$$1\ \dfrac{2\ \dfrac{3\ \dfrac{\dfrac{\neg D\ ^2 \quad \dfrac{C \supset D\ ^1 \quad C\ ^3}{D}\ _{\supset \text{E}}}{\bot}\ _{\neg \text{E}}}{\neg C}\ _{\neg \text{I}}}{\neg D \supset \neg C}\ _{\supset \text{I}}}{(C \supset D) \supset (\neg D \supset \neg C)}\ _{\supset \text{I}}$$

Let's call the deduction on the left δ_1. From the deduction on the right, we get one of

$$((A \wedge B) \supset A) \supset (\neg A \supset \neg(A \wedge B))$$

by replacing C everywhere by $A \wedge B$ and D everywhere by A—no new thinking is required.

$$1\ \dfrac{2\ \dfrac{3\ \dfrac{\dfrac{\neg A\ ^2 \quad \dfrac{(A \wedge B) \supset A\ ^1 \quad A \wedge B\ ^3}{A}\ _{\supset \text{E}}}{\bot}\ _{\neg \text{E}}}{\neg(A \wedge B)}\ _{\neg \text{I}}}{\neg A \supset \neg(A \wedge B)}\ _{\supset \text{I}}}{((A \wedge B) \supset A) \supset (\neg A \supset \neg(A \wedge B))}\ _{\supset \text{I}}$$

Let's call this deduction δ_2. Now we get a deduction of $\neg A \supset \neg(A \wedge B)$ simply by combining these with $\supset E$:

$$\dfrac{\begin{array}{c} \vdots\ \delta_2 \\ \vdots \\ ((A \wedge B) \supset A) \supset (\neg A \supset \neg(A \wedge B)) \end{array} \quad \begin{array}{c} \vdots\ \delta_1 \\ \vdots \\ (A \wedge B) \supset A \end{array}}{\neg A \supset \neg(A \wedge B)}\ _{\supset \text{E}}$$

The complete deduction looks like this:

$$\dfrac{1\ \dfrac{2\ \dfrac{3\ \dfrac{\dfrac{\neg A\ ^2 \quad \dfrac{(A \wedge B) \supset A\ ^1 \quad A \wedge B\ ^3}{A}\ _{\supset \text{E}}}{\bot}\ _{\neg \text{E}}}{\neg(A \wedge B)}\ _{\neg \text{I}}}{\neg A \supset \neg(A \wedge B)}\ _{\supset \text{I}}}{((A \wedge B) \supset A) \supset (\neg A \supset \neg(A \wedge B))}\ _{\supset \text{I}} \quad 4\ \dfrac{\dfrac{A \wedge B\ ^4}{A}\ _{\wedge \text{E}}}{(A \wedge B) \supset A}\ _{\supset \text{I}}}{\neg A \supset \neg(A \wedge B)}\ _{\supset \text{E}}$$

(We have renumbered the discharge labels on the right so that all inferences have unique discharge labels.)

If we already know how to prove $(A \wedge B) \supset A$ and $(C \supset D) \supset (\neg D \supset \neg C)$, this deduction requires the least amount of effort. We simply reuse the two deduction we already have and apply a single rule ($\supset$E) to arrive at the desired conclusion. That last rule, however, follows an application of $\supset I$—the last inference of δ_2. So it is not direct in the sense specified above. But we can also give a direct deduction, e.g., this one:

$$\cfrac{1\ \cfrac{2\ \cfrac{\cfrac{\neg A^{\,1}\quad \cfrac{A \wedge B^{\,2}}{A}\,{\scriptstyle\wedge E}}{\bot}\,{\scriptstyle\neg E}}{\neg(A \wedge B)}\,{\scriptstyle\neg I}}{\neg A \supset \neg(A \wedge B)}\,{\scriptstyle\supset I}}$$

This deduction is direct, and shorter than the indirect deduction given above (four inferences compared to eight). However, in general, indirect deductions can be *much* shorter than the shortest direct deduction.

There is another way we can compose deductions. Suppose we have a deduction δ_1 of some formula B from the open assumptions Γ, and another deduction δ_2 of A from the open assumption B. Then we can combine them into a single deduction of A from the open assumptions Γ by replacing every occurrence of the assumption B in δ_2 by the complete deduction δ_1. For instance, consider:

$$\cfrac{D \quad \neg E}{D \wedge \neg E}\,{\scriptstyle\wedge I} \qquad \cfrac{1\ \cfrac{\cfrac{D \wedge \neg E}{\neg E}\,{\scriptstyle\wedge E}\quad \cfrac{D \supset E^{\,1}\quad \cfrac{D \wedge \neg E}{D}\,{\scriptstyle\wedge E}}{E}\,{\scriptstyle\supset E}}{\bot}\,{\scriptstyle\neg E}}{\neg(D \supset E)}\,{\scriptstyle\neg I}$$

Deduction δ_1 on the left is a deduction of $D \wedge \neg E$ from $\{D, \neg E\}$; deduction δ_2 on the right is a deduction of $\neg(D \supset E)$ from open assumption $D \wedge \neg E$. If we replace both undischarged assumptions in δ_2 of the form $D \wedge \neg E$ by δ_1 we get:

$$\cfrac{1\ \cfrac{\cfrac{\cfrac{D \quad \neg E}{D \wedge \neg E}\,{\scriptstyle\wedge I}}{\neg E}\,{\scriptstyle\wedge E}\quad \cfrac{D \supset E^{\,1}\quad \cfrac{\cfrac{D \quad \neg E}{D \wedge \neg E}\,{\scriptstyle\wedge I}}{D}\,{\scriptstyle\wedge E}}{E}\,{\scriptstyle\supset E}}{\bot}\,{\scriptstyle\neg E}}{\neg(D \supset E)}\,{\scriptstyle\neg I}$$

In the resulting deduction, the $\wedge I$ inferences in which δ_1 ends are now followed immediately by $\wedge E$ inferences. So grafting δ_1 onto δ_2 has resulted in detours in the combined deduction. Again, a direct deduction is also possible:

$$\cfrac{1\ \cfrac{\neg E \quad \cfrac{D \supset E^{\,1}\quad D}{E}\,{\scriptstyle\supset E}}{\bot}\,{\scriptstyle\neg E}}{\neg(D \supset E)}\,{\scriptstyle\neg I}$$

In these examples, the direct deduction is shorter than the one with detours. This is in general not the case, as we'll see later: there are cases where deductions with detours are much shorter than the shortest direct deductions.

Since we'll use this operation of "grafting" deductions onto open assumptions a lot later, we'll introduce a notation for it now: if δ_1 has end-formula B, then $\delta_2[\delta_1/B]$ is the result of replacing every occurrence of an open assumption B in δ_2 by the complete deduction δ_1 (renumbering assumption discharge labels in δ_1 if necessary, to ensure δ_1 and δ_2 have no discharge labels in common).[1] It's clear that the open assumptions of $\delta_2[\delta_1/B]$ are exactly those of δ_2 except B, together with all open assumptions of δ_1. (We will prove this in Lemma 4.8.)

Constructing deductions using the "walk backwards" method of Section 3.2 always yields deductions without detours. If you look back to the examples in that section, you can verify that not once is the major premise of an elimination rule the conclusion of an introduction rule. But in complicated cases we often construct deductions by stringing together previously proved results, as in the case of the example deductions above, either by combining deductions using ⊃E or by grafting one deduction onto another. This accords with the practice of giving proofs in mathematics, where one often reuses results already proved in the proofs of new results. When we do this, we may introduce detours.

In the examples before, we saw that we could give a direct deduction for the same result that we first gave an indirect deduction of. Was this just a coincidence, or are detours in principle avoidable? Can everything that can be proved at all, be proved directly? As a proof-theoretic question we might pose it as: Is it the case that if A can be deduced from Γ, then A can be deduced from Γ without detours? A direct, constructive way of proving that the answer is yes is to show that every deduction (of A from Γ) can be transformed into a deduction (also of A from Γ) without detours, by eliminating detours one at a time. This method of eliminating detours from natural deductions is called *normalization*, and a deduction without detours is called *normal*.

Proof theorists prefer this method. An alternative would be to show that, once we know that a deduction exists, there is a way to always find a normal deduction, but disregard the specific structure of the given (non-normal) deduction. For instance, we can search for a normal deduction of A from open assumptions Γ in some systematic way.[2] This assumes we know, of course, that a normal deduction always exists if a deduction exists at all, which requires a separate proof. The normalization theorem provides one proof of this, but it can also be shown using model-theoretic methods—such a proof would be non-constructive, however.

The approach based on proof search is the basis for work in automated theorem proving (although natural deduction systems do not lend themselves easily to such search methods). In addition to relying on a non-constructive proof of existence of normal deductions, it has the disadvantage that it obscures the relationship between less direct (non-normal) and direct (normal) deductions. In particular, if we do not show how to transform a non-normal deduction to obtain a normal

---

[1] Although we use similar notation, this of course should not be confused with $A[c/t]$ or $\delta[c/t]$, the result of substituting a term t for a free variable c in a formula or deduction, as introduced in Definition 2.31 and Section 3.6.

[2] There are better and worse ways to do this. A good way to do it would be to make the procedure behind our heuristic for finding deductions precise and deterministic. A less good way would be to write a program that lists all possible normal deductions and stops once we find one that happens to have end-formula A and open assumptions Γ.

deduction, it becomes very hard to study how other measures of proof complexity relate (e.g., whether normal deductions are longer or shorter than non-normal deductions). In short, the direct method provides more information.

You will find some perhaps confusing terminology in the literature concerning normalization. A proof that merely yields the existence of normal deductions is often called a *normal form theorem*. It might claim, for instance, that if A has a deduction from Γ, it has a normal deduction. The proof of this result might not need to exploit the specific structure of the non-normal deduction to find the normal deduction. A proof of a *normalization* theorem successively transforms a deduction of A from Γ into a normal deduction. This is the direct method we mentioned above, and the method we will follow in our proofs below. The gold standard is a *strong normalization* proof. The difference between a (weak) normalization and a strong normalization proof lies in the order in which individual transformation steps are applied. In a weak normalization proof, we follow a particular order, which is shown to terminate in a normal deduction. But sometimes transformation steps could be applied in a different order. A strong normalization proof shows that *any* order in which these steps can be applied terminates in a normal form—and perhaps the normal form is the same in all possible cases. Strong normalization results are a lot harder to prove, and they go beyond the scope of this introductory book.[3] Via the so-called Curry-Howard isomorphism, they form the basis of a recent and important application of normalization results in programming language semantics and security (see Pierce, 2002; Sørensen and Urzyczyn, 2006; Wadler, 2015; Zach, 2019c).

4.2   Double induction

Before we proceed with the proof that deductions in **NM**, **NJ**, and **NK** can always be transformed into normal deductions, we have to revisit the principle of induction. So far, our inductive proofs have used induction on a single measure (e.g., length or size). In order to prove that all deductions have a certain property, we showed that deductions *of size n* have that property, for all n. The introduction of the measure n allowed us to use the principle of induction. Our approach was always to show (a) that deductions of size 0 have the property, and that, (b) if all deductions of size $< n$ have the property, so do deductions of size n. The inductive step, (b), consists in a proof that all deductions of size n have the property in which we appeal to the relevant inductive hypothesis, namely that any deduction of size $< n$ can be assumed to have the property in question.

The normalization theorem (and many other proof-theoretic results) is proved by induction on not one, but two measures. Here's the general principle: Suppose we want to show that $P(n, m)$ holds, for all n and m. A proof that $P(n, m)$ holds

---

[3] In some cases they do not hold; cf. n. 6 on p. 73.

for all n and m by double induction on n and m is a proof that follows the pattern below:

1. *Induction basis:* Prove that $P(0,0)$ holds.

2. Assume the *inductive hypothesis:* Suppose $P(k,\ell)$ holds whenever

    (a) $k < n$ or
    (b) $k = n$ but $\ell < m$.

3. *Inductive step:* Give a proof that $P(n,m)$ holds provided the induction hypothesis holds.

Why does this work? Remember that simple induction works because every descending sequence of numbers must eventually reach 0. Suppose $P(n)$ didn't hold for all n—then there must be some counterexample m_1. The proof of the inductive step tells us that there is some $m_2 < m_1$ such that, if $P(m_2)$ then $P(m_1)$. Since $P(m_1)$ is false, so is $P(m_2)$. But now applying the proof of the inductive step again, we find an $m_3 < m_2$ with $P(m_3)$ false, and so on. This descending sequence $m_1 > m_2 > m_3 \ldots$ must eventually reach 0—and $P(0)$ must also be false. But the proof of the induction basis shows that $P(0)$ is true, a contradiction. So there can be no m_1 for which $P(m_1)$ is false.

Double induction works by the same principle, except we now consider sequences of pairs of numbers. The proof of the inductive step shows that if $P(k,\ell)$ holds for all k and ℓ which satisfy either (a) or (b), then $P(n,m)$ holds. So, if $P(n,m)$ is false, then there are k and ℓ where either (a) $k < n$ or (b) $k = n$ and $\ell < m$, and $P(k,\ell)$ is false. So if $P(n_1,m_1)$ is false, we get a sequence $\langle n_1,m_1 \rangle$, $\langle n_2,m_2 \rangle$, $\langle n_3,m_3 \rangle$, ..., where each $P(n_i,m_i)$ is false, and either (a) $n_{i+1} < n_i$ or (b) $n_{i+1} = n_i$ and $m_{i+1} < m_i$. But such a sequence must eventually reach $\langle 0,0 \rangle$! For as long as $\langle n_i,m_i \rangle$ is followed by pairs $\langle n_j,m_j \rangle$ with $n_j = n_i$, the second coordinate m_j must decrease—and it can only do so at most m_i times before it reaches 0. And then, n_j has to decrease, and so forth.

It is perhaps easiest to see when visualizing all the possible pairs $\langle n_i,m_i \rangle$ in a grid:

$$\langle 0,0 \rangle \quad \langle 0,1 \rangle \quad \langle 0,2 \rangle \quad \langle 0,3 \rangle \quad \langle 0,4 \rangle \quad \ldots$$

$$\langle 1,0 \rangle \quad \langle 1,1 \rangle \quad \langle 1,2 \rangle \quad \langle 1,3 \rangle \quad \langle 1,4 \rangle \quad \ldots$$

$$\langle 2,0 \rangle \quad \langle 2,1 \rangle \quad \langle 2,2 \rangle \quad \langle 2,3 \rangle \quad \langle 2,4 \rangle \quad \ldots$$

$$\langle 3,0 \rangle \quad \langle 3,1 \rangle \quad \langle 3,2 \rangle \quad \langle 3,3 \rangle \quad \langle 3,4 \rangle \quad \ldots$$

$$\langle 4,0 \rangle \quad \langle 4,1 \rangle \quad \langle 4,2 \rangle \quad \langle 4,3 \rangle \quad \langle 4,4 \rangle \quad \ldots$$

$$\vdots \qquad \vdots \qquad \vdots \qquad \vdots \qquad \vdots \qquad \ddots$$

If $\langle n_i,m_i \rangle$ is anywhere in this grid, then $\langle n_{i+1},m_{i+1} \rangle$ is either in a previous line— when (a) holds—or in the same line, but to the left of $\langle n_i,m_i \rangle$—when (b) holds. In the same line, we only have m_i spaces to the left of $\langle n_i, m_i \rangle$ before we have to jump to a previous line. And in that previous line, we only have some finite number of places to the left, etc. Eventually we must reach the first line, and there we also have only finitely many places to the left before we reach $\langle 0, 0 \rangle$. Note, however, that when we jump to a previous line, we might land *anywhere* in that line—very far to the right. Still, however far to the right we land, there are only a finite number of places to go left before we have to jump to yet another previous line.

Problem 4.1. Suppose you have a chocolate bar divided into $n \times m$ squares. You want to divide it into individual squares by breaking along the lines. How many times do you have to break it? Show, by induction on $\langle n, m \rangle$ that the answer is $nm - 1$.

Let's consider a relatively simple example of a proof-theoretic result proved by double induction. Recall the $\perp_J$ rule:

$$\frac{\overset{\vdots}{\perp}}{B} \perp_J$$

The conclusion B may be any formula. The result we'll prove is that it is possible to restrict the rule to atomic B.

Proposition 4.2. *If A has an **NK**- or **NJ**-deduction from Γ, then it has an **NK**- or **NJ**-deduction, respectively, from the same assumptions in which every $\perp_J$ application has an atomic sentence as its conclusion.*

Proof. We show that for every deduction δ there is another deduction δ^* of the same formula from the same open assumptions in which every application of $\perp_J$ has an atomic conclusion. The basic idea is to remove first the applications of $\perp_J$ in which the conclusion is of greatest logical complexity and then to proceed in an orderly way to the removal of those applications of $\perp_J$ of lower complexity. But the process is complicated by the fact that a removal of an application of $\perp_J$ for a complex formula might increase the number of applications of $\perp_J$ for formulas of lower complexity. And to deal with this we will need double induction to keep track of the complexity of the formulas to which $\perp_J$ is applied and of the total number of applications of $\perp_J$ for formulas of any given complexity. Let us make this more precise.

Let the *degree* of a formula be the number of logical symbols in it (see Definition 2.5). Let $n(\delta)$ be the maximal degree of the conclusions of $\perp_J$ inferences, i.e., the largest number n such that some $\perp_J$ inference has a conclusion B of degree n. Let $m(\delta)$ be the number of applications of $\perp_J$ with non-atomic conclusions of maximal degree (that is, of degree $n(\delta)$). If $n(\delta) = 0$ then automatically $m(\delta) = 0$ since $n(\delta) = 0$ means that all applications of $\perp_J$ have atomic conclusions (the degree of atomic formulas is 0). We use double induction on $n(\delta), m(\delta)$.

Induction basis: $n(\delta) = m(\delta) = 0$. This means that there are no applications of $\perp_J$ inferences where the conclusion is not atomic, i.e., our deduction is already of the required form. Thus, in this case, there is nothing to prove.

Inductive hypothesis: Suppose all deductions δ' where (a) either $n(\delta') < n(\delta)$ or (b) $n(\delta') = n(\delta)$ and $m(\delta') < m(\delta)$ have the required property (they can be transformed into deductions $(\delta')^*$ with the same end-formula and open assumptions as δ' but with only atomic $\perp_J$ applications).

Inductive step: If $n(\delta) = 0$ then $m(\delta) = 0$. So we can assume that $n(\delta) > 0$ and $m(\delta) > 0$, i.e., there is at least one application of $\perp_J$ with conclusion B where the degree of $B = n(\delta)$. We pick a topmost such $\perp_J$ inference. There is a sub-deduction δ_1 of δ that ends in that inference, and δ has the form:

$$\genfrac{}{}{}{}{\vdots\;\delta_1}{} \quad \frac{\perp}{B}\;{\perp_J} \quad \genfrac{}{}{}{}{\vdots}{A}$$

Since the application of $\perp_J$ is topmost of the applications with conclusion of maximum degree $n(\delta)$, no application of $\perp_J$ in δ_1 can have degree $\geq n(\delta)$.

We now distinguish cases according to the form of B.

If $B \equiv \neg C$, let δ_1' be:

$$\genfrac{}{}{}{}{\vdots\;\delta_1}{} \quad \frac{\perp}{\neg C}\;{\neg\mathrm{I}}$$

If $B \equiv C \wedge D$, let δ_1' be:

$$\frac{\dfrac{\perp}{C}\;{\perp_J} \qquad \dfrac{\perp}{D}\;{\perp_J}}{C \wedge D}\;{\wedge\mathrm{I}}$$

If $B \equiv C \supset D$, let δ_1' be:

$$\frac{\dfrac{\perp}{D}\;{\perp_J}}{C \supset D}\;{\supset\mathrm{I}}$$

If $B \equiv \forall x\, C(x)$, let δ_1' be:

$$\vdots\, \delta_1$$

$$\frac{\dfrac{\bot}{C(a)}\ \bot_J}{\forall x\, C(x)}\ \forall\text{I}$$

where a is a free variable that does not occur in δ_1.

We leave the cases $B \equiv A \vee B$ and $B \equiv \exists x\, C(x)$ as exercises.

Now replace the sub-deduction of δ ending in B by δ_1'. The resulting deduction δ' is correct: the only inferences that have changed are those in δ_1' following $\bot_J$, and these are all correct. The end-formula of δ_1' is still B. The end-formula of δ has not changed, and neither have the open assumptions. (None of the δ_1' deductions have discharged assumptions that weren't discharged in δ_1, nor introduced open assumptions.)

However, either $n(\delta') < n(\delta)$ or $n(\delta') = n(\delta)$ and $m(\delta') < m(\delta)$. To see this, first note that the only $\bot_J$ inference in δ' that is not also a $\bot_J$ inference in δ with the same conclusion is the one we have considered in the definition of δ_1'. All the $\bot_J$ inferences below the indicated B remain the same. In δ', the $\bot_J$ inference with conclusion B has been replaced: In the case where $B \equiv \neg C$, it has been removed entirely. In all other cases, it has been replaced with $\bot_J$ inferences where the conclusion has degree $< n(\delta)$. The premise of each of these new $\bot_J$ inferences is the end-formula $\bot$ of δ_1. In one case, namely when $B \equiv C \wedge D$, this results in two copies of δ_1 where the original proof only had one. Since δ_1 may contain $\bot_J$ inferences, the new deduction δ' will contain two such inferences for every inference in δ_1. However, since the $\bot_J$ inference we replaced was a topmost one of degree $n(\delta)$, none of these inferences can be of degree $n(\delta)$.

If the $\bot_J$ inference δ we considered was the only one in δ of degree $n(\delta)$, we have just removed it or replaced it with $\bot_J$ inferences of lower degree. So, in δ' the conclusions of $\bot_J$ inferences are now all of lower degree than B, i.e., $n(\delta') < n(\delta)$. If it was not the only $\bot_J$ inference in δ of degree $n(\delta)$, we have replaced one $\bot_J$ inference of degree $n(\delta)$ with others of degree $< n(\delta)$. So the number of $\bot_J$ inferences of degree $n(\delta)$ has been decreased by 1, i.e., $n(\delta') = n(\delta)$ and $m(\delta') < m(\delta)$. The inductive hypothesis applies to δ': there is a deduction $(\delta')^*$ with same end-formula and open assumptions in which all applications of $\bot_J$ have atomic conclusions. $\qquad\square$

Problem 4.3. Complete the proof of Proposition 4.2 by verifying the cases for $\vee$ and $\exists$.

Problem 4.4. Apply the proof of Proposition 4.2 to transform the following deduction into one in which $\bot_J$ only has atomic conclusions:

$$\dfrac{\neg A \quad A}{\dfrac{\dfrac{\bot}{B \quad \dfrac{\bot}{B \supset \bot}\bot_J}}{\dfrac{\bot}{A \wedge B}\bot_J} \supset_E} \neg_E$$

At each step, give the deduction δ, and indicate the pair $\langle n(\delta), m(\delta) \rangle$.

4.3 Normalization for $\wedge, \supset, \neg, \forall$

We begin with a case in which it is relatively easy to define when a deduction counts as normal. Suppose we are only allowed the rules for $\wedge, \supset, \neg$, and $\forall$ (i.e., we are dealing with **NM** without $\vee$ or $\exists$). The other connectives add complexities that obscure the main ideas of the proof. Hence, we first present the proof for this fragment. We will prove the normalization theorem for full **NJ** in Section 4.6.

Definition 4.5. Let us call a sequence of occurrences of formulas $A_1, \ldots, A_k$ in a deduction δ a *branch* iff

1. A_1 is an assumption,

2. A_k is the end-formula,

3. for every $i = 1, \ldots, k-1$: A_i is a premise of an inference of which A_{i+1} is the conclusion.

The branches in our last example,

$$\dfrac{\dfrac{\neg A_a^2 \quad \dfrac{(A \wedge B) \supset A_g^1 \quad A \wedge B_i^3}{A_h}\supset_E}{\dfrac{\dfrac{\bot_b}{\neg(A \wedge B)_c}\neg_I}{\dfrac{\neg A \supset \neg(A \wedge B)_d}{\dfrac{((A \wedge B) \supset A) \supset (\neg A \supset \neg(A \wedge B))_e}{}}\supset_I}\supset_I}{\neg A \supset \neg(A \wedge B)_f} \quad \dfrac{\dfrac{\dfrac{A \wedge B_j^1}{A_k}\wedge_E}{(A \wedge B) \supset A_l}\supset_I}{}\supset_E} \neg_E$$

are: $abcdef, ghbcdef, ihbcdef, jklf$.

Definition 4.6. An occurrence of a formula in δ is called a *cut* if it is both the conclusion of an I-rule and the major premise of an E-rule.

Cut formulas are sometimes also called "maximal formulas." For instance, in our example above, formula occurrence e is a cut, and in the deductions below, $A \supset B$ and $\forall x\, A(x)$ are cuts:

$$\dfrac{\dfrac{\overset{\overset{A^1}{\vdots}}{B}}{A \supset B}\supset_I \quad \overset{\vdots}{A}}{B}\supset_E \qquad \dfrac{\dfrac{\overset{\vdots}{A(c)}}{\forall x\, A(x)}\forall_I}{A(t)}\forall_E$$

Cut formulas are "detours" in the sense discussed above. Not every inefficient, circuitous deduction involves a cut. For instance, the deduction

$$\frac{\dfrac{A \wedge A}{A} \wedge_{\mathrm{E}} \quad \dfrac{A \wedge A}{A} \wedge_{\mathrm{E}}}{A \wedge A} \wedge_{\mathrm{I}}$$

inefficiently deduces $A \wedge A$ from open assumptions $A \wedge A$, however, it does not involve a cut. Cuts are particularly circuitous since they involve formulas that are not sub-formulas of the open assumptions or of the end-formula. If we avoid cuts we can avoid deductions that use such formulas. As we will see in Section 4.4, normal deductions have the sub-formula property: every formula in such a deduction is a sub-formula of the end-formula or of an open assumption.

The definition of a cut ensures that if a cut is of the form $A \wedge B$, it is the conclusion of a $\wedge_{\mathrm{I}}$ and the premise of a $\wedge_{\mathrm{E}}$ inference; if it is $A \supset B$ it is the conclusion of a $\supset_{\mathrm{I}}$ and major premise of a $\supset_{\mathrm{E}}$ inference; and if it is of the form $\forall x\, A(x)$, it is the conclusion of a $\forall_{\mathrm{I}}$ and premise of a $\forall_{\mathrm{E}}$ inference. In fact, since a cut is the conclusion of an I-rule, there can be no atomic CUTS. The requirement that the cut be the *major* premise of an E-rule is crucial. For instance, in

$$\frac{(A \wedge B) \supset C \quad \dfrac{A \quad B}{A \wedge B} \wedge_{\mathrm{I}}}{C} \supset_{\mathrm{E}}$$

$A \wedge B$ is the conclusion of an I-rule and the minor premise of an E-rule. But this is not—at least not obviously—a detour in the same way $A \supset B$ was a detour in the preceding example, or in which $A \wedge B$ was a detour in our very first example in this chapter, i.e.:

$$\frac{\dfrac{A \quad B}{A \wedge B} \wedge_{\mathrm{I}}}{A} \wedge_{\mathrm{E}}$$

It is also not clear that such combinations of inferences can be avoided or simplified. What we're after is avoiding situations where an I-rule introduces a logical operator which the following E-rule eliminates.

Now the question is: can (all) cuts be removed from every deduction? It seems the answer should be yes: In the case of a cut $A \wedge B$, the deduction ending in the left premise is already a deduction of the conclusion.

$$\frac{\dfrac{\overset{\vdots}{\underset{\delta_1}{\vdots}}\ A \quad \overset{\vdots}{\underset{\delta_2}{\vdots}}\ B}{A \wedge B} \wedge_{\mathrm{I}}}{A} \wedge_{\mathrm{E}} \qquad \Rightarrow \qquad \overset{\vdots}{\underset{\delta_1}{\vdots}}\ A$$

In the case of a cut $A \supset B$, the cut can be avoided by replacing every occurrence of the assumption A in the left sub-deduction with the entire sub-deduction ending in the minor premise A of the $\supset$E inference.

$$
\begin{array}{c}
\begin{array}{c}
\overset{A^1}{\underset{\displaystyle \vdots\, \delta_1}{}} \\[2pt]
\cfrac{B}{A \supset B}\,\supset\!\mathrm{I}\,\,1 \qquad \overset{\displaystyle \vdots\, \delta_2}{} \\[6pt]
\cfrac{ A }{B}\,\supset\!\mathrm{E}
\end{array}
\end{array}
\quad\Rightarrow\quad
\left.\begin{array}{c}
\vdots\, \delta_2 \\
A \\
\vdots\, \delta_1 \\
B
\end{array}\right\}\delta_1[\delta_2/A]
$$

And in the case of a cut of the form $\forall x\, A(x)$, we can obtain the conclusion $A(t)$ by substituting t for c in the sub-deduction ending in the premise $A(c)$.

$$
\begin{array}{c}
\Gamma \\
\vdots\, \delta_1 \\
\cfrac{A(c)}{\forall x\, A(x)}\,\forall\mathrm{I} \\
\cfrac{}{A(t)}\,\forall\mathrm{E}
\end{array}
\qquad\Rightarrow\qquad
\begin{array}{c}
\Gamma \\
\vdots\, \delta_1[t/c] \\
A(t)
\end{array}
$$

Problem 4.7. Consider the following deductions. Identify the cuts in them and use the ideas above to avoid them. Repeat until no cuts remain.

$$
\cfrac{\cfrac{\cfrac{A^1 \quad A^1}{A \wedge A}\,\wedge\mathrm{I}}{A \supset (A \wedge A)}\,\supset\!\mathrm{I}\,\,1 \qquad \cfrac{B \supset A \quad B}{A}\,\supset\!\mathrm{E}}{A \wedge A}\,\supset\!\mathrm{E}
$$

$$
\cfrac{\cfrac{\cfrac{\cfrac{\cfrac{\forall x\, A(x)}{A(b)}\,\forall\mathrm{E} \quad \cfrac{\forall x\, A(x)}{A(b)}\,\forall\mathrm{E}}{A(b) \wedge A(b)}\,\wedge\mathrm{I}}{\forall y(A(y) \wedge A(y))}\,\forall\mathrm{I}}{A(c) \wedge A(c)}\,\forall\mathrm{E}}{A(c)}\,\wedge\mathrm{E}
$$

Of course, we should be careful and verify that when we avoid detours in the way described above, the deductions remain correct. The crucial case is the one that arises in the treatment of cuts of the form $A \supset B$: can we really just "graft" the deduction of A onto the assumptions A? We've done it before in an example in the introduction to this section, and introduced a notation for it: $\delta_1[\delta_2/A]$, the result of grafting a deduction δ_2 with end-formula A onto every open assumption in δ_1 of the form A. When quantifiers and eigenvariables are involved, we have to be careful and require that the open assumptions of δ_2 do not contain free variables which are used as eigenvariables in δ_1. Furthermore, when we apply this "grafting" operation, we will be careful to only replace some occurrences of the assumption A (those which have label i, let's say). Let's prove now that this actually works. (Although in this section we've so far excluded deductions with

∨ᴇ and ∃ᴇ from consideration, the result holds also for them, and we prove it for the general case.)

Lemma 4.8. *Suppose δ_1 is a deduction of B from open assumptions Γ_1, and δ_2 is a deduction of A from open assumptions Γ_2 with no occurrence of an eigenvariable of δ_1. Then there is a deduction $\delta_1[\delta_2/A^i]$ of B, in which every occurrence of an open assumption A labelled i in δ_1 has been replaced by δ_2. The open assumptions of $\delta_1[\delta_2/A^i]$ are among $\Gamma_1 \cup \Gamma_2$. A is an open assumption of $\delta_1[\delta_2/A^i]$ only if A is an open assumption of δ_2, or δ_1 has an open assumption A not labelled i.*

Proof. We will assume that δ_1 and δ_2 have no discharge labels in common. This can always be assured by, e.g., renumbering the discharge labels in δ_2 (say, by adding k to every label in it, where k is the largest label occurring in δ_1). In particular, this means that no assumption of the form A is labelled i in δ_2.

We prove the lemma by induction on the size n of δ_1.

Induction basis. If $n = 0$, then δ_1 is just the formula B which is also an open assumption. We have two cases: either B is an open assumption of the form A labelled i, or it is not. If it is not, there are no open assumptions A labelled i to be replaced. In this case, $\delta_1[\delta_2/A^i]$ just is δ_1. Clearly the open assumptions of δ_1 are among $\Gamma_1 \cup \Gamma_2$ and A is an open assumption of δ_1 only if B is an open assumption of the form A not labelled i.

Now suppose B is an open assumption of the form A labelled i. Since B is the only open assumption of δ_1, this means B and A are identical formulas. Then $\delta_1[\delta_2/A^i]$ is δ_2, which is then also a deduction of B. Its open assumptions are those of δ_2, i.e., contained in $\Gamma_1 \cup \Gamma_2$. Trivially, A is open in $\delta_1[\delta_2/A^i] = \delta_2$ only if it is open in δ_2.

Inductive step. Suppose δ_1 has size $n > 0$. Then δ_1 ends in an inference I. Consider a premise C of that inference; suppose the sub-deduction ending in premise C is δ' from open assumptions Γ':

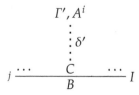

The size of δ' is $< n$, so the inductive hypothesis applies. There is a correct deduction $\delta'[\delta_2/A^i]$ of C in which every occurrence of the open assumption A labelled i in δ' has been replaced by δ_2. Its open assumptions are among $\Gamma' \cup \Gamma_2$, and A is an open assumption only if δ' contains an open assumption not labelled i or if $A \in \Gamma_2$.

To yield $\delta_1[\delta_2/A^i]$, we combine the results of replacing assumptions in the sub-deductions of the premises by applying the same inference I:

$$
\begin{array}{c}
\Gamma' \\
\vdots \\
\delta'[\delta_2/A^i] \\
\vdots \\
\end{array}
$$

$$
j \; \dfrac{\cdots \quad C \quad \cdots}{B} \; I
$$

By inductive hypothesis, if δ' ended in a formula C, so does $\delta'[\delta_2/A^i]$. Thus, the premises of I in $\delta_1[\delta_2/A^i]$ are the same as they are in δ_1.

Now suppose D is an open assumption of $\delta_1[\delta_2/A^i]$. Then it is an open assumption of one of the sub-deductions $\delta'[\delta_2/A^i]$. By inductive hypothesis, the open assumptions of $\delta'[\delta_2/A^i]$ are among $\Gamma' \cup \Gamma_2$. Now δ' may have an open assumption D^j which, in δ_1, is discharged by I. In this case $D \in \Gamma'$ but possibly $D \notin \Gamma_1$. However, any occurrence of D labelled j in δ' has a corresponding occurrence in $\delta'[\delta_2/A^i]$, and these are all discharged in $\delta_1[\delta_2/A^i]$ by I. Thus, the open assumptions of $\delta_1[\delta_2/A^i]$ are among $\Gamma_1 \cup \Gamma_2$. (Note that since δ_2 has no discharge labels in common with δ_1, no assumption of δ_2 is labelled j, and so I does not discharge any assumptions resulting from replacing A by δ_2.)

The premises and conclusion of I in $\delta_1[\delta_2/A^i]$ are the same as they are in δ_1. No eigenvariable of δ_1 occurs in δ_2, in particular, the open assumptions in Γ_2 contain no eigenvariables of δ_1. Thus, the inference I is correct, and no eigenvariable condition in the new deduction is violated.

The claim about A follows directly from the inductive hypothesis. Suppose A is an open assumption of $\delta_1[\delta_2/A^i]$. Since no inference can introduce open assumptions (only discharge them), A is an open assumption of at least one sub-deduction $\delta_1[\delta_2/A^i]$ ending in a premise of I. By inductive hypothesis, A is an open assumption of δ' not labelled i, or A is an open assumption of δ_2. In the former case, it is an open assumption of δ_1 not labelled i.   □

For instance, suppose δ_1 ends in $\forall_I$, i.e., it has the form on the left below. The resulting deduction is given on the right:

$$
\begin{array}{cc}
& \Gamma_2 \\
& \vdots \\
& \delta_2 \\
& \vdots \\
\Gamma', A^i & \dfrac{\Gamma' \quad A}{} \\
\vdots & \vdots \\
\delta' & \delta' \\
\vdots & \vdots \\
\dfrac{C(a)}{\forall x\, C(x)}\; \forall_I & \dfrac{C(a)}{\forall x\, C(x)}\; \forall_I
\end{array}
$$

Since a does not occur in Γ', Γ'', or Γ_2, the eigenvariable conditions are satisfied. The entire sub-deduction ending in $C(a)$ on the right is $\delta'[\delta_2/A^i]$.

If δ_1 ends in $\exists$E, the situation is the following:

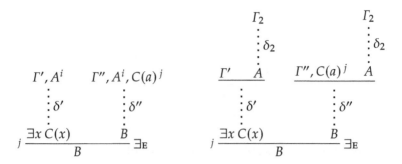

The eigenvariable conditions are again satisfied. The sub-deductions ending in $\exists x\, C(x)$ and B are $\delta'[\delta_2/A^i]$ and $\delta''[\delta_2/A^i]$, respectively.

The question of whether a deduction remains correct when manipulated in this way is, however, not enough. We also have to convince ourselves that when a cut is removed in the way envisaged, we don't inadvertently introduce new cuts. If we cannot guarantee this, we may end up removing a cut only to generate a new one, and the procedure never ends. Let's see how this may happen.

Suppose our deduction δ contains a cut $A \wedge B$, then the part of δ containing the cut ends with

By carefully choosing the cut, we can even make sure that δ_1 and δ_2 are normal. For instance, we may insist that we remove rightmost, topmost cuts first. So let's suppose that $A \wedge B$ is a rightmost, topmost cut and δ_1 and δ_2 are therefore normal, i.e., do not contain further cuts. Suppose we replace this entire sub-deduction of δ with just:

$$\vdots\ \delta_1$$
$$A$$

We have now removed the cut $A \wedge B$. However, the overall number of cuts in the resulting deduction may still be n, i.e., even though one cut has been removed, a new one appeared. This may happen if A is itself introduced, at the last inference of δ_1, by an I rule, and the next inference below A in δ is an E-rule. For example, A might be of the form $C \supset D$, and the part of δ where it appears plus the inference below it actually look like this:

$$
\begin{array}{c}
C^{1} \\
\vdots\, \delta_1 \\
\dfrac{D}{C \supset D}\, {}^{\supset\mathrm{I}} \qquad \vdots\, \delta_2
\end{array}
$$

Removing the cut on $A \wedge B$ as above then would yield:

This contains a new cut on $C \supset D$!

This kind of phenomenon can't be avoided. But what we should notice is that the new cut itself is less complex than the old cut: $(C \supset D)$ is a sub-formula of $(C \supset D) \wedge B$. So if we don't reduce the number of cuts, at least we reduce the complexity of cuts in δ. So to show that all cuts can be removed, we have to proceed by induction not on the number of cuts—which would be the most obvious approach—but by some measures that takes both the number and complexity of cuts into account. These are the *cut degree* of a deduction (the maximal degree of its cuts)[4] and the *cut rank* (the number of cuts of maximal degree).

Definition 4.9. The *cut degree* $d(\delta)$ of a deduction δ is the maximal degree of its cuts. The *rank* $r(\delta)$ of δ is the number of cuts in δ that have degree $d(\delta)$.

We proceed by double induction on the pair $\langle d(\delta), r(\delta) \rangle$ ordered lexicographically, that is $\langle d(\delta), r(\delta) \rangle < \langle d(\delta'), r(\delta') \rangle$ if either $d(\delta) < d(\delta')$, or $d(\delta) = d(\delta')$ and $r(\delta) < r(\delta')$. For instance, $\langle 5, 2 \rangle < \langle 5, 7 \rangle < \langle 8, 2 \rangle$. In other words, a deduction δ counts as "less than" a deduction δ' if either the cut degree of δ is less than that of δ' (δ's cuts are all less complex than the most complex CUTS of δ'), or the cut degrees of δ and δ' are equal but δ contains fewer cuts of maximal degree. Obviously, δ is normal if $d(\delta) = r(\delta) = 0$: it has no cuts at all. If we proceed by double induction on this ordering of deductions, we suppose that all deductions with a lower value than $\langle d(\delta), r(\delta) \rangle$ can be transformed to normal form, and use this to show that δ itself can be reduced to normal form.

---

[4] Recall from Section 2.3 that the degree of a formula is the number of logical symbols in it.

Theorem 4.10. *Suppose δ is a deduction of A from open assumptions Γ using only the rules for ∧, ⊃, ¬, and ∀. Then there is a normal deduction δ* also of A from open assumptions contained in Γ.*

Proof. By double induction on $\langle d(\delta), r(\delta) \rangle$.

Induction basis: If $\langle d(\delta), r(\delta) \rangle = \langle 0, 0 \rangle$, there is nothing to prove: δ is already normal.

Inductive step: Now suppose δ contains at least one cut. Since δ is a tree, and cuts are formula occurrences in that tree, there must be at least one cut with no cuts of the same or higher degree above them. Specifically, there must be a cut (formula occurrence that is both conclusion of ɪ-rule and premise of ᴇ-rule) such that the sub-deduction ending in the premise of the ɪ-rule contains no cut of same or higher degree. Pick one of them, say, the rightmost, topmost cut in δ of degree $d(\delta)$. We distinguish cases according to the form of the cut.

Suppose the cut is $B \land C$, then the part of δ containing the cut ends with

$$
\begin{array}{cc}
\vdots\, \delta_1 & \vdots\, \delta_2 \\
B & C \\
\hline
\multicolumn{2}{c}{B \land C} \\
\multicolumn{2}{c}{B}
\end{array}
\quad
\begin{array}{l}
\\ \\ {\scriptstyle \land\textsc{i}} \\ {\scriptstyle \land\textsc{e}}
\end{array}
$$

We replace this entire sub-deduction of δ with just

$$
\begin{array}{c}
\vdots\, \delta_1 \\
B
\end{array}
$$

Let us call the new deduction δ*. Clearly the end-formula of δ* is the same as that of δ, and the open assumptions of δ* are a subset of those of δ. The cuts in δ_1 remain cuts, but all of them have degree $< d(\delta)$, since otherwise there would be a cut of maximal degree higher in the proof than the cut we are reducing, and we assumed that was a topmost cut. We may have introduced a cut at the end of this deduction: if so, the cut is B. But the degree of B is less than the degree of $B \land C$, so it is not a cut of degree $d(\delta)$. The other cuts in the parts of δ that were not replaced have not changed. Altogether, the number of cuts of degree $d(\delta)$ in the new deduction is one less than in δ, i.e., $r(\delta) - 1$. If $r(\delta)$ was $= 1$, we have just removed the only cut of degree $d(\delta)$, so the maximal degree of cuts in δ*, i.e., $d(\delta^*)$ is $< d(\delta)$. If $r(\delta) > 1$, then $d(\delta^*) = d(\delta)$ (since we have introduced no cuts of higher degree than $d(\delta)$), but $r(\delta^*) = r(\delta) - 1 < r(\delta)$. Thus $\langle d(\delta^*), r(\delta^*) \rangle < \langle d(\delta), r(\delta) \rangle$. By the inductive hypothesis, δ* can be reduced to a normal deduction.

Suppose the cut is $B \supset C$, then the part of δ containing the cut ends with

$$
\begin{array}{c}
B^i \\
\vdots\; \delta_1 \\
\dfrac{C}{B \supset C}\; \supset_I \quad \vdots\; \delta_2 \\
\end{array}
$$

$$
i\;\dfrac{\dfrac{\begin{array}{c} B^i \\ \vdots\; \delta_1 \\ C \end{array}}{B \supset C}\; \supset_I \qquad \begin{array}{c} \vdots\; \delta_2 \\ B \end{array}}{C}\; \supset_E
$$

We replace this entire sub-deduction of δ with just

$$
\begin{array}{c}
\vdots\; \delta_1[\delta_2/B^i] \\
\vdots \\
C
\end{array}
$$

Let's call the resulting deduction δ^*. The end-formula of δ^* is the same as that of δ. By Lemma 4.8, the open assumptions of δ^* are a subset of those of δ.

Any cuts in δ_1 and δ_2 remain unchanged, although since δ_2 was grafted onto every undischarged assumption of the form B in δ_1, there now might be many more of them. However, all the cuts in δ_2 were of lower degree than $d(\delta)$, otherwise the cut being reduced would not be a rightmost, topmost cut. New cuts may have appeared: either the last inference of δ_1 with C as conclusion was an introduction, and the inference below C in δ^* is an elimination, or the last inference of δ_2 with B as conclusion was an introduction, and one of the undischarged assumptions of the form B in δ_1 is the major premise of an elimination rule. But these would be cuts of C or B, respectively, hence of lower degree than the cut $B \supset C$ we just eliminated. Again, the new deduction δ^* either has one fewer cut of degree $d(\delta)$, so $d(\delta^*) = d(\delta)$ and $r(\delta^*) < r(\delta)$, or it has no cuts of degree $d(\delta)$ at all anymore, in which case $d(\delta^*) < d(\delta)$. The inductive hypothesis applies; δ^* can be converted to a normal deduction.

We leave the case where the cut-formula is $\neg B$ as an exercise (Problem 4.11 below).

Suppose the cut is $\forall x\, B(x)$; the cut occurs in δ as:

$$
\begin{array}{c}
\Gamma \\
\vdots\; \delta_1 \\
\dfrac{\dfrac{B(c)}{\forall x\, B(x)}\; \forall_I}{B(t)}\; \forall_E
\end{array}
$$

By Theorem 3.20, there is a deduction δ_2 of $B[t/c]$ from $\Gamma[t/c]$. Because the eigenvariable condition on the $\forall_I$ inference requires that c does not occur in the open assumptions Γ, $\Gamma[t/c] = \Gamma$. We replace the sub-deduction of δ with just:

$$
\begin{array}{c}
\Gamma \\
\vdots \\
\delta_2 \\
\vdots \\
B(t)
\end{array}
$$

Let's call the resulting deduction δ^*. The end-formula of δ^* is the same as that of δ. The open assumptions of δ^* are the same as those of δ (since the assumptions in δ_2 are the same as those in δ_1). The cuts in δ_2 are the same as those in δ_1, with the possible exception that the free variable c is replaced by t. This is seen by inspecting the proof of Theorem 3.20: the structure of the deduction δ_2 is exactly the same as that of δ; only eigenvariables are renamed and t is substituted for c in some formulas in the deduction. This does not change the degree of formulas in the deduction, and hence also not the degrees of any cuts. Since the substitution is uniform, no new cuts are introduced.

All the cuts in δ_1 were of lower degree than $d(\delta)$. New cuts may have appeared: If the the last inference of δ_1 with $B(c)$ as conclusion was an introduction, then the last inference of δ_2 with $B(t)$ as conclusion is also an introduction. If now the inference below $B(t)$ in δ^* is an elimination, $B(t)$ is a cut. But this new cut has a lower degree than the cut $\forall x\, B(x)$ we just eliminated. If the eliminated cut on $\forall x\, B(x)$ was the only cut of degree $d(\delta)$, $d(\delta^*) < d(\delta)$. Otherwise, $d(\delta^*) = d(\delta)$ but $r(\delta^*) < r(\delta)$, and the inductive hypothesis applies. $\square$

Problem 4.11. Extend the preceding proof to cover the rules $\neg\mathrm{I}$ and $\neg\mathrm{E}$. What does a cut of the form $\neg B$ involve? Write out the missing case in the proof of Theorem 4.10. How can it be avoided? Identify the cut in the following example, and find the corresponding normal deduction by applying the appropriate reduction steps.

$$
\dfrac{
\dfrac{
1\dfrac{\dfrac{\neg A^1 \quad A^2}{\bot}\neg\mathrm{E}}{\neg\neg A}\neg\mathrm{I} \quad \neg A
}{
2\dfrac{\bot}{\neg A}\neg\mathrm{I}
}\neg\mathrm{E}
}{}
$$

Let's now consider an example. We begin with the deduction δ:

$$
\dfrac{
1\dfrac{
2\dfrac{
3\dfrac{
\neg A^2 \quad \dfrac{(A \wedge B) \supset A^1 \quad A \wedge B^3}{A}\supset\mathrm{E}
}{\bot}\neg\mathrm{E}
}{\dfrac{\neg(A \wedge B)}{\neg A \supset \neg(A \wedge B)}\supset\mathrm{I}}\neg\mathrm{I}
}{\boxed{((A \wedge B) \supset A) \supset (\neg A \supset \neg(A \wedge B))}}\supset\mathrm{I}
\quad
4\dfrac{\dfrac{A \wedge B^4}{A}\wedge\mathrm{E}}{(A \wedge B) \supset A}\supset\mathrm{I}
}{\neg A \supset \neg(A \wedge B)}\supset\mathrm{E}
$$

This is the deduction of $\neg A \supset \neg(A \wedge B)$ we constructed in Section 4.1. This deduction contains one cut on a formula of degree 7 (indicated by a box). Thus it has cut degree $d(\delta) = 7$ and cut rank $r(\delta) = 1$, since there is only one cut of degree 7. It is reduced by taking the sub-deduction ending in the right premise $(A \wedge B) \supset A$ of $\supset$E and "plugging it" into the assumptions (in this case, just one) discharged by the $\supset$I rule labelled 1, and removing the $\supset$I and $\supset$E rules at the bottom. We obtain δ^*:

$$
\cfrac{\neg A^{\,2} \quad \cfrac{\quad \cfrac{\cfrac{\cfrac{A \wedge B^{\,4}}{A}\;{\scriptstyle\wedge\text{E}}}{\boxed{(A \wedge B) \supset A}}\;4\;{\scriptstyle\supset\text{I}} \quad A \wedge B^{\,3}}{A}\;{\scriptstyle\supset\text{E}}}{\cfrac{\bot}{\neg(A \wedge B)}\;3\;{\scriptstyle\neg\text{I}}}\;{\scriptstyle\neg\text{E}}}{\neg A \supset \neg(A \wedge B)}\;2\;{\scriptstyle\supset\text{I}}
$$

This deduction no longer contains any cuts of degree 7, but it does contain a new cut of degree 2, which is now the cut of maximal degree. Thus, the cut degree $d(\delta^*)$ is now $2 < d(\delta)$. The cut rank has remained the same, since there is only one cut of degree 2. This cut can again be eliminated by replacing the assumption discharged by the $\supset$I rule labelled 4 by the sub-deduction ending in the minor premise of the following $\supset$E rule, and removing the cut. The resulting deduction δ^{**} is normal:

$$
\cfrac{\neg A^{\,2} \quad \cfrac{A \wedge B^{\,3}}{A}\;{\scriptstyle\wedge\text{E}}}{\cfrac{\cfrac{\bot}{\neg(A \wedge B)}\;3\;{\scriptstyle\neg\text{I}}}{\neg A \supset \neg(A \wedge B)}\;2\;{\scriptstyle\supset\text{I}}}\;{\scriptstyle\neg\text{E}}
$$

It is, coincidentally, the normal deduction we found before by a direct construction. It is not in general the case that normalization yields the shortest direct deduction, and different non-normal deductions with the same end-formula and from the same open assumptions can normalize to different normal deductions. However, even though our proof required a specific order in which cuts were removed, it so happens that they can be removed in any order and still yield a normal form—the same one in every case. Specifically, we do not have to deal with the topmost, rightmost cut of maximal degree first, or begin with a cut of maximal degree at all. This result, called "strong normalization," is a lot harder to prove though. It was first proved for **NM** and **NJ** by Prawitz (1971).[5]

It is useful to think of our proof by double induction as a proof that a certain procedure always terminates. The procedure is given by the proof itself: find a rightmost, topmost cut of maximal degree. Apply the transformation described in the inductive step—which case is used depends on the main connective of the formula occurring as the cut. Then apply the procedure to the resulting deduction: find a rightmost, topmost cut of maximal degree, apply the transformation in

---

[5] See also Cagnoni (1977) for a different strong normalization proof of **NM**.

the inductive step, and so on. Each deduction of an end-formula E from a set of assumptions Γ is associated with a pair of numbers. For each specific non-normal deduction δ (associated with the pair of numbers $\langle d(\delta), r(\delta)\rangle$) with end-formula E and assumptions Γ, the process of removing a rightmost, topmost cut yields a different deduction of E from assumptions included in Γ. In each case in the inductive step we have shown that the new deduction either has the same degree but lower rank, or has lower degree (that is, the degree-rank pair of the new deduction is smaller, in the lexicographical ordering, than the degree-rank pair of the preceding deduction). If we indicate the individual step of removing a single cut by *, we obtain a sequence of deductions $\delta, \delta^*, \delta^{**}, \ldots$, associated with a descending sequence of pairs,

$$\langle d(\delta), r(\delta)\rangle, \quad \langle d(\delta^*), r(\delta^*)\rangle, \quad \langle d(\delta^{**}), r(\delta^{**})\rangle, \ldots$$

After finitely many steps, this sequence of ordered pairs must reach $\langle 0, 0\rangle$, that is, a normal deduction of E from a subset of Γ. In the example we have discussed, the sequence corresponding to our removal of cuts was $\langle 7, 1\rangle, \langle 2, 1\rangle, \langle 0, 0\rangle$.

Problem 4.12. Normalize the deduction displayed below by removing the cuts one by one starting with the rightmost, topmost cut of maximal degree. At each stage of the normalization process, display the new transformed deduction. Throughout the process state the values of $d(\delta)$ and $r(\delta)$ (i.e., the ordered pair $\langle d(\delta), r(\delta)\rangle$) where δ stands for any of the deductions in the sequence.

$$
\cfrac{
\cfrac{\cfrac{\cfrac{C \wedge B^{1}}{B} \wedge_{\mathrm{E}}}{(C \wedge B) \supset B} \supset_{\mathrm{I}} \quad
\cfrac{A \quad \cfrac{\cfrac{C \quad B}{C \wedge B} \wedge_{\mathrm{I}}}{A \wedge (C \wedge B)} \wedge_{\mathrm{I}}}{\cfrac{C \wedge B}{C \wedge B} \wedge_{\mathrm{E}}}
}{B} \supset_{\mathrm{E}}
$$

4.4   The sub-formula property

The normalization theorem shows that all deductions can be transformed into normal deductions and shows how to effectively obtain the normal deduction from the original deduction. Normal deductions are direct in the sense that they don't contain cuts. They are also, and more importantly, direct in another sense: Every formula occurring in a normal deduction is a sub-formula of either the end-formula or an open assumption. In order to prove this—and also because it is independently interesting—we will first consider what normal deductions look like more generally.

We defined the notion of sub-formula already in Section 2.3. Later, in Section 2.10.1, we introduced $\bot$ as a special atomic formula, as well as the quantifiers in Section 2.13. Let's extend our definition of sub-formulas to formulas including $\bot, \forall$, and $\exists$ now.

Definition 4.13. The sub-formulas of a formula A are defined inductively as follows:

1. *Basis clause:* If A is atomic, A is the only sub-formula of A.

2. *Inductive clauses:* If A is of the form $\neg B$, the sub-formulas of A are A itself, $\bot$, and the sub-formulas of B.

3. If A is of the form $(B \wedge C)$, $(B \vee C)$, or $(B \supset C)$, then the sub-formulas of A are A itself and the sub-formulas of B and C.

4. If A is of the form $\exists x\, B(x)$ or $\forall x\, B(x)$, and $B(x)$ is $B[x/a]$, then the sub-formulas of A are A itself and the sub-formulas of all formulas $B[t/a]$, t any term.

5. *Extremal clause:* Nothing else is a sub-formula of A.

Problem 4.14. For each of the following, write out a list of all sub-formulas:

1. $\neg\neg\neg A$.

2. $\bot \vee \neg\bot$.

3. $((A \wedge B) \supset C) \supset (A \supset (B \supset C))$.

The definition of sub-formulas coincides with our previous definition for formulas not involving $\bot$, $\vee$ and $\exists$. When constructing quantified formulas, recall, we start with a formula B containing a free variable a and form $\forall x\, B(x)$ by replacing every occurrence of a in B by x. So $B(x)$ is not a formula, although $\forall x\, B(x)$ and $B(a)$ are. For this reason, we can't say that $B(x)$ is a sub-formula of $\forall x\, B(x)$. For maximal generality, and in order to use the concept "sub-formula" in the following results, we count not just $B(a)$ but any formula $B[t/a]$ as a sub-formula of $\forall x\, B(x)$.

In a normal deduction, no conclusion of an I-inference can be the major premise of an E-inference. One might at first glance think that on any branch of a normal deduction (cf. Definition 4.5), all E-inferences must occur above all I-inferences. This is, however, not the case. The reason is that a conclusion of an I-inference may well be the *minor* premise of an E-inference. Since $\supset$E and $\neg$E are the only E-inferences considered so far that have a minor premise, this means the conclusion of an I-inference may be the minor premise of an $\supset$E- or $\neg$E-inference. For instance,

$$
\cfrac{
  (A \wedge B) \supset C_a \qquad
  \cfrac{
    \cfrac{A_e{}^2 \qquad B_f{}^1}{A \wedge B_g}\ \wedge\text{I}
  }{}
}{
  \cfrac{
    \cfrac{
      \cfrac{C_b}{B \supset C_c}\ {}^1\ \supset\text{I}
    }{}
  }{A \supset (B \supset C)_d}\ {}^2\ \supset\text{I}
}\ \supset\text{E}
$$

is a normal deduction, in which the conclusion $A \wedge B$ of a $\wedge$I inference is the minor premise of an $\supset$E rule. In the branch $egbcd$ beginning in the assumption A and ending with the end-formula, an I-inference precedes (stands above) an E-inference.

Definition 4.15. A *thread* in a deduction δ is a sequence of formula occurrences $A_1, \ldots, A_k$ in which:

1. A_1 is an assumption,

2. A_i is a premise of an I-inference or a major premise of an E-inference of which A_{i+1} is the conclusion, for all $i < k$, and

3. A_k is either the end-formula or the minor premise of an $\supset$E- or $\neg$E-inference.

So a thread is like a branch, except it stops when we hit a minor premise of $\supset$E or $\neg$E. Why stop when we reach a minor premise? Because once we pass to the conclusion of the inference in question, then the information contained in the minor premise is lost (the minor premise itself is not a sub-formula of the conclusion).

In the preceding deduction, the branch *abcd* is also a thread: (1) *a* is an assumption; (2) *a* is the major premise of an E-inference of which *b* is the conclusion, *b* is a premise of an I-inference of which *c* is the conclusion, *c* is a premise of an I-inference of which *d* is the conclusion; and (3) *d* is the end-formula. *eg* is also a thread: (1) *e* is an assumption; (2) *e* is a premise of an I-inference of which *g* is the conclusion; (3) *g* is the minor premise of an $\supset$E-inference. Similarly, *fg* is a thread. But not every branch is a thread. For instance, the branch *egbcd* is not a thread, since *g* is not a premise of an I-inference nor the major premise of an E-inference (even though it is a premise of an inference of which *b* is the conclusion).

Proposition 4.16. *In any thread* $A_1, \ldots, A_k$ *in a normal deduction* δ, *if* A_i *is the conclusion of an I-rule, every* A_j *with* $j > i$ *is the conclusion of an I-rule.*

Proof. Suppose not. Then there is some $j > i$ where all A_ℓ with $\ell = i, \ldots, j$ are conclusions of I-rules, and A_{j+1} is not, i.e., it is the conclusion of an E-rule. Since A_{j+1} is the conclusion of the inference with A_j as premise, that means that A_j is a premise of an E-rule. In other words, we have:

$$\cdots \quad \frac{\frac{\vdots}{A_j}\ \text{I} \quad \cdots}{A_{j+1}}\ \text{E}$$

What are the possibilities? Suppose A_j is the conclusion of $\wedge$I. Then we can observe the following:

1. A_j cannot be the premise of $\wedge$E, since the deduction is normal.

2. A_j cannot be the major premise of $\supset$E, since the major premise of $\supset$E has the form $B \supset C$, and the conclusion of $\wedge$I has the form $B \wedge C$—and a formula can't have both $\supset$ and $\wedge$ as main operator.

3. A_j cannot be the major premise of $\neg$E or $\forall$E, for similar reasons.

4. A_j cannot be the minor premise of $\supset$E or $\neg$E, since otherwise the thread would end with A_j (cf. Definition 4.15) and there is no A_{j+1} on the thread.

By parallel reasoning, we can check that A_j can't be the conclusion of $\supset$I, $\neg$I, or of $\forall$I: it can't also be a major premise of the E-rule for the same logical operator since the deduction is normal, it can't be the major premise of the E-rule for a different operator because it can't have two different main operators, and it can't be the minor premise of $\supset$E or $\neg$E, since the thread would otherwise end at A_j. So in each case we obtain a contradiction. □

In other words, a thread in a normal deduction passes from an assumption through a (possibly empty) sequence of major premises of elimination rules and then through a (possibly empty) sequence of introduction rules. In particular this means that in every thread in a normal deduction, there is some m such that for all $j > m$, A_j is the conclusion of an I-inference, and for every $j < m$, A_j is the major premise of an E-inference. That m is the least index such that A_m is a premise of an I-inference. We might call $A_1, \ldots, A_{m-1}$ the *analytic* part of the thread (premises of E-inferences), $A_{m+1}, \ldots, A_k$ the *synthetic* part (conclusions of I-inferences), and A_m the *minimum formula* of the thread.

Now every thread is obviously an initial segment of a branch through δ. If the thread is not just an initial segment of a branch, but an entire branch, the thread ends in the end-formula of δ. And in fact there is always a thread that ends in the end-formula in any (and thus in any normal) deduction. This can be seen by starting from the end-formula and proceeding upward, going through only the major premises of the inferences leading to each formula thus reached. So the end-formula of any proof belongs to a thread. More generally:

Proposition 4.17. *Every formula in a deduction δ belongs to some thread.*

Proof. Induction basis: If the size of δ is 0, then δ consists of a single assumption, which obviously belongs to a thread.

Inductive step: Assume δ has size greater than 0. Then δ ends in an inference with conclusion A. We can assume, by inductive hypothesis, that all the formulas of the deduction appearing in the sub-deduction(s) leading to the premise(s) of the last inference belong to some thread. Consider the conclusion A of the inference. If A is obtained by an I-rule then, A belongs to the thread to which its major premise (or one of its major premises) belongs (and they belong to some thread by inductive hypothesis). The same holds if A is obtained by $\wedge$E. If A is obtained by

⊃E or ¬E, then A belongs to the thread to which the major premise of the inference belongs (and the latter must belong to a thread by inductive hypothesis).   □

Definition 4.18 (Sub-formula property). A deduction δ has the *sub-formula property* if every formula in δ is a sub-formula of the end-formula or of an open assumption.

One important upshot of the normalization theorem (Theorem 4.10) is that every deduction can be transformed into a normal form which has the sub-formula property. That normal deductions have the sub-formula property is surprisingly difficult to prove rigorously.

Here's the basic idea: we have just shown that every formula A in a deduction δ belongs to some thread. So we are done if we can show that every formula in a thread is either a sub-formula of an open assumption or of the end-formula of δ. We'll first show that the formula must be a sub-formula of the assumption that starts the thread, or of the last formula in the thread. Of course, a thread may not begin with an *open* assumption—the assumption that begins it may be discharged—and it need not end with the *end*-formula. In that case we have to show that the discharged assumption and the last formula of the thread are themselves sub-formulas of an open assumption or of the end-formula of the entire deduction. We do this by induction on the "order" of a thread. The order of a thread measures how many ⊃E and ¬E inferences (and hence other threads) we pass through as we proceed downward to the end-formula.

Proposition 4.19. *Every formula on a thread $A_1, \ldots, A_k$ in a normal deduction δ is a sub-formula of either A_1 or A_k.*

Proof. Recall that by Proposition 4.16, there is an m, $1 \le m \le k$, such that every A_j with $j < m$ is the major premise of an E-rule, and every A_j with $j > m$ is the conclusion of an I-rule.

The conclusion of an E-rule ∧E, ⊃E, ¬E, ∀E is always a sub-formula of its major premise as well. The premise of an I-rule is a sub-formula of its conclusion. (Recall that ⊥ is always a sub-formula of ¬A. This is why the conclusion ⊥ of ¬E is a sub-formula of its major premise ¬A, and the premise ⊥ of ¬I a sub-formula of the conclusion ¬A.) In other words, all A_j with $j < m$ (the analytic part) are sub-formulas of A_1, and all A_j with $j > m$ (the synthetic part) are sub-formulas of A_k, and the minimum formula A_m is a sub-formula of both A_1 and A_k.   □

As mentioned before the proposition just established is not sufficient to give us what we want for two reasons. The first is that the assumption A_1 might have been discharged and thus it is not one of the open assumptions of δ. The second is that the end-formula of the thread need not be the end-formula of the deduction δ.

However, we can exploit the above result to gain more information on those threads that end in the end-formula (i.e., threads that are also branches). What we would like to show is that if A occurs in a thread $A_1, \ldots, A_k$ in a normal deduction δ *where A_k is the end-formula*, then A is either a sub-formula of A_k or a sub-formula of an open assumption of δ.

Proposition 4.20. *If A occurs in a thread $A_1, \ldots, A_k$ ending with the end-formula of a normal deduction δ, then A is either a sub-formula of the end-formula A_k or of an open assumption of δ.*

Proof. By the previous Proposition 4.19, A is a sub-formula of A_k (and thus of the end-formula of δ) or a sub-formula of A_1. If A_1 is not discharged, it remains an open assumption in the original deduction δ. Then A is a sub-formula of an open assumption of δ and we are done.

Now suppose A_1 is discharged. The discharging inference can only be $\supset$I or $\neg$I, since those are the only inference rules we are currently taking into account which allow assumptions to be discharged. The assumption(s) discharged by an inference must lie in the sub-deduction ending in the premise of the inference. Conversely, the inference discharging an assumption must lie below that assumption, i.e., on the branch beginning with that assumption and ending in the end-formula of the deduction.

Because deductions are trees, for every assumption there is exactly one branch containing that assumption. And since the thread $A_1, \ldots, A_k$ ends in the end-formula of δ and proceeds from premises of inferences to their conclusions, it is also a branch. Consequently, if A_1 is discharged in δ, then it is discharged by an $\supset$I or $\neg$I inference on the thread $A_1, \ldots, A_k$.

Let A_m be the minimum formula of the thread. If A_1 is an assumption which is discharged by an $\supset$I or $\neg$I rule the conclusion of which is also on the thread—say, it is A_j for some $j \leq k$—then A_1 is a sub-formula of A_j. And as A_j is the conclusion of an I-inference, j must be $> m$, i.e., it must belong to the synthetic part of the thread, where each formula is a sub-formula of A_k. So if A_1 is discharged on the thread, it is a sub-formula of A_k. $\square$

Notice that our argument in the last part also establishes the following:

Proposition 4.21. *If $A_1, \ldots, A_k$ is a thread in a normal deduction δ, and A_1 is discharged on the thread, then it is a sub-formula of A_k.*

If A lies on a thread that does not end in the end-formula (i.e., it is not a branch), we still know that A is a sub-formula of the end-formula of the thread or of the assumption A_1 that begins it. And if A_1 is discharged on the thread, it, and hence also A, is a sub-formula of A_k. But what if A_1 is discharged on some other thread? Consider the following example:

$$
\cfrac{(B \supset ((A \land B)) \supset C \quad \cfrac{2 \ \cfrac{\cfrac{A^1 \quad B^2}{A \land B} \land \text{I}}{B \supset (A \land B)} \supset \text{I}}{}}{\cfrac{C}{1 \ \ A \supset C}} \supset \text{E}
$$

The formula B is discharged on the same thread to which it belongs, whereas A is discharged on a different thread than the one to which it belongs. However, A is still a sub-formula of the conclusion of the inference that discharges it, and eventually of the end-formula.

Let's pause here to consider our example from p. 122.

$$\cfrac{(A \wedge B) \supset C_a \quad \cfrac{A_e{}^2 \quad B_f{}^1}{A \wedge B_g} \wedge\mathrm{I}}{\cfrac{1 \ \cfrac{C_b}{B \supset C_c} \supset\mathrm{I}}{2 \ \dfrac{}{A \supset (B \supset C)_d} \supset\mathrm{I}}} \supset\mathrm{E}$$

We had one thread that is also a branch: $abcd$. In this case, the minimum formula A_m is formula b: every formula after it is the conclusion of an I-inference, every formula before it is the major premise of an E-inference. And indeed, it and all formulas before it in the thread are sub-formulas of the assumption a on the thread. Moreover, it and all formulas after it are sub-formulas of the end-formula. In this example, the assumption on the thread is not discharged by $\supset\mathrm{I}$. But suppose it were, e.g., by another $\supset\mathrm{I}$ at the end. Then we'd have:

$$\cfrac{(A \wedge B) \supset C_a{}^3 \quad \cfrac{A_e{}^2 \quad B_f{}^1}{A \wedge B_g} \wedge\mathrm{I}}{\cfrac{1 \ \cfrac{C_b}{B \supset C_c} \supset\mathrm{I}}{\cfrac{2 \ \dfrac{}{A \supset (B \supset C)_d} \supset\mathrm{I}}{3 \ \dfrac{}{((A \wedge B) \supset C) \supset (A \supset (B \supset C))_h} \supset\mathrm{I}}}} \supset\mathrm{E}$$

Here, a thread that's also a branch is $abcdh$. The relevant minimum formula A_m in this thread is again formula b. It and all formulas after it are sub-formulas of the end-formula h. The assumption a is now discharged, and is not a sub-formula of an open assumption on the branch. But since it is discharged by $\supset\mathrm{I}$, it is a sub-formula of the conclusion of that $\supset\mathrm{I}$ inference, namely h. So formula a is also a sub-formula of the end-formula.

Formula e (A) lies on the thread eg, which does not end in the end-formula. But A is also a sub-formula of the end-formula of the deduction. It is so for two reasons: First, since the thread eg only contains a single $\wedge\mathrm{I}$ inference, formula e lies in the synthetic part of the thread, and so is a sub-formula of the end-formula g of the thread. This formula, as minor premise of an $\supset\mathrm{E}$-rule, is a sub-formula of the major premise, namely formula a, which we've just shown is a sub-formula of the end-formula. The other reason is that even though formula e is not discharged on the thread it lies on, the conclusion of the inference which does discharge it (formula d) lies on a thread that is also a branch. And the conclusion of that $\supset\mathrm{I}$ contains the assumption it discharges (formula e) as a sub-formula. The same goes for formula f. So, in this example at least, the formulas on threads that don't end in the end-formula are sub-formulas of formulas on such a thread, and our previous result establishes that these are in turn sub-formulas of the end-formula (or of an open assumption, although this case does not apply here as all assumptions are discharged).

To deal with the general case of formulas not on a thread through the end-formula, we have to proceed by induction. We need to assign some measure to threads along which we can proceed. We'll want the case we've already established to be the induction basis. So let us say that the order of a thread is 0 if the thread ends in the end formula. If a thread does not end in the end-formula, it ends in the minor premise of an $\supset$E or $\neg$E inference. The major premise of the same inference and thus also the conclusion belong to another thread. If a thread has order n, then the threads which end in minor premises of $\supset$E or $\neg$E the conclusion of which lies on the thread of order n, have order $n + 1$. In other words, a thread of order 0 does not end in the minor premise of $\supset$E or $\neg$E but instead ends in the end-formula. It may contain major premises of $\supset$E or $\neg$E inferences, and the minor premises of these are where other threads stop. These will have order 1. These in turn may contain major premises of $\supset$E or $\neg$E inferences, the minor premises of which mark the end of yet other threads. These will have order 2, etc. So the order of a thread measures the distance of that thread from a thread of order 0 by counting steps from a minor to a major premise of an $\supset$E or $\neg$E rule.

In our preceding example, the thread $abcdh$ ends in the end-formula, so is of order 0. The threads eg and fg are of order 1. Here is an example of a deduction with threads of order 2:

$$
\cfrac{(B \supset C) \supset C_a \quad \cfrac{A \supset (B \supset C)_d \quad \cfrac{\cfrac{A \wedge B_f \,^1}{A_g} \wedge \text{E}}{B \supset C_e} \supset \text{E}}{B \supset C_e}\supset \text{E}}{\cfrac{C_b}{{}^1 \ (A \wedge B) \supset C_c} \supset \text{I}}
$$

Thread abc is of order 0, thread de of order 1, and thread fg of order 2.

Problem 4.22. Show that the above is a proper definition, i.e., show:

1. If A lies on a thread of order n, then every thread through A has order n.

2. Every thread is assigned an order in this way.

Theorem 4.23. *Normal deductions have the sub-formula property.*

Proof. Now we proceed by induction on the order n of a thread containing A.

Induction basis: If $n = 0$, we have already established that A is a sub-formula of the end-formula or of the assumption of the thread, if that assumption is open (Proposition 4.20).

Inductive step: Suppose that $n > 0$. In that case, A is on a thread $A_1, \ldots, A_k$ ending in some formula A_k which is the minor premise of an $\supset$E or $\neg$E inference.

This thread is a thread in a normal deduction and so Proposition 4.16 applies: there is some $m \leq k$ such that every A_j with $j < m$ is the major premise of an E-rule and every A_j with $j > m$ is the conclusion of an I-rule. By the same considerations as in the proof of Proposition 4.19, A is a sub-formula of A_k if A is A_j for some $j \geq m$, or of the assumption A_1 of the thread if A is A_j for some $j < m$ (since every A_j, for $j < m$, is the result of an elimination rule).

In the first case, A_k is the minor premise of $\supset$E or $\neg$E. The major premise is either $A_k \supset C$ or $\neg A_k$, and lies on a thread of order $n - 1$. So A_k is a sub-formula of a formula on a thread of order $n - 1$. The inductive hypothesis applies: all formulas on that thread are sub-formulas of the end-formula or of an open assumption. So A is a sub-formula of the end-formula or of an open assumption, since the relation of being a sub-formula is transitive.

In the second case, A is a sub-formula of the assumption A_1 of the thread. If the assumption remains open throughout the deduction, A is a sub-formula of a formula in Γ, and so we are done. If it is discharged, it is discharged by an $\supset$I or $\neg$I inference and A_1 and hence A is a sub-formula of the conclusion of that inference. If the discharging inference lies on the same thread as A_1, then A_1 and hence A is a sub-formula of A_k (we recorded this as Proposition 4.21). As in the first case, A_k (and hence A) then is a sub-formula of a formula on a thread of order $n - 1$. If A_1 is discharged below A_k, A_1 is a sub-formula of a formula on a thread of order $< n$, namely of the conclusion of the $\supset$I or $\neg$I inference which discharges it. In either case, the inductive hypothesis applies to that thread, and so A is a sub-formula of either the end-formula or of an open assumption. □

Consider our example of a normal deduction again. It has three threads, the formulas labelled $abcd$; the formulas labelled eg; and the formulas labelled fg.

$$\cfrac{(A \wedge B) \supset C_a \qquad \cfrac{A_e{}^2 \quad B_f{}^1}{A \wedge B_g}\wedge\text{I}}{\cfrac{\cfrac{C_b}{B \supset C_c}1 \supset\text{I}}{A \supset (B \supset C)_d}2 \supset\text{I}}\supset\text{E}$$

The thread $abcd$ has order 0, and the threads eg and fg have order 1 (they end in the minor premise of $\supset$E). All formulas on the thread $abcd$ are sub-formulas of the end-formula or of the open assumption beginning that thread, i.e., formula a. The formulas in the threads eg and fg are all sub-formulas of the last formula in that thread, i.e., formula g. It, being the minor premise of an $\supset$E rule, is a sub-formula of the major premise of that $\supset$E rule, i.e., of formula a. And since a occurs in a thread of order 0 it must be a sub-formula of the end-formula or of the open assumption that begins that thread. (In this case, it *is* the assumption that begins the thread.) Note, however, that the assumptions e and f are also discharged below the threads (eg and fg) in which they occur. They thus are sub-formulas of the conclusions (formulas d and c, respectively) of the $\supset$I inferences that discharge them, which lie on a thread of order 0. In our final example above, formula f on the thread fg (of order 2) is discharged on the thread abc of order 0, and hence a sub-formula of the end-formula. (Had it been discharged on the thread de of order 1, the inductive hypothesis would still have applied and we could have concluded that it must be a sub-formula of the end-formula or of an open assumption.)

The sub-formula property has a number of interesting consequences. The most important is that a natural deduction system in which every deduction can be converted into a normal deduction, and in which normal deductions have the sub-formula property, is consistent. For if it were inconsistent, there would be a deduction of $\bot$ (from no open assumptions). There would then also be a normal deduction of $\bot$ from no open assumptions. But there cannot be such a deduction. In fact, by the sub-formula property, every formula in such a deduction would have to be a sub-formula of the end-formula $\bot$. But only $\bot$ itself is a sub-formula of $\bot$. So the deduction cannot contain any I-rules, as $\bot$ can't be the conclusion of an I-rule, and it can't contain any E-rules since $\bot$ can't be the major premise of an E-rule. So we have:

Corollary 4.24. *NM without rules for* $\vee$ *and* $\exists$ *is consistent.*

By a slight generalization of our reasoning, we obtain also:

Corollary 4.25. *In NM without rules for* $\vee$ *and* $\exists$ *there cannot be a deduction ending in an atomic formula unless that formula is a sub-formula of an open assumption.*

In Theorem 4.42 we will prove that full **NJ** also has the sub-formula property. By similar reasoning we can then establish consistency for **NJ**.

4.5   The size of normal deductions

Normal deductions are simpler than non-normal ones in that they contain no detours. Importantly, and perhaps surprisingly, they can however be more complex in that they are *much longer*. We'll look at a simple example to make this clear.

Let $A(n)$ be the formulas defined by $A(1) = p$ and $A(n + 1) = (A(n) \wedge A(n))$. So $A(2) = (p \wedge p)$, $A(3) = ((p \wedge p) \wedge (p \wedge p))$, etc. In general, $A(n)$ is a conjunction of 2^n p's. Now obviously, $p \supset A(n)$ is a tautology for all n. Let us consider normal and (some) non-normal deductions of $p \supset A(n)$.

Normal proofs of $p \supset A(2)$, $p \supset A(3)$, etc., are structurally very simple:

$$
1\;\dfrac{\dfrac{p^1 \quad p^1}{A(2)}\wedge\text{I}}{p \supset A(2)}\supset\text{I}
\qquad\qquad
1\;\dfrac{\dfrac{\dfrac{p^1 \quad p^1}{A(2)}\wedge\text{I} \quad \dfrac{p^1 \quad p^1}{A(2)}\wedge\text{I}}{A(3)}\wedge\text{I}}{p \supset A(3)}\supset\text{I}
$$

$$\frac{\dfrac{\dfrac{p^1 \quad p^1}{A(2)}\wedge\text{I} \quad \dfrac{p^1 \quad p^1}{A(2)}\wedge\text{I}}{A(3)}\wedge\text{I} \qquad \dfrac{\dfrac{p^1 \quad p^1}{A(2)}\wedge\text{I} \quad \dfrac{p^1 \quad p^1}{A(2)}\wedge\text{I}}{A(3)}\wedge\text{I}}{1\ \dfrac{A(4)}{p \supset A(4)}\supset\text{I}}\wedge\text{I}$$

These deductions have sizes 2, 4, 8, and have 2, 4, 8 assumptions. In general, a normal deduction of $p \supset A(n+1)$ consists of a normal deduction of $p \supset A(n)$ where each assumption p has been replaced by

$$\frac{p^1 \quad p^1}{A(2)}\wedge\text{I}$$

and each p other than the antecedent of $p \supset A(n)$ has been replaced by $A(2)$. In fact, although we will not prove this, there are no smaller normal proofs. Thus, the size of a normal deduction of $p \supset A(n)$ is at least 2^{n-1}.

There is, however, another easy way of producing deductions of $p \supset A(n)$. Consider the following deduction δ of size 5:

$$1\ \cfrac{2\ \cfrac{3\ \cfrac{\cfrac{C \supset D^2 \quad \cfrac{B \supset C^1 \quad B^3}{C}\supset\text{E}}{D}\supset\text{E}}{B \supset D}\supset\text{I}}{(C \supset D) \supset (B \supset D)}\supset\text{I}}{(B \supset C) \supset ((C \supset D) \supset (B \supset D))}\supset\text{I}$$

Obviously, we can replace B, C, D, with any formulas in this deduction, and the size is the same. Let's abbreviate the conclusion of this deduction with $R(B, C, D)$. Suppose we have deductions δ_1 and δ_2 of $B \supset C$ and $C \supset D$, then we can produce a deduction δ of $B \supset D$ as follows:

$$\cfrac{\begin{matrix}\vdots\ \delta_1 \\ C \supset D\end{matrix} \quad \cfrac{\begin{matrix}\vdots\ \delta_2 \\ B \supset C\end{matrix} \quad \cfrac{\begin{matrix}\vdots\ \delta \\ (B \supset C) \supset ((C \supset D) \supset (B \supset D))\end{matrix}}{(C \supset D) \supset (B \supset D)}\supset\text{E}}{B \supset D}}{}\supset\text{E}$$

The size of this deduction is the sum of the sizes of δ_1 and δ_2 plus 7.

Now consider the deduction of $p \supset A(2)$, i.e., of $p \supset (p \wedge p)$ above. It remains a deduction if we replace p by $A(2)$, by $A(3)$, etc., and these deductions all have the same size: 2. In other words, for any n, there is a deduction of $A(n) \supset A(n+1)$ of size 2:

$$1\ \frac{\dfrac{A(n)^1 \quad A(n)^1}{A(n+1)}\wedge\text{I}}{A(n) \supset A(n+1)}\supset\text{I}$$

So, if we chain together a deduction of $A(1) \supset A(2)$ and one of $A(2) \supset A(3)$ in the above way, we get a deduction of $A(1) \supset A(3)$, of size $2 + 2 + 7 = 11$. Chain this together with a deduction of $A(3) \supset A(4)$ to get a deduction of $A(1) \supset A(4)$ of size $11 + 2 + 7 = 20$. Chain this together with a deduction of $A(4) \supset A(5)$, and we get a deduction of $A(1) \supset A(5)$ of size $20 + 2 + 7 = 29$, etc. In general, a deduction of $A(1) \supset A(n)$ obtained this way has size $2 + 9(n - 2) = 9n - 16$ (if $n \geq 3$). For $n = 3$, 4, 5, and 6, the normal deductions of $A(1) \supset A(n)$ are smaller (4 vs. 11, 8 vs. 20, 16 vs. 29, 32 vs. 38), but for $n > 6$ the non-normal deductions are soon dramatically smaller: 64 vs. 47, 128 vs. 56, 256 vs. 65, 512 vs. 74, 1024 vs. 83, etc.[6]

4.6 Normalization for **NJ**

We now proceed to discuss the general case of normalization for **NJ**. The remaining rules that may appear in a deduction are those for $\vee$, $\exists$, and the absurdity rule $\perp_J$. The latter, intuitively, generates additional detours. Instead of using $\perp_J$ to introduce a formula only to apply an elimination rule to it, you can always directly apply $\perp_J$ to obtain the conclusion of the elimination rule. For instance, instead of

$$\frac{\dfrac{\vdots}{\perp}}{\dfrac{A \wedge B}{A} \wedge_E} \perp_J \qquad \text{deduce} \qquad \dfrac{\vdots}{\perp} \perp_J \atop A$$

instead. And just like an $\wedge_E$ immediately following an $\wedge_I$ is a detour, so are $\vee_E$ following $\vee_I$ and $\exists_E$ following $\exists_I$. As we will see, such detours can also be eliminated.

The rules for $\vee$ and $\exists$ create additional difficulties because in their presence, a major premise of an elimination rule may follow the conclusion of an introduction rule without doing so *immediately*. For instance, we might have a situation like the following in a deduction:

[6] More striking results as well as rigorous proofs that no short normal deductions are possible are given by Orevkov (1982, 1993); see Troelstra and Schwichtenberg (2000, §6.11) for a presentation in natural deduction.

Here we intuitively have a detour: first $D \wedge E$ is inferred from D (and E), and then further down this $\wedge$ is eliminated, to obtain D again. But the conclusion of the $\wedge_I$ rule is not directly the premise of the $\wedge_E$ rule. We deal with this complication by generalizing the notion of a cut in **NJ**-deductions.

Definition 4.26. A *cut segment* in a deduction δ in **NJ** is a sequence $A_1, \ldots, A_n$ of formula occurrences in δ where

1. A_1 is the conclusion of an I-rule or of $\perp_J$,

2. A_i is a minor premise of a $\vee_E$ or $\exists_E$ inference, and A_{i+1} is its conclusion (for $1 \le i < n$), and

3. A_n is the major premise of an E-rule or of $\perp_J$.

The length of the cut segment is n.

This is a direct generalization of the previous definition: a cut according to the old definition is a cut segment of length 1 according to the new definition in which A_1 is the conclusion of an I-inference, and also (since $n = 1$ and $A_1 = A_n$) the major premise of an E-rule. Note that in $\vee_E$ and $\exists_E$, the conclusion formula is the same as the minor premises, so in a cut segment, all A_i are occurrences of the same formula. Hence we can again define the degree of a cut segment as the degree of the formulas A_i (and the degree $d(\delta)$ of a deduction as the maximal degree of its cut segments), or 0 if there are no cut segments. A *maximal cut segment* is one of degree $d(\delta)$. In the preceding example, the three boxed formula occurrences constitute a cut segment of length 3.

The proof that cut segments can be eliminated from deductions again proceeds by an induction on $\langle d(\delta), r^*(\delta) \rangle$. It is now not enough to count the number of cut segments of degree $d(\delta)$. Instead, we must consider the collective length of these cut segments: the *rank* $r^*(\delta)$ is now *the sum of* the lengths of all cut segments of degree $d(\delta)$. A normal deduction, again, is one without cut segments: for a normal deduction $d(\delta) = r^*(\delta) = 0$. (See the example starting on p. 143.)

In order to ensure that the inductive hypothesis applies when we transform cut segments, we must judiciously choose which cut segment to work on. In the proof of Theorem 4.10, it was enough to pick a topmost, rightmost cut. Now, we have to be a bit more careful.

Definition 4.27. Let a *highest cut segment* be a cut segment $A_1, \ldots, A_n$ in δ such that:

1. A_i has maximal degree $d(\delta)$,

2. no cut segment of maximal degree ends above A_1,

3. if A_n is a major premise of an E-inference, B one of its minor premises, then B is not on or below a maximal cut segment.

Proposition 4.28. *Every deduction δ has a highest cut segment, if it has a cut segment at all.*

Proof. Among a set of maximal cuts segments we can always find one that satisfies condition 2: For every maximal cut segment, consider the first formula occurrence in it. One of these must be topmost, i.e., none of the other occurrences of formulas in a maximal segment is above it. Pick the corresponding cut segment.

Let $A_1^1, \ldots, A_{n_1}^1$ (or for short: A_i^1) be a maximal cut segment satisfying condition 2. If A_i^1 does not already satisfy condition 3, we define cut segments A_i^2, $A_i^3, \ldots$, until we find one that does: If A_i^j satisfies condition 3, we are done. If A_i^j violates condition 3, $A_{n_j}^j$, the last formula occurrence in it, must be the major premise of an E-inference with a minor premise B, and B is on or below a cut segment of maximal degree. Among these maximal cut segments, we can again find one that satisfies condition 2. Let that be A_i^{j+1}. Each formula occurrence A_i^{j+1} lies either below the last formula occurrence $A_{n_j}^j$ of the previous cut segment, or else in a sub-deduction not containing any A_i^j (namely, in the sub-deduction ending in B). So, the occurrences A_i^{j+1} of formulas that make up the new cut segment are different occurrences than any of $A_i^1, \ldots, A_i^j$. Since δ contains only finitely many formula occurrences, eventually we must find a maximal cut segment A_i^k that satisfies conditions 2 and 3. □

Here is a simple example that illustrates the process of finding a highest maximal cut segment:

We see two cut segments, one consists of the single occurrence of the formula $A \lor B_a$, the other consists of the two labelled occurrences of $A \land C_b$. The cut segment $A \lor B_a$ is of maximal degree and has no cut-segments, of maximal degree or otherwise, above it. Thus, it satisfies conditions 1 and 2. It does not satisfy 3, however, since it is the major premise of an ∨E-inference, and one of the minor premises of that inference lies on the maximal cut segment $A \land C_b$. So, to arrive at a highest cut segment, we consider $A \land B_b$ and all the cut segments ending above it (e.g., any that are contained in the sub-deduction δ_1 ending in C). Among those, we pick one that satisfies condition 2. If it satisfies condition 3, we have found a highest maximal cut segment, otherwise we keep going.

Theorem 4.29. *Every deduction δ in **NJ** can be transformed into a normal deduction.*

Proof. By induction on $\langle d(\delta), r^*(\delta)\rangle$. By Lemma 3.18, we may assume that no two $\forall$I or $\exists$E inferences have the same eigenvariable, and that every eigenvariable used for such an inference only occurs above that inference.

Induction basis: If $d(\delta) = 0$ and $r^*(\delta) = 0$, δ is already normal.

Inductive step: First suppose that $d(\delta) = 0$ and $r^*(\delta) > 0$. Then the cut segment consists of a formula A of degree 0. No I-rule can have a formula of degree 0 as conclusion, so the first formula occurrence in the cut segment must be the conclusion of $\perp_J$. On the other hand, no E-rule can have a formula of degree 0 as major premise, so the last formula occurrence A_n in the cut segment must be the premise of $\perp_J$, i.e., an occurrence of $\perp$. Since $A_1, \ldots, A_n$ are all occurrences of the same formula, they must all be occurrences of $\perp$. Let δ_1 be the sub-deduction of δ that ends in the conclusion A_1 of the first $\perp_J$ inference, and δ_2 be the sub-deduction that ends in conclusion of the second $\perp_J$ with A_n as premise. Clearly, δ_1 is a sub-deduction of δ_2. Then δ_2 has the form given on the left below; we replace it by the sub-deduction on the right.

$$
\left.\begin{array}{c}
\vdots\ \delta_1 \\
\dfrac{\perp}{\perp}\perp_J \\
\vdots \\
\dfrac{\perp}{C}\perp_J
\end{array}\right\}\delta_2 \quad \Rightarrow \quad
\begin{array}{c}
\vdots\ \delta_1 \\
\vdots \\
\dfrac{\perp}{C}\perp_J
\end{array}
$$

For instance, if the length of the segment is 1, i.e., the conclusion of $\perp_J$ is also the premise of $\perp_J$, we have the simple situation:

$$
\begin{array}{c}
\vdots\ \delta_1 \\
\dfrac{\perp}{\dfrac{\perp}{C}\perp_J}\perp_J
\end{array}
\quad \Rightarrow \quad
\begin{array}{c}
\vdots\ \delta_1 \\
\dfrac{\perp}{C}\perp_J
\end{array}
$$

No new cut segments are introduced in this replacement, since δ_1 remains unchanged, and if C is the start of a cut segment in the resulting deduction δ^*, it already was in the original deduction δ. In particular, no new cut segments of degree $> d(\delta)$ were introduced. We have removed a cut segment of maximal degree of length at least 1. If it was the only cut segment in δ, then $r^*(\delta^*) = 0$, and we have arrived at a normal deduction. If not, at least the sum of the lengths of maximal cut segments in the new deduction δ^* is less than that in δ, i.e., $r^*(\delta^*) < r^*(\delta)$, and the inductive hypothesis applies.

Now suppose that $d(\delta) > 0$. Among the cuts of maximal degree $d(\delta)$, pick a highest one.

We consider the possible cases, concentrating on the sub-deduction of δ that ends in the conclusion of the last E-rule of the highest cut segment considered.

In each case, we replace this sub-deduction with another, and show that the deduction δ^* resulting from δ in this way is simpler, i.e., either $d(\delta^*) < d(\delta)$ or $d(\delta^*) = d(\delta)$ but $r^*(\delta^*) < r^*(\delta)$. In carrying out this replacement, the cut segment A is replaced either with a shorter one consisting also of A, or with possibly longer cut segments, which however consist of a sub-formula of A, and hence of a formula of lower degree. We call this transformation a *conversion*. In the process, we might introduce cut segments unrelated to A, and we will verify that these are all also of lower degree.

In some cases the conversion is very simple, namely if the last E-rule in the cut segment is a $\vee$E or $\exists$E inference in which in one of the (or the, respectively) sub-deductions ending in the minor premise(s) the indicated assumption is not in fact discharged, (See Section 3.2.3 for an example.) If the last inference is $\vee$E (and the left disjunct is not discharged in the sub-deduction leading to the left minor premise), the conversion is this:

$$
\begin{array}{ccccc}
& & & C^{\,i} & \\
\vdots\,\delta_1 & \vdots\,\delta_2 & \vdots\,\delta_3 & & \Rightarrow \qquad \vdots\,\delta_2 \\
{}_i\dfrac{B \vee C \quad D \quad D}{D}\;\vee\text{E} & & & & D
\end{array}
$$

If instead the right premise is not discharged in the deduction ending in the right minor premise, we instead replace with δ_3, of course.

If the last inference is $\exists$E, the conversion is:

$$
\begin{array}{ccc}
\vdots\,\delta_1 & \vdots\,\delta_2 & \Rightarrow \qquad \vdots\,\delta_2 \\
\dfrac{\exists x\,B(x) \quad D}{D}\;\exists\text{E} & & D
\end{array}
$$

In both cases, the resulting deduction clearly contains no new cut segments, so the inductive hypothesis applies.

The preceding conversions are often called *simplification conversions*.

Now consider the remaining cases, and *first assume that the length of the cut segment is 1*.

In this case, the cut segment consists of a single formula occurrence A which is either the conclusion of an I-rule or of $\perp_J$, and at the same time the major premise of an E-rule. (Recall that we have already dealt with the case where the degree of A is 0, so A cannot be $\perp$, and the cut segment cannot end in $\perp_J$.) We further distinguish cases according to the main operator of A. In each case, we replace the deduction ending in the conclusion of the E-rule with another deduction in which the cut segment A is replaced by one, or possibly multiple, cut segments of lower degree. Since these conversions directly remove a detour, we call them collectively *reductions* or *detour conversions*. We will have to verify that any new cut segments introduced are of lower degree than A.

If *A is the conclusion of $\perp_J$ and the major premise of an E-rule*, the detour conversions have the following forms, according to the form of *A*:

1. *A* is $B \land C$:

$$
\cfrac{\cfrac{\cfrac{\vdots\ \delta_1}{\perp}}{B \land C}\ \perp_J}{B}\ \land\text{E}
\qquad \Rightarrow \qquad
\cfrac{\cfrac{\vdots\ \delta_1}{\perp}}{B}\ \perp_J
$$

Similarly if the conclusion of $\land$E is *C*; then we conclude *C* by $\perp_J$.

2. *A* is $B \supset C$:

$$
\cfrac{\cfrac{\cfrac{\vdots\ \delta_1}{\perp}}{B \supset C}\ \perp_J \qquad \cfrac{\vdots\ \delta_2}{B}}{C}\ \supset\text{E}
\qquad \Rightarrow \qquad
\cfrac{\cfrac{\vdots\ \delta_1}{\perp}}{C}\ \perp_J
$$

3. *A* is $\forall x\, B(x)$:

$$
\cfrac{\cfrac{\cfrac{\vdots\ \delta_1}{\perp}}{\forall x\, B(x)}\ \perp_J}{B(t)}\ \forall\text{E}
\qquad \Rightarrow \qquad
\cfrac{\cfrac{\vdots\ \delta_1}{\perp}}{B(t)}\ \perp_J
$$

In these three cases, no new cut segments can be introduced, except perhaps a cut segment consisting of a sub-formula of *A*, which is now the conclusion of $\perp_J$. Any such cut segment is of lower degree than *A*, so does not affect the degree or rank of the new deduction.

4. *A* is $B \lor C$:

$$
\cfrac{\cfrac{\cfrac{\vdots\ \delta_1}{\perp}}{{}_i\ B \lor C}\ \perp_J \qquad \cfrac{B^i\ \ \vdots\ \delta_2}{D} \qquad \cfrac{C^i\ \ \vdots\ \delta_2}{D}}{D}\ \lor\text{E}
\qquad \Rightarrow \qquad
\left.\begin{array}{c}\vdots\ \delta_1 \\[2pt] \cfrac{\perp}{B}\ \perp_J \\[4pt] \vdots\ \delta_2 \\[2pt] D \end{array}\right\}\ \delta_2[\delta_1'/B^i]
$$

5. A is $\exists x\, B(x)$:

$$
\cfrac{\cfrac{\vdots\ \delta_1}{\cfrac{\bot}{\exists x\, B(x)}\ \bot_J}\qquad \cfrac{\overset{\displaystyle B(a)^i}{\vdots\ \delta_2}}{D}}{C}\ \exists_E
\qquad\Rightarrow\qquad
\left.\cfrac{\cfrac{\cfrac{\vdots\ \delta_1}{\bot}}{B(a)}\ \bot_J}{\cfrac{\vdots\ \delta_2}{D}}\right\}\delta_2[\delta_1'/B(a)^i]
$$

(with i labeling the discharge at $\exists x\, B(x)$)

In the last two cases, δ_1' is, respectively,

$$
\cfrac{\cfrac{\vdots\ \delta_1}{\bot}}{B}\ \bot_J
\qquad\text{or}\qquad
\cfrac{\cfrac{\vdots\ \delta_1}{\bot}}{B(a)}\ \bot_J
$$

We have assumed that all $\forall_I$ and $\exists_E$ inferences use distinct eigenvariables, and that these eigenvariables only occur above the respective inferences. That means that any free variables in open assumptions of δ_1 are not used as eigenvariables in δ_2, and so no eigenvariable conditions of $\forall_I$ or $\exists_E$ inferences in δ_2 are violated by replacing open assumptions of δ_2 by the deduction δ_1.

Here is the first time we make use of the fact that the cut segment considered is not just of maximal degree, but a highest cut segment. Since it satisfies condition 2 of Definition 4.27, the sub-deduction δ_1 ending in the premise of $\bot_J$ cannot contain a cut segment of maximal degree. This is important: the assumptions discharged by the $\forall_E$ or $\exists_E$ rule have been replaced in δ^* by sub-deductions δ_1', which includes δ_1 as a sub-deduction. Since these assumptions may have more than one occurrence, any cut segment contained in δ_1 can have multiple copies in δ^*. If such a cut segment had maximal degree, we would have potentially increased the rank. The new $\bot_J$ inferences in δ_1' are potential beginnings of new cut segments. These, however, are cut segments involving the formula B, C, or $B(a)$. Since these are sub-formulas of A, any new cut segments have lower degree than $d(\delta)$.

We list the remaining detour conversions, when A is the conclusion of an I-rule, for all possible cases of A:

1. A is $B \wedge C$:

$$
\cfrac{\cfrac{\cfrac{\vdots\ \delta_1}{B}\qquad\cfrac{\vdots\ \delta_2}{C}}{B \wedge C}\ \wedge_I}{B}\ \wedge_E
\qquad\Rightarrow\qquad
\cfrac{\vdots\ \delta_1}{B}
$$

Similarly if the conclusion of $\wedge_E$ is C; then we replace with δ_2.

2. A is $B \supset C$:

$$
\begin{array}{c}
B^{\,i} \\
\vdots\ \delta_1 \\
\vdots \\
C \\
\hline
i\ \overline{B \supset C}\ \ \supset\!\text{\scriptsize I} \quad \vdots\ \delta_2 \\
\underline{\hspace{2cm}}\ \ B \\
\qquad C
\end{array}
\ \ \ \supset\!\text{\scriptsize E}
\qquad\Rightarrow\qquad
\left.
\begin{array}{c}
\vdots\ \delta_2 \\
\vdots \\
B \\
\vdots\ \delta_1 \\
\vdots \\
C
\end{array}
\right\}\delta_1[\delta_2/B^{\,i}]
$$

3. A is $\forall x\, B(x)$:

$$
\begin{array}{c}
\Gamma \\
\vdots\ \delta_1 \\
\vdots \\
\underline{B(c)}\ \ \forall\text{\scriptsize I} \\
\underline{\forall x\, B(x)}\ \ \forall\text{\scriptsize E} \\
B(t)
\end{array}
\qquad\Rightarrow\qquad
\begin{array}{c}
\Gamma \\
\vdots\ \delta_1[t/c] \\
\vdots \\
B(t)
\end{array}
$$

As you see, the cases where A is the conclusion of a $\wedge$I, $\supset$I, or $\forall$I rule (and hence A is of the form $B \wedge C$, $B \supset C$, or $\forall x\, B(x)$, respectively) are treated the same as in the proof of Theorem 4.10, when we treated normalization of deductions involving only the rules for $\wedge$, $\supset$, $\neg$, and $\forall$. (The case of $\neg$ will be Problem 4.30.) The assumption about eigenvariables again guarantees that the resulting deduction in case (2) is correct: no free variable occurring in an open assumption of δ_2 can be an eigenvariable of an $\exists$E-inference in δ_1. In case (3), we make use of Theorem 3.20.

We have to verify that we have not introduced any new cuts of maximal degree. How could new cuts have been introduced? One way is that the sub-deduction has been replaced by a deduction which ends in an I- or $\perp_J$-inference. This can happen if δ_1 (or δ_2) ends in such an inference. For instance, in case (1), if δ_1 ends in an I-inference concluding B, the new deduction δ_1 might now contain a cut segment beginning with the indicated occurrence of B. However, all these potential new cut segments involve sub-formulas of A (B, C, and $B(t)$, respectively). Thus they are not of maximal degree.

In case (2), another other kind of potential new cut might appear: In δ_1, B is an assumption which is replaced, in δ^*, by a copy of the deduction δ_2 of B. If δ_2 ends in an I- or $\perp_J$ inference, the (possibly multiple occurrences) of B may now be parts of cut segments. These, however, are also of smaller degree than the original cut segment, since B is a sub-formula of A.

In this last case, the assumption B occurs perhaps multiple times, so the deduction δ_2 has perhaps been multiplied. However, the cut segment we consider was highest. Condition 3 of Definition 4.27 guarantees that the sub-deduction δ_2 contains no cut segment of maximal degree, so no new such cut segments have been introduced.

4. A is $B \vee C$:

$$
\begin{array}{ccc}
\begin{array}{c}
\vdots \; \delta_1 \\
\dfrac{\dfrac{B}{B \vee C} \; {\rm V_I}}{i} \quad
\end{array}
&
\begin{array}{cc}
B^{\,i} & C^{\,i} \\
\vdots \; \delta_2 & \vdots \; \delta_3 \\
D & D
\end{array}
\\
\dfrac{}{D} \; {\rm V_E}
\end{array}
\quad\Rightarrow\quad
\left.
\begin{array}{c}
\vdots \; \delta_1 \\
B \\
\vdots \; \delta_2 \\
D
\end{array}
\right\} \delta_2[\delta_1/B^i]
$$

Similarly if the premise of $\rm V_I$ is C, then we combine δ_1 with $\delta_3[\delta_1/C^i]$.

If A is the conclusion of $\rm V_I$, A has the form $B \vee C$. The cuts in δ_2 (or δ_3) remain as they are. Any cut segment in δ_1 is now potentially multiplied. However, since we selected a highest cut segment to reduce, the cut segments in δ_1 are of degree $< d(\delta)$. Potential new cuts are cuts on B, so also have lower degree than $d(\delta)$.

5. A is $\exists x\, B(x)$:

$$
\begin{array}{cc}
\begin{array}{c}
\vdots \; \delta_1 \\
\dfrac{\dfrac{B(t)}{\exists x\, B(x)} \; \exists_I}{i}
\end{array}
&
\begin{array}{c}
B(c)^{\,i} \\
\vdots \; \delta_2 \\
D
\end{array}
\\
\dfrac{}{D} \; \exists_E
\end{array}
\quad\Rightarrow\quad
\left.
\begin{array}{c}
\vdots \; \delta_1 \\
B(t) \\
\vdots \; \delta_2[t/c] \\
D
\end{array}
\right\} (\delta_2[t/c])[\delta_1/B(t)^i]
$$

Here, A has the form $\exists x\, B(x)$. The new deduction is obtained by applying Theorem 3.20, i.e., substituting t for c in δ_2. This yields a deduction of D from $B(t)$ instead of $B(c)$. The assumptions $B(t)$ which correspond to the assumptions $B(c)$ which were discharged in δ_2 are then replaced by (copies of) the deduction δ_1 of $B(t)$. Since the free variables in open assumptions of δ_1 do not occur in δ_2, no eigenvariable conditions are violated in the new deduction.

The cut segments in δ_2 remain unchanged. Since the reduced cut is highest, all cut segments in δ_1 are of degree $< d(\delta)$. Potential new cuts segments consist of occurrences of $B(t)$, so have lower degree than $d(\delta)$. The sum of lengths of cuts of degree $d(\delta)$ is reduced by 1, since we have removed one such cut segment of length 1.

Note that in the last two cases, there can be no new cut segments involving D: if D is part of a cut segment in δ^*, then D already belongs to a cut segment in δ, and the conversion has merely reduced the length of this cut segment.

In every case considered above, we have verified that no new cut segments of degree $d(\delta)$ or higher have been introduced in δ^*. The cut segment of length 1 on A in δ has been removed. If it was the only cut segment of degree $d(\delta)$, then $d(\delta^*) < d(\delta)$. Otherwise, we have decreased the collective length of cut segments of degree $d(\delta)$ by at least 1, i.e., $r^*(\delta^*) < r^*(\delta)$. Thus, the inductive hypothesis applies.

Now consider the case where *the length of the cut segment is* > 1. Then the cut segment ends in an E-rule, where the major premise of the E-rule is the cut-formula A. Because the cut segment has length > 1, that occurrence of A is the conclusion of an $\lor$E or of an $\exists$E inference in which A is the minor premise. In these cases, we move the E-rule upward to apply directly to the minor premise of $\lor$E or $\exists$E. These *permutation conversions* take the following forms:

$$
\cfrac{{}_i\,D \lor E \quad \cfrac{\begin{array}{c}D^i \\ \vdots\;\delta_2 \\ A\end{array} \quad \begin{array}{c}E^i \\ \vdots\;\delta_3 \\ A\end{array}}{A}\;{\scriptstyle \lor E}}{C}\;{\star\text{E} \quad \cdots}
\quad \Rightarrow \quad
\cfrac{{}_i\,D \lor E \quad \cfrac{\begin{array}{c}D^i \\ \vdots\;\delta_2 \\ A\end{array}\;\cdots}{C}\;{\star\text{E}} \quad \cfrac{\begin{array}{c}E^i \\ \vdots\;\delta_3 \\ A\end{array}\;\cdots}{C}\;{\star\text{E}}}{C}\;{\scriptstyle \lor E}
$$

Here, $\star$E is the E-rule that ends the cut segment, and $\cdots$ represents the sub-deductions ending in the minor premises (if any). As a special case, the cut segment, say, begins with $\supset$I and ends with $\supset$E. Then A is of the form $B \supset C$, and the sub-deduction ending in the conclusion of $\supset$E takes the form

$$
\cfrac{\cfrac{{}_i\,D \lor E \quad \begin{array}{c}D^i \\ \vdots\;\delta_2 \\ B \supset C\end{array} \quad \begin{array}{c}E^i \\ \vdots\;\delta_3 \\ B \supset C\end{array}}{B \supset C}\;{\scriptstyle \lor E} \quad \begin{array}{c} \vdots\;\delta_4 \\ B\end{array}}{C}\;{\scriptstyle \supset E}
$$

The conversion replaces this deduction with

$$
\cfrac{{}_i\,D \lor E \quad \cfrac{\cfrac{\begin{array}{c}D^i \\ \vdots\;\delta_2 \\ B \supset C\end{array} \quad \begin{array}{c} \vdots\;\delta_4 \\ B\end{array}}{C}\;{\scriptstyle \supset E}}{} \quad \cfrac{\cfrac{\begin{array}{c}E^i \\ \vdots\;\delta_3 \\ B \supset C\end{array} \quad \begin{array}{c} \vdots\;\delta_4 \\ B\end{array}}{C}\;{\scriptstyle \supset E}}{}}{C}\;{\scriptstyle \lor E}
$$

Any cut segment that ends with $\star$E has now been reduced in length by 1. All other cut segments remain the same. Since we have picked a highest cut, no cut segment (partly) contained in the deductions leading to the minor premise(s) of the $\star$E-inference (i.e., in δ_4 in the example) is of degree $d(\delta)$. (Had this not been the case, e.g., had δ_4 contained a cut segment of degree $d(\delta)$, then $r^*(\delta^*)$ may well have been increased, since that cut segment now occurs in two copies of δ_4 and so would count double!) Thus, the length of at least one cut segment of degree $d(\delta)$ has been reduced, and no new cut segments of degree $d(\delta)$ or higher have been introduced. Hence, $d(\delta^*) = d(\delta)$ but $r^*(\delta^*) < r^*(\delta)$ and the inductive hypothesis applies.

In the other case, when the cut segment ends in $\exists$E, the conversion is:

$$
\begin{array}{c}
\quad\quad D(c)^i \\
\vdots\, \delta_1 \quad \vdots\, \delta_2 \\
i\, \dfrac{\exists x\, D(x) \quad A}{\dfrac{A}{C} \,\star\mathrm{E}} \exists\mathrm{E} \quad \cdots
\end{array}
\quad \Rightarrow \quad
\begin{array}{c}
\quad\quad D(c)^i \\
\quad\quad \vdots\, \delta_2 \\
\vdots\, \delta_1 \\
i\, \dfrac{\exists x\, D(x) \quad \dfrac{A \quad \cdots}{C} \,\star\mathrm{E}}{C} \exists\mathrm{E}
\end{array}
$$

Here the verification that $d(\delta^*) = d(\delta)$ and $r^*(\delta^*) < r^*(\delta)$ is even simpler, since the cut segments in the sub-deductions leading to the minor premise of the $\star$E-rule are not multiplied. We should note, however, that it is essential here that eigenvariables in δ must be distinct and only occur above the corresponding eigenvariable inference ($\exists$E in this case). If they were not, then the eigenvariable c of the $\exists$E inference might also occur in the open assumptions of the sub-deductions indicated by $\cdots$, and then the eigenvariable condition of $\exists$E in δ^* would be violated.   □

Problem 4.30. Give the detour conversion for $\perp_J$ or $\neg$I followed by $\neg$E.

Problem 4.31. Write down the permutation conversions for the cases where the cut segment ends in $\vee$E followed by $\vee$E.

We mentioned at the end of Section 4.1 that the normalization theorem can be strengthend to a strong normalization theorem. In our proof, we applied (in a particular order) transformations of sub-deductions of our deduction which either simplified the deduction, shortened cut segments, or removed cut segments of length 1 (directly removed detours). These so-called simplification, permutation, and detour conversions (or contractions) are listed in Appendix E.2. Now one may ask: is the order we gave (based on picking a highest cut segment of maximal degree) the only possible one? One might think not any order will result in a normal deduction (e.g., we relied on the fact that we were picking highest cut segments to ensure that no new cut segments of maximal degree were being introduced). Call a sequence of deductions $\delta_1, \delta_2, \ldots$, in which δ_{i+1} results from δ_i by applying one of the conversions in Appendix E.2, a *conversion sequence*. Of course, if no conversion can be applied to a deduction, the deduction is normal. We have the following strong normalization theorem that shows that any sequence of conversion steps eventually results in a normal deduction.

Theorem 4.32 (Prawitz, 1971). *There are no infinite conversion sequences; every conversion sequence ends in a normal deduction.*

The proof, as mentioned, goes beyond the scope of this book.

4.7   An example

Suppose you're asked to show that $A \supset C$ can be deduced from $(A \supset C) \vee (B \supset C)$ and $A \supset B$ in **NJ**. Consider the first assumption, which is a disjunction. You notice that from the first disjunct, you easily get $(A \supset B) \supset (A \supset C)$, and from the second disjunct you get the same formula using a previously proved formula of the form $(B \supset C) \supset ((A \supset B) \supset (A \supset C))$. Once you have $(A \supset B) \supset (A \supset C)$, you can get $A \supset C$ by $\supset$E from $A \supset B$. So you set up the following deduction:

$$
\cfrac{
(A \supset C) \vee (B \supset C) \quad
\cfrac{A \supset C^1}{(A \supset B) \supset (A \supset C)}\;{\supset}\text{\scriptsize I} \quad
\cfrac{(B \supset C) \supset ((A \supset B) \supset (A \supset C)) \quad B \supset C^1}{(A \supset B) \supset (A \supset C)}\;{\supset}\text{\scriptsize E}
}{
\cfrac{\dfrac{(A \supset B) \supset (A \supset C)}{A \supset C}\;\vee\text{\scriptsize E} \qquad A \supset B}{A \supset C}\;{\supset}\text{\scriptsize E}
}\;{}_1
$$

with δ_1 above $(B \supset C) \supset$.

The deduction δ_1 of the required formula is:

$$
\cfrac{
B \supset C^2 \quad
\cfrac{
\dfrac{A \supset B^3 \quad A^4}{B}\;{\supset}\text{\scriptsize E}
}{
\cfrac{\dfrac{C}{A \supset C}\;{}_4\,{\supset}\text{\scriptsize I}}{(A \supset B) \supset (A \supset C)}\;{}_3\,{\supset}\text{\scriptsize I}
}\;{\supset}\text{\scriptsize E}
}{
(B \supset C) \supset ((A \supset B) \supset (A \supset C))
}\;{}_2\,{\supset}\text{\scriptsize I}
$$

If we abbreviate $(A \supset B) \supset (A \supset C)$ by D, the entire deduction looks like this:

$$
\cfrac{
(A \supset C) \vee (B \supset C) \quad
\cfrac{A \supset C^1}{\boxed{D}}\;{\supset}\text{\scriptsize I} \quad
\cfrac{
\cfrac{\dfrac{A \supset B^3 \quad A^4}{B}\;{\supset}\text{\scriptsize E} \;\; B \supset C^2}{\cfrac{\dfrac{C}{A \supset C}\;{}_4\,{\supset}\text{\scriptsize I}}{D}\;{}_3\,{\supset}\text{\scriptsize I}}
}{\boxed{(B \supset C) \supset D}}\;{}_2\,{\supset}\text{\scriptsize I} \quad B \supset C^1
}{
\cfrac{\boxed{D}}{A \supset C}
}
$$

The deduction δ contains two cut segments: one of length 2 with cut-formula D (of degree 3), and one of length 1 with cut-formula $(B \supset C) \supset D$ (of degree 5). (Note that the D in the right minor premise of $\vee$E is not part of a cut segment, since it is not the conclusion of an I-rule.) Hence, $d(\delta) = 5$ and $r^*(\delta) = 1$. We pick a highest cut segment of maximal degree to reduce first: that's the cut on $(B \supset C) \supset D$. To reduce it, we replace the sub-deduction ending in D (on the left below) with the one on the right:

$$
\cfrac{
\cfrac{
B \supset C^2 \quad \cfrac{A \supset B^3 \quad A^4}{B}\supset_E
}{
\cfrac{\cfrac{\cfrac{C}{A \supset C}\supset_I}{D}\supset_I}{(B \supset C) \supset D}\supset_I
}\quad B \supset C^1
}{D}\supset_E
\qquad \Rightarrow \qquad
\cfrac{B \supset C^1 \quad \cfrac{A \supset B^3 \quad A^4}{B}\supset_E}{\cfrac{\cfrac{C}{A \supset C}\supset_I}{D}\supset_I}\supset_E
$$

We've replaced the assumption discharged in the last $\supset_I$, namely $B \supset C^2$, by the sub-deduction ending in the minor premise of the $\supset_E$ rule, namely $B \supset C^1$, and removed the last two $\supset_I$ and $\supset_E$ inferences.

The entire deduction turns into the deduction δ^*:

The new deduction has the same cut segment with cut-formula D, but now has another one, going through the right minor premise of the $\vee_E$ inference, also of length 2. (The right minor premise of $\vee_E$ is now the conclusion of an $\supset_I$ rule, whereas it was not in the original deduction.) So $d(\delta^*) = 3$ and $r^*(\delta^*) = 4$ since we have 2 cut segments of length 2 (which happen to share a formula occurrence).

Both cut segments are of maximal degree, there are no cut segments above them, and also no cuts above the minor premise $A \supset B$ of the $\supset_E$ rule. So both cut segments are "highest" in the sense of Definition 4.27. Since both cut segments involve the same $\vee_E$ inference, the result of applying the permutation conversion is the same. We transform the deduction into δ^{**} where the $\supset E$ rule is applied to the minor premises of the $\vee_E$ rule instead of to its conclusion:

We still have two cut segments of degree 3, but each one has length 1, so $r^*(\delta^{**}) = 2 < r^*(\delta^*)$. Both of them count as highest cut segments. The first reduction conversion applied to the left cut segment yields δ^{***} with $r^*(\delta^{***}) = 1$:

$$\cfrac{(A \supset C) \vee (B \supset C) \qquad \cfrac{B \supset C^1 \qquad \cfrac{4 \ \cfrac{\cfrac{A \supset B^3 \quad A^4}{B} \supset_E}{C} \supset_I}{3 \ \cfrac{A \supset C}{\boxed{D}} \supset_I}}{A \supset C^1} \qquad \cfrac{\qquad\qquad A \supset B}{A \supset C} \supset_E}{A \supset C}_{\ VE} \ _1$$

We reduce the remaining cut segment and obtain a normal deduction:

$$\cfrac{(A \supset C) \vee (B \supset C) \quad A \supset C^1 \quad \cfrac{B \supset C^1 \quad \cfrac{4 \ \cfrac{\cfrac{A \supset B \quad A^4}{B} \supset_E}{C} \supset_I}{A \supset C}}{}}{A \supset C} \ _1 \quad {}_{VE}$$

4.8   The sub-formula property for **NJ**

A normal deduction is one that does not contain any cut segments: this is our definition of a normal deduction. But this definition is merely negative: it tells us what a normal deduction *cannot* be like. What does this tell us about what normal deductions *do* look like? We must answer this question in order to prove that normal deductions have the *sub-formula property*: a deduction has the sub-formula property if every formula occurrence in it is a sub-formula of the end-formula or else a sub-formula of an open assumption.

We begin by considering the form of branches of a normal deduction δ. Recall that a branch of δ is a sequence of formula occurrences $A_1, \ldots, A_n$ where A_1 is an assumption (open or discharged in δ), each A_i is a premise of an inference of which A_{i+1} is the conclusion, and A_n is the end-formula of δ. In other words, we obtain a branch by beginning at an assumption of δ and proceeding downwards from premises to conclusions until we arrive at the end-formula. Obviously, every formula in δ is on at least one branch but may be on more than one branch, and the end-formula is on every branch.

A cut segment is a part of every branch that passes through the first formula occurrence in the segment. If there are no cut segments in the deduction, it tells us something about branches $A_1, \ldots, A_n$: beginning with A_1, as long as we don't encounter a minor premise of an E-rule, we pass through inferences in the following order: first the formulas in the branch are major premises of E-rules, then a premise of $\perp_J$, then premises of I-rules, until we arrive either at the end-formula or the minor premise of an E-rule. Each one of these parts may be empty, but if they are not, they occur in the preceding order.

To see this, consider an initial segment $A_1, \ldots, A_k$ of a branch, where A_k is either the end-formula or the minor premise of an $\supset_E$ or $\neg_E$-rule, and no A_j with $j < k$ is a minor premise of an $\supset_E$ or $\neg_E$-rule.

1. No major premise of an E-rule may appear following $\perp_J$ (otherwise we'd have a cut segment of length 1).

2. No major premise of an E-rule may follow after a premise of an I-rule (that would also be a cut segment of length 1).

3. No $\perp_J$ can follow an I-rule (the conclusion of an I-rule is never $\perp$).

4. No $\perp_J$ can follow a $\perp_J$ (otherwise we'd have a cut segment of length 1).

So, if A_i is the premise of an I-rule, no $\perp_J$ or major premises of E-rules may follow below it (by (2) and (3)). If A_i is the premise of $\perp_J$, no major premise of E-rules nor another $\perp_J$ inference may follow below it.

Let us call a part of a branch just considered a *thread* and record what we've just discovered:

Definition 4.33. A *thread* in a deduction is an initial segment $A_1, \ldots, A_k$ of a branch where:

1. A_1 is an assumption,

2. A_i ($1 \le i < k$) is a premise of $\perp_J$, an I-inference, or a major premise of an E-inference, and A_{i+1} is the conclusion of the same inference.

3. A_k is either the end-formula of the deduction, or a minor premise of an E-rule.

(This is a generalization of the definition of 'thread' given in Definition 4.15 for the case where the deduction can also contain $\perp_J$, VE and ∃E rules.)

Proposition 4.34. *If $A_1, \ldots, A_k$ is a thread in a normal deduction, then there is an m, $1 \le m \le k$, such that:*

1. *each A_i with $i < m$ is the major premise of an E-inference;*

2. *A_m is the premise of $\perp_J$ or of an I- inference, unless $m = k$;*

3. *each A_i with $m < i < k$ is the premise of an I-inference.*

Consider the normal deduction from our last example. We'll add another ⊃I rule at the bottom so that the features of the proof of the sub-formula property come out more clearly.

$$
\cfrac{(A \supset C) \vee (B \supset C)_a \quad \cfrac{A \supset C_d^1}{} \quad \cfrac{B \supset C_e^1 \quad \cfrac{\cfrac{A \supset B_h^5 \quad A_j^4}{B_i} \supset E}{C_f} \supset E}{4 \; \cfrac{}{A \supset C_g} \supset I}}{\cfrac{A \supset C_b}{5 \; \cfrac{}{(A \supset B) \supset (A \supset C)_c} \supset I}} \; VE
\quad 1
$$

The threads are: abc, d, efg, hi, and j. The A_m's of Proposition 4.34 are b, d, f, h, and j, respectively.

The purpose of considering threads is to derive the sub-formula property from facts about the structure of threads. Previously, when we considered the sub-formula property for normal deductions without ∨ and ∃ rules, the sub-formula property followed, more or less, just from the fact that threads can be divided into an E-part and an I-part, and then the minimum formula is a sub-formula of either an assumption (since the E-rules pass from major premises to conclusions which are sub-formulas of these premises), or of the end-formula (since I-rules pass from premises to conclusions of which the premises are sub-formulas). But in the presence of ∨E and ∃E this is no longer true: the conclusion of these rules is not, in general, a sub-formula of the major premise. Hence, in the case of full **NJ**, we have to consider a different, more complicated notion. However, the conclusion of an ∨E or ∃E inference *is* a sub-formula of its minor premise (they are identical formulas), so the sub-formula property is only violated when going from minor premise to conclusion in an ⊃E or ¬E inference. So, we'll have to define something analogous to threads that traces the sub-formula relationships in a deduction. That is the notion of a *path*:

Definition 4.35. A *path* in a deduction is a sequence of formula occurrences A_1, ..., A_k where:

1. A_1 is either an open assumption or an assumption discharged by ⊃I or ¬I,

2. A_k is either the end-formula of the deduction or a minor premise of an ⊃E or ¬E rule,

3. if A_i is the major premise of an ∨E or ∃E inference, A_{i+1} is an assumption discharged by that inference,

4. if A_i is

   (a) a premise of ⊥$_I$,

   (b) a premise of an I-inference,

   (c) a major premise of an E-inference other than ∨E or ∃E,

   (d) a minor premise of ∨E or ∃E,

   then A_{i+1} is the conclusion of the same inference.

In our example, the paths are $adbc, aefgbc, hi$, and j. Only hi and j start in an assumption discharged by ⊃I, the others start with open assumptions. The first two paths end in the end-formula, the latter two in the minor premise of an ⊃E rule. We see that unlike threads, a path does not just go from premises to conclusions, but also may pass from a major premise of ∨E or ∃E to an assumption above a minor premise of that inference.

Along a path we may encounter sequences of formula occurrences in which one is the minor premise of a ∨E or ∃E inference and the following formula occurrence is its conclusion (hence, they are occurrences of the same formula). Let us call such a sequence a *segment* of the path. If we count a formula which is not both the conclusion of ∨E or ∃E and a minor premise of ∨E or ∃E as a segment of length 1, any path completely divides into segments. Just like we were able to divide threads into an E-part followed by an I-part, with a minimum formula in the middle, we can divide a path into an E-part and an I-part, with a minimum segment in the middle:

Proposition 4.36. *Any path $A_1, \ldots, A_k$ in a deduction can be divided into segments A_1^i, $\ldots, A_{n_i}^i$, i.e.,*

$$A_1 = A_1^1, \ldots, A_{n_1}^1, \ldots, A_1^\ell, \ldots, A_{n_\ell}^\ell = A_k.$$

If the deduction is normal then there is an m, $1 \leq m \leq \ell$, such that,

1. *each $A_{n_i}^i$ with $i < m$ is the major premise of an E-rule with A_1^{i+1} its conclusion or, in the case of ∨E and ∃E, an assumption discharged by the inference;*

2. *$A_{n_m}^m$ is the premise of $\bot_J$, or of an I-rule with A_1^{m+1} its conclusion, or is A_k;*

3. *each $A_{n_i}^i$ with $m < i < k$ is the premise of an I-rule with A_1^{i+1} its conclusion.*

$A_1^m, \ldots, A_{n_m}^m$ is called the minimum segment *of the path. All formula occurrences in the minimum segment are occurrences of the same formula, the* minimum formula.

Proof. The argument proceeds just as the argument for threads above: if it were not the case, then the deduction would not be normal. The only difference is that now we also consider the possibility that there is a cut segment $A_1^i, \ldots, A_{n_i}^i$ on the path, not just a cut-formula.

The proposition claims that as we follow a path, we encounter segments in the following order:

1. segments where the last formula occurrence is a major premise of an E-rule;

2. a segment the last formula of which is the premise of $\bot_J$;

3. segments where the last formula is a premise of an I-rule.

If this is the case, then the segment that ends in a premise of $\bot_J$ or, if there is no such segment, the first segment that ends in a premise of an I-inference, is the minimum segment.

So suppose this were false: what are the possibilities?

1. We have a segment that ends in the premise of an I-rule followed by a segment that ends in the major premise of an E-rule: then the second segment would be $A_1^i, \ldots, A_{n_i}^i$ where A_1^i is the conclusion of an I-inference and $A_{n_i}^i$ is the major premise of an E-inference.

2. We have a segment that ends in the premise of an I-rule followed by a segment that ends in the premise of $\bot_J$: then the second segment would be $A_1^i, \ldots, A_{n_i}^i$ where A_1^i is the conclusion of an I-rule and $A_{n_i}^i$ is the premise of $\bot_J$.

3. We have a segment that ends in the premise of an $\perp_J$ followed by a segment that also ends in the premise of $\perp_J$: then the second segment would be A_1^i, ..., $A_{n_i}^i$ where A_1^i is the conclusion of $\perp_J$ and $A_{n_i}^i$ is the premise of another $\perp_J$.

4. We have a segment that ends in the premise of an $\perp_J$ followed by a segment that ends in the major premise of an E-rule: then the second segment would be $A_1^i, \ldots, A_{n_i}^i$ where A_1^i is the conclusion of $\perp_J$ and $A_{n_i}^i$ is the major premise of an E-inference.

In each case we see that the second segment would satisfy the definition of a cut segment. But since the deduction is normal, it does not contain any cut segments.

That A_1^{i+1} is the conclusion of an I-inference with premise $A_{n_i}^i$, or the conclusion of an E-inference other than $\vee$E and $\exists$E with $A_{n_i}^i$ as major premise, or an assumption discharged by an $\vee$E or $\exists$E inference with $A_{n_i}^i$ as major premise follows simply from the definition of a path.   □

Consider our example deduction again. Since there is only one $\vee$E rule, the only segments of length > 1 are db and gb. The remaining formula occurrences are all segments of length 1.

The path $adbc$ divides into segments a, db, c and db is the minimum segment. The path $aefgbc$ divides into segments a, e, f, gb, c, and f is the minimum segment. The path hi divides into segments h and i, and i is the minimum segment. The path j only has one segment, j, which is also the minimum segment.

Proposition 4.37. *In a path $A_1, \ldots, A_k$ in a normal deduction, every formula A in it is either a sub-formula of A_1 or of A_k.*

Proof. If A occurs before the minimum segment, it is a sub-formula of A_1, since all inferences leading to A along the path either go from major premises of E-rules other than $\vee$E and $\exists$E to conclusions of those rules, or from major premises of $\vee$E or $\exists$E to assumptions discharged by them, or go from minor premises of $\vee$E or $\exists$E to their conclusions.

If A is the minimum formula or occurs after it, all inferences after A in the path are I-inferences, and so go from premises of I-rules to their conclusions, of which the premises are sub-formulas, or go from minor premises of $\vee$E or $\exists$E to their conclusions.   □

Let's see how this plays out in our example. In the first path, $adbc$, db is the minimum segment and $A \supset C$ is the minimum formula. The minimum formula is a sub-formula of the first and the last formula of the path, in this case, it is a sub-formula of both an open assumption and of the end-formula. In the path $aefgbc$, f is the minimum segment, and so C is the minimum formula. Formula f, i.e., C is a sub-formula of e (the assumption $B \supset C$) which in turn is a sub-formula of a, the assumption that starts the segment and the major premise of the $\vee$E inference that discharges e. Formula f is also a sub-formula of formulas g and b, i.e., $A \supset C$, which is in turn a sub-formula of the end-formula c. In path hi, i is a sub-formula of h which starts the path; and in j, of course, j is the assumption that starts the path.

We now show that every formula belongs to at least one path. The previous proposition then entails that every formula in a deduction is a sub-formula of the assumption that starts the path it is on, or of the final formula in the path.

Proposition 4.38. *Every formula in δ must be on at least one path.*

Proof. By induction on the size of δ.

Induction basis: If the size of δ is 0, δ is just an assumption by itself. This single formula is a path.

Inductive step: Suppose the size of δ is > 0. Then δ ends in an inference, and the sub-deductions ending in these premises are deductions of smaller size than δ. Hence, the inductive hypothesis applies to them. Each premise of that last inference lies on a path, and since it is the end-formula of the sub-deduction ending in the premise, it is the last formula on a path. Any path ending in a premise of $\perp_J$, an I-rule, the major premise of an $\supset$E or $\neg$E rule, or in a minor premise of a $\vee$E or $\exists$E rule, can be extended by its conclusion A to a path in δ which contains A. Every other path in the sub-deductions ending in a premise of the last inference is still a path in δ.   □

The proposition tells us that every formula in a deduction lies on a path and hence is a sub-formula of the assumption at the beginning of the path or of the final formula of the path. However, paths need not start in *open* assumptions and need not end in the *end-formula of the entire deduction*, so we don't have the full sub-formula property yet. (a) A path may start with an assumption discharged by $\supset$I or $\neg$I (this is the only case: assumptions discharged by $\vee$E or $\exists$E cannot be the first formula in a path). (b) A path may end in a minor premise of $\supset$E or $\neg$E. We have to show that these assumptions or final formulas of the path are sub-formulas of some open assumption or of the end-formula of the deduction. We do this by an inductive proof using an ordering of paths.

To deal with (a) we have to consider paths that begin in assumptions discharged by $\supset$I or $\neg$I. If the $\supset$I or $\neg$I inference happens to be located on the same path as A, then that inference must be in the I-part of the path. Formulas in the I-part of the path are sub-formulas of the final formula of the path. The assumption discharged by $\supset$I or $\neg$I is a sub-formula of the conclusion of the $\supset$I or $\neg$I inference. Thus, if a formula A on a path is a sub-formula of an assumption discharged by $\supset$I or $\neg$I on the same path, it is also a sub-formula of the final formula of the path. However, the $\supset$I or $\neg$I inference may not be on the same path. In this case we have to make sure that the $\supset$I or $\neg$I inference where the assumption is discharged lies on a path for which we can (inductively) assume that all its formulas are sub-formulas of open assumptions or of the end-formula. To deal with (b), we have to relate the final formula of the path, say, B, to the major premise of the $\supset$E or $\neg$E inference of which B is the minor premise. That major premise is $B \supset C$ or $\neg B$. Any path through it must continue through the conclusion of the $\supset$E or $\neg$E inference. So if we could assume that the sub-formula property holds for paths running through $B \supset C$ or $\neg B$, we would be done. This suggests that we should proceed by induction in a way that the inductive hypothesis guarantees the sub-formula property for major premises of $\supset$I and $\neg$I.

To summarize: we want to assign numbers to paths—call them the *order* of a path—such that the first formula (if it is discharged by a $\supset$I or $\neg$I not on the path) and the final formula (if it is the minor premise of $\supset$E or $\neg$E) of a path of order n are sub-formulas of a formula in a path of order $< n$.

First, note that if a path ends in a formula B, the entire path is contained within the sub-deduction that ends in B. B may be the end-formula, or it may be the minor premise of an $\supset$E or $\neg$E inference. If we consider the sub-deduction that ends in that minor premise, the path is also a path in that sub-deduction, and it ends in the end-formula of that sub-deduction.

This leads us to the following definition:

Definition 4.39. We assign an *order* to every path in δ inductively as follows:

1. *Base clause:* The order of a path that ends in the end-formula is 0.

2. *Inductive clause:* Now consider a path that does not end in the end-formula, but in a minor premise of $\supset$E or $\neg$E, and such that the major premise of the $\supset$E or $\neg$E lies on a branch to which an order o has been assigned. The order of the path that ends in the minor premise is $o + 1$.

This assigns an order to every path in δ. The paths that end in the end-formula have order 0. The paths that end in minor premises of $\supset$E or $\neg$E with major premises on paths through the end-formula have order 1. The paths that end in minor premises of $\supset$E or $\neg$E with major premise on a path of order 1 have order 2, etc.

Lemma 4.40. *If a formula A occurs on a path of order o in a normal deduction δ, it is either a sub-formula of the first formula of the path, of the end-formula, or of a formula on a path of order $o - 1$.*

Proof. If $o = 0$, the path ends in the end-formula, and the claim of the lemma follows by Proposition 4.37. If $o > 0$, then the path ends in a minor premise B of $\supset$E or $\neg$E. By the same proposition, A is a sub-formula of B or of the first formula in the path. If the former, the major premise is $B \supset C$ or $\neg B$, and A is a sub-formula of it. But the major premise occurs on a path of order $o - 1$, by the definition of order of paths. $\square$

In our example, the paths of order 0 are $adbc$ and $aefgbc$ since they both end in the end-formula c. As we've seen, all formulas on those paths are either sub-formulas of the end-formula c or of the assumption a. The remaining paths are hi and j. hi is contained in the sub-deduction ending in i, and j in the sub-deduction that is just j by itself; both of these end in minor premises of $\supset$E. In the sub-deduction consisting of h, i, and j, the path hi ends in the end-formula, so in that sub-deduction, it has order 0. This means it has order 1 in the entire deduction. The path j considered as a path in the deduction consisting only of j has order 0. Consequently, it has order 1 in the sub-deduction consisting of g, h, and i, which in turn means it has order 2 in the entire deduction. On the path hi, i is a sub-formula both of the first formula on the path, h, but also of e, which lies on a path of order 0—that's because i is the minor premise of an $\supset$E rule, and the major premise lies on the path $aefgbc$. On the path j, the formula j is a sub-formula both of the first formula of the path but also of the major premise of the $\supset$E rule that ends the path, namely of h—and h lies on the path hi of order 1.

Lemma 4.41. *If the first formula A_1 of a path of order o is discharged by $\supset$I or $\neg$I in δ, the conclusion of that $\supset$I or $\neg$I inference either lies on the same path, or on a path of order $< o$.*

Proof. Suppose the assumption formula in question is not discharged on the path. Then it must be discharged on a branch between the end-formula A_k of the path and the end-formula of δ, since an $\supset$I or $\neg$I inference can only discharge assumptions in the sub-deductions ending in the premise of that deduction. But all formulas between A_k and the end-formula of δ lie on paths of order $< o$.   □

Theorem 4.42. *If δ is a normal **NJ**-deduction of A from Γ, then every formula in δ is a sub-formula of A or of a formula in Γ.*

Proof. We proceed by induction on the order o of a path to which an occurrence of a formula B belongs.

Induction basis: If the order is 0, B belongs to a path that ends in the end-formula. So B is a sub-formula of the end-formula or, if not, a sub-formula of the first formula in the path. By Lemma 4.41, since there are no paths of order < 0, if the first formula of the path is discharged by $\supset$I or $\neg$I, the conclusion of that inference is on the same path. Since that conclusion appears in the I-part of the path, it is a sub-formula of the end-formula of the path, i.e., the end-formula of δ.

Inductive step: Suppose the order of the path is > 0. B is a sub-formula of the end-formula of the path or of its first formula, by Proposition 4.37. If the first formula is an assumption discharged on the path, B is a sub-formula of the end-formula of the path. That end-formula is a minor premise of an $\supset$E or $\neg$E inference, and thus a sub-formula of the major premise. The major premise lies on a path of order $< o$, so by inductive hypothesis it is a sub-formula either of A or of Γ. The remaining case is that B is a sub-formula of the assumption of δ that is the first formula in the path, and that assumption is discharged by an $\supset$I or $\neg$I inference not on the path. By Lemma 4.41, the conclusion of the $\supset$I or $\neg$I inference discharging it lies on a path of order $< o$, and the inductive hypothesis applies. So B is a sub-formula of A or of Γ.   □

Let's apply the considerations of the proof to our example from p. 146. So far we have figured out that all the formulas on paths of order 0 must be sub-formulas either of the end-formula c or of the (open) assumption a that starts those paths. We know that j (the sole formula on the path of order 2) is either a sub-formula of a path of order 1 or a sub-formula of an assumption (in fact, it is both). And we also know that every formula on the path of order 1, i.e., hi, is a sub-formula either of the assumption that starts that path or of a path of order 0: The formula i satisfies both, and h is a sub-formula of the assumption that starts that path. Both j and h are assumptions discharged by $\supset$I, in both cases these $\supset$I inferences lie outside the path that these assumptions start. In fact, they lie on the path $aefgbc$. Since they fall into the I-part of that path, the conclusions are sub-formulas of the last formula on that path—in this case, of the end-formula. Thus, the assumptions h and j are sub-formulas of the end-formula, and consequently all formulas on the E-part of the paths hi and j must be sub-formulas of the end-formula.

Corollary 4.43. *NJ is consistent.*

Proof. The argument is the same as for Corollary 4.24: Only $\bot$ itself is a sub-formula of $\bot$. But $\bot$ can't be the conclusion of an I-rule, and it can't be the major premise of an E-rule. It can be the conclusion of $\bot_J$, but then the premise of that is also $\bot$. Since $\bot_J$ does not discharge any assumptions, no deduction involving only $\bot_J$ can be a deduction from no assumptions. □

We also obtain a normalization theorem for **NM** in its full language:

Corollary 4.44. *Every deduction in NM can be transformed into a normal deduction.*

Proof. Every deduction in **NM** is also one in **NJ**, so the proof of Theorem 4.29 established that the deduction can be transformed into a normal deduction. By inspecting the individual steps it is clear that if no $\bot_J$ is present in the original deduction, no $\bot_J$ is introduced during the procedure. Thus the resulting normal deduction (and all intermediate deductions) are deductions in **NM**. □

Corollary 4.45. *NM is consistent.*

Corollary 4.46. *If there is an NM- or NJ-proof of $A \vee B$ then there is a proof either of A or of B (in NM or NJ, respectively).*

Proof. Suppose there is a deduction (and hence a normal deduction by Theorem 4.29 or Corollary 4.44) of $A \vee B$. By Proposition 4.36, a normal deduction of $A \vee B$ must end with either $\vee$I or $\bot_J$. A proof has no open assumptions, and $\bot_J$ discharges no assumptions. So the proof cannot end with $\bot_J$, as otherwise we would have a deduction of $\bot$ from no open assumptions, contradicting Corollary 4.43 and Corollary 4.45. So it must end with $\vee$I. The sub-deduction ending in the premise is a deduction either of A from no open assumptions, or of B. □

We are now in a position to prove some of the independence results from Section 2.11 proof-theoretically.

Corollary 4.47. *NJ does not prove $A \lor \neg A$ for all A.*

Proof. If it did, by Corollary 4.46, **NJ** would prove either A or $\neg A$. If A is atomic, this is impossible. Take a proof of p and replace p everywhere in it by $\bot$. The result is still a proof, but by Corollary 4.43, **NJ** does not prove $\bot$. If there were a normal proof of $\neg p$, it would have to end either in $\bot_J$ or in $\neg$I. In the first case, we'd have a proof of $\bot$, again contradicting Corollary 4.43. In the latter case, we would have a deduction δ of $\bot$ from open assumption p. Now replace p everywhere by $A \supset A$, and every assumption $A \supset A$ by a proof thereof. Again, this would give us a proof of $\bot$, contradicting Corollary 4.43. □

Of course, it follows that **NM** does not prove all instances of $A \lor \neg A$ either, since a proof in **NM** is also a proof in **NJ**.

Corollary 4.48. *In **NM** there cannot be a deduction ending in an atomic formula unless that formula is a sub-formula of an open assumption.*

Corollary 4.49. *NM does not prove $(A \land \neg A) \supset B$ for all A and B.*

Proof. Clearly **NM** proves $(A \land \neg A) \supset B$ if, and only if, **NM** proves B from open assumptions A and $\neg A$. Suppose we had a normal deduction of B from A and $\neg A$, with A and B atomic (and neither equal to $\bot$). Take any path ending in B in such a deduction. By Proposition 4.36, since **NM** does not contain $\bot_J$, B must be the minimum formula. But the only possible conclusion of an elimination rule applied to the assumptions A and $\neg A$ is $\bot$. □

4.9 Normalization for **NK**

Normalization in **NK** is harder to prove than in **NJ**, and it has historically taken quite a while to find a simple proof. There are a number of difficulties compared to normalization for **NJ**. The first one is apparent from a comparison of $\bot_K$ with $\bot_J$. An application of $\bot_J$ followed by an elimination rule is a cut: it generates a detour in the deduction, and a failure of the sub-formula property. Such a detour can be eliminated easily, e.g.:

$$\frac{\dfrac{\vdots}{\dfrac{\bot}{B \land C}\bot_J}}{B}\land\text{E} \qquad \Rightarrow \qquad \dfrac{\vdots}{\dfrac{\bot}{B}\bot_J}$$

The classical rule $\perp_K$ can generate the same kinds of detour, but it is not as easily eliminated. If we simply try what we did for $\perp_J$ we'd have:

$$\neg(B \wedge C)^i \qquad\qquad\qquad \neg(B \wedge C)^i$$

$$i \; \frac{\dfrac{\perp}{B \wedge C} \perp_K}{B} \wedge_E \qquad \Rightarrow \qquad *i \; \frac{\perp}{B} \perp_K$$

The problem is that on the right, the application of $\perp_K$ is incorrect because the conclusion does not match the discharged assumption anymore. To fix this, we also have to change the assumptions discharged, but of course the deduction below the original assumption $\neg(B \wedge C)$ relies for its correctness on the shape of that assumption—if we just replace it by $\neg B$, the next inference below it will generally become incorrect. So we have to replace the assumption $\neg(B \wedge C)$ by a deduction of it, a deduction in which instead $\neg B$ appears as an assumption which can be discharged by the $\perp_K$ rule. Statman (1974) proposed the following conversion:

$$\neg(B \wedge C)^i \qquad\qquad\qquad k \; \frac{\neg B^i \quad \dfrac{\dfrac{B \wedge C^k}{B} \wedge_E}{}}{\dfrac{\perp}{\neg(B \wedge C)} \neg_I} \neg_E$$

$$i \; \frac{\dfrac{\perp}{B \wedge C} \perp_K}{B} \wedge_E \qquad \Rightarrow \qquad\qquad i \; \frac{\perp}{B} \perp_K$$

The problem here is that although we have removed the cut on $B \wedge C$, we may have introduced a *new* cut: instead of being an assumption, $\neg(B \wedge C)$ now is the conclusion of $\neg_I$, and it may well be followed in the deduction δ by an $\neg_E$ rule. This potential new cut is of higher degree than the one removed. (The $\perp_K$ inference now concludes B, which if B is the major premise of an E-rule is now a new cut, but at least a cut of lower degree.) Similar, and worse, concerns arise for the other inferences. Consider the case where the cut is $B \vee C$, and $\perp_K$ is followed by $\vee_E$.

$$\neg(B \vee C)^i \quad B^j \quad C^j \qquad\qquad\qquad\qquad B^j \quad C^j$$

$$j \; \frac{\dfrac{i \; \dfrac{\perp}{B \vee C} \perp_K \quad D \quad D}{}}{D} \vee_E \qquad \Rightarrow \qquad k \; \frac{\neg D^i \quad \dfrac{j \dfrac{B \vee C^k \quad D \quad D}{D} \vee_E}{}}{\dfrac{\perp}{\neg(B \vee C)} \neg_I} \neg_E$$

$$i \; \frac{\perp}{D} \perp_K$$

The $\perp_K$ inference now concludes D, and if D is the major premise of an E-rule, it is a new cut. But while in the other cases this potential new cut is on a sub-formula of the original cut, and hence a cut of lower degree, D is not related to $B \vee C$ at all, and may have much higher degree than the original cut. The same issue arises with a cut being a major premise of $\exists$E.

To solve these issues, we follow the proof given by Andou (1995). First of all, observe that since our rules allow but do not require discharge, an application of $\perp_K$ in a deduction may not actually discharge an assumption. Such an application can obviously be replaced by $\perp_J$. From now on, we assume that any application of $\perp_K$ in fact discharges an assumption.

Problem 4.50. Show by induction that every deduction in **NK** can be transformed into one in which every application of $\perp_K$ discharges at least one assumption.

Furthermore, we can assume that δ has the following property: If an assumption $\neg A$ is discharged by an application of $\perp_K$ in δ (we'll call $\neg A$ a $\perp_K$-assumption in this case), then it is the major premise of an $\neg$E-inference. This can always be guaranteed by replacing a $\perp_K$-assumption $\neg A^i$ which is not a major premise of $\neg$E by the short deduction

$$
\cfrac{\cfrac{\neg A^{\,i} \quad A^{\,k}}{\perp}\neg\mathrm{E}}{k \; \cfrac{\perp}{\neg A}}\neg\mathrm{I}
$$

where k is a discharge label not already used in δ. From now on we assume that any $\perp_K$-assumption $\neg A$ is the major premise of $\neg$E.

With these conventions, we can now state the additional *classical conversions* necessary to remove detours where a formula A is the conclusion of $\perp_K$ and the major premise of an elimination rule $\star$E. The general pattern is this:

In contrast to the manipulations involved in Theorem 4.10 and Theorem 4.29, here we do not just replace assumptions by deductions of them, we replace entire sub-deductions. Also, sub-deductions involving different occurrences of the assumption $\neg A$ with discharge label i are replaced by different sub-deductions, since the deduction δ'_1 may be different for different occurrences of $\perp_K$-assumptions $\neg A^i$. A suitable generalization of Lemma 4.8 guarantees that this manipulation always results in a correct deduction, provided in the deduction δ the eigenvariables of $\forall$I and $\exists$E inferences are distinct and only occur above them, as guaranteed by Lemma 3.18.

For the proof we need appropriate definitions of "cut" and "cut segment," similar to the definitions in Section 4.6.

Definition 4.51. A *cut* in δ is an occurrence of a formula A which is

1. the conclusion of an I-rule, $\perp_J$, $\perp_K$, $\vee$E, or $\exists$E, and

2. the major premise of an E-rule.

A deduction in **NK** is *normal* if it has no cuts.

Definition 4.52. A *segment* in a deduction δ in **NK** is a sequence $A_1, \ldots, A_n$ of formula occurrences in δ where

1. A_1 is not the conclusion of $\perp_K$, $\vee$E, $\exists$E,

2. A_i is (for $1 \leq i < n$)

   (a) a minor premise of a $\vee$E or $\exists$E inference, and A_{i+1} is its conclusion, or

   (b) the minor premise of a $\neg$E inference the major premise of which is a $\perp_K$-assumption, and A_{i+1} is the conclusion of the $\perp_K$ discharging the major premise,

3. A_n is a cut, i.e., the major premise of an E-rule.

The *length* of the segment $A_1, \ldots, A_n$ is n. The *degree* of the segment is the degree of A_n.

Note that these notions of cut and segment are similar to the notions of cut and cut segment used in our proofs for **NJ**, but not identical. But the idea is still almost the same. Segments are sequences of formula occurrences in a deduction that connect a major premise of an $\star$E-rule to the beginning of a potential detour (e.g., an $\star$I or $\perp_K$ inference) via a sequence of intermediate identical formula occurrences. Cuts are formula occurrences in a deduction that keep it from being normal, and which the conversions remove. Here, cuts can both be direct detours as well as the ending of a segment that ends in $\vee$E or $\exists$E, to which a permutation conversion can be applied.[7]

Proposition 4.53. *Any cut A is the last formula in a segment.*

Proof. If A is the conclusion of an I-rule or of $\perp_J$, and the major premise of an E-rule, then A is its own segment of length 1. If it is the major premise of an E-rule and the conclusion of an $\vee$E or $\exists$E rule, it is the last formula in the segments we obtain by following formulas upwards, going from conclusions of $\vee$E and $\exists$E to a

---

[7] The main difference between Andou's and Prawitz's definition (Definition 4.26) is that here an $\vee$E concluding the major premise D of an E-rule is *always* a cut, even if the D is not (at the beginning of the segment) the conclusion of an I- or $\perp_K$ inference.

minor premise, and from conclusions of $\perp_K$ to the minor premise of the $\neg$E rules of which the assumptions discharged by $\perp_K$ are the major premise.   □

Definition 4.54. If A is a cut in δ, then $d(A)$ is its degree. Let the *rank* $r(A)$ of A be the maximum length of a segment to which A belongs, and the *height* $h(A)$ be the number of inferences below A in δ.

Definition 4.55. We introduce four measures on δ:

1. $d(\delta)$ is the maximal degree of cuts occurring in δ;

2. $r(\delta)$ is the maximal rank of cuts of degree $d(\delta)$ in δ;

3. $s(\delta)$ is the number of cuts of maximal rank among those of maximal degree $d(\delta)$ ("maximal cuts" for short), and

4. $h(\delta)$ is the sum of the heights of maximal cuts.

Definition 4.56. Let a *highest cut* be a cut A in δ such that:

1. A is a maximal cut,

2. no maximal cut lies above the first formula in a segment ending in A,

3. if A is a major premise of an inference, B one of its minor premises, then B is not on or below a segment ending in a maximal cut.

Proposition 4.57. *Every deduction δ has a highest cut, if it has a cut at all.*

Proof. Among a set of maximal cuts we can always find one that satisfies condition (2): For every maximal cut, consider the formulas which begin segments ending in those cuts. One of these must be such that no other occurrence of a formula in a segment ending in a maximal cut is above it; pick the corresponding cut. Let A_1 be a maximal cut satisfying (2). If A_1 does not already satisfy (3), we define $A_2, A_3, \ldots$, until we find one that does: If A_i satisfies (3), we are done. If A_i violates (3), it must be the major premise of an inference with a minor premise B, and B is on or below a segment ending in a maximal cut A. Among these maximal cuts, we can again find one that satisfies (2). Let that be A_{i+1}. Clearly, A_{i+1} must lie below A_i, or in a sub-deduction not containing A_i. So, A_{i+1} is different from A_1, $\ldots, A_i$. Since δ contains only finitely many maximal cuts, eventually we must find an A_k that satisfies (1)–(3).   □

You'll recall the way double induction works from Section 4.2. We show that $P(0,0)$ holds and also show that supposing $P(k, \ell)$ holds whenever either $k < n$, or $k = n$ but $\ell < m$, then it holds for $P(n, m)$. We applied this in our proofs of normalization. There, we assigned pairs of numbers $\langle k, \ell \rangle$ to proofs, and showed by double induction that every proof assigned $\langle n, m \rangle$ can be transformed to one in normal form. First we showed that if a proof has $\langle 0, 0 \rangle$ assigned to it, then it already is normal. Then we showed that if we have a proof that's assigned

$\langle n, m \rangle$, we can transform it into a new proof by applying a permutation or detour conversion. We showed that this new proof gets assigned $\langle k, \ell \rangle$, and that either $k < n$ or $k = n$ and $\ell < m$. Our proof of normalization for **NK** is the same, except instead of pairs of numbers we use quadruples of numbers $\langle d, r, s, h \rangle$. We show that any proof assigned $\langle d, r, s, h \rangle$ can be transformed into a new proof δ^*, and the new proof is assigned $\langle d', r', s', h' \rangle$ where

(a) $d' < d$, or

(b) $d' = d$ and $r' < r$, or

(c) $d' = d$ and $r' = r$ and $s' < s$, or

(d) $d' = d$ and $r' = r$ and $s' = s$ and $h' < h$.

We'll see how and why this "principle of quadruple induction" works in Chapter 8, and you might want to revisit this proof after reading that chapter. For now you can think of it as an extension of the principle of double induction, and take it on faith that it works the same way.

Theorem 4.58. *Every deduction δ in **NK** can be converted into a normal deduction.*

Proof. Induction basis: If $d(\delta) = r(\delta) = s(\delta) = h(\delta) = 0$, then δ has no maximal cuts, so is already normal.

Inductive step: If δ is not normal, there is a cut, and of course a maximal cut. Pick a highest cut, say, A. (By Proposition 4.57, a highest cut exists.) We transform δ to δ^* by applying the appropriate conversion to the sub-deduction ending in the conclusion of the inference of which A is the premise. In each case we have to verify that the overall complexity of δ^* is less than that of δ.

1. If a detour conversion of **NJ** applies, we obtain a new deduction in which at least one maximal cut has been removed and replaced with cuts of smaller degree. Let us look at one of the cases, e.g., $A = B \supset C$:

$$
\begin{array}{c}
\begin{array}{c}
B^i \\
\vdots\ \delta_1 \\
\dfrac{C}{i\ \ B \supset C}\ \supset\!\mathrm{I} \quad \begin{array}{c} \vdots\ \delta_2 \\ B \end{array} \\
\dfrac{}{\quad\quad C\quad\quad}\ \supset\!\mathrm{E}
\end{array}
\quad\Rightarrow\quad
\left.
\begin{array}{c}
\vdots\ \delta_2 \\
B \\
\vdots\ \delta_1 \\
C
\end{array}
\right\} \delta_1[\delta_2/B^i]
\end{array}
$$

The cut on $A = B \supset C$ was a highest cut, so δ_2 does not contain a maximal cut. Thus, even though in δ^* the subdeduction δ_2 has been multiplied as many times as there are occurrences of B discharged in δ_1, the cuts in δ_2 that have been added this way are not maximal cuts. The only potential new cuts that have been introduced are cuts on B, which have lower degree than A. The cut on $B \supset C$ has disappeared, so the number of maximal cuts is now $s(\delta^*) = s(\delta) - 1$. Of course, if $s(\delta) = 0$, i.e., the cut on A was the *only* maximal cut, then δ^* contains only cuts of degree $d(\delta)$ of lower rank (then $r(\delta^*) < r(\delta)$). If in addition $r(\delta^*) = 1$, then A was the only cut of degree $d(\delta)$, and so $d(\delta^*) < d(\delta)$.

2. If a permutation or simplification conversion applies, we have a new deduction in which all segments belonging to A have been shortened or eliminated altogether, respectively. Thus, the new occurrences of A are no longer maximal cuts. For instance, suppose we applied the following conversion:

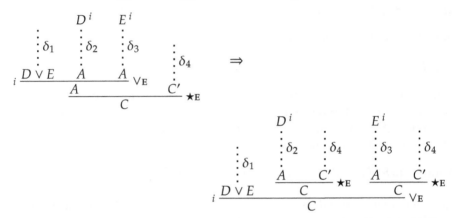

Here $\star$E is an E-rule with minor premise C'. A was a maximal cut in δ and at least one of the segments ending in A was of length $r(\delta)$. In the new deduction, the length of every segment ending in A has decreased by 1 (the $\vee$E inference is no longer part of any of those segments). Thus, although the corresponding occurrences of A in δ^* may still be cuts, their rank is $< r(\delta)$. The part of the deduction indicated by δ_4 may contain additional cuts, which have been duplicated in δ^*. But, since the cut A was a highest cut, no cut in δ_4 can be maximal. Thus, duplicating δ_4 has not duplicated any *maximal* cuts.

We cannot rule out that there is a cut G which ends a segment that includes C' (e.g., G might be C). We can, however, rule out that G is maximal: since A was a highest cut, and A and C' are major and minor premises of $\star$E, C' is not on a segment ending in a maximal cut. Since A is a cut of maximal degree, $d(G) \leq d(A)$. What if $d(G) = d(A)$ and G has rank $r(\delta) - 1$, that is, the length of the longest segment that ends in G is just 1 short of the maximum rank? Then in δ^* we now have at least two segments ending in G of length $r(\delta)$ (since there is an additional $\vee$E inference between C' and G). This would make G a maximal cut in δ^*, and overall the number of maximal cuts would not have been reduced.

In this case we rely on the last measure, the sum of the heights of all maximal cuts. It is easy to see that every maximal cut of δ except A corresponds to a maximal cut of δ^*, and the height of all of these is the same in δ as in δ^*. The height of A in δ, however, is greater than the height of G (the only new maximum cut) in δ^*. Consequently, $h(\delta^*) < h(\delta)$.

If this scenario does not apply (i.e., there is no cut G that ends a segment including the minor premise of the $\star$E inference), we have decreased the number of maximal cuts, or (if A was the only maximal cut) decreased the maximum rank of cuts of degree $d(\delta)$, or (if that rank was already 1) are only left with cuts of degree $< d(\delta)$.

3. Finally, let us consider the cases where A is a classical cut, i.e., the conclusion of $\perp_K$.

(a) A is $B \wedge C$:

$$
\begin{array}{c}
\vdots \delta_1' \\
\cfrac{\neg(B \wedge C)^i \quad B \wedge C}{\perp} \text{\scriptsize ¬E} \\
\vdots \delta_1 \\
i\ \cfrac{\cfrac{\perp}{B \wedge C}\, \perp_K}{B} \text{\scriptsize ∧E}
\end{array}
\qquad \Rightarrow \qquad
\begin{array}{c}
\vdots \delta_1' \\
\cfrac{\neg B^i \quad \cfrac{B \wedge C}{B}\text{\scriptsize ∧E}}{\perp} \text{\scriptsize ¬E} \\
\vdots \delta_1 \\
i\ \cfrac{\perp}{B}\, \perp_K
\end{array}
$$

(b) A is $B \supset C$:

$$
\begin{array}{c}
\vdots \delta_1' \\
\cfrac{\neg(B \supset C)^i \quad B \supset C}{\perp} \text{\scriptsize ¬E} \\
\vdots \delta_1 \qquad\qquad \vdots \delta_2 \\
\cfrac{i\ \cfrac{\perp}{B \supset C}\, \perp_K \qquad\qquad B}{C} \text{\scriptsize ⊃E}
\end{array}
\qquad \Rightarrow \qquad
\begin{array}{c}
\vdots \delta_1' \qquad \vdots \delta_2 \\
\cfrac{\neg C^i \quad \cfrac{B \supset C \quad B}{C}\text{\scriptsize ⊃E}}{\perp} \text{\scriptsize ¬E} \\
\vdots \delta_1 \\
i\ \cfrac{\perp}{C}\, \perp_K
\end{array}
$$

(c) A is $\neg B$: exercise.

(d) A is $\forall x\, B(x)$:

$$
\begin{array}{c}
\vdots \delta_1' \\
\cfrac{\neg\forall x\, B(x)^i \quad \forall x\, B(x)}{\perp} \text{\scriptsize ¬E} \\
\vdots \delta_1 \\
i\ \cfrac{\cfrac{\perp}{\forall x\, B(x)}\, \perp_K}{B(t)} \text{\scriptsize ∀E}
\end{array}
\qquad \Rightarrow \qquad
\begin{array}{c}
\vdots \delta_1' \\
\cfrac{\neg B(t)^i \quad \cfrac{\forall x\, B(x)}{B(t)}\text{\scriptsize ∀E}}{\perp} \text{\scriptsize ¬E} \\
\vdots \delta_1 \\
i\ \cfrac{\perp}{B(t)}\, \perp_K
\end{array}
$$

In each of the preceding cases, A is a highest cut in δ. The formula preceding A in the segments it ends is the last formula A of δ_1' (the minor premise of the $\neg$E inference the major premise of which is discharged in the $\perp_K$ inference). In δ^*, the last formula of δ_1' may now be a cut, e.g., if it is the conclusion of an I-rule or of $\vee$E or $\exists$E. Since the cut is also on A, the degree of these new cuts is the same as the cut we have reduced. But the segments of which these cuts are the last formulas are shorter (by 1) than the segments ending in the original A in δ. In the case where $A = B \supset C$ we have also multiplied the deduction δ_2 by grafting it onto possibly multiple assumption parts. But, since the cut A was highest, δ_2 contains no maximal cuts. The multiplication of δ_2 therefore does not increase the number of maximal cuts. Finally, the conclusion of $\perp_K$ in δ^* is now B, C, or $B(t)$. It is possible that this formula is the major premise of an E-rule and thus would be a new cut. But as this new cut is on a sub-formula of A, it is a cut of degree $< d(\delta)$, i.e., not a maximal cut. Overall, we have decreased the number of maximal cuts by 1, i.e., $s(\delta^*) < s(\delta)$ (or, if there was only one maximal cut, decreased the rank or degree).

(e) A is $B \vee C$:

(f) A is $\exists x\, B(x)$:

In the last two cases, the conclusion of the $\perp_K$ inference has been replaced by D.

In δ^*, we have removed the cut A ending in the conclusion of $\perp_K$, so that's one less cut of degree $d(\delta)$ and rank $r(\delta)$. The end-formula $B \vee C$ (or $\exists x\, B(x)$) of δ'_1 in δ^* is possibly a new cut. If it is, the segments ending in it are again shorter (by 1) than the segments ending in A in δ. Hence, these new cuts do not increase the number of maximal cuts. Because we picked a highest cut, no maximal cut can lie in δ_2 or δ_3 (if they had, we would have multiplied them by copying δ_2—and possibly δ_3—to multiple assumption-parts).

There is one additional case, similar to the case of permutative conversions. It is possible that D is part of a segment of length $r(\delta) - 1$ ending in a cut G of degree $d(\delta)$. This would result in G now being a maximal cut in δ^*, since the segments would be made longer by the occurrence of D now serving as conclusion of $\vee$E and minor premise of $\neg$E. However, the height of this new maximal cut G is less than $h(A)$, and the heights of every cut other than A in δ is the same in δ^*. Hence, $h(\delta^*) < h(\delta)$.

Again, if the last scenario does not apply, either $s(\delta^*) < s(\delta)$, or else $r(\delta^*) < r(\delta)$, or else $d(\delta^*) < d(\delta)$.

In each case, we have obtained a proof δ^* with measure $\langle d(\delta^*), r(\delta^*), s(\delta^*), h(\delta^*) \rangle$ less than that assigned to δ, so the inductive hypothesis applies. $\square$

Problem 4.59. Spell out the case in the proof of Theorem 4.58 for a classical cut of the form $\neg B$.

Problem 4.60. The proof of Theorem 4.58 assumed that every assumption discharged by $\perp_K$ is the major premise of an $\neg$E inference. Show that the conversions used in the proof produce a new deduction with this property if the original deduction satisfies it.

Andou (2003) and David and Nour (2003) showed that the conversions used here are strongly normalizing. A different and slightly more complicated normalization procedure was given by Stålmarck (1991).

As noted in Section 3.4, there are natural deduction systems for classical logic other than **NK**. Seldin (1989) established normalization for the system using the ND rule. For multiple-conclusion natural deduction, normalization was proved by Cellucci (1992) and Parigot (1992).

In reasoning classically we often make use of the principle of excluded middle: for any formula A, we can prove $A \vee \neg A$. We can use this fact in deductions in **NK**, e.g., the following deduction of $\neg\neg A \supset A$:

$$\cfrac{\neg(A \vee \neg A)^1 \quad \cfrac{A^2}{A \vee \neg A}\ \text{VI}\left.\right\}\delta_1''}{\cfrac{\bot}{2\ \boxed{\neg A}}\ \neg\text{I}}\ \neg\text{E}$$

We have two cuts, the first a regular detour ($\neg A$) and the second a classical cut concluding $A \vee \neg A$. The latter is maximal, but as the conversions are strongly normalizing the order in which we reduce them does not matter. We will first apply the classical conversion (case e) to the maximum cut on $A \vee \neg A$ to obtain the deduction below. The role of $B \vee C$ is played by $A \vee \neg A$; the role of D is played by A. Each of the two occurrences of $A \vee \neg A$ serving as minor premises of $\neg\text{E}$ in the original deduction will now be major premises of the $\vee\text{E}$ inference. Recall that different assumptions discharged by $\bot_K$ are treated differently, depending on which of δ_1'' and δ_1''' plays the role of δ_1'.

This has generated two new cuts on $A \vee \neg A$, which we eliminate by applying the intuitionistic detour conversions. This results in:

We eliminate the remaining cut $\neg A$:

$$
\cfrac{
  \cfrac{\neg A^1 \qquad \cfrac{\neg\neg A^5 \qquad \cfrac{6\ \cfrac{\neg A^3 \quad A^6}{\cfrac{\bot}{\neg A}\,{}_{\neg\mathrm{I}}}}{\bot}\,{}_{\neg\mathrm{E}}}{\cfrac{\bot}{A}\,\bot_J}}{\bot}{}_{\neg\mathrm{E}}
}{
  \cfrac{\neg A^1 \qquad 3\ \cfrac{\bot}{A}\,\bot_K}{\bot}\,{}_{\neg\mathrm{E}}
}
$$

$$
5\ \cfrac{1\ \cfrac{\bot}{A}\,\bot_K}{\neg\neg A \supset A}\,\supset\mathrm{I}
$$

It will be instructive to apply the detours in a different order than the one used in the proof. If we first eliminate the cut on $\neg A$, we get:

Now apply the $\bot_K/\vee\mathrm{E}$ classical conversion: replace the the two assumptions $\neg(A \vee \neg A)$ by $\neg A$, and apply the $\vee\mathrm{E}$ to the minor premises of the $\neg\mathrm{E}$ inferences.

This has introduced two new $\vee$I/$\vee$E-detours. (Note that we also had to relabel some assumptions.) Reducing the top detour results in:

$$
\cfrac{\cfrac{\neg A^1 \qquad \cfrac{\cfrac{\neg\neg A^5 \quad \cfrac{6\cfrac{\neg A^3 \quad A^6}{\bot}\neg E}{\neg A}\neg I}{\cfrac{\bot}{A}\bot_J}}{A}\neg E}{\cfrac{\neg A^1 \qquad \cfrac{3\cfrac{\bot}{A}\bot_K}{\boxed{A \vee \neg A}}\vee I \qquad A^7 \qquad \cfrac{\neg\neg A^5 \quad \neg A^7}{\cfrac{\bot}{A}\bot_J}\neg E}{7 \quad A}\vee E}{1\cfrac{\bot}{A}\bot_K}\neg E}{5\cfrac{}{\neg\neg A \supset A}\supset I}
$$

We reduce the remaining detour to arrive at a normal deduction:

$$
\cfrac{\cfrac{\neg A^1 \qquad \cfrac{\neg A^1 \qquad \cfrac{\neg\neg A^5 \quad \cfrac{6\cfrac{\neg A^3 \quad A^6}{\bot}\neg E}{\neg A}\neg I}{\cfrac{\bot}{A}\bot_J}}{A}\neg E}{3\cfrac{\bot}{A}\bot_K}\neg E}{1\cfrac{\bot}{A}\bot_K}\supset I
}{5\cfrac{}{\neg\neg A \supset A}}
$$

Compare this to the result of our first conversion sequence: the results are the same! This is no accident, as Andou (2003) showed.

Problem 4.61. Transform the following deduction into normal form:

$$
\cfrac{\cfrac{\neg((A \supset B) \vee \neg(A \supset B))^2 \\ \vdots \delta_1 \\ 2\cfrac{\bot}{(A \supset B) \vee \neg(A \supset B)}\bot_K \qquad \cfrac{A \supset B^1}{(A \supset B) \vee (B \supset A)}\vee I \qquad \cfrac{\neg(A \supset B)^1 \quad 4\cfrac{\cfrac{\neg A^3 \quad A^4}{\bot}\neg E}{\cfrac{\bot}{B}\bot_J} \\ \cfrac{}{A \supset B}\supset I}{\cfrac{3\cfrac{\bot}{A}\bot_K}{\cfrac{B \supset A}{(A \supset B) \vee (B \supset A)}\vee I}\neg E}}{1 \qquad (A \supset B) \vee (B \supset A)}\vee E
$$

(The deduction δ_1 would be carried out like the deduction of $A \vee \neg A$ on p. 86, with $A \supset B$ replacing A.)

In Definition 4.26, we generalized the notion of cut introduced in Definition 4.6 to that of a *cut segment*. Prawitz's proof of weak normalization for **NJ** had the aim of showing that every deduction can be transformed into one without cut segments. A cut segment begins with the conclusion of an I-rule and ends with the major premise of an E-rule. The elimination of cut segments is all that is required to derive the sub-formula property for **NJ**. For the proof of normalization in **NJ** (Theorem 4.29), we defined a number of possible transformations. First we had detour conversions, which transform sub-deductions ending in a detour (a cut in the original sense), that is to say, sub-deductions ending in an E-rule the major premise of which is an I-rule. Such sub-deductions were transformed into deductions that end only in cuts on sub-formulas, if they end in cuts at all, so these transformations reduced the degree of a cut segment. Second, we had permutation conversions. These transform sub-deductions ending in an E-rule the major premise of which is the conclusion of VE or ∃E. Such transformations shortened cut segments, and so decreased the rank of the deduction.

Our proof of normalization for **NK** used a slightly different notion of normal deduction. First of all, we introduced new classical conversions. Secondly, in the context of this section, we counted a deduction as normal if no conversion (permutation, detour, or classical) can be applied to it. If a deduction in **NJ** is normal in this sense, it is also normal in the sense of Section 4.6, but not vice versa. For instance, if the minor premise of an VE rule is an assumption, and the conclusion is the major premise of another E-rule, a permutation conversion can be applied, so the proof is not normal in the new sense. But it is already normal in the old sense, since the minor premise of the E-rule is not the conclusion of an I-rule, and so this does not constitute a cut segment in the sense of Definition 4.26.

Andou's normalization proof presented in this section produces a deduction that is normal in the sense that no conversions can be applied to it. In such a deduction, no formula is both the premise or result of an I-rule (or $\perp_K$ rule) and also the major premise of an E-rule, so there are no "detours" and the sub-formula property can be shown to hold. It would be possible to define cut segments the way we did in Section 4.6, define normal deduction as one without such cut segments, and adapt the proof to show that a deduction in **NK** can always be transformed into such a deduction. The proof would be more complicated, since we would have to keep track of which possible permutation conversions have to be applied to reduce the length of segments, and which ones we could leave as is (e.g., segments not beginning with the conclusion of an I- or $\perp_J$ rule such as those ending in VE followed by another E-rule where the minor premise of VE is an assumption). Conversely, it is possible to adapt the proof of Theorem 4.29 to show that a deduction can always be transformed into one in which no simplification, permutation, or detour conversion can be applied.

The sub-formula property holds for normal **NK** proofs in the following slightly attenuated form: every formula A in a normal **NK**-deduction is a sub-formula of the end-formula or of an open assumption or, if A is of the form $\neg B$, then B is such a sub-formula.

5

The sequent calculus

5.1   The language of the sequent calculus

In addition to developing the classical and intuitionistic natural deduction calculi,
NK and **NJ**, Gentzen also introduced—and studied—a second pair of calculi for
classical and intuitionistic logic based on the notion of *sequents*. A sequent consists
of two sequences of formulas separated by a *sequent arrow*, e.g., $A, A \supset B \Rightarrow B$.
Gentzen took the idea of devising a calculus which operates not on single formulas
but on sequents from Hertz (1929). However, Hertz did not consider rules for
logical operators. Sequents display in Gentzen an appealing symmetry, which is
broken only by the difference between classical and intuitionistic sequents: classical
sequents allow any number of formulas on the right hand side of $\Rightarrow$ as well as on
the left. In intuitionistic sequents, there may be at most one formula to the right
of $\Rightarrow$. After Gentzen, the notion of a sequent was varied by other proof theorists.
Gentzen's notion of a sequent is that of a pair of ordered lists of formulas arranged
in a determinate way, whereas some of the authors after Gentzen conceive of
sequents as pairs of sets of formulas; others as pairs of multisets of formulas. Yet
others use "one-sided" sequents, where formulas that would belong to the left
side of a Gentzen-style sequent are all negated. In what follows we will use the
original notion of a sequent as introduced by Gentzen.

What distinguishes the calculus of sequents from the natural deduction calculus
(and from the axiomatic calculus in the Hilbertian formulation) is not the structure
of the formulas but rather the structure of sequents and the formulation of the
inference rules as rules operating on sequents. For this reason we begin with the
presentation of the concept of sequent. A sequent is an expression of the form

$$A_1, \ldots, A_n \Rightarrow B_1, \ldots, B_m$$

where each A_i and B_j is a formula, and $\Rightarrow$ is the "sequent arrow," or just "arrow."
We will use uppercase Greek letters, like Γ, Δ, Σ, Π, Φ, Θ, Λ, Ξ, as meta-variables
for finite lists or *sequences* of formulas like $A_1, \ldots, A_n$. If Γ is the sequence A_1,
$\ldots, A_n$, and Δ is the sequence $B_1, \ldots, B_m$, then the sequent above will be simply

An Introduction to Proof Theory: Normalization, Cut-Elimination, and Consistency Proofs.
Paolo Mancosu, Sergio Galvan, and Richard Zach, Oxford University Press. © Paolo Mancosu,
Sergio Galvan and Richard Zach 2021. DOI: 10.1093/oso/9780192895936.003.0005

written as $\Gamma \Rightarrow \Delta$. Sequences may be empty or contain a single formula, and may contain more than one occurrence of the same formula.

Definition 5.1 (Sequent). If Γ and Δ are sequences of formulas, then

$$\Gamma \Rightarrow \Delta$$

is a *sequent*.

The sequence of formulas on the left is called the *antecedent* of the sequent; the sequence on the right is called the *succedent* of the sequent.

The sequences on either side may be empty, so

$$\Rightarrow \Delta \qquad \Gamma \Rightarrow \qquad \qquad \Rightarrow$$

are also sequents. " $\Rightarrow$ " is called the *empty sequent*.

We will use a comma not just to separate the formulas in a sequence Γ, but also to indicate concatenation of sequences or of sequences with formulas. So for instance, Γ, Δ stands for

$$A_1, \ldots, A_n, B_1, \ldots, B_m.$$

If C is a formula, then C, Γ and Γ, C mean

$$C, A_1, \ldots, A_n \quad \text{and} \quad A_1, \ldots, A_n, C$$

respectively.

The intuitive meaning of a sequent $\Gamma \Rightarrow \Delta$ is roughly this: if all the formulas in the antecedent are true, at least one formula in the succedent is true. If Γ is empty, it means that one of the formulas in Δ is true. If Δ is empty, it means that at least one of the formulas in Γ is false.

Here's a translation of sequents into formulas that makes this precise:

1. $n \neq 0, m \neq 0$: $A_1, \ldots, A_n \Rightarrow B_1, \ldots, B_m$ has the meaning of

$$(A_1 \wedge \cdots \wedge A_n) \supset (B_1 \vee \cdots \vee B_m).$$

2. $n = 0, m \neq 0$: $\Rightarrow B_1, \ldots, B_m$ has the meaning of

$$B_1 \vee \cdots \vee B_m$$

3. $n \neq 0, m = 0$: $A_1, \ldots, A_n \Rightarrow$ has the meaning of

$$\neg(A_1 \wedge \cdots \wedge A_n).$$

4. $n = 0, m = 0$: $\Rightarrow$, i.e., the empty sequent, means the contradiction $\bot$ (see Section 2.10.1).

Proofs in the sequent calculus are trees of sequents, where the topmost sequents are of the form $A \Rightarrow A$ (the *axioms*), and every other sequent in the tree follows from the sequents immediately above it by one of the rules of inference given below. A sequent is *provable* (or *derivable*) in **LK** if there is a proof of it in the calculus **LK**.

As will become clear after we present the axioms and rules, each line of a derivation must be valid and thus sequent systems are more like axiomatic ones than natural deduction systems.

5.2   Rules of **LK**

5.2.1   *Axioms*

Definition 5.2. Sequents of the type $A \Rightarrow A$, where A is any formula, are *axioms*.

The axioms are initial sequents. One has to start from them to obtain a proof. They are, so to speak, the sequents that turn on the inferential engine.

5.2.2   *Structural rules*

The first set of rules do not deal with logical symbols, but simply with the arrangement of formulas in the antecedent or succedent of a sequent. There are three pairs: Weakening (sometimes also called thinning or dilution) allows us to add a formula on either the far left or far right side of a sequent. Contraction allows us to replace two occurrences of the same formula next to each other by a single occurrence, on either the far left or the far right side of a sequent. Interchange (sometimes called exchange or permutation) allows us to switch two neighboring formulas. Each has two versions, one for the left and one for the right side of a sequent. In proofs, we will often indicate any number of structural inferences by double lines.

Weakening

$$\frac{\Gamma \Rightarrow \Theta}{A, \Gamma \Rightarrow \Theta} \ \text{WL} \qquad \frac{\Gamma \Rightarrow \Theta}{\Gamma \Rightarrow \Theta, A} \ \text{WR}$$

Contraction

$$\frac{A, A, \Gamma \Rightarrow \Theta}{A, \Gamma \Rightarrow \Theta} \ \text{CL} \qquad \frac{\Gamma \Rightarrow \Theta, A, A}{\Gamma \Rightarrow \Theta, A} \ \text{CR}$$

Interchange

$$\frac{\Delta, A, B, \Gamma \Rightarrow \Theta}{\Delta, B, A, \Gamma \Rightarrow \Theta} \ \text{IL} \qquad \frac{\Gamma \Rightarrow \Theta, A, B, \Lambda}{\Gamma \Rightarrow \Theta, B, A, \Lambda} \ \text{IR}$$

In the preceding rules the formulas A and B are called the *principal formulas* of the rule. In some contexts we will drop explicit mention of whether the structural rules have been applied to the left side or the right side of the sequent (or both). In those cases we will simply use w, c, and i.[1]

5.2.3   Cut

The CUT rule is counted among the structural rules, but it has a special status: it's the only rule that does not have a left and right version.

$$\frac{\Gamma \Rightarrow \Theta, A \qquad A, \Delta \Rightarrow \Lambda}{\Gamma, \Delta \Rightarrow \Theta, \Lambda} \text{ CUT}$$

The formula A in the cut that disappears in the conclusion is called the *cut-formula*.

5.2.4   Operational rules

The logical inferences or, as Gentzen calls them, "operational" rules deal with the logical operators and quantifiers. They also come in pairs: for every operator or quantifier there is a left rule, where the conclusion sequent contains a formula with it as main operator in the antecedent, and a corresponding right rule. The $\wedge$L and $\vee$R each have two versions (similarly to the $\vee$I and $\wedge$E rules in **NM**, **NJ**, and **NK**).

Conjunction

$$\frac{A, \Gamma \Rightarrow \Theta}{A \wedge B, \Gamma \Rightarrow \Theta} \wedge \text{L} \qquad\qquad \frac{\Gamma \Rightarrow \Theta, A \qquad \Gamma \Rightarrow \Theta, B}{\Gamma \Rightarrow \Theta, A \wedge B} \wedge \text{R}$$

$$\frac{B, \Gamma \Rightarrow \Theta}{A \wedge B, \Gamma \Rightarrow \Theta} \wedge \text{L}$$

When it will be necessary to distinguish between the two variants we will indicate them with $\wedge_1$L for the first and $\wedge_2$L for the second.

Disjunction

$$\frac{A, \Gamma \Rightarrow \Theta \qquad B, \Gamma \Rightarrow \Theta}{A \vee B, \Gamma \Rightarrow \Theta} \vee \text{L} \qquad\qquad \frac{\Gamma \Rightarrow \Theta, A}{\Gamma \Rightarrow \Theta, A \vee B} \vee \text{R}$$

$$\frac{\Gamma \Rightarrow \Theta, B}{\Gamma \Rightarrow \Theta, A \vee B} \vee \text{R}$$

---

[1] The three structural rules are fundamental also in the context of other logics, for instance in Curry's combinatory logic, where they are denoted differently. In that context, weakening is denoted by K, contraction by W, and interchange is denoted by C (see Curry, Hindley, and Seldin, 1972).

When it will be necessary to distinguish between the two variants we will indicate them with $\vee_1 R$ for the first and $\vee_2 R$ for the second.

Conditional

$$\frac{\Gamma \Rightarrow \Theta, A \quad B, \Delta \Rightarrow \Lambda}{A \supset B, \Gamma, \Delta \Rightarrow \Theta, \Lambda} \supset L \qquad \frac{A, \Gamma \Rightarrow \Theta, B}{\Gamma \Rightarrow \Theta, A \supset B} \supset R$$

Negation

$$\frac{\Gamma \Rightarrow \Theta, A}{\neg A, \Gamma \Rightarrow \Theta} \neg L \qquad \frac{A, \Gamma \Rightarrow \Theta}{\Gamma \Rightarrow \Theta, \neg A} \neg R$$

Universal quantifier

$$\frac{A(t), \Gamma \Rightarrow \Theta}{\forall x\, A(x), \Gamma \Rightarrow \Theta} \forall L \qquad !\, \frac{\Gamma \Rightarrow \Theta, A(a)}{\Gamma \Rightarrow \Theta, \forall x\, A(x)} \forall R$$

The rule $\forall R$ qualified by the exclamation mark ! is a critical rule. This means that it can only be applied if the indicated free variable a in the premise (the *eigenvariable* of the rule) does not appear in the lower sequent, i.e., it does not occur in Γ, Θ, or in $\forall x\, A(x)$.

Existential quantifier

$$!\, \frac{A(a), \Gamma \Rightarrow \Theta}{\exists x\, A(x), \Gamma \Rightarrow \Theta} \exists L \qquad \frac{\Gamma \Rightarrow \Theta, A(t)}{\Gamma \Rightarrow \Theta, \exists x\, A(x)} \exists R$$

The rule $\exists L$ qualified by the exclamation mark ! is also a critical rule. It too is subject to the restriction that the free variable a indicated in the premise of the rule (the *eigenvariable*) does not appear in Γ, Θ, or $\exists x\, A(x)$.

The eigenvariable condition on $\forall R$ and $\exists L$ is necessary to avoid proofs of sequents that are not valid. For instance, if we did not have the condition on $\forall R$, the following would be a proof of $\exists x\, A(x) \supset \forall x\, A(x)$:

$$\frac{\dfrac{\dfrac{A(a) \Rightarrow A(a)}{A(a) \Rightarrow \forall x\, A(x)} \forall R}{\exists x\, A(x) \Rightarrow \forall x\, A(x)} \exists L}{\Rightarrow \exists x\, A(x) \supset \forall x\, A(x)} \supset R$$

The eigenvariable condition is violated because a occurs in the antecedent of the conclusion of the $\forall R$ "inference."

Here's an example where the eigenvariable condition on $\exists$L is violated:

$$\cfrac{\cfrac{\cfrac{A(a) \Rightarrow A(a)}{\exists x\, A(x) \Rightarrow A(a)}\,\exists\text{L}}{\Rightarrow \exists x\, A(x) \supset A(a)}\,\supset\text{R}}{\Rightarrow \forall y(\exists x\, A(x) \supset A(y))}\,\forall\text{R}$$

There can be no correct proof of $\Rightarrow \forall y(\exists x\, A(x) \supset A(y))$ since it is also not valid.

Note that the eigenvariable condition also excludes some "proofs" of valid sequents as incorrect. For instance, the "proof" of $\forall x\, A(x) \Rightarrow \forall x\, A(x)$ on the left is incorrect, since the eigenvariable condition on $\forall$R is violated. The proof on the right, however, is correct.

$$\cfrac{\cfrac{A(a) \Rightarrow A(a)}{A(a) \Rightarrow \forall x\, A(x)}\,\forall\text{R}}{\forall x\, A(x) \Rightarrow \forall x\, A(x)}\,\forall\text{L} \qquad \cfrac{\cfrac{A(a) \Rightarrow A(a)}{\forall x\, A(x) \Rightarrow A(a)}\,\forall\text{L}}{\forall x\, A(x) \Rightarrow \forall x\, A(x)}\,\forall\text{R}$$

1. As can easily be seen, each operational rule introduces a new logical sign either on the left or on the right. The formula newly introduced by the rule is said to be the *principal formula* of the rule. The immediate sub-formula(s) of the principal formula occurring in the premises are called the *auxiliary formula(s)*. The formulas in Γ and Θ are called the *side formulas*.

2. The rules of interchange and contraction are necessary on account of the fact that the antecedent and the succedent are sequences of formulas. If, by contrast, $\Gamma, \Delta, \Phi, \Pi, \ldots$ were sets of formulas, then such rules would become redundant. However, if $\Gamma, \Delta, \Phi, \Pi, \ldots$ were to be understood as *multi-sets*—namely as sets which allow for the multiple occurrence of the same formula in the same set— then the rule of contraction would still be necessary.

3. It is worth mentioning that in all the rules the sequences $\Gamma, \Delta, \Phi, \Pi, \ldots$ may be empty. These sequences of formulas are called the *context* of the rule. One should also remember that in a sequent the whole antecedent and the whole succedent may be empty. As already pointed out earlier, the empty antecedent means that the succedent holds unconditionally, whereas the empty succedent means that the assumptions making up the antecedent are contradictory.

Problem 5.3. Using the translation of sequents into formulas given in Section 5.1 and setting all the side formulas of the sequents equal to the single formula C, show by means of truth-tables that $\supset$L and $\supset$R translate into logically valid inferences.

Definition 5.4. A *proof in LK* is a finite tree of sequents with an end-sequent inductively defined as follows.

1. *Basis clause:* Any axiom sequent $A \Rightarrow A$ is a proof with $A \Rightarrow A$ as both initial sequent and end-sequent.

2. *Inductive clauses:* Rules with one premise: Let a proof π_1 with end-sequent $\Gamma_1 \Rightarrow \Delta_1$ be given. If $\Gamma_1 \Rightarrow \Delta_1$ is the upper sequent of a one premise rule R that has $\Gamma \Rightarrow \Delta$ as conclusion then:

$$\begin{array}{c} \vdots\, \pi_1 \\ \dfrac{\Gamma_1 \Rightarrow \Delta_1}{\Gamma \Rightarrow \Delta}\ R \end{array}$$

   is a proof with end-sequent $\Gamma \Rightarrow \Delta$.

3. Rules with two premises: Let π_1 and π_2 be two proofs having as end-sequents $\Gamma_1 \Rightarrow \Delta_1$ and $\Gamma_2 \Rightarrow \Delta_2$, respectively. If $\Gamma_1 \Rightarrow \Delta_1$ and $\Gamma_2 \Rightarrow \Delta_2$ are the upper sequents of a two premise rule R that has $\Gamma \Rightarrow \Delta$ as end-sequent, then:

$$\begin{array}{cc} \vdots\, \pi_1 & \vdots\, \pi_2 \\ \dfrac{\Gamma_1 \Rightarrow \Delta_1 \quad \Gamma_2 \Rightarrow \Delta_2}{\Gamma \Rightarrow \Delta}\ R \end{array}$$

   is a proof with $\Gamma \Rightarrow \Delta$ as end-sequent.

4. *Extremal clause:* Nothing else is a proof.

Definition 5.5. A sequent is *intuitionistic* if its succedent contains at most one formula. A *proof in LJ* is a proof that only consists of intuitionistic sequents.

Definition 5.6. A *proof in LM* is a proof in LJ that does not use the wR rule.

The intuitionistic calculus **LJ** is nothing other than a restriction of the calculus **LK**. More precisely, we obtain **LJ** from **LK** by imposing the restriction that the succedent of any sequent must be constituted by at most one formula. We will give more details on **LJ** later. **LM** is a further restriction of **LJ**, and is the sequent calculus system for minimal logic.[2] It is perhaps worth pointing out that by the definition, every proof in **LM** is also one in **LJ**, and every proof in **LJ** is also one in **LK**.

In the following we will use π (with possible subscripts or superscripts) to denote proofs. A proof is also often called a derivation. We will indicate the inference rule used in given inference on the right of the inference line (indicated by R in Definition 5.4). However, when a sequent is obtained from the one above it using one or more structural (weakening, interchange, contraction) inferences, we often leave out the rule name(s) and use a double inference line.

---

[2] Gentzen's original papers did not include **LM**, just as they didn't include **NM**. **LM** is due to Johansson (1937).

When we argue about proofs in the sequent calculus, we will often use induction, just as we did in Chapters 3 and 4. In order to do this, we have to define a measure on proofs, which we will also call their *size* (cf. Definition 3.13).

Definition 5.7. The *size* of a proof π is defined inductively as follows:

1. If π consists only of an axiom, the size of π is 0.

2. If π ends in an inference, the size of π is 1 plus the sum of the sizes of the sub-proofs ending in these premises.

5.3   Constructing proofs in **LK**

Proofs in **LK** are relatively easy to find. All the rules for the logical operators have a very nice property: if we know that a sequent is obtained by such a rule, i.e., an operational rule, we also know what its premises must be. So finding a proof for a sequent is a matter of applying the rules in reverse. Let's start with a simple example: $A \wedge B \Rightarrow B \wedge A$. This may be obtained either by $\wedge$L or $\wedge$R. Let's pick the latter. This is what the inference must look like:

$$\frac{A \wedge B \Rightarrow B \qquad A \wedge B \Rightarrow A}{A \wedge B \Rightarrow B \wedge A} \wedge\text{R}$$

Each of the premises, in turn, is a possible conclusion of a $\wedge$L inference. $\wedge$L, however, comes in two versions: In one, the $A \wedge B$ on the left is obtained from an A on the left, in the other, from a B. The left premise, $A \wedge B \Rightarrow B$ can therefore be obtained from either $A \Rightarrow B$ or $B \Rightarrow B$. The latter, but not the former, is an axiom. The right premise, on the other hand, can be obtained from the axiom $A \Rightarrow A$. This gives us the proof:

$$\frac{\dfrac{B \Rightarrow B}{A \wedge B \Rightarrow B} \wedge\text{L} \qquad \dfrac{A \Rightarrow A}{A \wedge B \Rightarrow A} \wedge\text{L}}{A \wedge B \Rightarrow B \wedge A} \wedge\text{R}$$

If we had instead started looking for a proof by applying $\wedge$L in reverse, we would have run into trouble. For we have to pick one of the two versions of $\wedge$L. We'd end up with one of:

$$\frac{\dfrac{A \Rightarrow B \qquad A \Rightarrow A}{A \Rightarrow B \wedge A} \wedge\text{R}}{A \wedge B \Rightarrow B \wedge A} \wedge\text{L} \qquad\qquad \frac{\dfrac{B \Rightarrow B \qquad B \Rightarrow A}{B \Rightarrow B \wedge A} \wedge\text{R}}{A \wedge B \Rightarrow B \wedge A} \wedge\text{L}$$

In each case, our search only finds one axiom but leaves another sequent that's not an axiom. To avoid situations like this, we have to be a bit careful. This is where the contraction rule comes in handy. From top to bottom, contraction allows us to replace two occurrences of the same formula by one. From bottom to top, it allows us to duplicate a formula:

$$\frac{A \wedge B, A \wedge B \Rightarrow B \wedge A}{A \wedge B \Rightarrow B \wedge A} \text{CL}$$

Suppose we now apply ∧L "backwards:"

$$\frac{\dfrac{A, A \wedge B \Rightarrow B \wedge A}{A \wedge B, A \wedge B \Rightarrow B \wedge A} \text{ ∧L}}{A \wedge B \Rightarrow B \wedge A} \text{ CL}$$

The formula $A \wedge B$ still occurs on the left side of the top sequent. However, we can use the interchange rule to maneuver the remaining $A \wedge B$ to the outside position. After this is done, we can apply the other version of ∧L in reverse. The result is that ∧ no longer occurs on the left side of the top sequent.

$$\frac{\dfrac{\dfrac{\dfrac{B, A \Rightarrow B \wedge A}{A \wedge B, A \Rightarrow B \wedge A} \text{ ∧L}}{A, A \wedge B \Rightarrow B \wedge A} \text{ IL}}{A \wedge B, A \wedge B \Rightarrow B \wedge A} \text{ ∧L}}{A \wedge B \Rightarrow B \wedge A} \text{ CL}$$

If we now apply ∧R, we end up with:

$$\frac{\dfrac{\dfrac{\dfrac{\dfrac{B, A \Rightarrow B \quad B, A \Rightarrow A}{B, A \Rightarrow B \wedge A} \text{ ∧R}}{A \wedge B, A \Rightarrow B \wedge A} \text{ ∧L}}{A, A \wedge B \Rightarrow B \wedge A} \text{ IL}}{A \wedge B, A \wedge B \Rightarrow B \wedge A} \text{ ∧L}}{A \wedge B \Rightarrow B \wedge A} \text{ CL}$$

The sequents $B, A \Rightarrow B$ and $B, A \Rightarrow A$ are not, strictly speaking, axioms. But each one can be obtained from an axiom by weakening (and interchange). And this is generally the case: if $\Gamma \Rightarrow \Delta$ contains some formula A on both sides, it can be proved from the axiom $A \Rightarrow A$ just by weakenings and interchanges. Thus, we've obtained another proof:

$$\frac{\dfrac{\dfrac{\dfrac{\dfrac{\dfrac{B \Rightarrow B}{A, B \Rightarrow B} \text{ WL}}{B, A \Rightarrow B} \text{ IL} \quad \dfrac{\dfrac{A \Rightarrow A}{B, A \Rightarrow A} \text{ WL}}{} }{B, A \Rightarrow B \wedge A} \text{ ∧R}}{A \wedge B, A \Rightarrow B \wedge A} \text{ ∧L}}{A, A \wedge B \Rightarrow B \wedge A} \text{ IL}}{\dfrac{A \wedge B, A \wedge B \Rightarrow B \wedge A}{A \wedge B \Rightarrow B \wedge A} \text{ CL}} \text{ ∧L}$$

So, when searching for a proof, in a case where we apply ∧L backwards, always first copy the conjunction using CL first. Also, when you've arrived at a sequent containing some formula on both the left and the right, you can prove it from an axiom using just weakenings and interchanges.

The same principle applies also to disjunction on the right: since ∨R comes in two versions, always copy the disjunction using CR first; then apply ∨R in reverse;

then use IR; then apply the other version of $\lor$R. For instance, to find a proof of $\Rightarrow A \lor \neg A$, start with

$$\frac{\dfrac{\dfrac{\dfrac{\dfrac{\Rightarrow A, \neg A}{\Rightarrow A, A \lor \neg A}\ \lor\text{R}}{\Rightarrow A \lor \neg A, A}\ \text{IR}}{\Rightarrow A \lor \neg A, A \lor \neg A}\ \lor\text{R}}{\Rightarrow A \lor \neg A}\ \text{CR}}{}$$

The topmost sequent $\Rightarrow A, \neg A$ contains a negated formula on the right, so apply $\neg$R in reverse. The corresponding premise is $A \Rightarrow A$, an axiom.

A similar challenge arises when you want to apply $\supset$L in reverse: the rule combines the premise sequents, so in reverse we would have to decide how to split up the antecedent into Γ and Δ and the succedent into Θ and Λ. But, thanks to contraction and interchange, we don't have to. If your sequent is $A \supset B, \Gamma \Rightarrow \Theta$, use contraction and interchange in reverse to first get $A \supset B, \Gamma, \Gamma \Rightarrow \Theta, \Theta$, and then use Γ as Δ and Θ as Λ.

Here's a simple example. Suppose you want to prove $A \supset B \Rightarrow \neg A \lor B$. If you want to reduce $A \supset B$ first, you'd first use contraction to get a second copy of $\neg A \lor B$. Then for the application of $\supset$L in reverse, Γ and Δ are empty, and Θ and Λ are both $\neg A \lor B$:

$$\frac{\dfrac{\Rightarrow \neg A \lor B, A \qquad B \Rightarrow \neg A \lor B}{A \supset B \Rightarrow \neg A \lor B, \neg A \lor B}\ \supset\text{L}}{A \supset B \Rightarrow \neg A \lor B}\ \text{CR}$$

The further development would yield this proof:

$$\frac{\dfrac{\dfrac{\dfrac{\dfrac{A \Rightarrow A}{\Rightarrow A, \neg A}\ \neg\text{R}}{\Rightarrow A, \neg A \lor B}\ \lor\text{R}}{\Rightarrow \neg A \lor B, A}\ \text{IR} \qquad \dfrac{B \Rightarrow B}{B \Rightarrow \neg A \lor B}\ \lor\text{R}}{A \supset B \Rightarrow \neg A \lor B, \neg A \lor B}\ \supset\text{L}}{A \supset B \Rightarrow \neg A \lor B}\ \text{CR}$$

In order to apply an operational rule backward, the principal formula must be either on the far left or the far right of the sequent. Just as in the cases above, using one or more interchanges, you can always maneuver a formula in the sequent to the outside position. Since this may require a number of interchanges, we abbreviate any number of structural inferences in a proof with a double line. Say you wanted to prove $\Rightarrow \neg(A \land B), A, B$. You'd want to apply $\neg$R in reverse to $\neg(A \land B)$, but the A and B are in the way. So first, interchanges, then $\neg$R:

$$\frac{\dfrac{A \land B \Rightarrow A, B}{\Rightarrow A, B, \neg(A \land B)}\ \neg\text{R}}{\Rightarrow \neg(A \land B), A, B}\ \text{IR}$$

The procedure for systematically developing proofs from the bottom up in fact is always guaranteed to find a proof, if one exists. Quantifiers, however, introduce a degree of indeterminacy and guesswork. For if, say, the succedent contains $\exists x\, A(x)$, we know we should apply $\exists$R backward, but not what the instance $A(t)$ in the premise should be. In this case we must also be careful not to make choices that in the further construction create problems with the eigenvariable condition. For instance, suppose we are looking for a proof of $\exists x\, A(x) \Rightarrow \exists x(A(x) \vee B)$. If we start by applying $\exists$R backwards, we have to pick some term, e.g., a free variable a:

$$\frac{\exists x\, A(x) \Rightarrow A(a) \vee B}{\exists x\, A(x) \Rightarrow \exists x(A(x) \vee B)}\ \exists\text{R}$$

But now we are in trouble. We can't pick the same a for the application of $\exists$L, for it would violate the eigenvariable condition if we did this:

$$\frac{\dfrac{A(a) \Rightarrow A(a) \vee B}{\exists x\, A(x) \Rightarrow A(a) \vee B}\ \exists\text{L}}{\exists x\, A(x) \Rightarrow \exists x(A(x) \vee B)}\ \exists\text{R}$$

On the other hand, if we pick some other variable, say, b, we'd have:

$$\frac{\dfrac{A(b) \Rightarrow A(a) \vee B}{\exists x\, A(x) \Rightarrow A(a) \vee B}\ \exists\text{L}}{\exists x\, A(x) \Rightarrow \exists x(A(x) \vee B)}\ \exists\text{R}$$

The eigenvariable condition is satisfied, but $A(b) \Rightarrow A(a) \vee B$ is not provable: neither $A(b) \Rightarrow A(a)$ nor $A(b) \Rightarrow B$ is an axiom.

The solution is to always first reduce using a rule with the eigenvariable condition, i.e., $\forall$R or $\exists$L, and use a new variable. When reducing using a rule without eigenvariable condition, i.e., $\forall$L or $\exists$R, use a variable that already occurs in the sequent. In our case, this produces:

$$\frac{\dfrac{\dfrac{A(a) \Rightarrow A(a)}{A(a) \Rightarrow A(a) \vee B}\ \forall\text{R}}{A(a) \Rightarrow \exists x(A(x) \vee B)}\ \exists\text{R}}{\exists x\, A(x) \Rightarrow \exists x(A(x) \vee B)}\ \exists\text{L}$$

In some cases, we have to apply $\exists$L or $\forall$R backwards more than once using the same formula, with different variables. So to be safe, you should use the same trick we used for $\wedge$L and $\vee$R: use contraction to keep a copy of $\exists x\, A(x)$ or $\forall x\, A(x)$ in the premise of $\exists$L or $\forall$R, respectively.

5.4   The significance of CUT

As you can see, the procedure we have for finding proofs will produce proofs that never use the CUT rule. So what is the CUT rule good for? In the first instance, it is useful for *combining* proofs. Here are two example proofs, generated by the proof search method.

Proof π_1 is this:

$$
\cfrac{
  \cfrac{
    \cfrac{
      \cfrac{
        \cfrac{
          \cfrac{
            \cfrac{A \Rightarrow A}{\Rightarrow A, \neg A} \; \neg\text{R}
          }{\Rightarrow \neg A, A} \; \text{IR} \qquad B \Rightarrow B
        }{A \supset B \Rightarrow \neg A, B} \; \supset\text{L}
      }{A \supset B \Rightarrow \neg A, \neg A \vee B} \; \vee_2\text{R}
    }{A \supset B \Rightarrow \neg A \vee B, \neg A} \; \text{IR}
  }{A \supset B \Rightarrow \neg A \vee B, \neg A \vee B} \; \vee_1\text{R}
}{A \supset B \Rightarrow \neg A \vee B} \; \text{CR}
$$

Proof π_2 is this:

$$
\cfrac{
  \cfrac{
    \cfrac{
      \cfrac{
        \cfrac{\cfrac{A \Rightarrow A}{\neg A, A \Rightarrow} \; \neg\text{L}}{A, \neg A \Rightarrow} \; \text{IL}
      }{A \wedge \neg B, \neg A \Rightarrow} \; \wedge\text{L}
    }{\neg A \Rightarrow \neg(A \wedge \neg B)} \; \neg\text{R}
    \qquad
    \cfrac{
      \cfrac{\cfrac{B \Rightarrow B}{\neg B, B \Rightarrow} \; \neg\text{L}}{A \wedge \neg B, B \Rightarrow} \; \wedge\text{L}
    }{B \Rightarrow \neg(A \wedge \neg B)} \; \neg\text{R}
  }{\neg A \vee B \Rightarrow \neg(A \wedge \neg B)} \; \vee\text{L}
}{}
$$

Now suppose you needed a proof of $A \supset B \Rightarrow \neg(A \wedge \neg B)$. Rather than laboriously generating a proof from scratch you could just use CUT:

$$
\cfrac{
  \begin{matrix} \vdots \\ \pi_1 \\ \vdots \end{matrix} \quad A \supset B \Rightarrow \neg A \vee B
  \qquad
  \begin{matrix} \vdots \\ \pi_2 \\ \vdots \end{matrix} \quad \neg A \vee B \Rightarrow \neg(A \wedge \neg B)
}{A \supset B \Rightarrow \neg(A \wedge \neg B)} \; \text{CUT}
$$

The CUT rule allows us to combine proofs, much like the process of "grafting" a deduction of A from Γ onto all open assumptions of the form A allowed us to combine proofs in natural deduction (see Section 4.1). CUT also allows us to simulate inferential steps we might want to make. For instance, suppose we had proofs π_3, π_4 of $\Rightarrow A$ and $\Rightarrow A \supset B$. There should be a way to turn these proofs into one proof of $\Rightarrow B$—we would expect **LK** to let us use *modus ponens,* just as we can use it in axiomatic derivations and in natural deduction. Now there is clearly no way to apply *operational* inferences alone to these two sequents to obtain $\Rightarrow B$. In every inference rule of **LK** other than CUT, all formulas that appear in the premises also appear (perhaps as sub-formulas) in the conclusion. So any sequent generated from $\Rightarrow A$ and $\Rightarrow A \supset B$ will have A and $A \supset B$ as sub-formulas—but $\Rightarrow B$ does not. However, we can derive the sequent $A \supset B, A \Rightarrow B$, and then use CUT to obtain a proof of $\Rightarrow B$:

$$
\cfrac{
  \begin{matrix} \vdots \\ \pi_3 \\ \vdots \end{matrix} \; \Rightarrow A
  \qquad
  \cfrac{
    \begin{matrix} \vdots \\ \pi_4 \\ \vdots \end{matrix} \; \Rightarrow A \supset B
    \qquad
    \cfrac{A \Rightarrow A \qquad B \Rightarrow B}{A \supset B, A \Rightarrow B} \; \supset\text{L}
  }{A \Rightarrow B} \; \text{CUT}
}{\Rightarrow B} \; \text{CUT}
$$

You can see how the CUT rule here generates a similar kind of phenomenon in **LK**-proofs as what we called "cuts" in the proof of the normalization theorem: an introduction of $\supset$ (at the end of π_4) followed by an "elimination." Of course in **LK**, there are no elimination rules, but the CUT rule does eliminate a formula from the premises. Proofs with CUT inferences are indirect in much the same way as deductions with cuts (i.e., detours) are.

Because a deduction has to be copied for every open assumption it is grafted onto, the combined deduction will often be much larger than the two deductions being combined. Combining two **LK**-proofs using CUT, however, results in a proof that just adds one additional sequent (the conclusion of CUT) to the two proofs. In natural deduction, the multiplication of deductions can be avoided by first discharging them using $\supset$I, and then using the resulting conditional as the major premise of $\supset$E—introducing thereby a cut in the sense of natural deduction.

The combined proof above using CUT on π_1 and π_2 consists of 20 sequents. There is a shorter proof of $A \supset B \Rightarrow \neg(A \wedge \neg B)$ that does not use CUT. In general, however, proofs with CUT can be *much* shorter than the shortest proofs without, just like natural deductions including detours can be much shorter than the shortest direct deduction (see Section 4.5 and Boolos, 1984).

We will prove a result in Chapter 6—Gentzen's cut-elimination theorem—that shows that the use of CUT can always be avoided in proofs in **LK**, **LJ**, and **LM**. It plays a similar role as the normalization theorem did for natural deduction: there is always a direct proof if there is a proof at all.

Problem 5.8. Describe the structure cut-free **LK**-proofs of the formulas $p \supset A(n)$ of Section 4.5. How many inferences does each contain? How would you give shorter proofs using CUT?

5.5   Examples of proofs

In the present section we will derive in **LK** some logical laws of interest. In some of these proofs we will employ the CUT rule. After the proof of the cut-elimination theorem we will see how to use the elimination procedure delineated in the proof in order to eliminate the cuts used in such proofs.

Remark.   It is important to note that there is never only one way to derive the laws or rules. In particular, as will become apparent from the cut-elimination theorem, all the proofs in which use is made of the CUT rule are transformable into proofs in which this rule is not used. Therefore, later in this chapter you will use the CUT rule, which faithfully mirrors what happens in informal reasoning, and which is also a rule in natural deduction ($\supset$E). After the proof of the cut-elimination theorem, in Chapter 6 some of these proofs will be used as a starting point to show how the theorem works.

General axioms.   We call a sequent of the form

$$\Gamma, A, \Delta \Rightarrow \Sigma, A, \Theta$$

a *general axiom*.

General axioms can be obtained from the axiom $A \Rightarrow A$ through successive applications of WL and WR and possibly IL and IR.[3]

Note.   In the following proofs, the initial sequents are at times general axioms. We will not indicate this explicitly and we treat them at the same level of the simple axioms.

5.5.1   *Law of the Pseudo-Scotus LPS* (ex contradictione sequitur quodlibet)

$$A, \neg A \Rightarrow B$$

Proof.

$$\frac{\dfrac{\dfrac{\dfrac{A \Rightarrow A}{A \Rightarrow A, B} \text{ WR}}{A \Rightarrow B, A} \text{ IR}}{\neg A, A \Rightarrow B} \text{ ¬L}}{A, \neg A \Rightarrow B} \text{ IL}$$

Here is an alternative proof that shows that **LJ** proves the theorem.

$$\frac{\dfrac{\dfrac{\dfrac{A \Rightarrow A}{\neg A, A \Rightarrow} \text{ ¬L}}{A, \neg A \Rightarrow} \text{ IL}}{A, \neg A \Rightarrow B} \text{ WR}}$$

Note that this proof is not valid in **LM**. Why?

5.5.2   *Paradox of implication PI*

$$A \Rightarrow B \supset A$$

Proof.

$$\frac{B, A \Rightarrow A}{A \Rightarrow B \supset A} \text{ ⊃R}$$

---

[3] In **LJ** and **LM**, a general axiom would be of the form $\Gamma, A, \Delta \Rightarrow A$.

5.5.3   Law of double negation (classical) LDNC

$$\neg\neg A \Rightarrow A$$

Proof.

$$
\frac{\dfrac{A \Rightarrow A}{\Rightarrow A, \neg A}\ \scriptstyle\neg R}{\neg\neg A \Rightarrow A}\ \scriptstyle\neg L
$$

$\square$

Note that this is not a proof of LDNC in **LJ**. Why?

5.5.4   Law of double negation (intuitionistic) LDNI

$$A \Rightarrow \neg\neg A$$

Proof.

$$
\frac{\dfrac{A \Rightarrow A}{\neg A, A \Rightarrow}\ \scriptstyle\neg L}{A \Rightarrow \neg\neg A}\ \scriptstyle\neg R
$$

$\square$

Note that this rule is also derivable in **LJ** and **LM**. Why?

5.5.5   Tertium non datur *or excluded middle TND*

$$\Rightarrow A \vee \neg A$$

Proof.

$$
\frac{\dfrac{\dfrac{\dfrac{\dfrac{A \Rightarrow A}{\Rightarrow A, \neg A}\ \scriptstyle\neg R}{\Rightarrow A, A \vee \neg A}\ \scriptstyle\vee_2 R}{\Rightarrow A \vee \neg A, A}\ \scriptstyle IR}{\Rightarrow A \vee \neg A, A \vee \neg A}\ \scriptstyle\vee_1 R}{\Rightarrow A \vee \neg A}\ \scriptstyle CR
$$

$\square$

Note that this is not a proof in **LJ**. Why?

5.5.6   Law of non-contradiction LNC

$$\Rightarrow \neg(A \wedge \neg A)$$

Proof.

$$\cfrac{\cfrac{\cfrac{\cfrac{\cfrac{\cfrac{A \Rightarrow A}{\neg A, A \Rightarrow} \neg L}{A \wedge \neg A, A \Rightarrow} \wedge_2 L}{A, A \wedge \neg A \Rightarrow} IL}{A \wedge \neg A, A \wedge \neg A \Rightarrow} \wedge_1 L}{A \wedge \neg A \Rightarrow} CL}{\Rightarrow \neg(A \wedge \neg A)} \neg R$$

□

Note that this law is also derivable in **LJ** and **LM**.

5.5.7   *Law of* Modus Ponens *LMP*

$$A \supset B, A \Rightarrow B$$

Proof.

$$\cfrac{A \Rightarrow A \quad B \Rightarrow B}{A \supset B, A \Rightarrow B} \supset I$$

□

Note that this law is also derivable in **LJ** and **LM**. Why?

5.5.8   *Law of transitivity of implication*

$$C \supset A, A \supset B \Rightarrow C \supset B$$

Proof.

$$\cfrac{\cfrac{\cfrac{\cfrac{C \Rightarrow C}{C \Rightarrow C, A} WR}{C \Rightarrow A, C} IR \quad A \Rightarrow A}{C \supset A, C \Rightarrow A, A} \supset L}{\cfrac{C \supset A, C \Rightarrow A}{C \supset A, C \Rightarrow A} CR} \quad \cfrac{\cfrac{\cfrac{\cfrac{A \Rightarrow A}{A \Rightarrow A, B} WR}{A \Rightarrow B, A} IR \quad B \Rightarrow B}{A \supset B, A \Rightarrow B, B} \supset L}{\cfrac{A \supset B, A \Rightarrow B}{A, A \supset B \Rightarrow B} CR} IL}{\cfrac{\cfrac{C \supset A, C, A \supset B \Rightarrow B}{C, C \supset A, A \supset B \Rightarrow B} IL}{C \supset A, A \supset B \Rightarrow C \supset B} \supset R} CUT$$

□

Problem 5.9. The law of transitivity of implication is also derivable in the minimal sequent calculus **LM**. But the proof above is not a proof in **LM**. Can you see why? Give an alternative proof in **LM**.

5.5.9   *Laws of transformation of the quantifiers*

(a) $\exists x\, A(x) \Rightarrow \neg\forall x\neg A(x)$

Proof.

$$\cfrac{\cfrac{\cfrac{\cfrac{A(a) \Rightarrow A(a)}{\neg A(a), A(a) \Rightarrow} \neg\text{L}}{\forall x\,\neg A(x), A(a) \Rightarrow} \forall\text{L}}{A(a) \Rightarrow \neg\forall x\,\neg A(x)} \neg\text{R}}{\exists x\, A(x) \Rightarrow \neg\forall x\,\neg A(x)} \exists\text{L}$$

Note that here and in the following proofs the critical conditions for quantifiers rules are satisfied. This is a proof in **LJ** and **LM**, so the law is intuitionistically valid and valid in minimal logic.

(b) $\neg\forall x\,\neg A(x) \Rightarrow \exists x\, A(x)$

Proof.

$$\cfrac{\cfrac{\cfrac{\cfrac{\cfrac{A(a) \Rightarrow A(a)}{A(a) \Rightarrow \exists x\, A(x)} \exists\text{R}}{\Rightarrow \exists x\, A(x), \neg A(a)} \neg\text{R}}{\Rightarrow \exists x\, A(x), \forall x\,\neg A(x)} \forall\text{R}}{\neg\forall x\,\neg A(x) \Rightarrow \exists x\, A(x)} \neg\text{L}}$$

This is *not* a proof in **LJ**. Indeed, the law is not intuitionistically valid.

(c) $\neg\forall x\, A(x) \Rightarrow \exists x\neg A(x)$

Proof.

$$\cfrac{\cfrac{\cfrac{\cfrac{A(a) \Rightarrow \neg A(a)}{\neg A(a) \Rightarrow \exists x\,\neg A(x)} \exists\text{R}}{\Rightarrow \exists x\neg A(x), \neg\neg A(a)} \neg\text{R} \qquad \cfrac{\cfrac{\cfrac{A(a) \Rightarrow A(a)}{\Rightarrow A(a), \neg A(a)} \neg\text{R}}{\neg\neg A(a) \Rightarrow A(a)} \neg\text{L}}{}}{\Rightarrow \exists x\neg A(x), A(a)} \text{CUT}}{\cfrac{\Rightarrow \exists x\neg A(x), \forall x\, A(x)}{\neg\forall x\, A(x) \Rightarrow \exists x\neg A(x)} \neg\text{L}} \forall\text{R}}$$

Like b, this is not an intuitionistically valid principle, and the proof is not a proof in **LJ**.

(d) $\neg\exists x\, A(x) \Rightarrow \forall x\,\neg A(x)$

Proof.

$$
\cfrac{
\cfrac{A(a) \Rightarrow A(a)}{A(a) \Rightarrow \exists x\, A(x)} \exists \text{R}
\qquad
\cfrac{
\cfrac{
\cfrac{\exists x\, A(x) \Rightarrow \exists x\, A(x)}{\neg\exists x\, A(x), \exists x\, A(x) \Rightarrow} \neg\text{L}
}{\exists x\, A(x), \neg\exists x\, A(x) \Rightarrow} \text{IL}
}{\,}
}{
\cfrac{
\cfrac{
\cfrac{A(a), \neg\exists x\, A(x) \Rightarrow}{\neg\exists x\, A(x) \Rightarrow \neg A(a)} \neg\text{R}
}{\neg\exists x\, A(x) \Rightarrow \forall x \neg A(x)} \forall\text{R}
}{\,}
} \text{CUT}
$$

□

This *is* a proof in **LJ**, and even in **LM**, and therefore d is intuitionistically and minimally valid.

Problem 5.10. Give proofs of the following sequents. For all but (4) you should give a proof in **LJ**.

1. $A \vee B \Rightarrow B \vee A$
2. $A \supset B, A \supset C, (B \wedge C) \supset D \Rightarrow A \supset D$
3. $A \supset C, B \supset C \Rightarrow (A \vee B) \supset C$
4. $\neg A \supset B, \neg A \supset \neg B \Rightarrow A$
5. $A \supset B, A \supset \neg B \Rightarrow \neg A$
6. $A \wedge (B \vee C) \Rightarrow (A \wedge B) \vee (A \wedge C)$

Problem 5.11. Give proofs of the following sequents relating $\vee, \wedge, \neg$ with $\supset$. For (5)–(8), (11), and (12), you should give a proof in **LJ**.

1. $A \supset B \Rightarrow \neg A \vee B$
2. $\neg A \supset B \Rightarrow A \vee B$
3. $A \supset \neg B \Rightarrow \neg A \vee \neg B$
4. $\neg A \supset \neg B \Rightarrow A \vee \neg B$
5. $A \vee B \Rightarrow \neg A \supset B$
6. $\neg A \vee B \Rightarrow A \supset B$
7. $A \vee \neg B \Rightarrow \neg A \supset \neg B$
8. $\neg A \vee \neg B \Rightarrow A \supset \neg B$
9. $\neg(A \wedge \neg B) \Rightarrow A \supset B$
10. $\neg(\neg A \wedge \neg B) \Rightarrow \neg A \supset B$
11. $\neg(A \wedge B) \Rightarrow A \supset \neg B$
12. $\neg(\neg A \wedge B) \Rightarrow \neg A \supset \neg B$

Problem 5.12. Give proofs of the following sequents (De Morgan's Laws). For (1)–(4), you should give a proof in **LJ**.

1. $A \vee B \Rightarrow \neg(\neg A \wedge \neg B)$

2. $\neg A \vee B \Rightarrow \neg(A \wedge \neg B)$

3. $A \vee \neg B \Rightarrow \neg(\neg A \wedge B)$

4. $\neg A \vee \neg B \Rightarrow \neg(A \wedge B)$

5. $\neg(\neg A \wedge \neg B) \Rightarrow A \vee B$

6. $\neg(A \wedge \neg B) \Rightarrow \neg A \vee B$

7. $\neg(\neg A \wedge B) \Rightarrow A \vee \neg B$

8. $\neg(A \wedge B) \Rightarrow \neg A \vee \neg B$

Problem 5.13. Give proofs of the following derived rules, i.e., show that the bottom sequent can be derived from the top sequent(s) in **LK**. For (3) and (4), give proofs in **LJ**.

1. $$\frac{\Gamma, \neg A \Rightarrow}{\Gamma \Rightarrow A}$$

2. $$\frac{\Gamma, \neg A \Rightarrow A}{\Gamma \Rightarrow A}$$

3. $$\frac{\Gamma, A \Rightarrow \neg A}{\Gamma \Rightarrow \neg A}$$

4. $$\frac{\Gamma, A \Rightarrow \neg B}{\Gamma, B \Rightarrow \neg A}$$

5. $$\frac{\Gamma, A \Rightarrow B \qquad \Delta, \neg A \Rightarrow B}{\Gamma, \Delta \Rightarrow B}$$

Problem 5.14. Give proofs in **LJ** of the following rules involving $\vee$, $\wedge$, $\supset$.

1. $$\frac{A, B, \Gamma \Rightarrow C}{A \wedge B, \Gamma \Rightarrow C}$$

2. $$\frac{A \wedge B, \Gamma \Rightarrow C}{A, B, \Gamma \Rightarrow C}$$

3. $$\frac{\Gamma \Rightarrow A \supset B}{A, \Gamma \Rightarrow B}$$

Problem 5.15. Give proofs in **LK** of the following rules involving $\vee$, $\wedge$, $\supset$.

1. $$\frac{\Gamma \Rightarrow \Theta, A, B}{\Gamma \Rightarrow \Theta, A \vee B}$$

2. $$\frac{\Gamma, \neg A \Rightarrow \Theta \qquad \Delta, B \Rightarrow \Theta}{\Gamma, \Delta, A \supset B \Rightarrow \Theta}$$

Problem 5.16. Give proofs of the following sequents, except for the last, in **LJ**. For the last sequent give a proof in **LK**.

1. $(D \forall \wedge) \ \forall x(A(x) \wedge B(x)) \Rightarrow \forall x \, A(x) \wedge \forall x \, B(x)$

2. (D∀⊃) $\forall x(A(x) \supset B(x)) \Rightarrow \forall x\, A(x) \supset \forall x\, B(x)$

3. (D∃∧) $\exists x(A(x) \wedge B(x)) \Rightarrow \exists x\, A(x) \wedge \exists x\, B(x)$

4. (D∃∨) $\exists x(A(x) \vee B(x)) \Rightarrow \exists x\, A(x) \vee \exists x\, B(x)$

5. (E∀∧) $\forall x\, A(x) \wedge \forall x\, B(x) \Rightarrow \forall x(A(x) \wedge B(x))$

6. (E∀∨) $\forall x\, A(x) \vee \forall x\, B(x) \Rightarrow \forall x(A(x) \vee B(x))$

7. (E∃∨) $\exists x\, A(x) \vee \exists x\, B(x) \Rightarrow \exists x(A(x) \vee B(x))$

8. (E∃⊃) $\exists x\, A(x) \supset \exists x\, B(x) \Rightarrow \exists x(A(x) \supset B(x))$

5.6   Atomic logical axioms

In Chapter 7, we'll need to know that logical initial sequents can be restricted to atomic formulas. We'll prove this fact here, as a simple example of proof transformation.

Proposition 5.17. *If π is a proof, there is a proof π′ with the same end-sequent in which all logical initial sequents are atomic.*

Proof. It is clearly enough to show that every initial sequent $A \Rightarrow A$ can be derived from just atomic initial sequents. Then any proof π can be transformed to the proof π′ with the same end-sequent by replacing every complex initial sequent $A \Rightarrow A$ by its proof from atomic initial sequents.

To prove the claim, we proceed by induction on the degree of A, i.e., number of logical operators and quantifiers in A.

Induction basis: If the degree of A is 0, A is already atomic, and there is nothing to prove.

Inductive step: Suppose that the degree of A is > 0. We distinguish cases according to the main logical operator of A. We treat the cases where that operator is $\wedge$ and $\vee$, the other cases are left as exercises.

Suppose A is of the form $B \wedge C$. We have to show that $B \wedge C \Rightarrow B \wedge C$ can be derived from atomic initial sequents. By inductive hypothesis, $B \Rightarrow B$ and $C \Rightarrow C$ can be so derived using proofs π_B and π_C, since B and C contain fewer logical operators than $B \wedge C$. We use them to construct the proof of $B \wedge C \Rightarrow B \wedge C$ from atomic initial sequents as follows:

$$
\cfrac{\cfrac{\cfrac{\cfrac{\cfrac{\cfrac{\cfrac{\vdots\ \pi_B}{B \Rightarrow B}\ \text{WL}}{B, C \Rightarrow B} \quad \cfrac{\cfrac{\vdots\ \pi_C}{C \Rightarrow C}\ \text{WL}}{B, C \Rightarrow C}}{B, C \Rightarrow B \wedge C}\ \wedge\text{R}}{B \wedge C, C \Rightarrow B \wedge C}\ \wedge\text{L}}{C, B \wedge C \Rightarrow B \wedge C}\ \text{IL}}{B \wedge C, B \wedge C \Rightarrow B \wedge C}\ \wedge\text{L}}{B \wedge C \Rightarrow B \wedge C}\ \text{CL}
$$

Now suppose A is of the form $\forall x\, B(x)$. $B(a)$ contains fewer logical operators than $\forall x\, B(x)$, so by induction hypothesis, there is a proof $\pi_{B(a)}$ of $B(a) \Rightarrow B(a)$ from atomic initial sequents. We obtain a proof of $\forall x\, B(x) \Rightarrow \forall x\, B(x)$ thus:

$$
\vdots\, \pi_{B(a)}
$$

$$
\cfrac{\cfrac{B(a) \Rightarrow B(a)}{\forall x\, B(x) \Rightarrow B(a)}\ \forall_L}{\forall x\, B(x) \Rightarrow \forall x\, B(x)}\ \forall_R
$$

Note that because we use $\forall_L$ first, the premise of $\forall_R$ no longer contains the proper variable a in the side formulas, and so the eigenvariable condition is satisfied. We cannot use $\forall_R$ first, since then the eigenvariable condition would not be satisfied: a would still occur in the conclusion $B(a) \Rightarrow \forall x\, B(x)$. $\qquad\square$

Problem 5.18. Complete the proof of Proposition 5.17 by verifying the remaining cases.

5.7   Lemma on variable replacement

It will be useful to know that proofs of a sequent $\Gamma \Rightarrow \Delta$ containing a variable a can be transformed into proofs of $\Gamma' \Rightarrow \Delta'$, where Γ' results from Γ simply by replacing every occurrence of a by t. One might think that this can be achieved by replacing every occurrence of a in the entire proof π by t.

Lemma 5.19 (Variable replacement). *Suppose $\pi(a)$ is a proof, t is a term not containing any eigenvariables of $\pi(a)$, and a is a variable that is not used as an eigenvariable of an inference in $\pi(a)$. Then $\pi(t)$, which results from $\pi(a)$ by replacing every occurrence of a by t, is a correct proof.*

Proof. The proof is by induction on the number of inferences in $\pi(a)$.

Induction basis: If $\pi(a)$ is an initial sequent of the form $A(a) \Rightarrow A(a)$, then $\pi(t)$ is $A(t) \Rightarrow A(t)$ which is also an initial sequent.

Inductive step: Now suppose $\pi(a)$ contains at least one inference. Suppose, for instance, that $\pi(a)$ ends in $\wedge_R$. Then $\pi(a)$ has the following form:

$$
\cfrac{\displaystyle \vdots\, \pi_1(a) \qquad\qquad \vdots\, \pi_2(a)}{}
$$

$$
\cfrac{\Gamma(a) \Rightarrow \Theta(a), A(a) \qquad \Gamma(a) \Rightarrow \Theta(a), B(a)}{\Gamma(a) \Rightarrow \Theta(a), A(a) \wedge B(a)}\ \wedge_R
$$

By assumption, a is a free variable that is not used as an eigenvariable of any inference in π_1 or π_2, and t does not contain an eigenvariable of π_1 or π_2. So the induction hypothesis applies to the sub-proofs ending in the premises: $\pi_1(t)$ and

$\pi_2(t)$ ending in $\Gamma(t) \Rightarrow \Theta(t), A(t)$ and $\Gamma(t) \Rightarrow \Theta(t), B(t)$, respectively, are correct inferences. But then $\pi(t)$, i.e.,

$$
\frac{
\begin{array}{cc}
\vdots\,\pi_1(t) & \vdots\,\pi_2(t) \\
\Gamma(t) \Rightarrow \Theta(t), A(t) & \Gamma(t) \Rightarrow \Theta(t), B(t)
\end{array}
}{\Gamma(t) \Rightarrow \Theta(t), A(t) \wedge B(t)} \wedge\mathrm{R}
$$

is also correct.

The important cases are $\forall$R and $\exists$L. We consider the case of $\forall$R, the other is analogous.

If $\pi(a)$ ends in $\forall$R, it has the following form:

$$
\frac{
\begin{array}{c}
\vdots\,\pi_1(a) \\
\Gamma(a) \Rightarrow \Theta(a), A(b)
\end{array}
}{\Gamma(a) \Rightarrow \Theta(a), \forall x\, A(x)} \forall\mathrm{R}
$$

The inductive hypothesis applies to $\pi_1(a)$: if we replace every occurrence of a in $\pi_1(a)$ by t, we obtain a correct proof $\pi(t)$. What is its end-sequent? If we replace every occurrence of a in $\Gamma(a)$ and $\Theta(a)$ by t, we obtain $\Gamma(t)$ and $\Theta(t)$, respectively. Since b is an eigenvariable of π, it does not occur in t, by assumption. So $\Gamma(t)$ and $\Theta(t)$ do not contain b. Also by assumption, a is not used as an eigenvariable in π, so a and b must be different variables. If a does not occur in $A(b)$ at all, replacing a by t just results in the same formula $A(b)$. If it does, it is really of the form $A(a,b)$, and we pass from a formula $A(a,b)$ to $A(t,b)$—occurrences of b in it remain untouched. The corresponding quantified formula $\forall x\, A(t,x)$ does not contain b, since $\forall x\, A(a,x)$ did not, and t does not contain b either. So, if we consider $\pi(t)$, i.e.,

$$
\frac{
\begin{array}{c}
\vdots\,\pi_1(t) \\
\Gamma(t) \Rightarrow \Theta(t), A(t,b)
\end{array}
}{\Gamma(t) \Rightarrow \Theta(t), \forall x\, A(t,x)} \forall\mathrm{R}
$$

we see that the eigenvariable condition for b is satisfied, so the last inference is also correct.

The rule $\exists$L is treated analogously. We leave it and the other cases as exercises. $\square$

Problem 5.20. Complete the proof of Lemma 5.19 by verifying the remaining cases.

Corollary 5.21. *Suppose a proof π ends in $\forall$R or $\exists$L with eigenvariable a, contains no other $\forall$R or $\exists$L inference with eigenvariable a, and b is a variable not occurring in π. Then the result of replacing a by b throughout π is a proof of the same end-sequent.*

Proof. Consider the case where π ends in $\forall$R (the case where it ends in $\exists$L is analogous). Then π has the form

$$
\begin{array}{c}
\vdots\; \pi_1(a) \\
\vdots \\
\dfrac{\Gamma \Rightarrow \Theta, A(a)}{\Gamma \Rightarrow \Theta, \forall x\, A(x)}\; \forall \text{R}
\end{array}
$$

Since $\pi_1(a)$ contains no $\forall$R or $\exists$L inferences with eigenvaraible a, a is not used as an eigenvariable in $\pi_1(a)$. So by Lemma 5.19, we can replace a by b throughout $\pi_1(a)$. By adding a $\forall$R inference, we obtain the proof $\pi(b)$:

$$
\begin{array}{c}
\vdots\; \pi_1(b) \\
\vdots \\
\dfrac{\Gamma \Rightarrow \Theta, A(b)}{\Gamma \Rightarrow \Theta, \forall x\, A(x)}\; \forall \text{R}
\end{array}
$$

Since b does not occur in π, it also does not occur in $\pi_1(a)$, in particular, not in Γ or Θ. By the eigenvariable condition of the last $\forall$R inference, a does not occur in Γ, Θ, or $A(x)$, so replacing a by b in it results in the same sequent. Hence the eigenvariable condition is also satisfied for the $\forall$R inference in $\pi(b)$. $\qquad\square$

Definition 5.22. A proof in **LK** is *regular* if every eigenvariable is the eigenvariable of a single $\forall$R or $\exists$L inference and occurs only above that one inference.

Proposition 5.23. *Every proof π can be transformed into a regular proof π' of the same end-sequent by replacing eigenvariables only.*

Proof. By induction on the number n of $\exists$L and $\forall$R inferences in π in which the eigenvariable used is also eigenvariable of another inference or occurs below it.

Induction basis: If n is 0, then every eigenvariable is the eigenvariable of only a single inference, so the proof is regular.

Inductive step: Suppose $n > 0$. There is a topmost $\exists$L or $\forall$R inference with an eigenvariable a which is also used in another inference or which occurs below it. Consider the sub-proof π_1 ending in this inference: it is a proof that ends in $\exists$L or $\forall$R and contains no other eigenvariable inferences with eigenvariable a. So Corollary 5.21 applies, and we can replace the sub-proof π_1 by π_1' where we replace a everywhere in π_1 by some free variable b not in π. In the resulting proof, the last inference of π_1' is an inference with eigenvariable b which is not the eigenvariable of any other inference and which does not occur below it. Eigenvariables of other inferences have not been changed. So the number of $\exists$L and $\forall$R inferences in which the eigenvariable also serves as eigenvariable of another inference or occurs below it is $< n$, and the inductive hypothesis applies. $\qquad\square$

Problem 5.24. Consider the proof

$$
\cfrac{
\cfrac{A(a) \Rightarrow A(a)}{\cfrac{\forall x\, A(x) \Rightarrow A(a)}{\forall x\, A(x) \Rightarrow \forall x\, A(x)}\; \text{VR}}\; \text{VL}
\qquad
\cfrac{
\cfrac{A(a) \Rightarrow A(a)}{\cfrac{\forall x\, A(x) \Rightarrow A(a)}{\cfrac{\forall x\, A(x) \Rightarrow \forall x\, A(x)}{\forall x\, A(x) \Rightarrow \forall x\, A(x), A(a)}\; \text{WR}}\; \text{VR}}\; \text{VL}
}
}{\forall x\, A(x) \Rightarrow \forall x\, A(x), A(a)}\; \text{CUT}
$$

Apply the method in the proof of Proposition 5.23 to find the corresponding regular proof.

Corollary 5.25. *If π is a regular proof of a sequent $\Gamma(a) \Rightarrow \Delta(a)$ containing a, t is a term not containing eigenvariables of π, then $\pi(t)$ is a proof of $\Gamma(t) \Rightarrow \Delta(t)$.*

Proof. Since π is regular, a is not used as an eigenvariable in it (otherwise it would occur below the corresponding eigenvariable inference in π). The result follows by Lemma 5.19. □

Problem 5.26. Consider the following proof $\pi(c)$:

$$
\pi_1(a,b)\begin{cases}\end{cases}
\quad
\cfrac{
\cfrac{
\cfrac{
\cfrac{A(a,b) \Rightarrow A(a,b)}{\cfrac{A(a,b) \wedge C(c) \Rightarrow A(a,b)}{\forall x(A(x,b) \wedge C(c)) \Rightarrow A(a,b)}\; \text{VL}}\; \wedge\text{L}
\quad
\cfrac{C(c) \Rightarrow C(c)}{\cfrac{A(a,b) \wedge C(c) \Rightarrow C(c)}{\forall x(A(x,b) \wedge C(c)) \Rightarrow C(c)}\; \text{VL}}\; \wedge\text{L}
}{\forall x(A(x,b) \wedge C(c)) \Rightarrow A(a,b) \wedge C(c)}\; \wedge\text{R}
}{\forall x(A(x,b) \wedge C(c)) \Rightarrow \exists y(A(a,y) \wedge C(c))}\; \exists\text{R}
}{\cfrac{\exists y\forall x(A(x,y) \wedge C(c)) \Rightarrow \exists y(A(a,y) \wedge C(c))}{\exists y\forall x(A(x,y) \wedge C(c)) \Rightarrow \forall x\exists y(A(x,y) \wedge C(c))}\; \text{VR}}\; \exists\text{L}
$$

The inferences marked ! are critical (have eigenvariables). Note that the proof is regular.

1. Test your understanding of Corollary 5.21 by replacing the free variable b in $\pi_1(a,b)$ by another free variable d. Verify each inference in the resulting proof.

2. Test your understanding of Corollary 5.25 by replacing every occurrence of c in $\pi(c)$ by d. Verify that the result is a correct proof of

$$\exists y\forall x(A(x,y) \wedge C(d)) \Rightarrow \forall x\exists y(A(x,y) \wedge C(d)).$$

What happens if you replace c by a?

5.8 Translating NJ to LJ

Leaving J_1 as well as the classical systems **NK** and **LK** aside for the moment, we'll consider the relationship between **NJ** and **LJ** separately first. In an **NJ** deduction, a single formula A is deduced from a set of formulas Γ, the open assumptions of the deduction. In **LJ**, the result of a proof is a sequent of the form $\Gamma \Rightarrow A$. One might interpret such a sequent as the statement that A can be deduced from assumptions Γ. This interpretation suggests that there should be a natural translation of **LJ** deductions of the sequent $\Gamma \Rightarrow A$ to **NJ**-proofs of A from undischarged assumptions Γ, and vice versa. And in fact there is.[4] The translation from **NJ** to **LJ** is given already by Gentzen (1935c, §V 4).

Definition 5.27. If δ is an **NJ**-deduction, let As(δ) denote the set of all undischarged assumptions in δ. If Γ is a sequence of formulas, we write $\Gamma = $ As(δ) if every formula in Γ is an undischarged assumption of δ, and every undischarged assumption of δ occurs in Γ at least once. We write As$(\delta) \subseteq \Gamma$ if every formula in As(δ) occurs in Γ at least once.

Theorem 5.28. *If A has an **NJ**-deduction δ, then $\Gamma \Rightarrow A$ is provable in **LJ**, where $\Gamma = $ As(δ).*

Proof. We'll define a mapping P of **NJ**-deductions to **LJ**-proofs inductively, i.e., we'll define it for the smallest possible **NJ**-deductions (i.e., assumption formulas by themselves) and then show how to find the corresponding **LJ**-proof $P(\delta)$ for longer **NJ**-deductions δ based on the last inference in δ and the result of the mapping of the sub-deductions ending in the premises of the last inference. We do this so that the resulting proof $P(\delta)$ has end-sequent $\Gamma \Rightarrow A$ where A is the end-formula of δ and $\Gamma = $ As(δ). If δ is a deduction of A without undischarged assumptions, then Γ is empty and $P(\delta)$ is an **LJ**-proof of $\Rightarrow A$. If the end-formula of δ is $\bot$, then the end-sequent of $P(\delta)$ is $\Gamma \Rightarrow$.

$P(\delta)$ is inductively defined as follows:

1. *Basis clause:* If δ is just the assumption A by itself, let $P(\delta)$ be the proof consisting only of the initial sequent $A \Rightarrow A$. Obviously, $A = $ As(δ).

2. *Inductive clauses:* Suppose the last inference of δ is $\wedge$I. Then A is $B \wedge C$, δ has the form on the left, and $P(\delta)$ is defined as the **LJ**-proof on the right:

$$
\begin{array}{ccc}
\begin{array}{cc} \vdots \gamma & \vdots \gamma' \\ B & C \end{array} \\ \hline B \wedge C \end{array} {\scriptstyle \wedge\text{I}} \qquad \Rightarrow \qquad
\cfrac{\cfrac{\vdots P(\gamma)}{\cfrac{\Gamma \Rightarrow B}{\Gamma, \Gamma' \Rightarrow B}} \qquad \cfrac{\vdots P(\gamma')}{\cfrac{\Gamma' \Rightarrow C}{\Gamma, \Gamma' \Rightarrow C}}}{\Gamma, \Gamma' \Rightarrow B \wedge C} {\scriptstyle \wedge\text{R}}
$$

---

[4] A similar relationship holds between **NK** and **LK**, but because **LK**-sequents can have more than one formula on the right side but **NK** deductions are still deductions of single formulas, the translation is a bit more complicated to describe. The corresponding translations between **NM** and **LM**, however, are just like those for **NJ** and **LJ**, leaving out the treatment of wR and $\bot_J$.

(Recall our convention that double inference lines abbreviate multiple structural inferences. In this case we successively use WL to add all the formulas in Γ to the antecedent of the premises of the $\wedge$R inference.)

3. Suppose the last inference of δ is $\vee$I. Then A is $B \vee C$, δ has the form on the left, and $P(\delta)$ is defined as the LJ-proof on the right:

$$\frac{\begin{array}{c} \vdots \gamma \\ B \end{array}}{B \vee C}\ \vee\text{I} \qquad \Rightarrow \qquad \frac{\begin{array}{c} \vdots P(\gamma) \\ \Gamma \Rightarrow B \end{array}}{\Gamma \Rightarrow B \vee C}\ \vee\text{R}$$

If the premise of $\vee$I is C, then the right-hand side of the end-sequent of $P(\gamma)$ is also C, of course.

4. Now suppose the last inference of δ is $\supset$I. Then A is $B \supset C$, δ has the form on the left, and $P(\delta)$ is defined as the LJ-proof on the right:

$$\frac{\begin{array}{c} B^{\,i} \\ \vdots \\ \vdots \gamma \\ \vdots \\ C \end{array}}{B \supset C}\ i\ \supset\text{I} \qquad \Rightarrow \qquad \frac{\dfrac{\begin{array}{c} \vdots P(\gamma) \\ \Gamma \Rightarrow C \end{array}}{B, \Gamma' \Rightarrow C}\ *}{\Gamma' \Rightarrow B \supset C}\ \supset\text{R}$$

In this case, the last inference of δ can discharge assumptions of the form B which occur in γ. Recall that discharging of assumptions is permissive: $\supset$I need not discharge all assumptions of the form B, and in fact may not discharge any. This includes the case where γ has no open assumptions of the form B at all. If that is the case, then $As(\delta) = As(\gamma)$. Then we let $\Gamma' = \Gamma$ and derive $B, \Gamma' \Rightarrow C$ by WL in (*). If γ does in fact have assumptions B labeled by i (i.e., $B \in As(\gamma)$) and all of them are discharged (i.e., $B \notin As(\delta)$), then we derive (*) $B, \Gamma' \Rightarrow C$ from $\Gamma \Rightarrow C$ by interchanges and contractions so that Γ' no longer contains B. In either case, $\Gamma' = As(\delta)$.

5. Suppose the last inference of δ is $\forall$I. Then A is $\forall x\, B(x)$, δ has the form on the left, and $P(\delta)$ is defined as the LJ-proof on the right:

$$\frac{\begin{array}{c} \vdots \gamma \\ B(c) \end{array}}{\forall x\, B(x)}\ \forall\text{I} \qquad \Rightarrow \qquad \frac{\begin{array}{c} \vdots P(\gamma) \\ \Gamma \Rightarrow B(c) \end{array}}{\Gamma \Rightarrow \forall x\, B(x)}\ \forall\text{R}$$

The eigenvariable condition on $\forall$I guarantees that c does not occur in $\forall x\, B(x)$ or $As(\gamma)$, hence it also doesn't occur in Γ and the eigenvariable condition on $\forall$R at the end of $P(\gamma)$ is satisfied.

6. Suppose the last inference of δ is $\neg$I: exercise.

7. Suppose the last inference of δ is $\exists$I. If δ is the proof on the left, $P(\delta)$ is defined as the proof on the right:

$$
\frac{
\begin{array}{c}
\vdots\, \gamma \\
B(t)
\end{array}
}{\exists x\, B(x)}\ \exists\text{I}
\qquad \Rightarrow \qquad
\frac{
\begin{array}{c}
\vdots\, P(\gamma) \\
\Gamma \Rightarrow B(t)
\end{array}
}{\Gamma \Rightarrow \exists x\, B(x)}\ \exists\text{R}
$$

8. Now we consider the cases where δ ends in an elimination rule. The first case is where the end-formula of δ is the conclusion of $\wedge$E. If δ is the proof on the left, $P(\delta)$ is defined as the proof on the right:

$$
\frac{
\begin{array}{c}
\vdots\, \gamma \\
A \wedge B
\end{array}
}{A}\ \wedge\text{E}
\qquad \Rightarrow \qquad
\frac{
\begin{array}{c}
\vdots\, P(\gamma) \\
\Gamma \Rightarrow A \wedge B
\end{array}
\qquad
\dfrac{A \Rightarrow A}{A \wedge B \Rightarrow A}\ \wedge\text{L}
}{\Gamma \Rightarrow A}\ \text{CUT}
$$

9. Suppose δ ends in $\vee$E, i.e., δ has the form:

$$
\text{i}\ \frac{
\begin{array}{ccc}
 & B^i & C^i \\
\vdots\, \gamma & \vdots\, \theta & \vdots\, \lambda \\
B \vee C & A & A
\end{array}
}{A}\ \vee\text{E}
$$

If B is not discharged in θ or C not in λ (i.e., θ or λ do not contain assumptions of the form B or C labeled i, respectively), we obtain an **LJ**-proof $P(\delta)$ by

$$
\frac{
\begin{array}{c}
\vdots\, P(\theta) \\
\Theta \Rightarrow A
\end{array}
}{\Gamma \Rightarrow A}\ \text{WL}
\qquad \Rightarrow \qquad
\frac{
\begin{array}{c}
\vdots\, P(\lambda) \\
\Lambda \Rightarrow A
\end{array}
}{\Gamma \Rightarrow A}\ \text{WL}
$$

respectively. By inductive hypothesis, $\Theta = \text{As}(\theta)$. If, as we assume, the $\vee$E inference does not discharge assumptions of the form B, then $\Theta \subseteq \text{As}(\delta)$, and we can us WL to add the remaining formulas in $\Gamma = \text{As}(\delta)$. Similarly for λ and assumptions of the form C.

Otherwise, we can assume that B is discharged in θ and C in λ. Then we translate δ to **LJ** as follows:

$$
\cfrac{
\begin{array}{c}\vdots P(\gamma)\\ \Gamma \Rightarrow B \vee C\end{array}
\quad
\cfrac{
\cfrac{\begin{array}{c}\vdots P(\theta)\\ \Theta \Rightarrow A\end{array}}{B,\Theta',\Lambda' \Rightarrow A}
\quad
\cfrac{\begin{array}{c}\vdots P(\lambda)\\ \Lambda \Rightarrow A\end{array}}{C,\Theta',\Lambda' \Rightarrow A}
}{B \vee C, \Theta', \Lambda' \Rightarrow A}\ \text{VL}
}{\Gamma,\Theta',\Lambda' \Rightarrow A}\ \text{CUT}
$$

The structural inferences indicated by the double lines interchange and contract the B's appearing in Θ (and the C's in Λ) so that these come to stand at the far left of the sequent, and Θ' (Λ') are like Θ (Λ) except without B (C). In addition, we weaken with Λ' in the antecedent of the left premise and with Θ' in the antecedent of the right premise of the VL rule so that the side formulas in the premises match, as required by VL.

10. Suppose δ ends in $\supset$E:

$$
\cfrac{
\begin{array}{cc}\vdots \gamma & \vdots \theta\\ B \supset A & B\end{array}
}{A}\ \supset\text{E}
$$

We translate it to **LJ** as follows:

$$
\cfrac{
\begin{array}{c}\vdots P(\gamma)\\ \Gamma \Rightarrow B \supset A\end{array}
\quad
\cfrac{\begin{array}{cc}\begin{array}{c}\vdots P(\theta)\\ \Theta \Rightarrow B\end{array} & A \Rightarrow A\end{array}}{B \supset A, \Theta \Rightarrow A}\ \supset\text{L}
}{\Gamma, \Theta \Rightarrow A}\ \text{CUT}
$$

11. Suppose the last inference of δ is $\neg$E: exercise.

12. Suppose the last inference in δ is $\forall$E. Then A is $B(t)$. If δ is the proof on the left, $P(\delta)$ is defined as the proof on the right:

$$
\cfrac{\begin{array}{c}\vdots \gamma\\ \forall x\, B(x)\end{array}}{B(t)}\ \forall\text{E}
\qquad \Rightarrow \qquad
\cfrac{
\begin{array}{c}\vdots P(\gamma)\\ \Gamma \Rightarrow \forall x\, B(x)\end{array}
\quad
\cfrac{B(t) \Rightarrow B(t)}{\forall x\, B(x) \Rightarrow B(t)}\ \forall\text{L}
}{\Gamma \Rightarrow B(t)}\ \text{CUT}
$$

13. Suppose the last inference in δ is $\exists E$. If δ is the proof on the left, $P(\delta)$ is defined as the proof on the right:

$$
\cfrac{
\begin{array}{cc}
\begin{array}{c} B(c)^{\,i} \\ \vdots\,\gamma \\ \exists x\,B(x) \end{array} &
\begin{array}{c} \vdots\,\theta \\ A \end{array}
\end{array}
}{\displaystyle {}_{i}\ \frac{}{A}}\ \exists E
\qquad\Rightarrow\qquad
\cfrac{
\begin{array}{cc}
\begin{array}{c} \vdots\,P(\gamma) \\ \Gamma \Rightarrow \exists x\,B(x) \end{array} &
\cfrac{\cfrac{\cfrac{\vdots\,P(\theta)}{\Theta \Rightarrow A}}{B(c),\Theta' \Rightarrow A}\ \text{IL, CL}}{\exists x\,B(x),\Theta' \Rightarrow A}\ \exists L
\end{array}
}{\Gamma,\Theta' \Rightarrow A}\ \text{CUT}
$$

Let Θ' be Θ with all occurrences of $B(c)$ removed. We can derive $B(c), \Theta' \Rightarrow A$ from $\Theta \Rightarrow A$ using interchanges and contractions.

Since c is the eigenvariable of the last inference $\exists E$ in δ, c does not occur in any formulas in $As(\theta)$ other than $B(c)$, hence it does not occur in Θ'. It also occurs in neither A nor $\exists x\,B(x)$. So the eigenvariable condition on $\exists L$ is satisfied.

In the special case where $B(c)$ does not occur as an assumption in θ, $B(c)$ is not among the open assumptions of θ. Then $P(\theta)$ is a proof of $\Theta \Rightarrow A$, where Θ does not contain $B(c)$, and we obtain a proof of $\Gamma \Rightarrow A$ by WL inferences.

14. Lastly, a deduction δ ending in $\bot_J$ is translated to **LJ** as follows:

$$
\cfrac{\vdots\,\gamma}{\cfrac{\bot}{A}}\ \bot_J
\qquad\Rightarrow\qquad
\cfrac{\cfrac{\vdots\,P(\gamma)}{\Gamma \Rightarrow}}{\Gamma \Rightarrow A}\ \text{WR}
\qquad\qquad \square
$$

Problem 5.29. Complete the proof of Theorem 5.28 by extending the mapping P to cover the cases where δ ends with $\neg I$ or $\neg E$.

Problem 5.30. Extend the translation P to **NK**. Suppose δ ends in:

$$
\cfrac{
\begin{array}{c}
\neg A^{\,i} \\
\vdots\,\gamma \\
\end{array}
}{\displaystyle {}_{i}\ \frac{\bot}{A}}\ \bot_K
$$

How would you define $P(\delta)$ on the basis of $P(\gamma)$? You'll need to use inferences that are **LK** but not **LJ** inferences. Where does this happen?

5.9   Translating **LJ** to **NJ**

It is possible to reverse the translation, and obtain **NJ**-deductions of A from assumptions among Γ from an **LJ**-proof of $\Gamma \Rightarrow E$ as well, and thus obtain:

Theorem 5.31. *If $\Gamma \Rightarrow A$ has an **LJ**-proof π, then there is a deduction $D(\pi)$ of A in **NJ** with $\mathrm{As}(D(\pi)) \subseteq \Gamma$.*

Proof. By Proposition 5.23, we may assume that π is regular. We define a mapping of regular **LJ**-proofs π with end-sequent $\Gamma \Rightarrow [A]$ to **NJ**-deductions $D(\pi)$ of A (or $\bot$, if the succedent of the end-sequent is empty) with $\mathrm{As}(D(\pi)) \subseteq \Gamma$ by induction. (In this case it is easier to require just $\subseteq$ and not $=$.)

1. *Basis clause:* If π consists only of the initial sequent $A \Rightarrow A$, let $D(\pi)$ be the deduction consisting only of the assumption A.

Inductive clauses: We first treat all the cases where the **LJ**-proof π ends in a right rule. In each case, we take the translation of the sub-proof(s) ending in the premise(s), and add an ı-rule for the corresponding operator.

2. If π ends in $\wedge$R, A is $B \wedge C$, π is of the form given on the left, and $D(\pi)$ is defined as on the right:

$$
\cfrac{\begin{array}{cc} \vdots\,\gamma & \vdots\,\theta \\ \Gamma \Rightarrow B & \Gamma \Rightarrow C \end{array}}{\Gamma \Rightarrow B \wedge C}\,\wedge\text{R} \qquad \Rightarrow \qquad \cfrac{\begin{array}{cc} \vdots\,D(\gamma) & \vdots\,D(\theta) \\ B & C \end{array}}{B \wedge C}\,\wedge\text{I}
$$

The open assumptions of $D(\pi)$ are the assumptions of $D(\gamma)$ together with those of $D(\theta)$. As these are each $\subseteq \Gamma$, we have $\mathrm{As}(D(\pi)) \subseteq \Gamma$.

3. Suppose π ends in $\vee$R:

$$
\cfrac{\begin{array}{c} \vdots\,\gamma \\ \Gamma \Rightarrow B \end{array}}{\Gamma \Rightarrow B \vee C}\,\vee\text{R} \qquad \Rightarrow \qquad \cfrac{\begin{array}{c} \vdots\,D(\gamma) \\ B \end{array}}{B \vee C}\,\vee\text{I}
$$

4. Suppose π ends in $\supset$R:

$$
\cfrac{\begin{array}{c} \vdots\,\gamma \\ B, \Gamma \Rightarrow C \end{array}}{\Gamma \Rightarrow B \supset C}\,\supset\text{R} \qquad \Rightarrow \qquad i\,\cfrac{\begin{array}{c} B^{\,i} \\ \vdots\,D(\gamma) \\ C \end{array}}{B \supset C}\,\supset\text{I}
$$

The deduction ending in the premise C is obtained from $D(\gamma)$ by labelling (discharging) all open assumptions of the form B in it with the label i. This guarantees that $\mathrm{As}(D(\pi)) \subseteq \Gamma$ (even if Γ contains B).

5. Suppose π ends in ¬R: exercise.

6. Suppose π ends in ∀R:

$$
\begin{array}{ccc}
\begin{array}{c}
\vdots\, \gamma \\[2pt]
\hline
\dfrac{\Gamma \Rightarrow B(c)}{\Gamma \Rightarrow \forall x\, B(x)}\ \forall_R
\end{array}
&
\Rightarrow
&
\begin{array}{c}
\vdots\, D(\gamma) \\[2pt]
\dfrac{B(c)}{\forall x\, B(x)}\ \forall_I
\end{array}
\end{array}
$$

Since c must obey the eigenvariable condition in the ∀R inference, it cannot occur in Γ or $\forall x\, B(x)$. Since $\mathrm{As}(D(\gamma)) \subseteq \Gamma$, c can therefore also not occur in any open assumptions of $D(\gamma)$, and so the eigenvariable condition on ∀I is also satisfied.

7. Suppose π ends in ∃R:

$$
\begin{array}{ccc}
\begin{array}{c}
\vdots\, \gamma \\[2pt]
\hline
\dfrac{\Gamma \Rightarrow B(t)}{\Gamma \Rightarrow \exists x\, B(x)}\ \exists_R
\end{array}
&
\Rightarrow
&
\begin{array}{c}
\vdots\, D(\gamma) \\[2pt]
\dfrac{B(t)}{\exists x\, B(x)}\ \exists_I
\end{array}
\end{array}
$$

Now consider the cases where the **LJ**-proof π ends in a left rule. In these cases, we use a corresponding elimination rule where the major premise is an assumption. $[A]$ indicates that A may occur on the right side of the end-sequent of π, or the right side of the sequent may be empty. In the former case, the end-formula A^* of $D(\pi)$ is A; in the latter, A^* is $\bot$.

8. If π ends in ∧L, it is of the form given on the left. If $B \notin \mathrm{As}(D(\gamma))$, let $D(\pi) = D(\gamma)$. If B is an open assumption in $D(\gamma)$, let $D(\pi)$ be as given on the right:

$$
\begin{array}{ccc}
\begin{array}{c}
\vdots\, \gamma \\[2pt]
\hline
\dfrac{B, \Gamma \Rightarrow [A]}{B \wedge C, \Gamma \Rightarrow [A]}\ \wedge_L
\end{array}
&
\Rightarrow
&
\begin{array}{c}
\dfrac{B \wedge C}{B}\ \wedge_E \\[4pt]
\vdots\, D(\gamma) \\[2pt]
A^*
\end{array}
\end{array}
$$

The deduction on the right is obtained by writing $B \wedge C$ atop each open assumption of the form B in $D(\gamma)$ as a premise of an ∧E rule. Thus, $D(\pi)$ no longer has any open assumptions of the form B, but does have open assumptions of the form $B \wedge C$, i.e., $\mathrm{As}(D(\pi)) \subseteq B \wedge C, \Gamma$. Recall that π is regular, i.e., eigenvariables in γ occur only above the corresponding eigenvariable inference. In particular, no eigenvariable of γ occurs in $B \wedge C$. Hence no eigenvariable of $D(\gamma)$ occurs in the new open assumption $B \wedge C$, so all eigenvariable conditions in $D(\gamma)$ are still satisfied.

9. Suppose π ends in $\lor$L:

$$
\dfrac{\genfrac{}{}{0pt}{}{\vdots\,\gamma}{B,\Gamma\Rightarrow[A]} \quad \genfrac{}{}{0pt}{}{\vdots\,\theta}{C,\Gamma\Rightarrow[A]}}{B\lor C,\Gamma\Rightarrow[A]}\ \lor\text{L} \qquad\Rightarrow\qquad {}_i\dfrac{B\lor C \quad \genfrac{}{}{0pt}{}{B^i}{\genfrac{}{}{0pt}{}{\vdots\,D(\gamma)}{A^*}} \quad \genfrac{}{}{0pt}{}{C^i}{\genfrac{}{}{0pt}{}{\vdots\,D(\theta)}{A^*}}}{A^*}\ \lor\text{E}
$$

In $D(\pi)$ obtained this way, the assumptions of the form B in $D(\gamma)$ and those of the form C in $D(\theta)$ are labeled i and thus discharged. The major premise $B\lor C$ of the $\lor$E rule is a new open assumption.

10. Suppose π ends in $\supset$L:

$$
\dfrac{\genfrac{}{}{0pt}{}{\vdots\,\gamma}{\Gamma\Rightarrow B} \quad \genfrac{}{}{0pt}{}{\vdots\,\theta}{C,\Delta\Rightarrow[A]}}{B\supset C,\Gamma,\Delta\Rightarrow[A]}\ \supset\text{L} \qquad\Rightarrow\qquad \left.\dfrac{B\supset C \quad \genfrac{}{}{0pt}{}{\vdots\,D(\gamma)}{B}}{\genfrac{}{}{0pt}{}{C}{\genfrac{}{}{0pt}{}{\vdots\,D(\theta)}{A^*}}}\ \supset\text{E}\right\}\delta
$$

Here, the deduction δ ending in C has open assumptions among $B\supset C,\Gamma$. $D(\theta)$ has open assumptions C and Δ. By replacing each assumption C in $D(\theta)$ by the deduction δ ending in C, there are no longer open assumptions of the form C in $D(\theta)$. Together, $\mathrm{As}(D(\pi))\subseteq B\supset C,\Gamma,\Delta$. Again, because π was regular, no eigenvariable of θ occurs in $B\supset C$ or in γ (and consequently also not in $D(\gamma)$). So the eigenvariables of $D(\theta)$ do not occur in any of the new open assumptions. (Specifically: the conditions of Lemma 4.8 apply, and we can write $D(\pi)$ as $D(\theta)[\delta/C]$. The lemma guarantees that this is a correct deduction.)

11. Suppose π ends in $\neg$L: exercise.

12. Suppose π ends in $\forall$L:

$$
\dfrac{\genfrac{}{}{0pt}{}{\vdots\,\gamma}{B(t),\Gamma\Rightarrow[A]}}{\forall x\,B(x),\Gamma\Rightarrow[A]}\ \forall\text{L} \qquad\Rightarrow\qquad \genfrac{}{}{0pt}{}{\dfrac{\forall x\,B(x)}{B(t)}\ \forall\text{E}}{\genfrac{}{}{0pt}{}{\vdots\,D(\gamma)}{A^*}}
$$

Again, the deduction on the right is obtained by replacing every occurrence of an open assumption $B(t)$ in $D(\gamma)$ by a deduction of $B(t)$ from an open assumption $\forall x\,B(x)$. $\forall x\,B(x)$ contains no free variables not already in $B(t)$, so no eigenvariable inference in $D(\gamma)$ is disturbed by adding $\forall x\,B(x)$ as a new open assumption.

13. Suppose π ends in $\exists$L:

$$\cfrac{\vdots\,\gamma}{\cfrac{B(c), \Gamma \Rightarrow [A]}{\exists x\, B(x), \Gamma \Rightarrow [A]}\ {}_{\exists \text{L}}} \qquad \Rightarrow \qquad {}_{i}\,\cfrac{\exists x\, B(x) \qquad \begin{matrix} B(c)^{\,i} \\ \vdots \\ D(\gamma) \\ A^* \end{matrix}}{A^*}\ {}_{\exists \text{E}}$$

Since by the eigenvariable condition, c doesn't occur in $\exists x\, A(x), \Gamma \Rightarrow [A]$ and $\text{As}(D(\gamma)) \subseteq B(c), \Gamma$, c cannot occur in open assumptions of $D(\gamma)$ other than $B(c)$. Hence the eigenvariable condition on $\exists$E is satisfied.

It remains to consider the structural rules.

14. If π ends in WL, CL, IL, and γ is the deduction of the premise, let $D(\pi) = D(\gamma)$. Note that if π ends in WL, i.e.,

$$\cfrac{\vdots\,\gamma}{\cfrac{\Gamma \Rightarrow [A]}{B, \Gamma \Rightarrow [A]}\ {}_{\text{WL}}}$$

the open assumptions of $D(\gamma)$ need not contain B. For this reason we only get that $\text{As}(D(\pi)) \subseteq \Gamma, B$ and not that $\text{As}(D(\pi)) = \Gamma, B$.

15. Suppose π ends in WR:

$$\cfrac{\vdots\,\gamma}{\cfrac{\Gamma \Rightarrow}{\Gamma \Rightarrow A}\ {}_{\text{WR}}} \qquad \Rightarrow \qquad \cfrac{\vdots\,D(\gamma)}{\cfrac{\perp}{A}}\ {}_{\perp_J}$$

16. Lastly, suppose π ends in a CUT:

$$\cfrac{\begin{matrix}\vdots\,\theta \\ \Gamma \Rightarrow B\end{matrix} \qquad \begin{matrix}\vdots\,\lambda \\ B, \Delta \Rightarrow [A]\end{matrix}}{\Gamma, \Delta \Rightarrow [A]}\ {}_{\text{CUT}} \qquad \Rightarrow \qquad \begin{matrix}\vdots\,D(\theta) \\ B \\ \vdots\,D(\lambda) \\ A^*\end{matrix}$$

In the resulting **NJ**-deduction, the open assumptions of the form B in $D(\lambda)$ are each replaced by the deduction $D(\theta)$ of B, i.e, it is $D(\lambda)[D(\theta)/B]$. Since π is regular, no eigenvariable of λ occurs in θ, and so also no eigenvariable of $D(\lambda)$ occurs in $D(\theta)$. Hence the conditions of Lemma 4.8 are satisfied, and $D(\lambda)[D(\theta)/B]$ is a correct proof. The open assumptions of the combined deduction are those of $D(\theta) \subseteq \Gamma$ and of $D(\lambda) \setminus \{B\} \subseteq \Delta$.

Note that the cases where π ends in CR or IR are impossible, because the premise of such an inference would contain at least two occurrences of a formula, i.e., it would not be an intuitionistic sequent.

The reason we require $\mathrm{As}(D(\pi)) \subseteq \Gamma$ and not $\mathrm{As}(D(\pi)) = \Gamma$ lies in case 14. There is no simple equivalent to WL in **NJ**; we can't easily add undischarged assumptions to an **NJ**-deduction. Since we take the translation of a proof

$$\cfrac{\cfrac{\begin{matrix}\vdots\\ \gamma \\ \vdots\end{matrix}}{\Gamma \Rightarrow [A]}}{B,\Gamma \Rightarrow [A]}\ \text{WL} \qquad \text{to be} \qquad \begin{matrix}\vdots\\ D(\gamma)\\ \vdots\\ A^*\end{matrix}$$

the assumptions of $D(\gamma)$ and hence of $D(\pi)$ need not contain B. If we really wanted to, we could make the translation more complicated to guarantee that B was an undischarged assumption of $D(\pi)$, e.g., as follows:

$$\cfrac{1\ \cfrac{\cfrac{\cfrac{A^1 \quad B}{A \wedge B}\ \wedge\text{I}}{A}\ \wedge\text{E}}{A \supset A}\ \supset\text{I} \qquad \cfrac{\begin{matrix}\vdots\\ D(\gamma)\\ \vdots\\ A\end{matrix}}{}}{A}\ \supset\text{E} \qquad\qquad \square$$

Problem 5.32. Complete the proof of Theorem 5.31 by extending the mapping D to cover the cases of ¬R and ¬L.

Problem 5.33. In Problem 5.13(1) we showed that the following rule is derivable in **LK**:
$$\cfrac{\Gamma, \neg A \Rightarrow}{\Gamma \Rightarrow A}\ \neg\neg$$
Consider the calculus **LJ** + ¬¬, i.e., the sequent calculus in which sequents are restricted to at most one formula on the right, but which allows inferences using the ¬¬ rule. If Θ is $A_1, \ldots, A_n$, let ¬Θ be $\neg A_n, \ldots, \neg A_1$. Show that if $\Gamma \Rightarrow \Theta, A$ has a proof in **LK**, then $\Gamma, \neg\Theta \Rightarrow A$ has a proof in **LJ** + ¬¬.

Problem 5.34. Show that if $\Gamma \Rightarrow A_1, \ldots, A_n$ has a proof in **LK**, so does $\Gamma \Rightarrow A_1 \vee \cdots \vee A_n$.

Problem 5.35. Extend the translation D of **LJ** proofs to **NJ**-deductions to **LJ** + ¬¬ proofs and **NK**-deductions. Use this and the preceding two problems to show that if $\Gamma \Rightarrow A_1, \ldots, A_n$ has an **LK**-proof, then $A_1 \vee \cdots \vee A_n$ has an **NK**-deduction from open assumptions among Γ.

6

The cut-elimination theorem

Gentzen proved a "main theorem" in his original paper on the sequent calculus, which we devote this chapter to. It concerns the CUT rule and establishes a claim we've already hinted at a number of times: if there is a proof of a sequent that uses CUT, there is also one without. In other words, CUTS can be eliminated from proofs in **LK**, **LJ**, and **LM**, and so the main theorem is perhaps more descriptively called the cut-elimination theorem. Gentzen's original German term for the "main theorem" was "*Hauptsatz*"—and the cut-elimination theorem is now often called "Gentzen's *Hauptsatz*" or simply "the *Hauptsatz*."

Gentzen proves the cut-elimination theorem in §3 of Section III of the *Investigations into Logical Deduction* (Gentzen, 1935b). However, he does not directly show how to eliminate the CUT rule. Rather, he obtains the result from the fact that another rule, the MIX rule, which is deductively equivalent to the CUT rule, is eliminable from **LK** and **LJ**. Following Gentzen (first section of 3.1), we will do the same.

The MIX rule is:

$$\frac{\Gamma \Rightarrow \Theta \qquad \Delta \Rightarrow \Lambda}{\Gamma, \Delta^* \Rightarrow \Theta^*, \Lambda} \text{ MIX}$$

where we assume that Δ and Θ each contain at least one occurrence of the formula M, called the MIX formula, and that Δ^* and Θ^* are obtained from Δ and Θ, respectively, by erasing every occurrence of M.

The equivalence of the CUT rule with the MIX rule, modulo the remaining rules of **LK**, follows from the following two results:

Proposition 6.1. *The MIX rule is derivable in **LK**.*

Proof. Let $\Gamma \Rightarrow \Theta$ and $\Delta \Rightarrow \Lambda$ be two derivable sequents in **LK**. Then the sequent $\Gamma, \Delta^* \Rightarrow \Theta^*, \Lambda$ is also derivable in **LK**. In fact:

$$\cfrac{\cfrac{\cfrac{\Gamma \Rightarrow \Theta}{\Gamma \Rightarrow \Theta^*, M, \ldots, M} \text{ IR}}{\Gamma \Rightarrow \Theta^*, M} \text{ CR} \qquad \cfrac{\cfrac{\Delta \Rightarrow \Lambda}{M, \ldots, M, \Delta^* \Rightarrow \Lambda} \text{ IL}}{M, \Delta^* \Rightarrow \Lambda} \text{ CL}}{\Gamma, \Delta^* \Rightarrow \Theta^*, \Lambda} \text{ CUT}$$

An Introduction to Proof Theory: Normalization, Cut-Elimination, and Consistency Proofs.
Paolo Mancosu, Sergio Galvan, and Richard Zach, Oxford University Press. © Paolo Mancosu, Sergio Galvan and Richard Zach 2021. DOI: 10.1093/oso/9780192895936.003.0006

Note that IR (respectively IL) moves all occurrences of M right (respectively left) and CR (respectively CL) is used if M occurs more than once.[1]   □

Proposition 6.2. *The* CUT *rule is derivable in* **LK** − CUT + MIX.

Proof. Let $\Gamma \Rightarrow \Theta, M$ and $M, \Delta \Rightarrow \Lambda$ be two sequents derivable in **LK** − CUT + MIX. Then $\Gamma, \Delta \Rightarrow \Theta, \Lambda$ is also derivable. In fact:

$$\frac{\dfrac{\Gamma \Rightarrow \Theta, M \qquad M, \Delta \Rightarrow \Lambda}{\Gamma, \Delta^* \Rightarrow \Theta^*, \Lambda} \text{ MIX}}{\Gamma, \Delta \Rightarrow \Theta, \Lambda} \text{ WL, WR, IL, IR}$$

Note that the last passage is to be carried out only if the occurrences of M eventually eliminated by the application of MIX have to be restored.   □

Given the equivalence of the CUT rule and the MIX rule, Gentzen can restate the cut-elimination theorem as follows.

Theorem 6.3 (Cut-elimination theorem, *Hauptsatz*). *The* MIX *rule is eliminable. That is, given any proof of* **LK** − CUT + MIX *it is possible to construct another proof with the same end-sequent but that does not contain any* MIX.

At this point Gentzen introduces a lemma (also stated in §3.1 of Gentzen, 1935b) in virtue of which any proof which ends with a MIX but not containing any other MIXes can be transformed into a proof of the same end-sequent without any MIX. From the lemma (our Lemma 6.10) it is easy to obtain the main theorem.

Indeed, given any proof let us consider the part of the proof that ends with the first MIX (if two MIXes are at the same level let us choose the rightmost MIX). This MIX is eliminable in virtue of the lemma and consequently the part of the proof ending with the first MIX can be replaced by a proof without MIX. One can proceed to remove that part of the proof containing the second application of MIX in the original proof. By the lemma, also this MIX can be eliminated. By iterating the process, it is therefore possible to eliminate, one by one, all the occurrences of MIX which appear in the original proof.

In our proof of the lemma we will follow Gentzen's exposition quite closely. However, we will rearrange and complete the proof so that the double induction at its core can be made explicit and fully clear. We will, after the introduction of a few preliminary definitions and before entering the proper details of the demonstration, begin with an outline of the proof. This outline will be helpful in understanding the structure of the proof. A summary of the proof is also provided in Appendix G. (The proof involves an unusually large number of lemmas and cases, so such a roadmap might be especially helpful here.)

---

[1] Recall that we indicate multiple applications of structural rules c, w, i by a double inference line.

We should note at this point that we are proving the cut-elimination theorem by transforming a given proof (possibly including CUT) into a cut-free proof. In this respect, the result is akin to the normalization results in natural deduction discussed in Section 4.1. There we distinguished normalization proofs which transform given proofs from normal form proofs, which merely establish the existence of normal deductions without transforming a given proof. In the case of the sequent calculus we can also consider proofs that every provable sequent has a cut-free proof, which don't proceed by transforming a given proof. In this context, one often speaks of "cut-free completeness:" The cut-free sequent calculus is complete in the sense that every valid sequent has a proof. Such proofs, using semantic methods, are fairly common. For instance, Takeuti (1987) contains cut-free completeness proofs for **LJ** and **LK** in addition to cut-elimination proofs.

6.1   Preliminary definitions

Recall that the degree of a formula is the number of logical symbols in it (see Section 2.3).

Definition 6.4 (Degree of a MIX). The degree of a MIX is the degree of the MIX formula of that MIX.

In the remaining definitions we will take into consideration proofs that end with a MIX but that do not contain other MIXes. For this reason, when the context is unambiguous, we will at times understand by "proof" a proof that ends up with a MIX and that does not contain other MIXes.

Definition 6.5 (Degree of a proof). The degree $dg(\pi)$ of a proof π ending with a MIX is given by the degree of the MIX with which it terminates.

Definition 6.6 (Right (left) branch). We call the sub-proof of π ending with the right premise (the left premise, respectively) of the MIX with which the proof ends the *right branch* (*left branch*) of π. Sometime we will use the symbols π_1 and π_2 to indicate the branches of a proof.

Definition 6.7 (Rank of a proof). The *left rank* $rk_l(\pi)$ of a proof π ending with a MIX is the maximum number of consecutive sequents which constitute a path that terminates with the left premise of the MIX and which contain the MIX formula in the succedent. A path is understood as a sequence of sequents in which the successor relation coincides with the relation of being one of the premises of the immediately following sequent.

The *right rank* $rk_r(\pi)$ of a proof π ending with a MIX is the maximum number of consecutive sequents which constitute a path that terminates with the right premise of the MIX and which contain the MIX formula in the antecedent.

The *rank* $rk(\pi)$ is the sum of its left and right rank, i.e., $rk(\pi) = rk_l(\pi) + rk_r(\pi)$.

Example 6.8. We box the MIX formula for clarity:

$$
\cfrac{
F(a) \Rightarrow F(a)
}{Fa \Rightarrow \boxed{\exists x\, F(x)}} \; \exists\text{R}
\qquad
\cfrac{
\cfrac{
\cfrac{\boxed{\exists x\, F(x)} \Rightarrow \exists x\, F(x)}{\neg \exists x\, F(x),\, \boxed{\exists x\, F(x)} \Rightarrow} \; \neg\text{L}
}{\boxed{\exists x\, F(x)},\, \neg \exists x\, F(x) \Rightarrow} \; \text{IL}
}{}
$$

$$
\frac{\qquad\qquad\qquad\qquad\qquad\qquad\qquad}{F(a),\, \neg \exists x\, F(x) \Rightarrow} \; \text{MIX}
$$

Here, the left rank $\mathrm{rk}_l(\pi)$ is 1, the right rank $\mathrm{rk}_r(\pi)$ is 3, so the rank $\mathrm{rk}(\pi) = 1 + 3 = 4$.

Example 6.9. In the proof below, M is taken to be the MIX formula and we assume that in the part of the proof indicated with ... there are no occurrences of M relevant for the rank.

$$
\frac{
\cfrac{\Gamma \Rightarrow \boxed{M}}{\cfrac{\Gamma \Rightarrow \boxed{M},A}{\neg A, \Gamma \Rightarrow \boxed{M}} \; \neg\text{L}} \; \text{WR}
\qquad
\cfrac{
\cfrac{\cfrac{\boxed{M} \Rightarrow}{\boxed{M} \Rightarrow D} \; \text{WR}}{\boxed{M} \Rightarrow D, B} \; \text{WR}
\qquad
\cfrac{\Rightarrow D, C}{\boxed{M} \Rightarrow D, C} \; \text{WL}
}{\boxed{M} \Rightarrow D, B \wedge C} \; \wedge\text{R}
}{\neg A, \Gamma \Rightarrow D, B \wedge C} \; \text{MIX}
$$

We have: $\mathrm{rk}_l(\pi) = 3$, $\mathrm{rk}_r(\pi) = 4$, so $\mathrm{rk}(\pi) = 3 + 4 = 7$.

Note that a proof ending in MIX must have at least rank 2. Indeed, it is necessary that the MIX formula occurs at least in the succedent of the left premise and in the antecedent of the right premise. That is, the left rank is always at least 1, and so is the right rank.

At this point we have at our disposal the fundamental notions for the formulation and the proof of the lemma. The notions of degree and rank of a proof (ending with a MIX) are of the essence. The proof of the lemma proceeds by a double induction on the degree of the MIX (primary induction) and on the rank of a proof (ending with a MIX). The aspects from which our presentation differs from Gentzen's original proof concern the rearrangement of the procedure and the completion of all the essential steps. In this way we will make the double induction more explicit than in the original proof. We have preferred to insist on the completeness for systematic and pedagogical reasons. You might want to take advantage of the complete presentation of the steps by first trying to work out the details yourself, and comparing them to the detailed exposition we give.

Lemma 6.10 (Main Lemma). *Any regular proof containing a MIX with a single application occurring in its final step can be transformed into a proof with the same end-sequent in which MIX is not applied. A proof that can be transformed in such a way is said to be reducible.*

Note that Proposition 5.23 guarantees that any proof can be transformed into a regular proof.

Let us begin with the structural schema of the proof. In this way we can grasp the idea of how the proof is articulated. Besides, since the proof proceeds by a double induction, the schema will bring to light the structure of the induction.

6.2   Outline of the lemma

As already mentioned, the proof proceeds by a double induction on the degree of the MIX and on the rank of the proof.

The meta-theoretical proposition which is established by induction is: "If π is an **LK**-proof containing as its last inference a single MIX of degree $dg(\pi)$ and with rank $rk(\pi)$, then there exists a proof π' of the same end-sequent, but without any application of MIX." Using the notion "reducible" introduced above we can shorten the statement to: "Every **LK**-proof π containing as its last inference a single MIX of degree $dg(\pi)$ and with rank $rk(\pi)$ is reducible."

What does it mean then to prove such a proposition by double induction? It means to prove that the proposition holds for proofs with minimal complexity (namely those with $dg(\pi) = 0$ and $rk(\pi) = 2$) and that, assuming by inductive hypothesis that the proposition holds for any proof "less complex than" π, it will also hold for π.

The complexity of π is given by its degree and rank. When is a proof "less complex" than another? We proceed as in Section 4.2, and order pairs $\langle d, r \rangle$ in the standard way there introduced. In other words, a proof π_1 is less complex than a proof π_2 iff either (a) $dg(\pi_1) < dg(\pi_2)$ or (b) $dg(\pi_1) = dg(\pi_2)$ but $rk(\pi_1) < rk(\pi_2)$.

The choice of the order of the parameters on which the induction will be carried out is important. Its choice depends on the "weight" which we attribute to the parameter. We will give more weight to the degree $dg(\pi)$. That the degree of a proof has more "weight" means that the decrease, even by a single unit, of $dg(\pi)$ implies that the inductive hypothesis holds for any value of the other parameter, that is the rank $rk(\pi)$. By contrast, it is not the case that the decrease of the rank implies that the conclusion of the inductive step holds for instances of greater degree.

Once this order has been fixed we can now make explicit how the inductive hypothesis in the proof will work. Here is the schema.

1. *Induction basis:* $dg(\pi) = 0$ and $rk(\pi) = 2$.

We prove that every proof of degree 0 and rank 2 is reducible. We will do this in Section 6.3.

2. *Inductive hypothesis:*

Every proof of lower complexity than π is reducible. In other words, π' is reducible, provided either (a) $dg(\pi') < dg(\pi)$ or (b) $dg(\pi') = dg(\pi)$ and $rk(\pi') < rk(\pi)$.

3. *Inductive step:*

Assume the inductive hypothesis is true. Then π is reducible.

The inductive step divides into two parts. The first part is the case when $\mathrm{rk}(\pi) = 2$, i.e., the rank of the MIX is minimal. A proof of lower complexity than π in this case is any proof π' with $\mathrm{dg}(\pi') < \mathrm{dg}(\pi)$. So, this case can be dealt with if we can show that π is reducible provided any proof of lower degree is reducible. We can exploit the fact that $\mathrm{rk}(\pi) = 2$. This means the MIX formula must be introduced by an inference rule on both sides of the MIX, and it cannot have more than one occurrence on either side. In other words, the MIX formula is the principal formula of a weakening, or of a R-rule on the left and a L-rule on the right. In these cases we can show that the same end-sequent can be derived using proofs that end in MIXes on the premises of these rules. Where the MIX-formula is the result of a weakening, we weaken instead with a sub-formula of the original MIX-formula. If the MIX-formula is the result of an operational rule, the new MIX-formulas are the side-formulas of the original operational inferences. The degree of the MIXes will be $< \mathrm{dg}(\pi)$, and we can apply the induction hypothesis. Simplified, the case looks something like this. We begin with a proof π:

$$
\begin{array}{cc}
\vdots\ \pi_1 & \vdots\ \pi_2 \\[4pt]
\dfrac{\Gamma \Rightarrow \Pi}{\Gamma \Rightarrow \Theta}\ R_1 \qquad & \dfrac{\Psi \Rightarrow \Lambda}{\Delta \Rightarrow \Lambda}\ R_2 \\[8pt]
\end{array}
$$
$$
\dfrac{}{\Gamma, \Delta^* \Rightarrow \Theta^*, \Lambda}\ \text{MIX}
$$

The MIX formula in this case is M, R_1 is a R-rule for the main operator of M, and R_2 the corresponding L-rule. For instance, M might be $\forall x\, B(x)$, R_1 is then $\forall$R and R_2 is $\forall$L. In this simplified scenario, both R_1 and R_2 have only one premise, and no auxiliary formulas in Γ or Λ, but this won't always be the case.

We now combine the proofs π_1 and π_2 directly using MIX, but with an immediate sub-formula M' of M as MIX-formula (e.g., $B(t)$):

$$
\begin{array}{cc}
\vdots\ \pi_1 & \vdots\ \pi_2 \\[4pt]
\dfrac{\Gamma \Rightarrow \Pi \qquad \Psi \Rightarrow \Lambda}{\Gamma, \Psi^* \Rightarrow \Pi^*, \Lambda}\ \text{MIX}
\end{array}
$$

Because $\mathrm{rk}(\pi) = 2$, M cannot appear in Π or Ψ. The inductive hypothesis applies to the new proof: there is a MIX-free proof of $\Gamma, \Psi^* \Rightarrow \Pi^*, \Lambda$ where Π^* and Ψ^* are Π and Ψ, respectively, without M'. We then show that using this proof we can obtain $\Gamma, \Delta^* \Rightarrow \Theta^*, \Lambda$ without MIX. Of course, there are numerous cases to verify. We deal with this case, i.e., the case where $\mathrm{rk}(\pi) = 2$, in Section 6.4.

The second case is when $\mathrm{rk}(\pi) > 2$. Here, the induction hypothesis applies in particular to all proofs π' with $\mathrm{dg}(\pi') = \mathrm{dg}(\pi)$ and $\mathrm{rk}(\pi') < \mathrm{rk}(\pi)$. Since $\mathrm{rk}(\pi) > 2$, either $\mathrm{rk}_r(\pi) > 1$ or $\mathrm{rk}_l(\pi) > 1$. This means that either the MIX-formula occurs in the antecedent of a premise of the rule leading to the right premise of the MIX, or it appears in the succedent of a premise of the rule leading to the left premise, respectively. We show that, regardless of which rules are applied, it is always possible to obtain the end-formula by an application of MIX of lower rank. Again, a very simplified schema is this: We begin with π where $\mathrm{rk}(\pi) > 2$, e.g., when $\mathrm{rk}_r(\pi) > 1$.

$$
\cfrac{\Gamma \Rightarrow \Theta \quad \cfrac{\Psi \Rightarrow \Lambda \quad \Delta \Rightarrow \Lambda}{\Delta \Rightarrow \Lambda}\ R}{\Gamma, \Delta^* \Rightarrow \Theta^*, \Lambda}\ \text{MIX}
$$

This means the MIX-formula M occurs in Ψ, the antecedent of a premise of the rule leading to the right premise of the MIX. We now consider the proof

$$
\cfrac{\Gamma \Rightarrow \Theta \quad \Psi \Rightarrow \Lambda}{\Gamma, \Psi^* \Rightarrow \Theta^*, \Lambda}\ \text{MIX}
$$

where the MIX-formula is again M. This proof has degree $\mathrm{dg}(\pi)$, since the MIX-formula is the same. It has rank $\mathrm{rk}(\pi) - 1$, since we have removed one sequent in any path leading to the right premise containing M in the antecedent (namely, we have removed $\Delta \Rightarrow \Lambda$). So, the inductive hypothesis applies to this proof, and we can assume that $\Gamma, \Psi^* \Rightarrow \Theta^*, \Lambda$ has a proof without MIX. We then verify that using this proof, we can obtain a MIX-free proof of $\Gamma, \Delta^* \Rightarrow \Theta^*, \Lambda$, typically by adding R below the end-formula, plus some structural inferences. In simple cases, when M is not the principal formula of R operating on the antecedent, this works. In more complicated cases, where M is the principal formula, we will then have re-introduced the MIX-formula M. For instance, we might have a MIX-free proof like this:

$$
\cfrac{\cfrac{\Gamma, \Psi^* \Rightarrow \Theta^*, \Lambda}{\Psi^*, \Gamma \Rightarrow \Theta^*, \Lambda}\ \text{IL}}{M, \Psi^*, \Gamma \Rightarrow \Theta^*, \Lambda}\ R
$$

However, since M does not appear in Ψ^*, we can consider another proof ending in a MIX on M:

$$
\cfrac{\Gamma \Rightarrow \Theta \quad \cfrac{\cfrac{\Gamma, \Psi^* \Rightarrow \Theta^*, \Lambda}{\Psi^*, \Gamma \Rightarrow \Theta^*, \Lambda}\ \text{IL}}{M, \Psi^*, \Gamma \Rightarrow \Theta^*, \Lambda}\ R}{\Gamma, \Psi^*, \Gamma \Rightarrow \Theta^*, \Theta^*, \Lambda}\ \text{MIX}
$$

This proof also has degree $dg(\pi)$ and rank $< rk(\pi)$, since its right rank is 1. The inductive hypothesis applies, and we obtain a MIX-free proof of $\Gamma, \Psi^*, \Gamma \Rightarrow \Theta^*, \Theta^*, \Lambda$. Using interchanges and contractions only, we obtain the end-sequent of π, i.e., $\Gamma, \Psi^* \Rightarrow \Theta^*, \Lambda$. We carry out this case in detail in Section 6.5.

This shows that all proofs of arbitrary degree and rank are reducible.

Gentzen does not make the inductive structure of the proof explicit. His proof of the Main Lemma is divided into section 3.11 (corresponding to $rk(\pi) = 2$) and section 3.12 (corresponding to $rk(\pi) > 2$). In our reconstruction we will refer to these sections and their subsections by prefixing the letter 'G'.

Let us now move on to the detailed proof. We remark that in the presentation the specific parts of the induction will be given with explicit reference to Gentzen's original work. This will allow the reader to see how the single pieces of his demonstration fit in the frame of the double induction that we have taken care to make fully explicit.

6.3   Removing MIXes directly

There is a case where the MIX ending π can be removed outright, namely when the MIX-formula is introduced immediately above the MIX either by weakening, or in an axiom, and on the side where it is so introduced, there is only one occurrence of the MIX formula. Let's see how this is done.

Lemma 6.11. *If π contains a single MIX as its last inference, and one premise of the MIX is an axiom, then there is a proof of the end-sequent without MIX.*

Proof. π has one of the following forms:

$$
\begin{array}{cc}
\vdots\ \pi_1 & \vdots\ \pi_1 \\
\dfrac{M \Rightarrow M \qquad \Delta \Rightarrow \Lambda}{M, \Delta^* \Rightarrow \Lambda}\ \text{MIX} & \dfrac{\Gamma \Rightarrow \Theta \qquad M \Rightarrow M}{\Gamma \Rightarrow \Theta^*, M}\ \text{MIX}
\end{array}
$$

In both cases the proofs can be transformed so as to obtain the same end-sequent without MIXes. Let us consider only the first case, and leave the second as an exercise.

$$
\dfrac{\dfrac{\dfrac{\vdots\ \pi_1}{\Delta \Rightarrow \Lambda}}{M, \ldots, M, \Delta^* \Rightarrow \Lambda}\ \text{IL}}{M, \Delta^* \Rightarrow \Lambda}\ \text{CL}
$$

Since π_1 contains no MIX, the new proof does not contain any MIX. (Gentzen covers this case in G.3.111 and G.3.112.) Note that in this case, M may be of any complexity. $\qquad\square$

Problem 6.12. Show how to eliminate the MIX in the second case.

Lemma 6.13. *If π contains a single MIX as its last inference, (a) the left premise of the MIX is a WR or the right premise is a WL inference, (b) the MIX formula M is the principal formula of the weakening, and (c) M does not occur in the succedent or antecedent, respectively, of the premises of the weakening, then there is a proof of the end-sequent without MIX.*

Proof. π has one of the following forms:

$$
\begin{array}{cc}
\genfrac{}{}{0pt}{}{\vdots\ \pi_1}{} & \\
\dfrac{\Gamma \Rightarrow \Theta}{\dfrac{\Gamma \Rightarrow \Theta, M}{\ }\ \text{WR}} & \vdots\ \pi_2
\end{array}
\qquad
\dfrac{\Gamma \Rightarrow \Theta, M \qquad \Delta \Rightarrow \Lambda}{\Gamma, \Delta^* \Rightarrow \Theta, \Lambda}\ \text{MIX}
$$

$$
\dfrac{\Gamma \Rightarrow \Theta \qquad \dfrac{\Delta \Rightarrow \Lambda}{M, \Delta \Rightarrow \Lambda}\ \text{WL}}{\Gamma, \Delta \Rightarrow \Theta^*, \Lambda}\ \text{MIX}
$$

By condition (c), the M introduced by the weakening is the only occurrence of M in the succedent Θ, M of the left premise of the MIX, or in the antecedent M, Δ of the right premise of the MIX, respectively. In particular, in the first case, Θ does not contain M and so $\Theta^* = \Theta$; in the second case Δ does not contain M and so $\Delta^* = \Delta$.

In both cases the proofs can be transformed so as to obtain the same end-sequent without MIXes.

$$
\dfrac{\dfrac{\dfrac{\dfrac{\Gamma \Rightarrow \Theta}{\Delta^*, \Gamma \Rightarrow \Theta}\ \text{WL}}{\Gamma, \Delta^* \Rightarrow \Theta}\ \text{IL}}{\Gamma, \Delta^* \Rightarrow \Theta, \Lambda}\ \text{WR}}{}
\qquad\qquad
\dfrac{\dfrac{\dfrac{\dfrac{\Delta \Rightarrow \Lambda}{\Gamma, \Delta \Rightarrow \Lambda}\ \text{WL}}{\Gamma, \Delta \Rightarrow \Lambda, \Theta^*}\ \text{WR}}{\Gamma, \Delta \Rightarrow \Theta^*, \Lambda}\ \text{IR}}{}
$$

(which corresponds to G.3.113.1)   (which corresponds to G.3.113.2)

Since π_1 and π_2 do not contain MIX, we have thus a proof with the same end-sequent and without MIX. Note that M may be of any complexity, which will allow us to use this lemma also in the proof of the inductive steps.[2]   □

Lemmas 6.11 and 6.13 immediately yield the basis of the induction on rank and degree of the MIX in π, i.e., the case where both rank and degree are minimal. The least possible degree is 0; the least possible rank is 2. We just have to verify that indeed when $\mathrm{dg}(\pi) = 0$ and $\mathrm{rk}(\pi) = 2$, there is a proof of the same end-sequent without MIX. When we consider all the possible cases of how the premises of the

---

[2] Care must be taken here in adapting the proof for **LJ**, where the WR inference is restricted. In that case, however, either Θ or Λ is empty. **LM** does not include the WR rule, so there cannot be a proof where the left premise of a MIX is the conclusion of a WR.

MIX could be obtained, we notice that all but the cases treated by Lemmas 6.11 and 6.13 are incompatible with $dg(\pi) = 0$ and $rk(\pi) = 2$.

Lemma 6.14. *If* $dg(\pi) = 0$ *and* $rk(\pi) = 2$, *then there is a proof of the same end-sequent without* MIX.

Proof. We consider the possible cases:

A. One of the premises of the MIX is an axiom. Then the conclusion follows by Lemma 6.11.

B. One of the premises is the result of weakening. Suppose the weakening occurs above the left premise. Since $rk_l(\pi) = 1$, M does not occur in the succedent of the premise of the weakening. This means that the weakening can only be WR with M as principal formula. Similarly, if the weakening occurs on the right, the inference must be WL with M as principal formula. Now the conditions of Lemma 6.13 are satisfied, and the conclusion follows.

C. One of the premises is the result of contraction or interchange. In that case, the MIX formula M must occur in the premise of that contraction or interchange. Since $rk_l(\pi) = rk_r(\pi) = 1$, this is impossible.

D. One of the premises is the result of an operational rule. Since we have that $rk(\pi) = 2$, both $rk_l(\pi) = 1$ and $rk_r(\pi) = 1$. This means that the MIX formula M can appear neither in the succedent of the sequents immediately above the left premise of the MIX nor in the antecedent of the sequents immediately above the right premise of the MIX. Thus in order to appear in the conclusion of an operational rule that ends in a premise of the MIX, the formula M would have to occur as its principal formula. But this is impossible since $dg(\pi) = 0$ and thus M cannot be a complex formula, something which would have to be the case if M were the principal formula of an operational rule.   □

It is useful to deal with the case in which M occurs not only as MIX formula but at the same time also in the antecedent of the left premise of the MIX, or in the succedent of the right premise of the MIX (G.3.121.1). This is not part of the induction basis, but will be used in the proof of Lemma 6.17.

Lemma 6.15. *If* π *ends in a* MIX, *and the* MIX *formula occurs in the antecedent of the left premise of the* MIX *or the succedent of the right premise of the* MIX, *then* π *is reducible.*

Proof. Consider the case where the MIX-formula M (which may be of any complexity) occurs in the antecedent of the left premise of the MIX. Then π is the proof below, where M occurs in Π.

$$\frac{\overset{\vdots}{\underset{\Pi \Rightarrow \Sigma}{\pi_1}} \quad \overset{\vdots}{\underset{\Delta \Rightarrow \Lambda}{\pi_2}}}{\Pi, \Delta^* \Rightarrow \Sigma^*, \Lambda} \text{ MIX}$$

In this case, we can replace the entire proof with:

$$
\begin{array}{c}
\vdots \; \pi_2 \\[4pt]
\dfrac{\varDelta \Rightarrow \varLambda}{M,\dots,M,\varDelta^* \Rightarrow \varLambda} \; \text{IL} \\[4pt]
\hline
\dfrac{M,\dots,M,\varDelta^* \Rightarrow \varLambda,\varSigma^*}{} \; \text{WR} \\[4pt]
\hline
\varPi,\varDelta^* \Rightarrow \varSigma^*,\varLambda \; \text{WL, I, C}
\end{array}
$$

Indeed, the presence of M in $\varPi$ guarantees that the derivability of $\varDelta \Rightarrow \varLambda$ can be transferred to the end-sequent through several applications of w, ı, and c in the antecedent and the succedent. But, thus transformed, the proof does not have any MIX, since π_2 did not contain a MIX.

The case where M occurs in $\varLambda$ is treated similarly. □

Let us look at an example. Consider the following proof:

$$
\dfrac{M,A \Rightarrow B,C \supset D,M \qquad A \vee C,M,B,M \Rightarrow A \vee B}{M,A,A \vee C,B \Rightarrow B,C \supset D,A \vee B} \; \text{MIX}
$$

It can be transformed in the following way:

$$
\dfrac{A \vee C,M,B,M \Rightarrow A \vee B}{M,A,A \vee C,B \Rightarrow B,C \supset D,A \vee B} \; \text{WL, IL, CL, WR, IR}
$$

6.4   Reducing the degree of MIX

Lemma 6.16. *Suppose* $\mathrm{dg}(\pi) > 0$ *and* $\mathrm{rk}(\pi) = 2$, *and assume that every proof* π' *with* $\mathrm{dg}(\pi') < \mathrm{dg}(\pi)$ *is reducible. Then there is a proof of the same end-sequent as* π *without* MIX.

Proof. We again distinguish cases.

A. One of the premises is an axiom. The result follows by Lemma 6.11.

B. One of the premises is a weakening. Since $\mathrm{rk}(\pi) = 2$, by the same reasons as in Lemma 6.14, case B, we can apply Lemma 6.13, and obtain a MIX-free proof of the same end-sequent.

C. One of the premises is a contraction or interchange. By the same reasons as in Lemma 6.14, case C, this case is ruled out by the assumption that $\mathrm{rk}(\pi) = 2$.

D. The remaining case is one where both premises are the results of operational rules. Since $\mathrm{rk}(\pi) = 2$, the MIX formula cannot occur in the succedent of any premise of the rule leading to the left premise of the MIX, nor in the antecedent of a premise of the rule leading to he right premise of the MIX. In other words, the two premises of the MIX are the result of the application of some operational rule in which M occurs as principal formula. There are six of them in accordance with the number of possibilities for the main logical operator of M (G.3.113.31–36).

D1. $M = A \wedge B$.

The final part of the proof has the following form:

$$
\cfrac{
  \cfrac{
    \cfrac{\vdots \pi_1}{\Gamma_1 \Rightarrow \Theta_1, A} \quad \cfrac{\vdots \pi_2}{\Gamma_1 \Rightarrow \Theta_1, B}
  }{\Gamma_1 \Rightarrow \Theta_1, A \wedge B} \wedge R
  \quad
  \cfrac{\cfrac{\vdots \pi_3}{A, \Gamma_2 \Rightarrow \Theta_2}}{A \wedge B, \Gamma_2 \Rightarrow \Theta_2} \wedge L
}{\Gamma_1, \Gamma_2 \Rightarrow \Theta_1, \Theta_2} \text{ MIX}
$$

It is worth mentioning that the application of MIX does not affect in any way Θ_1 or Γ_2 but only the elimination of the displayed MIX formula $A \wedge B$. How is this fact to be explained? It depends from the circumstance that since $\mathrm{rk}(\pi) = 2$, the right rank and the left rank are both equal to 1. Let us consider the left rank. The fact that it is equal to 1 implies that the MIX formula $A \wedge B$ occurs, with respect to the left branch of the MIX, only in the succedent of the premise of the MIX. On the other hand, the application of the operational rule $\wedge R$ does not induce any change in the part Θ_1 of the succedent of the same premise. For these reasons the MIX only involves the elimination of the formula $A \wedge B$ from the succedent of the left premise.

An analogous argument also holds for the right branch of the MIX and from the argument it follows that the MIX does not induce any change in Γ_2 but only involves the elimination of $A \wedge B$ from the antecedent of the corresponding premise. Hence we end up with the particular form of MIX in which there are no changes in either Θ_1 or Γ_2 but which only involves the elimination of the displayed MIX formula $A \wedge B$.

The proof can be transformed into a new proof as follows:

$$
\left.
\cfrac{
  \cfrac{
    \cfrac{\vdots \pi_1}{\Gamma_1 \Rightarrow \Theta_1, A} \quad \cfrac{\vdots \pi_3}{A, \Gamma_2 \Rightarrow \Theta_2}
  }{\Gamma_1, \Gamma_2^* \Rightarrow \Theta_1^*, \Theta_2} \text{ MIX}
}{\Gamma_1, \Gamma_2 \Rightarrow \Theta_1, \Theta_2} \text{ W, I}
\right\} \pi'
$$

The proof π' ends in a MIX applied to A and not to $A \wedge B$. Thus, $\mathrm{dg}(\pi') < \mathrm{dg}(\pi)$ (however, the rank may increase). By inductive hypothesis, there is a proof of $\Gamma_1, \Gamma_2^* \Rightarrow \Theta_1^*, \Theta_2$ without MIX. If we replace π' by that proof, we obtain a proof of the same end-sequent as π without MIX.

D2. $M = A \lor B$.

This implies, on account of considerations similar to those carried out in the preceding case, that the proof ends in the following way:

$$
\cfrac{
\cfrac{\vdots \; \pi_1}{\cfrac{\Gamma_1 \Rightarrow \Theta_1, A}{\Gamma_1 \Rightarrow \Theta_1, A \lor B} \; \text{VR}}
\qquad
\cfrac{\cfrac{\vdots \; \pi_2}{A, \Gamma_2 \Rightarrow \Theta_2} \qquad \cfrac{\vdots \; \pi_3}{B, \Gamma_2 \Rightarrow \Theta_2}}{A \lor B, \Gamma_2 \Rightarrow \Theta_2} \; \text{VL}
}{\Gamma_1, \Gamma_2 \Rightarrow \Theta_1, \Theta_2} \; \text{MIX}
$$

Then the proof can be transformed, as in the previous case, into the following one:

$$
\left.
\cfrac{
\cfrac{\cfrac{\vdots \; \pi_1}{\Gamma_1 \Rightarrow \Theta_1, A} \qquad \cfrac{\vdots \; \pi_2}{A, \Gamma_2 \Rightarrow \Theta_2}}{\Gamma_1, \Gamma_2^* \Rightarrow \Theta_1^*, \Theta_2} \; \text{MIX}
}{\Gamma_1, \Gamma_2 \Rightarrow \Theta_1, \Theta_2} \; \text{W, I}
\right\} \pi'
$$

and we obtain the same result.

D3. $M = A \supset B$.

For reasons analogous to those presented above the proof has the following form:

$$
\cfrac{
\cfrac{\cfrac{\vdots \; \pi_1}{A, \Gamma_1 \Rightarrow \Theta_1, B}}{\Gamma_1 \Rightarrow \Theta_1, A \supset B} \; \supset \text{R}
\qquad
\cfrac{\cfrac{\vdots \; \pi_2}{\Gamma \Rightarrow \Theta, A} \qquad \cfrac{\vdots \; \pi_3}{B, \Delta \Rightarrow \Lambda}}{A \supset B, \Gamma, \Delta \Rightarrow \Theta, \Lambda} \; \supset \text{L}
}{\Gamma_1, \Gamma, \Delta \Rightarrow \Theta_1, \Theta, \Lambda} \; \text{MIX}
$$

But then the proof can be transformed as follows:

$$
\left.
\cfrac{
\cfrac{\cfrac{\vdots \; \pi_2}{\Gamma \Rightarrow \Theta, A} \qquad \cfrac{\cfrac{\vdots \; \pi_1}{A, \Gamma_1 \Rightarrow \Theta_1, B} \qquad \cfrac{\vdots \; \pi_3}{B, \Delta \Rightarrow \Lambda}}{A, \Gamma_1, \Delta^* \Rightarrow \Theta_1^*, \Lambda} \; \text{MIX}}{\Gamma, \Gamma_1^*, \Delta^{**} \Rightarrow \Theta^*, \Theta_1^*, \Lambda} \; \text{MIX}
}{\Gamma_1, \Gamma, \Delta \Rightarrow \Theta_1, \Theta, \Lambda} \; \text{W, I}
\right\} \pi'
$$

Note that in Δ^* and in Θ_1^* the occurrences of B are erased, while in Γ_1^*, Δ^{**}, and Θ^*, the occurrences of A are erased.

The new proof has two MIXes in it. Consider π': its degree is smaller than that of the MIX in the original proof. Consequently, despite the fact that the rank could now be higher than in the original proof, by inductive hypothesis there is a proof with the same end-sequent $A, \Gamma_1, \Delta^* \Rightarrow \Theta_1^*, \Lambda$ without MIX. We can replace π' with that MIX-free proof:

$$
\left.
\begin{array}{c}
\vdots\ \pi_2 \qquad\qquad \vdots \\[2pt]
\dfrac{\Gamma \Rightarrow \Theta, A \qquad A, \Gamma_1, \Delta^* \Rightarrow \Theta_1^*, \Lambda}{\dfrac{\Gamma, \Gamma_1^*, \Delta^{**} \Rightarrow \Theta^*, \Theta_1^*, \Lambda}{\Gamma_1, \Gamma, \Delta \Rightarrow \Theta_1, \Theta, \Lambda}\ \text{W, I}}\ \text{MIX}
\end{array}
\right\}\pi''
$$

Now we can proceed to eliminate the lower MIX with the same method: π'' has degree less than π. By inductive hypothesis, there is a proof of

$$
\Gamma, \Gamma_1^*, \Delta^{**} \Rightarrow \Theta^*, \Theta_1^*, \Lambda
$$

without MIX. Replacing π'' with that proof results in a MIX-free proof of the end-sequent of π.

D4. $M = \neg A$.

For the usual reasons the proof ends as follows:

$$
\dfrac{\dfrac{\begin{array}{c}\vdots\ \pi_1 \\[2pt] A, \Gamma_1 \Rightarrow \Theta_1\end{array}}{\Gamma_1 \Rightarrow \Theta_1, \neg A}\ \neg\text{R} \qquad \dfrac{\begin{array}{c}\vdots\ \pi_2 \\[2pt] \Gamma_2 \Rightarrow \Theta_2, A\end{array}}{\neg A, \Gamma_2 \Rightarrow \Theta_2}\ \neg\text{L}}{\Gamma_1, \Gamma_2 \Rightarrow \Theta_1, \Theta_2}\ \text{MIX}
$$

Thus, it can be transformed as follows:

$$
\left.
\begin{array}{c}
\vdots\ \pi_2 \qquad\qquad \vdots\ \pi_1 \\[2pt]
\dfrac{\Gamma_2 \Rightarrow \Theta_2, A \qquad A, \Gamma_1 \Rightarrow \Theta_1}{\dfrac{\Gamma_2, \Gamma_1^* \Rightarrow \Theta_2^*, \Theta_1}{\Gamma_1, \Gamma_2 \Rightarrow \Theta_1, \Theta_2}\ \text{W, I}}\ \text{MIX}
\end{array}
\right\}\pi'
$$

Again, the proof π' is of lower degree than π even if the rank might go up. By inductive hypothesis, there is a MIX-free proof of the end-sequent.

D5. $M = \forall x\, F(x)$.

On the basis of the aforementioned considerations we have the following proof:

$$
\dfrac{\dfrac{\begin{array}{c}\vdots\ \pi_1(b) \\[2pt] \Gamma_1 \Rightarrow \Theta_1, F(b)\end{array}}{\Gamma_1 \Rightarrow \Theta_1, \forall x\, F(x)}\ \forall\text{R} \qquad \dfrac{\begin{array}{c}\vdots\ \pi_2 \\[2pt] F(t), \Gamma_2 \Rightarrow \Theta_2\end{array}}{\forall x\, F(x), \Gamma_2 \Rightarrow \Theta_2}\ \forall\text{L}}{\Gamma_1, \Gamma_2 \Rightarrow \Theta_1, \Theta_2}\ \text{MIX}
$$

It can be now be transformed as follows:

$$
\left.
\begin{array}{c}
\begin{array}{cc}
\vdots\, \pi_1(t) & \vdots\, \pi_2 \\[4pt]
\dfrac{\Gamma_1 \Rightarrow \Theta_1, F(t) \qquad F(t), \Gamma_2 \Rightarrow \Theta_2}{\Gamma_1, \Gamma_2^* \Rightarrow \Theta_1^*, \Theta_2}\ \text{MIX} \\[8pt]
\dfrac{}{\Gamma_1, \Gamma_2 \Rightarrow \Theta_1, \Theta_2}\ \text{W, I}
\end{array}
\end{array}
\right\} \pi'
$$

Of course, the proof that ends with $\Gamma_1 \Rightarrow \Theta_1, F(t)$ now contains t in place of b. But this is allowed by Corollary 5.25 on replacement of proper variables that we have already proved. Indeed, the critical condition concerning the legitimacy of the application of the rule $\forall R$ requires that the variable b does not occur either in Γ_1 or in Θ_1. Hence, the substitution of t for b in the proof of $\Gamma_1 \Rightarrow \Theta_1, F(b)$—substitution allowed by the corollary—does not affect either Γ_1 or Θ_1.

The new proof π' is of lower degree than the original one. Thus by inductive hypothesis, there exists a proof of the same sequent without MIX.

D6. $M = \exists x\, F(x)$.

For the usual reasons the final part of the proof has the following form:

$$
\begin{array}{cc}
\vdots\, \pi_1 & \vdots\, \pi_2(b) \\[4pt]
\dfrac{\Gamma_1 \Rightarrow \Theta_1, F(t)}{\Gamma_1 \Rightarrow \Theta_1, \exists x\, F(x)}\ \exists R & \dfrac{F(b), \Gamma_2 \Rightarrow \Theta_2}{\exists x\, F(x), \Gamma_2 \Rightarrow \Theta_2}\ \exists L \\[10pt]
\multicolumn{2}{c}{\dfrac{}{\Gamma_1, \Gamma_2 \Rightarrow \Theta_1, \Theta_2}\ \text{MIX}}
\end{array}
$$

According to the lemma of variable replacement, the proof can be transformed by the replacement of b by t as in the previous case:

$$
\left.
\begin{array}{c}
\begin{array}{cc}
\vdots\, \pi_1 & \vdots\, \pi_2(t) \\[4pt]
\dfrac{\Gamma_1 \Rightarrow \Theta_1, F(t) \qquad F(t), \Gamma_2 \Rightarrow \Theta_2}{\Gamma_1, \Gamma_2^* \Rightarrow \Theta_1^*, \Theta_2}\ \text{MIX} \\[8pt]
\dfrac{}{\Gamma_1, \Gamma_2 \Rightarrow \Theta_1, \Theta_2}\ \text{W, I}
\end{array}
\end{array}
\right\} \pi'
$$

and this yields the result.   □

It is worth pausing to consider how we have made progress in the cases considered so far. The crucial idea in the inductive step in each case was the transformation of the original proof into one of lower degree. The rank of the reduced proofs might be higher than that of the original proof, but this does not matter. How can this be accounted for? It is simply a consequence of the fact that the double measure of proofs "weighs" the degree more heavily than the rank. The following example may clarify the entire situation. The proof

$$
\cfrac{
\cfrac{
\cfrac{\cfrac{A \Rightarrow A}{B,A \Rightarrow A}\text{WL}}{A,B \Rightarrow A}\text{IL}
\qquad
\cfrac{\cfrac{B \Rightarrow B}{A,B \Rightarrow B}\text{WL}}{A,B \Rightarrow \boxed{A \wedge B}}\wedge\text{R}
}{A,B \Rightarrow \boxed{A \wedge B}}
\qquad
\cfrac{\cfrac{\cfrac{C \Rightarrow C}{A,C \Rightarrow C}\text{WL}}{\boxed{A \wedge B},C \Rightarrow C}\wedge\text{L}}{}
}{A,B,C \Rightarrow C}\text{MIX}
$$

is transformed into:

$$
\cfrac{
\cfrac{\cfrac{\cfrac{A \Rightarrow \boxed{A}}{B,A \Rightarrow \boxed{A}}\text{WL}}{A,B \Rightarrow \boxed{A}}\text{IL}
\qquad
\cfrac{\cfrac{C \Rightarrow C}{\boxed{A},C \Rightarrow C}\text{WL}}{}
}{A,B,C \Rightarrow C}\text{MIX}
$$

Now, the rank of $A \wedge B$ in the first proof is 2. The rank of A in the transformed proof is 4. However, the transformed proof has a global level of complexity that is lower, since the degree of the MIX has more "weight" than the rank.

6.5   Reducing the rank

Lemma 6.17. *Suppose π is a proof with $\mathrm{rk}(\pi) > 2$, and every proof π' with $\mathrm{dg}(\pi') < \mathrm{dg}(\pi)$ or $\mathrm{dg}(\pi') = \mathrm{dg}(\pi)$ and $\mathrm{rk}(\pi') < \mathrm{rk}(\pi)$ is reducible. Then there is a proof of the same end-sequent as π with equal or lower degree, and lower rank.*

The proof is divided into two general cases: the first case in which the right rank is > 1, and the second case in which the right rank is = 1 and the left rank > 1. Since the rank is > 2, either the right rank is > 1, or the right rank = 1 and the left rank is > 1. So these are the only cases to consider. We divide the result into two lemmas.

Lemma 6.18. *Suppose π is a proof with $\mathrm{rk}_r(\pi) > 1$, and every proof π' with $\mathrm{dg}(\pi') < \mathrm{dg}(\pi)$ or $\mathrm{dg}(\pi') = \mathrm{dg}(\pi)$ and $\mathrm{rk}(\pi') < \mathrm{rk}(\pi)$ is reducible. Then there is a proof of the same end-sequent as π with equal or lower degree, and lower rank.*

Proof. The proof has the following form:

$$
\cfrac{\overset{\vdots\ \pi_1}{\Pi \Rightarrow \Sigma} \qquad \overset{\vdots\ \pi_2}{\Delta \Rightarrow \Lambda}}{\Pi,\Delta^* \Rightarrow \Sigma^*,\Lambda}\text{MIX}
$$

The MIX formula occurring in Σ and in Δ might well be the result of the application of an operational rule.

By Lemma 6.15, we can assume that M does not occur in the antecedent of the left premise of the MIX, i.e., in Π. Since the right rank is > 1, the size of the sub-proof π_2 is > 1, i.e., the right premise of the MIX is not just an axiom, but the conclusion of an inference. We distinguish cases according to the type of this inference leading to the right premise of the MIX: it may be either a structural inference on the left, or some other inference with one premise, or an inference with two premises.

Case A.   The right premise is the conclusion of WL, IL, or CL (G3.121.21). In other words, the proof has the following form:

$$
\frac{\Pi \Rightarrow \Sigma \quad \dfrac{\Psi \Rightarrow \Theta}{\Xi \Rightarrow \Theta} \text{ WL or CL or IL}}{\Pi, \Xi^* \Rightarrow \Sigma^*, \Theta} \text{ MIX}
$$

We have two slightly different cases. If the principal formula of WL, CL, or IL is not M, then the proof can now be transformed into the following proof:

$$
\left.
\frac{\dfrac{\dfrac{\dfrac{\Pi \Rightarrow \Sigma \quad \Psi \Rightarrow \Theta}{\Pi, \Psi^* \Rightarrow \Sigma^*, \Theta} \text{ MIX}}{\Psi^*, \Pi \Rightarrow \Sigma^*, \Theta} \text{ IL}}{\Xi^*, \Pi \Rightarrow \Sigma^*, \Theta} \text{ WL or CL or IL}}{\Pi, \Xi^* \Rightarrow \Sigma^*, \Theta} \text{ IL}
\right\} \pi'
$$

If M is the principal formula of the inference leading to the right premise, then Ψ^* and Ξ^* are identical. Then consider the proof:

$$
\left.
\frac{\Pi \Rightarrow \Sigma \quad \Psi \Rightarrow \Theta}{\Pi, \Psi^* \Rightarrow \Sigma^*, \Theta} \text{ MIX}
\right\} \pi'
$$

The end-sequent is already identical to the end-sequent of π, i.e., $\Pi, \Xi^* \Rightarrow \Sigma^*, \Theta$.

The rank of the proof π' is smaller than that of the original proof, and its degree is unchanged. Indeed, the MIX has been transposed above the application of the structural rules. Thus, in virtue of the hypothesis, there exists a proof which has as end-sequent $\Pi, \Xi^* \Rightarrow \Sigma^*, \Theta$ with no MIXes.

Let us look at two examples for each of rule WL, CL, or IL, one with M as principal formula and one with a principal formula different from M.

1. *Example with* WL *applied to M (principal formula)*

$$\cfrac{\neg A, C \Rightarrow B, M \quad \cfrac{A \vee B, M, B, M \Rightarrow C}{M, A \vee B, M, B, M \Rightarrow C} \text{ WL}}{\neg A, C, A \vee B, B \Rightarrow B, C} \text{ MIX}$$

Transformation of the proof:

$$\cfrac{\neg A, C \Rightarrow B, M \quad A \vee B, M, B, M \Rightarrow C}{\neg A, C, A \vee B, B \Rightarrow B, C} \text{ MIX}$$

2. *Example with* WL *applied to a formula different from M*

$$\cfrac{A \Rightarrow M, \neg A \quad \cfrac{A \vee B, D, M, E, M \Rightarrow C}{A \wedge C, A \vee B, D, M, E, M \Rightarrow C} \text{ WL}}{A, A \wedge C, A \vee B, D, E \Rightarrow \neg A, C} \text{ MIX}$$

Transformation of the proof:

$$\cfrac{\cfrac{\cfrac{A \Rightarrow M, \neg A \quad A \vee B, D, M, E, M \Rightarrow C}{A, A \vee B, D, E \Rightarrow \neg A, C} \text{ MIX}}{A \wedge C, A, A \vee B, D, E \Rightarrow \neg A, C} \text{ WL}}{A, A \wedge C, A \vee B, D, E \Rightarrow \neg A, C} \text{ IL}$$

3. *Example with* IL *applied to M (principal formula)*

$$\cfrac{A \Rightarrow M, \neg A \quad \cfrac{M, A \vee B, D, M, E \Rightarrow C}{A \vee B, M, D, M, E \Rightarrow C} \text{ IL}}{A, A \vee B, D, E \Rightarrow \neg A, C} \text{ MIX}$$

Transformation of the proof:

$$\cfrac{A \Rightarrow M, \neg A \quad M, A \vee B, D, M, E \Rightarrow C}{A, A \vee B, D, E \Rightarrow \neg A, C} \text{ MIX}$$

4. *Example with* IL *applied to a formula different from M*

$$\cfrac{A \Rightarrow M, \neg A \quad \cfrac{A \vee B, D, M, E, M \Rightarrow C}{D, A \vee B, M, E, M \Rightarrow C} \text{ IL}}{A, D, A \vee B, E \Rightarrow \neg A, C} \text{ MIX}$$

Transformation of the proof:

$$\cfrac{\cfrac{A \Rightarrow M, \neg A \quad A \vee B, D, M, E, M \Rightarrow C}{A, A \vee B, D, E \Rightarrow \neg A, C} \text{ MIX}}{A, D, A \vee B, E \Rightarrow \neg A, C} \text{ IL}$$

5. *Example with* CL *applied to M (principal formula)*

$$\cfrac{A \Rightarrow M, \neg A \quad \cfrac{M, M, A \vee B, D, E \Rightarrow C}{M, A \vee B, D, E \Rightarrow C} \text{ CL}}{A, A \vee B, D, E \Rightarrow \neg A, C} \text{ MIX}$$

Transformation of the proof:

$$\dfrac{A \Rightarrow M, \neg A \qquad M, M, A \vee B, D, E \Rightarrow C}{A, A \vee B, D, E \Rightarrow \neg A, C} \text{ MIX}$$

6. *Example with* CL *applied to a formula different from* M

$$\dfrac{A \Rightarrow M, \neg A \qquad \dfrac{A \wedge B, A \wedge B, D, M, M, E \Rightarrow C}{A \wedge B, D, M, M, E \Rightarrow C} \text{ CL}}{A, A \wedge B, D, E, \Rightarrow \neg A, C} \text{ MIX}$$

Transformation of the proof:

$$\dfrac{A \Rightarrow M, \neg A \qquad A \wedge B, A \wedge B, D, M, M, E \Rightarrow C}{\dfrac{\dfrac{\dfrac{\dfrac{A, A \wedge B, A \wedge B, D, E \Rightarrow \neg A, C}{A \wedge B, A \wedge B, A, D, E \Rightarrow \neg A, C} \text{ IL}}{A \wedge B, A, D, E \Rightarrow \neg A, C} \text{ CL}}{A, A \wedge B, D, E \Rightarrow \neg A, C} \text{ IL}}} \text{ MIX}$$

Case B.   The rule that ends with the right premise of the MIX is a rule with *one* premise, but not WL, IL, CL (G.3.121.22).

There are nine cases to consider. However, it is useful to remark from the outset that the proof relative to each one of them depends on the specific rule in question. If the rule introduces a logical operator in the antecedent then in the proof one must distinguish between the case in which M coincides with the principal formula of the rule and the opposite case. This distinction is irrelevant if the rule concerns the introduction of a logical operator in the succedent, or is one of WR, IR, or CR. The reason for the methodological difference in the two cases consists in the fact that, if the operational rule is a left rule, the application of the rule affects the MIX and consequently how to treat it in order to eliminate it. A partial exception to this classification occurs with the ¬R, for it affects the MIX in a peculiar way so that one has to keep track, in the transformation of the original proof, only of the case in which M coincides with the formula A (and not with the principal formula of the rule that is ¬A) and the case in which it does not. All these circumstances will be recalled at the beginning of each case.

B1. Rule WR. If the right premise of the MIX is the conclusion of WR, the proof has the following form:

$$\dfrac{\vdots \pi_1 \qquad \dfrac{\vdots \pi_2}{\dfrac{\Gamma \Rightarrow \Omega_1}{\Gamma \Rightarrow \Omega_1, A} \text{ WR}}}{\Pi \Rightarrow \Sigma \qquad \qquad } $$
$$\dfrac{\Pi \Rightarrow \Sigma \qquad \dfrac{\dfrac{\Gamma \Rightarrow \Omega_1}{\Gamma \Rightarrow \Omega_1, A} \text{ WR}}{}}{\Pi, \Gamma^* \Rightarrow \Sigma^*, \Omega_1, A} \text{ MIX}$$

The proof can be transformed in the following way:

$$
\begin{array}{c}
\vdots \pi_1 \qquad \vdots \pi_2 \\
\dfrac{\Pi \Rightarrow \Sigma \qquad \Gamma \Rightarrow \Omega_1}{\dfrac{\Pi, \Gamma^* \Rightarrow \Sigma^*, \Omega_1}{\Pi, \Gamma^* \Rightarrow \Sigma^*, \Omega_1, A}} \begin{array}{l} \text{MIX on } M \\ \\ \text{WR} \end{array}
\end{array}
$$

The new proof has lower rank by 1 than the original proof. The MIX-formula is the same and so the degree is the same. By inductive hypothesis, there is a proof of the same end-sequent and without MIX.

B2. Rule IR.

B3. Rule CR. See Problem 6.20.

B4. Rule ∧L. Since this is a rule of introduction in the antecedent the final part of the proof can present two different forms, depending on the following opposite conditions:

(a) $M \neq A \wedge B$. Then

$$
\begin{array}{c}
\qquad\qquad\qquad \vdots \pi_2 \\
\vdots \pi_1 \qquad\qquad \dfrac{A, \Gamma \Rightarrow \Omega_1}{A \wedge B, \Gamma \Rightarrow \Omega_1} \wedge \text{L} \\
\dfrac{\Pi \Rightarrow \Sigma \qquad\qquad\qquad\qquad\quad}{\Pi, A \wedge B, \Gamma^* \Rightarrow \Sigma^*, \Omega_1} \text{MIX}
\end{array}
$$

In this case the proof can be transformed into:

$$
\begin{array}{c}
\vdots \pi_1 \qquad\qquad \vdots \pi_2 \\
\dfrac{\Pi \Rightarrow \Sigma \qquad A, \Gamma \Rightarrow \Omega_1}{\dfrac{\Pi, A, \Gamma^* \Rightarrow \Sigma^*, \Omega_1}{\dfrac{A, \Pi, \Gamma^* \Rightarrow \Sigma^*, \Omega_1}{\dfrac{A \wedge B, \Pi, \Gamma^* \Rightarrow \Sigma^*, \Omega_1}{\Pi, A \wedge B, \Gamma^* \Rightarrow \Sigma^*, \Omega_1}}}} \begin{array}{l} \text{MIX on } M \\ \\ \text{IL} \\ \\ \wedge \text{L} \\ \\ \text{IL} \end{array}
\end{array}
$$

if $M \neq A$, or if $M = A$, into:

$$
\begin{array}{c}
\vdots \pi_1 \qquad\qquad \vdots \pi_2 \\
\dfrac{\Pi \Rightarrow \Sigma \qquad A, \Gamma \Rightarrow \Omega_1}{\dfrac{\Pi, \Gamma^* \Rightarrow \Sigma^*, \Omega_1}{\dfrac{A \wedge B, \Pi, \Gamma^* \Rightarrow \Sigma^*, \Omega_1}{\Pi, A \wedge B, \Gamma^* \Rightarrow \Sigma^*, \Omega_1}}} \begin{array}{l} \text{MIX on } M \\ \\ \text{WL} \\ \\ \text{IL} \end{array}
\end{array}
$$

Clearly we have a proof of smaller rank and the same degree. Thus, in virtue of the inductive hypothesis (on rank), we have the result.

(b) $M = A \wedge B$. Then,

$$
\cfrac{\Pi \Rightarrow \Sigma \qquad \cfrac{\vdots \pi_2 \\ A, \Gamma \Rightarrow \Omega_1}{A \wedge B, \Gamma \Rightarrow \Omega_1} \text{\scriptsize AL}}{\Pi, \Gamma^* \Rightarrow \Sigma^*, \Omega_1} \text{\scriptsize MIX}
$$

with π_1 above $\Pi \Rightarrow \Sigma$.

In this case, we can obtain our objective by constructing a new proof which is built out of two pieces that have to be put one after the other. Let us construct the first piece by adding inferences to π_1 and π_2 as follows:

$$
\left.
\cfrac{\cfrac{\cfrac{\Pi \Rightarrow \Sigma \qquad A, \Gamma \Rightarrow \Omega_1}{\Pi, A, \Gamma^* \Rightarrow \Sigma^*, \Omega_1} \text{\scriptsize MIX on } A \wedge B}{A, \Gamma^*, \Pi \Rightarrow \Sigma^*, \Omega_1} \text{\scriptsize IL}}{A \wedge B, \Gamma^*, \Pi \Rightarrow \Sigma^*, \Omega_1} \text{\scriptsize AL}
\right\} \pi_3
$$

with π_1 above $\Pi \Rightarrow \Sigma$ and π_2 above $A, \Gamma \Rightarrow \Omega_1$.

We'll call this proof π_3. Note, first of all, that it is a correct proof, as all the sequents in it are derivable. Recall that $\text{rk}_r(\pi) > 1$. Thus the MIX formula $A \wedge B$ must occur also in Γ. This means that $A, \Gamma \Rightarrow \Omega_1$ can serve as right premise of a MIX on $A \wedge B$. The rank of the MIX in π_3 is the rank of the original MIX minus 1. The left rank of both is the same. On the right side, we have removed the conclusion of the $\wedge$L rule from the sub-proof ending in the right premise, so there is one fewer sequent above the right premise containing the MIX formula in the antecedent. (The $\wedge$L rule itself is still applied, but after the MIX.) As a consequence, the rank of the MIX has decreased, while the degree remains the same. Thus, by inductive hypothesis, there is a derivation π_4 of the end-sequent of the new MIX. Let us replace the sub-proof ending in the MIX in π_3 by π_4. We have:

$$
\left.
\cfrac{\cfrac{\cfrac{\vdots \pi_4 \\ \Pi, A, \Gamma^* \Rightarrow \Sigma^*, \Omega_1}{A, \Gamma^*, \Pi \Rightarrow \Sigma^*, \Omega_1} \text{\scriptsize IL}}{A \wedge B, \Gamma^*, \Pi \Rightarrow \Sigma^*, \Omega_1} \text{\scriptsize AL}}{}
\right\} \pi_3'
$$

Now we add the following additional steps to the end of π'_3 to obtain π_5:

$$
\cfrac{\Pi \Rightarrow \Sigma \qquad \cfrac{\cfrac{\cfrac{\Pi, A, \Gamma^* \Rightarrow \Sigma^*, \Omega_1}{A, \Gamma^*, \Pi \Rightarrow \Sigma^*, \Omega_1}\text{ IL}}{A \wedge B, \Gamma^*, \Pi \Rightarrow \Sigma^*, \Omega_1}\text{ ΛL}}{\Pi, \Gamma^*, \Pi^* \Rightarrow \Sigma^*, \Sigma^*, \Omega_1}}{\Pi, \Gamma^* \Rightarrow \Sigma^*, \Omega_1}
$$

with π_1 on the left, π_4 above, MIX on $A \wedge B$, and I, C below.

This derivation still contains a MIX. But its rank is equal to that of the left rank of the original MIX plus 1. The left branch has not been modified, so the left rank is the same. The right rank is 1, for M does not occur in the antecedent of the sequent immediately above the right premise. In the transformation of the original derivation, the outcome of having pushed the MIX upwards results in the removal of the formula M. By hypothesis, M does not occur in Π. So, M does not occur in the antecedent of the sequent above the right premise of the MIX, $A, \Gamma^*, \Pi \Rightarrow \Sigma^*, \Omega_1$. So, since the right rank of the new derivation is equal to 1 while the right rank of the original derivation is > 1, and the left ranks are equal, the rank of the new derivation is less than the rank of the original one. It follows that we can apply the inductive hypothesis on rank to the new derivation (as we have lower rank and the same degree). Thus, we again have a cut-free derivation of the end-sequent of the original derivation.

We've spelled this out in detail to make clear how the inductive hypothesis applies in this case. We could also consider the following complex proof consisting of both pieces together:

$$
\cfrac{\Pi \Rightarrow \Sigma \qquad \cfrac{\cfrac{\cfrac{\cfrac{\Pi \Rightarrow \Sigma \qquad A, \Gamma \Rightarrow \Omega_1}{\Pi, A, \Gamma^* \Rightarrow \Sigma^*, \Omega_1}\text{ MIX on } A \wedge B}{A, \Gamma^*, \Pi \Rightarrow \Sigma^*, \Omega_1}\text{ IL}}{A \wedge B, \Gamma^*, \Pi \Rightarrow \Sigma^*, \Omega_1}\text{ ΛL}}{\Pi, \Gamma^*, \Pi^* \Rightarrow \Sigma^*, \Sigma^*, \Omega_1}\text{ MIX on } A \wedge B}{\Pi, \Gamma^* \Rightarrow \Sigma^*, \Omega_1}\;I, C
$$

Then first apply the inductive hypothesis to the sub-proof ending in the top MIX. The result is π_5, to which the inductive hypothesis also applies, as discussed above.

B5. Rule ∨R. In this case, since we are dealing with the rule of introduction of disjunction in the succedent, the final part of the proof has a unique form, regardless of whether $M = A \vee B$ or $M \neq A \vee B$. The original proof is:

$$
\cfrac{\Pi \Rightarrow \Sigma \quad \cfrac{\begin{array}{c} \vdots \ \pi_2 \\ \Gamma \Rightarrow \Omega_1, A \end{array}}{\Gamma \Rightarrow \Omega_1, A \vee B} \vee \text{R}}{\Pi, \Gamma^* \Rightarrow \Sigma^*, \Omega_1, A \vee B} \text{ MIX}
$$

Then, regardless of whether $M = A \vee B$ or $M \neq A \vee B$, the proof can be transformed in the following way:

$$
\cfrac{\cfrac{\Pi \Rightarrow \Sigma \quad \Gamma \Rightarrow \Omega_1, A}{\Pi, \Gamma^* \Rightarrow \Sigma^*, \Omega_1, A} \text{ MIX on } M}{\Pi, \Gamma^* \Rightarrow \Sigma^*, \Omega_1, A \vee B} \vee \text{R}
$$

This is obviously a proof of rank lower by 1 than the original proof but of the same degree. Thus, in virtue of the inductive hypothesis (on rank) there exists a proof having the same original end-sequent and without MIX.

B6. Rule ¬R. In this case the proof has the following form in which we can have either $M = A$ or $M \neq A$:

$$
\cfrac{\Pi \Rightarrow \Sigma \quad \cfrac{\begin{array}{c} \vdots \ \pi_2 \\ A, \Gamma \Rightarrow \Omega_1 \end{array}}{\Gamma \Rightarrow \Omega_1, \neg A} \neg \text{R}}{\Pi, \Gamma^* \Rightarrow \Sigma^*, \Omega_1, \neg A} \text{ MIX}
$$

Let us suppose that $M \neq A$. Then the proof can be transformed as follows:

$$
\cfrac{\cfrac{\cfrac{\Pi \Rightarrow \Sigma \quad A, \Gamma \Rightarrow \Omega_1}{\Pi, A, \Gamma^* \Rightarrow \Sigma^*, \Omega_1} \text{ MIX on } M}{A, \Pi, \Gamma^* \Rightarrow \Sigma^*, \Omega_1} \text{ IL}}{\Pi, \Gamma^* \Rightarrow \Sigma^*, \Omega_1, \neg A} \neg \text{R}
$$

This proof has a rank lower by 1 than the original proof but with the same degree. Thus, in virtue of the inductive hypothesis, there exists a proof having the same end-sequent without MIX.

Alternatively, let us suppose that $M = A$. Then the transformation reads:

$$
\frac{
\frac{
\frac{\vdots \pi_1 \qquad \vdots \pi_2}{\Pi \Rightarrow \Sigma \qquad A, \Gamma \Rightarrow \Omega_1} \text{ MIX on } A
}{\Pi, \Gamma^* \Rightarrow \Sigma^*, \Omega_1}
}{
\frac{A, \Pi, \Gamma^* \Rightarrow \Sigma^*, \Omega_1}{\Pi, \Gamma^* \Rightarrow \Sigma^*, \Omega_1, \neg A} \text{ ¬R}
} \text{ WL}
$$

Even in this case we have a proof with rank lower by 1 than the original proof but with the same degree. Thus, also in this case, the conditions of the inductive hypothesis are satisfied and the result holds.

B7. Rule ¬L. As in the case of rule ∧L, we need to take into consideration the two different cases already in the formulation of the original proof. Each one of these cases must then be treated differently in order to obtain the result. The reason for the different treatment is the one we have already explained above.

(a) $M \neq \neg A$. Then

$$
\frac{
\Pi \Rightarrow \Sigma \qquad \dfrac{\Gamma \Rightarrow \Omega_1, A}{\neg A, \Gamma \Rightarrow \Omega_1} \text{ ¬L}
}{\Pi, \neg A, \Gamma^* \Rightarrow \Sigma^*, \Omega_1} \text{ MIX}
$$

Note again that since $\mathrm{rk}_r(\pi) > 1$, the MIX formula M must be present also in Γ.

The proof can be transformed into:

$$
\frac{
\frac{
\dfrac{\Pi \Rightarrow \Sigma \qquad \Gamma \Rightarrow \Omega_1, A}{\Pi, \Gamma^* \Rightarrow \Sigma^*, \Omega_1, A} \text{ MIX on } M
}{\neg A, \Pi, \Gamma^* \Rightarrow \Sigma^*, \Omega_1} \text{ ¬L}
}{\Pi, \neg A, \Gamma^* \Rightarrow \Sigma^*, \Omega_1} \text{ IL}
$$

We have a proof of rank smaller by 1 than the original proof but with the same degree. Thus, in virtue of the inductive hypothesis there exists a proof having the same end-sequent without MIX.

(b) $M = \neg A$. Then:

$$
\frac{
\Pi \Rightarrow \Sigma \qquad \dfrac{\Gamma \Rightarrow \Omega_1, A}{\neg A, \Gamma \Rightarrow \Omega_1} \text{ ¬L}
}{\Pi, \Gamma^* \Rightarrow \Sigma^*, \Omega_1} \text{ MIX}
$$

In this case the objective is reached by constructing a new proof assembled from two pieces. Let us construct the first piece:

$$\left.\cfrac{\cfrac{\cfrac{\vdots\,\pi_1 \qquad \vdots\,\pi_2}{\Pi \Rightarrow \Sigma \qquad \Gamma \Rightarrow \Omega_1, A}\ \text{MIX on } \neg A}{\cfrac{\Pi, \Gamma^* \Rightarrow \Sigma^*, \Omega_1, A}{\Gamma^*, \Pi \Rightarrow \Sigma^*, \Omega_1, A}\ \text{IL}}}{\neg A, \Gamma^*, \Pi \Rightarrow \Sigma^*, \Omega_1}\ \neg \text{L}\right\}\pi_3$$

The second piece is the following:

$$\cfrac{\cfrac{\vdots\,\pi_1}{\Pi \Rightarrow \Sigma \qquad \neg A, \Gamma^*, \Pi \Rightarrow \Sigma^*, \Omega_1}\ \text{MIX on } \neg A}{\cfrac{\Pi, \Gamma^*, \Pi^* \Rightarrow \Sigma^*, \Sigma^*, \Omega_1}{\Pi, \Gamma^* \Rightarrow \Sigma^*, \Omega_1}\ \text{I, C}}$$

By combining the first and the second piece we obtain:

$$\cfrac{\cfrac{\vdots\,\pi_1}{\Pi \Rightarrow \Sigma} \qquad \left.\cfrac{\cfrac{\cfrac{\vdots\,\pi_1 \qquad \vdots\,\pi_2}{\Pi \Rightarrow \Sigma \qquad \Gamma \Rightarrow \Omega_1, A}\ \text{MIX on } \neg A}{\cfrac{\Pi, \Gamma^* \Rightarrow \Sigma^*, \Omega_1, A}{\Gamma^*, \Pi \Rightarrow \Sigma^*, \Omega_1, A}\ \text{IL}}}{\neg A, \Gamma^*, \Pi \Rightarrow \Sigma^*, \Omega_1}\ \neg \text{L}\right\}\pi_3}{\cfrac{\Pi, \Gamma^*, \Pi^* \Rightarrow \Sigma^*, \Sigma^*, \Omega_1}{\Pi, \Gamma^* \Rightarrow \Sigma^*, \Omega_1}\ \text{I, C}}\ \text{MIX on } \neg A$$

At this point one can obviously repeat the argument used in the case of ∧L to show that the transformed proof has lower rank and the same degree and consequently the eliminability of both MIXes appearing in it.

B8. Rule ∀L. Also in this case we need to consider two sub-cases from the outset. Each one of them must also be treated differently.

(a) $M \neq \forall x\, F(x)$. Then:

$$\cfrac{\cfrac{\vdots\,\pi_1}{\Pi \Rightarrow \Sigma} \qquad \cfrac{\cfrac{\vdots\,\pi_2}{F(t), \Gamma \Rightarrow \Omega_1}}{\forall x\, F(x), \Gamma \Rightarrow \Omega_1}\ \forall \text{L}}{\Pi, \forall x\, F(x), \Gamma^* \Rightarrow \Sigma^*, \Omega_1}\ \text{MIX}$$

In this case the proof can be transformed into:

$$
\begin{array}{cc}
\vdots\,\pi_1 & \vdots\,\pi_2
\end{array}
$$

$$
\cfrac{
\cfrac{
\cfrac{
\cfrac{\Pi \Rightarrow \Sigma \qquad F(t), \Gamma \Rightarrow \Omega_1}{\Pi, F(t), \Gamma^* \Rightarrow \Sigma^*, \Omega_1} \text{ MIX on } M
}{F(t), \Pi, \Gamma^* \Rightarrow \Sigma^*, \Omega_1} \text{ IL}
}{\forall x\, F(x), \Pi, \Gamma^* \Rightarrow \Sigma^*, \Omega_1} \text{ VL}
}{\Pi, \forall x\, F(x), \Gamma^* \Rightarrow \Sigma^*, \Omega_1} \text{ IL}
$$

If by chance $M = F(t)$ we proceed instead as follows:

$$
\begin{array}{cc}
\vdots\,\pi_1 & \vdots\,\pi_2
\end{array}
$$

$$
\cfrac{
\cfrac{
\cfrac{\Pi \Rightarrow \Sigma \qquad F(t), \Gamma \Rightarrow \Omega_1}{\Pi, \Gamma^* \Rightarrow \Sigma^*, \Omega_1} \text{ MIX on } M
}{\forall x\, F(x), \Pi, \Gamma^* \Rightarrow \Sigma^*, \Omega_1} \text{ WL}
}{\Pi, \forall x\, F(x), \Gamma^* \Rightarrow \Sigma^*, \Omega_1} \text{ IL}
$$

We have a proof of lower rank by 1 than the original proof and of the same degree. Thus, in virtue of the hypothesis, there exists a proof having the same end-sequent without MIX.

(b) $M = \forall x\, F(x)$. Then:

$$
\vdots\,\pi_2
$$

$$
\cfrac{
\Pi \Rightarrow \Sigma \qquad \cfrac{F(t), \Gamma \Rightarrow \Omega_1}{\forall x\, F(x), \Gamma \Rightarrow \Omega_1} \text{ VL}
}{\Pi, \Gamma^* \Rightarrow \Sigma^*, \Omega_1} \text{ MIX}
$$

Again, since $\text{rk}_r(\pi) > 1$, the MIX formula $M = \forall x\, F(x)$ occurs in Γ.

In this case we attain our goal of constructing a new proof made out of two pieces assembled together. First of all, we construct the first piece. It is:

$$
\begin{array}{cc}
\vdots\,\pi_1 & \vdots\,\pi_2
\end{array}
$$

$$
\left.
\cfrac{
\cfrac{
\cfrac{\Pi \Rightarrow \Sigma \qquad F(t), \Gamma \Rightarrow \Omega_1}{\Pi, F(t), \Gamma^* \Rightarrow \Sigma^*, \Omega_1} \text{ MIX on } \forall x\, F(x)
}{F(t), \Gamma^*, \Pi \Rightarrow \Sigma^*, \Omega_1} \text{ IL}
}{\forall x\, F(x), \Gamma^*, \Pi \Rightarrow \Sigma^*, \Omega_1} \text{ VL}
\right\} \pi_3
$$

The second piece of the proof is the following:

$$
\frac{\Pi \Rightarrow \Sigma \qquad \dfrac{\forall x\, F(x), \Gamma^*, \Pi \Rightarrow \Sigma^*, \Omega_1}{\dfrac{\Pi, \Gamma^*, \Pi^* \Rightarrow \Sigma^*, \Sigma^*, \Omega_1}{\Pi, \Gamma^* \Rightarrow \Sigma^*, \Omega_1}\ \text{I, C}}}{} \quad \text{MIX on } \forall x\, F(x)
$$

By assembling them together we get:

$$
\frac{\Pi \Rightarrow \Sigma \qquad \dfrac{\ \dfrac{\ \dfrac{\Pi \Rightarrow \Sigma \quad \dfrac{F(t), \Gamma \Rightarrow \Omega_1}{}}{\Pi, F(t), \Gamma^* \Rightarrow \Sigma^*, \Omega_1}\ \text{MIX on } \forall x\, F(x)}{F(t), \Gamma^*, \Pi \Rightarrow \Sigma^*, \Omega_1}\ \text{IL}}{\forall x\, F(x), \Gamma^*, \Pi \Rightarrow \Sigma^*, \Omega_1}\ \text{VL}}{\dfrac{\Pi, \Gamma^*, \Pi^* \Rightarrow \Sigma^*, \Sigma^*, \Omega_1}{\Pi, \Gamma^* \Rightarrow \Sigma^*, \Omega_1}\ \text{I, C}}}{} \quad \text{MIX on } \forall x\, F(x)
$$

Obviously, one can repeat here the argument carried out in the first case to show the eliminability of both MIXes occurring in the transformed proof.

B9. Rule ∃L. Here it is again necessary to distinguish two cases in the original formulation of the proof. Each one of them must also be treated differently to obtain the result. Recall that our proof is assumed to be regular, that is the eigenvariable a of the ∃L inference only occurs above the ∃L inference. That means that a does not occur in Π or Σ. Of course, it does not occur in Γ or Ω_1, as otherwise the eigenvariable condition would be violated.

(a) $M \neq \exists x\, F(x)$. Then:

$$
\frac{\Pi \Rightarrow \Sigma \qquad \dfrac{F(a), \Gamma \Rightarrow \Omega_1}{\exists x\, F(x), \Gamma \Rightarrow \Omega_1}\ \exists \text{L}}{\Pi, \exists x\, F(x), \Gamma^* \Rightarrow \Sigma^*, \Omega_1}\ \text{MIX}
$$

In this case the proof can be transformed into:

$$
\begin{array}{c}
\begin{array}{cc}
\vdots \pi_1 & \vdots \pi_2 \\
\end{array} \\
\cline{1-1}
\dfrac{\Pi \Rightarrow \Sigma \qquad F(a), \Gamma \Rightarrow \Omega_1}{\Pi, F(a), \Gamma^* \Rightarrow \Sigma^*, \Omega_1} \text{ MIX on } M \\
\dfrac{}{F(a), \Pi, \Gamma^* \Rightarrow \Sigma^*, \Omega_1} \text{ IL} \\
\dfrac{}{\exists x\, F(x), \Pi, \Gamma^* \Rightarrow \Sigma^*, \Omega_1} \exists \text{L} \\
\dfrac{}{\Pi, \exists x\, F(x), \Gamma^* \Rightarrow \Sigma^*, \Omega_1} \text{ IL}
\end{array}
$$

The new proof is correct: the eigenvariable condition on $\exists$L is satisfied, by the observation resulting from the fact that the proof is regular made above. It has rank lower by 1 than the original proof and has the same degree. Thus, in virtue of the inductive hypothesis (on rank), there exists a proof having the same end-sequent without MIX.

We might be worried about the special case where $M = F(a)$: then the MIX would remove the $F(a)$ from the antecendent and the proof given above would not work. However, this case cannot arise since then $M = F(a)$ would have to occur in Γ. But if it did, the eigenvariable condition of the original $\exists$L would be violated.

(b) $M = \exists x\, F(x)$. Then:

$$
\dfrac{\Pi \Rightarrow \Sigma \qquad \dfrac{F(a), \Gamma \Rightarrow \Omega_1}{\exists x\, F(x), \Gamma \Rightarrow \Omega_1} \exists \text{L}}{\Pi, \Gamma^* \Rightarrow \Sigma^*, \Omega_1} \text{ MIX}
$$

Just as above, the objective is that of constructing a proof made out of two pieces that are assembled together. The first piece is the following:

$$
\left.
\begin{array}{c}
\dfrac{\Pi \Rightarrow \Sigma \qquad F(a), \Gamma \Rightarrow \Omega_1}{\Pi, F(a), \Gamma^* \Rightarrow \Sigma^*, \Omega_1} \text{ MIX on } \exists x\, F(x) \\
\dfrac{}{F(a), \Gamma^*, \Pi \Rightarrow \Sigma^*, \Omega_1} \text{ IL} \\
\dfrac{}{\exists x\, F(x), \Gamma^*, \Pi \Rightarrow \Sigma^*, \Omega_1} \exists \text{L}
\end{array}
\right\} \pi_3
$$

The eigenvariable condition is satisfied, since a does not occur in Π, Σ, Γ, or Ω_1.

Just as above, the MIX is applied correctly since, under the assumption that $\mathrm{rk}_r(\pi) > 1$, the MIX formula $\exists x\, F(x)$ also occurs in Γ.

The second piece is:

$$\frac{\Pi \Rightarrow \Sigma \qquad \exists x\, F(x), \Gamma^*, \Pi \Rightarrow \Sigma^*, \Omega_1}{\dfrac{\Pi, \Gamma^*, \Pi^* \Rightarrow \Sigma^*, \Sigma^*, \Omega_1}{\Pi, \Gamma^* \Rightarrow \Sigma^*, \Omega_1}\; \text{I, C}} \;\; \text{MIX on } \exists x\, F(x)$$

with derivation π_1 above $\Pi \Rightarrow \Sigma$.

Assembling the two pieces together we get:

$$\frac{\Pi \Rightarrow \Sigma \qquad \dfrac{\dfrac{\dfrac{\Pi \Rightarrow \Sigma \qquad F(a), \Gamma \Rightarrow \Omega_1}{\Pi, F(a), \Gamma^* \Rightarrow \Sigma^*, \Omega_1}\;\text{MIX on } \exists x\, F(x)}{F(a), \Gamma^*, \Pi \Rightarrow \Sigma^*, \Omega_1}\;\text{IL}}{\exists x\, F(x), \Gamma^*, \Pi \Rightarrow \Sigma^*, \Omega_1}\;\exists\text{L}}{\dfrac{\Pi, \Gamma^*, \Pi^* \Rightarrow \Sigma^*, \Sigma^*, \Omega_1}{\Pi, \Gamma^* \Rightarrow \Sigma^*, \Omega_1}\;\text{I, C}}\;\text{MIX on } \exists x\, F(x)$$

with π_1, π_2 as the upper derivations and π_3 bracketing the right branch.

We can now repeat the argument given for Rule $\wedge$L to show that both MIXes in the transformed proof can be eliminated.

B10. Rule $\supset$R. Since this is a rule of introduction in the succedent we need not keep track of the usual distinction. The proof has the following form:

$$\frac{\Pi \Rightarrow \Sigma \qquad \dfrac{A, \Gamma \Rightarrow \Omega_1, B}{\Gamma \Rightarrow \Omega_1, A \supset B}\;\supset\text{R}}{\Pi, \Gamma^* \Rightarrow \Sigma^*, \Omega_1, A \supset B}\;\text{MIX}$$

with π_1 above $\Pi \Rightarrow \Sigma$ and π_2 above $A, \Gamma \Rightarrow \Omega_1, B$.

It can be transformed in the following way, regardless of the distinction between $M \neq A \supset B$ and $M = A \supset B$:

$$\frac{\dfrac{\dfrac{\Pi \Rightarrow \Sigma \qquad A, \Gamma \Rightarrow \Omega_1, B}{\Pi, A, \Gamma^* \Rightarrow \Sigma^*, \Omega_1, B}\;\text{MIX on } M}{A, \Pi, \Gamma^* \Rightarrow \Sigma^*, \Omega_1, B}\;\text{IL}}{\Pi, \Gamma^* \Rightarrow \Sigma^*, \Omega_1, A \supset B}\;\supset\text{R}$$

with π_1, π_2 as the upper derivations.

If by chance $M = A$, insert a WL to introduce the formula A, between the MIX and the $\supset$R inference.

This is obviously a proof with rank lower by 1 than the original proof and of the same degree. Thus, in virtue of the inductive hypothesis (on rank) there exists a proof having the same end-sequent without MIX.

B11. Rule ∀R. Also in this case, just as in the previous rule, we do not need to make any distinctions. The proof has the following form:

$$
\cfrac{\Pi \Rightarrow \Sigma \qquad \cfrac{\Gamma \Rightarrow \Omega_1, F(a)}{\Gamma \Rightarrow \Omega_1, \forall x\, F(x)}\ \forall \text{R}}{\Pi, \Gamma^* \Rightarrow \Sigma^*, \Omega_1, \forall x\, F(x)}\ \text{MIX}
$$

Regardless of whether $M \neq \forall x\, F(x)$ or $M = \forall x\, F(x)$, the proof can be transformed as follows:

$$
\cfrac{\cfrac{\Pi \Rightarrow \Sigma \qquad \Gamma \Rightarrow \Omega_1, F(a)}{\Pi, \Gamma^* \Rightarrow \Sigma^*, \Omega_1, F(a)}\ \text{MIX on } M}{\Pi, \Gamma^* \Rightarrow \Sigma^*, \Omega_1, \forall x\, F(x)}\ \forall \text{R}
$$

Since π is regular, a only occurs above the ∀R-inference. So it does not occur in Π, Σ, Γ, or Ω_1, and the eigenvariable condition is satisfied.

We have clearly a proof of rank lower by 1 than the original proof and of the same degree. Hence, in virtue of the inductive hypothesis (on rank) there exists a proof having the same end-sequent without MIX.

B12. Rule ∃R. The case is analogous to the preceding one. The proof has the following form:

$$
\cfrac{\Pi \Rightarrow \Sigma \qquad \cfrac{\Pi \Rightarrow \Omega_1, F(t)}{\Gamma \Rightarrow \Omega_1, \exists x\, F(x)}\ \exists \text{R}}{\Pi, \Gamma^* \Rightarrow \Sigma^*, \Omega_1, \exists x\, F(x)}\ \text{MIX}
$$

Then, regardless of whether $M \neq \exists x\, F(x)$ or $M = \exists x\, F(x)$, the proof can be transformed as follows:

$$
\cfrac{\cfrac{\Pi \Rightarrow \Sigma \qquad \Gamma \Rightarrow \Omega_1, F(t)}{\Pi, \Gamma^* \Rightarrow \Sigma^*, \Omega_1, F(t)}\ \text{MIX on } M}{\Pi, \Gamma^* \Rightarrow \Sigma^*, \Omega_1, \exists x\, F(x)}\ \exists \text{R}
$$

As before, this is a proof of rank lower by 1 than the original rank but with the same degree. Hence, in virtue of the inductive hypothesis (on rank) there exists a proof having the same end-sequent without MIX.

Case C.   The rule that ends in the right premise of the MIX is an operational rule with *two* premises, namely $\wedge$R, $\vee$L, or $\supset$L (G.3.121.23).

Let us consider the three rules in order.

C1. Rule $\wedge$R (G.3.121.231). The final part of the proof is:

$$
\cfrac{\Pi \Rightarrow \Sigma \qquad \cfrac{\Gamma \Rightarrow \Theta, A \qquad \Gamma \Rightarrow \Theta, B}{\Gamma \Rightarrow \Theta, A \wedge B}\ {\scriptstyle\wedge\text{R}}}{\Pi, \Gamma^* \Rightarrow \Sigma^*, \Theta, A \wedge B}\ {\scriptstyle\text{MIX on } M}
$$

where it does not matter whether M is identical to $A \wedge B$ or not.

In both cases the proof can be transformed as follows:

$$
\cfrac{\cfrac{\Pi \Rightarrow \Sigma \qquad \Gamma \Rightarrow \Theta, A}{\Pi, \Gamma^* \Rightarrow \Sigma^*, \Theta, A}\ {\scriptstyle\text{MIX on } M} \qquad \cfrac{\Pi \Rightarrow \Sigma \qquad \Gamma \Rightarrow \Theta, B}{\Pi, \Gamma^* \Rightarrow \Sigma^*, \Theta, B}\ {\scriptstyle\text{MIX on } M}}{\Pi, \Gamma^* \Rightarrow \Sigma^*, \Theta, A \wedge B}\ {\scriptstyle\wedge\text{R}}
$$

The sub-proofs ending in the premises of the $\wedge$R inference have lower rank than the original one. Thus, in virtue of the inductive hypothesis (on rank), they can be transformed into proof without MIX. This results, overall, in a proof without MIX.

C2. Rule $\vee$L (G.3.121.232). The final part of the proof can present two forms:

(a) $M \neq A \vee B$. Then

$$
\cfrac{\Pi \Rightarrow \Sigma \qquad \cfrac{A, \Gamma \Rightarrow \Theta \qquad B, \Gamma \Rightarrow \Theta}{A \vee B, \Gamma \Rightarrow \Theta}\ {\scriptstyle\vee\text{L}}}{\Pi, A \vee B, \Gamma^* \Rightarrow \Sigma^*, \Theta}\ {\scriptstyle\text{MIX on } M}
$$

In this case the proof can be transformed as follows:

$$
\cfrac{\cfrac{\cfrac{\Pi \Rightarrow \Sigma \qquad A, \Gamma \Rightarrow \Theta}{\Pi, A^*, \Gamma^* \Rightarrow \Sigma^*, \Theta}\ {\scriptstyle\text{MIX on } M}}{A, \Pi, \Gamma^* \Rightarrow \Sigma^*, \Theta}\ {\scriptstyle\text{IL, WL}} \qquad \cfrac{\cfrac{\Pi \Rightarrow \Sigma \qquad B, \Gamma \Rightarrow \Theta}{\Pi, B^*, \Gamma^* \Rightarrow \Sigma^*, \Theta}\ {\scriptstyle\text{MIX on } M}}{B, \Pi, \Gamma^* \Rightarrow \Sigma^*, \Theta}\ {\scriptstyle\text{IL, WL}}}{A \vee B, \Pi, \Gamma^* \Rightarrow \Sigma^*, \Theta}\ {\scriptstyle\vee\text{L}}
$$

where if $A = M$ then A^* is empty and if $A \neq M$ then $A^* = A$; moreover, if $B = M$ then B^* is empty and if $B \neq M$ then $B^* = B$.

Clearly the MIXes occurring in the new proof are of lower rank than the MIX in the original proof. Therefore, in virtue of the inductive hypothesis there exists a proof of the same end-sequent without MIX.

(b) $M = A \vee B$. Then

$$
\cfrac{\Pi \Rightarrow \Sigma \qquad \cfrac{A,\Gamma \Rightarrow \Theta \qquad B,\Gamma \Rightarrow \Theta}{A \vee B,\Gamma \Rightarrow \Theta} \; \text{VL}}{\Pi,\Gamma^* \Rightarrow \Sigma^*,\Theta} \; \text{MIX on } A \vee B
$$

Now, on account of the fact that $M = A \vee B$, $M \neq A$ and $M \neq B$, the preceding proof can be modified as follows:

$$
\cfrac{\Pi \Rightarrow \Sigma \qquad \cfrac{\left.\cfrac{\cfrac{\Pi \Rightarrow \Sigma \quad A,\Gamma \Rightarrow \Theta}{\Pi,A,\Gamma^* \Rightarrow \Sigma^*,\Theta}\text{MIX}}{\boxed{A,\Pi,\Gamma^*} \Rightarrow \Sigma^*,\Theta}\text{IL}\right\}\pi' \quad \left.\cfrac{\cfrac{\Pi \Rightarrow \Sigma \quad B,\Gamma \Rightarrow \Theta}{\Pi,B,\Gamma^* \Rightarrow \Sigma^*,\Theta}\text{MIX}}{\boxed{B,\Pi,\Gamma^*} \Rightarrow \Sigma^*,\Theta}\text{IL}\right\}\pi''}{A \vee B,\Pi,\Gamma^* \Rightarrow \Sigma^*,\Theta}\text{VL}}{\cfrac{\Pi,\Pi,\Gamma^* \Rightarrow \Sigma^*,\Sigma^*,\Theta}{\Pi,\Gamma^* \Rightarrow \Sigma^*,\Theta}\text{I, C}}\text{MIX}
$$

where the MIX is always on $A \vee B$. Note that here, too, $A \vee B$ must occur also in Γ, since $\text{rk}_r(\pi) > 1$.

First consider the two sub-proofs π' and π''. We have $\text{rk}_l(\pi') = \text{rk}_l(\pi'') = \text{rk}_l(\pi)$ and $\text{rk}_r(\pi')$ and $\text{rk}_r(\pi'')$ are both $\leq \text{rk}_r(\pi) - 1$. Hence the inductive hypothesis applies to both sub-proofs, and we can replace them by MIX-free proofs of their end-sequents. We now have:

$$
\cfrac{\Pi \Rightarrow \Sigma \qquad \left.\cfrac{\cfrac{\cfrac{\Pi,A,\Gamma^* \Rightarrow \Sigma^*,\Theta}{\boxed{A,\Pi,\Gamma^*} \Rightarrow \Sigma^*,\Theta}\text{IL} \qquad \cfrac{\Pi,B,\Gamma^* \Rightarrow \Sigma^*,\Theta}{\boxed{B,\Pi,\Gamma^*} \Rightarrow \Sigma^*,\Theta}\text{IL}}{A \vee B,\Pi,\Gamma^* \Rightarrow \Sigma^*,\Theta}\text{VL}}{}\right\}\pi'''}{\cfrac{\Pi,\Pi,\Gamma^* \Rightarrow \Sigma^*,\Sigma^*,\Theta}{\Pi,\Gamma^* \Rightarrow \Sigma^*,\Theta}\text{I, C}}\text{MIX}
$$

We have assumed at the outset (by appealing to Lemma 6.15) that $A \vee B$ does not occur in Π. $A \vee B$ also occurs neither in Γ^* nor in Σ^*. Thus $A \vee B$ does not occur in the antecedents (displayed in the boxes) of the premises of the right premise of the remaining MIX. In other words, $\mathrm{rk}_r(\pi''') = 1$. Since $\mathrm{rk}_l(\pi''') = \mathrm{rk}_l(\pi)$, we have that $\mathrm{rk}(\pi''') < \mathrm{rk}(\pi)$. Thus the hypothesis applies: π''' is reducible, and we can replace it with a MIX-free proof.

C3. Rule $\supset$L (G.3.121.233). The final part of the proof looks as follows:

$$
\cfrac{
\begin{array}{c}\vdots\,\pi_1\\{}\end{array}\quad
\cfrac{
\cfrac{
\begin{array}{c}\vdots\,\pi_2\\ \Gamma \Rightarrow \Theta, A\end{array}\qquad
\begin{array}{c}\vdots\,\pi_3\\ B, \Delta \Rightarrow \Lambda\end{array}
}{A \supset B, \Gamma, \Delta \Rightarrow \Theta, \Lambda}\ {\supset}\mathrm{L}
}{}
}{\Pi, (A \supset B)^*, \Gamma^*, \Delta^* \Rightarrow \Sigma^*, \Theta, \Lambda}\ \mathrm{MIX}
$$

with left premise $\Pi \Rightarrow \Sigma$ (proof π_1).

where, of course, if $M \ne A \supset B$ then $(A \supset B)^* = A \supset B$, while if $M = A \supset B$, then $(A \supset B)^*$ is empty.

In order to transform the proof one must take into account that $\mathrm{rk}_r(\pi) > 1$ and that M must consequently appear in the antecedent of one of the premises of the $\supset$L inference leading to the right premise of the MIX. If $M = B$, B must also occur in Γ or in Δ. Otherwise, M would not occur in the antecedent of the right premise of the MIX, since the left-most occurrence of B in the antecedent B, Δ is immediately eliminated by the application of $\supset$L. Consequently, we can treat three different cases: (1) The MIX formula M occurs in both Γ and Δ, (2) M occurs in Γ but not in Δ, and (3) M occurs in Δ but not in Γ. Case (1) is Gentzen's G.3.121.233.1, and cases (2) and (3) are his G.3.121.233.2.

C3.1. M occurs in both Γ and Δ.

In this case the proof takes place in two stages. We construct first the following piece:

$$
\cfrac{
\cfrac{
\begin{array}{c}\vdots\,\pi_1\\ \Pi \Rightarrow \Sigma\end{array}\qquad
\begin{array}{c}\vdots\,\pi_2\\ \Gamma \Rightarrow \Theta, A\end{array}
}{\Pi, \Gamma^* \Rightarrow \Sigma^*, \Theta, A}\ \mathrm{MIX}
\qquad
\cfrac{
\cfrac{
\cfrac{
\begin{array}{c}\vdots\,\pi_1\\ \Pi \Rightarrow \Sigma\end{array}\qquad
\begin{array}{c}\vdots\,\pi_3\\ B, \Delta \Rightarrow \Lambda\end{array}
}{\Pi, B^*, \Delta^* \Rightarrow \Sigma^*, \Lambda}\ \mathrm{MIX}
}{B, \Pi, \Delta^* \Rightarrow \Sigma^*, \Lambda}\ \mathrm{IL, \ possibly\ WL}
}{}
}{A \supset B, \Pi, \Gamma^*, \Pi, \Delta^* \Rightarrow \Sigma^*, \Theta, \Sigma^*, \Lambda}\ {\supset}\mathrm{L}
$$

where $B^* = B$ if $M \ne B$ and B^* is empty if $M = B$.

Now we have two possibilities:

(a) $M \ne A \supset B$. In this case the result follows immediately by adding to the preceding proof the following piece:

$$
\cfrac{A \supset B, \Pi, \Gamma^*, \Pi, \Delta^* \Rightarrow \Sigma^*, \Theta, \Sigma^*, \Lambda}{\Pi, A \supset B, \Gamma^*, \Delta^* \Rightarrow \Sigma^*, \Theta, \Lambda}\ \mathrm{I, C}
$$

(b) $M = A \supset B$. Then we can construct the following proof:

$$
\cfrac{
\cfrac{
\vdots \pi_1 \\
\Pi \Rightarrow \Sigma \qquad A \supset B, \Pi, \Gamma^*, \Pi, \Delta^* \Rightarrow \Sigma^*, \Theta, \Sigma^*, \Lambda
}{
\Pi, \Pi, \Gamma^*, \Pi, \Delta^* \Rightarrow \Sigma^*, \Sigma^*, \Theta, \Sigma^*, \Lambda
} \text{ MIX}
}{
\Pi, \Gamma^*, \Delta^* \Rightarrow \Sigma^*, \Theta, \Lambda
} \text{ I, C}
$$

This can be added to the initial piece so as to obtain a proof having as a whole a lower rank. The reason is analogous to that given in the preceding case (namely, sub-case C2(b) where the right premise is obtained by VL and the principal formula is the MIX formula).

In both cases (a) and (b) we obtain proofs of lower rank. Thus, the inductive hypothesis is satisfied and the result follows.

C3.2. M is in Γ but not in Δ. Let thus the proof be:

$$
\cfrac{
\vdots \pi_1 \\
\Pi \Rightarrow \Sigma \qquad
\cfrac{
\Gamma \Rightarrow \Theta, A \qquad B, \Delta \Rightarrow \Lambda
}{
A \supset B, \Gamma, \Delta \Rightarrow \Theta, \Lambda
} \supset \text{L}
}{
\Pi, (A \supset B)^*, \Gamma^*, \Delta \Rightarrow \Sigma^*, \Theta, \Lambda
} \text{ MIX}
$$

where M only occurs in Γ and of course in Σ and, moreover, $(A \supset B)^*$ is empty if $M = A \supset B$, while $(A \supset B)^* = A \supset B$ if $M \neq A \supset B$.

Here too we start from the construction of the first piece of the proof. Then we will proceed according to whether $M = A \supset B$ or $M \neq A \supset B$.

$$
\cfrac{
\cfrac{
\vdots \pi_1 \qquad \vdots \pi_2 \\
\Pi \Rightarrow \Sigma \qquad \Gamma \Rightarrow \Theta, A
}{
\Pi, \Gamma^* \Rightarrow \Sigma^*, \Theta, A
} \text{ MIX} \qquad B, \Delta \Rightarrow \Lambda
}{
A \supset B, \Pi, \Gamma^*, \Delta \Rightarrow \Sigma^*, \Theta, \Lambda
} \supset \text{L}
$$

Now we must distinguish the two further sub-cases:

(a) $M \neq A \supset B$. Then we proceed as follows:

$$
\cfrac{
A \supset B, \Pi, \Gamma^*, \Delta \Rightarrow \Sigma^*, \Theta, \Lambda
}{
\Pi, A \supset B, \Gamma^*, \Delta \Rightarrow \Sigma^*, \Theta, \Lambda
} \text{ IL}
$$

Note $A \supset B$ is not the MIX formula and so still appears in the end-sequent. Clearly we have a derivation of lower rank than the original one.

(b) $M = A \supset B$. Then:

$$
\cfrac{\cfrac{\Pi \Rightarrow \Sigma \qquad A \supset B, \Pi, \Gamma^*, \Delta \Rightarrow \Sigma^*, \Theta, \Lambda}{\Pi, \Pi, \Gamma^*, \Delta \Rightarrow \Sigma^*, \Sigma^*, \Theta, \Lambda} \text{ MIX}}{\Pi, \Gamma^*, \Delta \Rightarrow \Sigma^*, \Theta, \Lambda} \text{ I, C}
$$

with π_1 above $\Pi \Rightarrow \Sigma$.

Note that $\Delta^* = \Delta$ since M does not occur in Δ and similarly $\Pi^* = \Pi$. Also in this case we have a proof of lower rank than the original one.

C3.3. M is in Δ and not in Γ. The final part of the proof now looks as follows:

$$
\cfrac{\Pi \Rightarrow \Sigma \qquad \cfrac{\Gamma \Rightarrow \Theta, A \qquad B, \Delta \Rightarrow \Lambda}{A \supset B, \Gamma, \Delta \Rightarrow \Theta, \Lambda} \supset\text{L}}{\Pi, (A \supset B)^*, \Gamma, \Delta^* \Rightarrow \Sigma^*, \Theta, \Lambda} \text{ MIX}
$$

with π_1 over $\Pi \Rightarrow \Sigma$, π_2 over $\Gamma \Rightarrow \Theta, A$, and π_3 over $B, \Delta \Rightarrow \Lambda$.

where M occurs only in Δ and of course in Σ and, as above, $(A \supset B)^*$ is empty if $M = A \supset B$, while $(A \supset B)^* = A \supset B$ if $M \neq A \supset B$.

Here too we proceed in two stages. First of all we construct the following piece:

$$
\cfrac{\Gamma \Rightarrow \Theta, A \qquad \cfrac{\cfrac{\Pi \Rightarrow \Sigma \qquad B, \Delta \Rightarrow \Lambda}{\Pi, B^*, \Delta^* \Rightarrow \Sigma^*, \Lambda} \text{ MIX}}{B, \Pi, \Delta^* \Rightarrow \Sigma^*, \Lambda} \text{ I, possibly WL}}{A \supset B, \Gamma, \Pi, \Delta^* \Rightarrow \Theta, \Sigma^*, \Lambda} \supset\text{L}
$$

with π_2 over $\Gamma \Rightarrow \Theta, A$, π_1 over $\Pi \Rightarrow \Sigma$, and π_3 over $B, \Delta \Rightarrow \Lambda$.

where $B^* = B$ if $M \neq B$ and B^* is empty if $M = B$.

Now we must distinguish two further sub-cases:

(a) $M \neq A \supset B$. Then we proceed as follows:

$$
\cfrac{A \supset B, \Gamma, \Pi, \Delta^* \Rightarrow \Theta, \Sigma^*, \Lambda}{\Pi, A \supset B, \Gamma, \Delta^* \Rightarrow \Sigma^*, \Theta, \Lambda} \text{ IL}
$$

Here note that since $M \neq A \supset B$, $(A \supset B)^* = A \supset B$.

(b) $M = A \supset B$. Then we proceed as follows:

$$
\cfrac{\cfrac{\Pi \Rightarrow \Sigma \qquad A \supset B, \Gamma, \Pi, \Delta^* \Rightarrow \Theta, \Sigma^*, \Lambda}{\Pi, \Gamma, \Pi, \Delta^* \Rightarrow \Sigma^*, \Theta, \Sigma^*, \Lambda} \text{ MIX}}{\Pi, (A \supset B)^*, \Gamma, \Delta^* \Rightarrow \Sigma^*, \Theta, \Lambda} \text{ I, C}
$$

with π_1 over $\Pi \Rightarrow \Sigma$.

Here note that since M is not in Γ, $\Gamma = \Gamma^*$ and $(A \supset B)^*$ is empty.

In all cases, the proofs we obtain are of lower rank. Thus, the conditions for the application of the hypothesis are satisfied and the result follows.   □

We've concluded the case where $\mathrm{rk}_r(\pi) > 1$. We may now assume that $\mathrm{rk}_r(\pi) = 1$ and $\mathrm{rk}_l(\pi) > 1$ (G.3.122).

Lemma 6.19. *Suppose π is a proof with $\mathrm{rk}_r(\pi) = 1$, $\mathrm{rk}_l(\pi) > 1$, and every proof π' with $\mathrm{dg}(\pi') < \mathrm{dg}(\pi)$ or $\mathrm{dg}(\pi') = \mathrm{dg}(\pi)$ and $\mathrm{rk}(\pi') < \mathrm{rk}(\pi)$ is reducible. Then there is a proof of the same end-sequent as π with equal or lower degree, and lower rank.*

Proof. The situation is of course symmetric to the case where $\mathrm{rk}_r(\pi) > 1$ (i.e., the proof of Lemma 6.18), except instead of considering cases according to the last inference leading to the right premise of the MIX, we consider cases according to the last inference leading to the left premise of the MIX. The difficult cases are now not the L-rules, but the R-rules, as these can potentially introduce the MIX-formula. There is however one case that we should carry out in detail, namely the rule ⊃L.

The proof ends in the following way:

$$
\cfrac{\cfrac{\vdots \pi_1 \qquad \vdots \pi_2}{\cfrac{\Gamma \Rightarrow \Theta, A \quad B, \Delta \Rightarrow \Lambda}{A \supset B, \Gamma, \Delta \Rightarrow \Theta, \Lambda}\, {\supset}\mathrm{L}} \qquad \cfrac{\vdots \pi_3}{\Sigma \Rightarrow \Pi}}{A \supset B, \Gamma, \Delta, \Sigma^* \Rightarrow \Theta^*, \Lambda^*, \Pi}\, \mathrm{MIX}
$$

For reasons symmetrical to those of the corresponding case C3 of Lemma 6.18 above, where the right premise of the MIX ends in ⊃L, we have three cases:

1. M occurs both in Θ and in Λ. The proof is then transformed as follows:

$$
\cfrac{\cfrac{\cfrac{\vdots \pi_1 \quad \vdots \pi_3}{\cfrac{\Gamma \Rightarrow \Theta, A \quad \Sigma \Rightarrow \Pi}{\cfrac{\Gamma, \Sigma^* \Rightarrow \Theta^*, A^*, \Pi}{\Gamma, \Sigma^* \Rightarrow \Theta^*, \Pi, A}\, \mathrm{IR,\ possibly\ WR}}\, \mathrm{MIX}} \qquad \cfrac{\cfrac{\vdots \pi_2 \quad \vdots \pi_3}{B, \Delta \Rightarrow \Lambda \quad \Sigma \Rightarrow \Pi}{B, \Delta, \Sigma^* \Rightarrow \Lambda^*, \Pi}\, \mathrm{MIX}}{\quad}}{A \supset B, \Gamma, \Sigma^*, \Delta, \Sigma^* \Rightarrow \Theta^*, \Pi, \Lambda^*, \Pi}\, {\supset}\mathrm{L}}{A \supset B, \Gamma, \Delta, \Sigma^* \Rightarrow \Theta^*, \Lambda^*, \Pi}\, \mathrm{I,\,C}
$$

where A^* is empty if $A = M$ and $A^* = A$ if $A \neq M$.

Since both MIXes are of lower rank than the original MIX, they can both be eliminated.

2. M occurs in Λ but not in Θ. Then, the proof can be transformed as follows:

$$
\cfrac{
  \cfrac{\vdots\;\pi_1}{\Gamma \Rightarrow \Theta, A}
  \quad
  \cfrac{
    \cfrac{\vdots\;\pi_2}{B, \Delta \Rightarrow \Lambda}
    \quad
    \cfrac{\vdots\;\pi_3}{\Sigma \Rightarrow \Pi}
  }{B, \Delta, \Sigma^* \Rightarrow \Lambda^*, \Pi}\;\text{MIX}
}{A \supset B, \Gamma, \Delta, \Sigma^* \Rightarrow \Theta, \Lambda^*, \Pi}\;\supset\text{L}
$$

Since we assumed that M is not in Θ, Θ^* and Θ are identical.

Even here the MIX can be eliminated in virtue of the hypothesis that proofs of lower rank are reducible.

3. M occurs in Θ but not in Λ. Exercise. ☐

Problem 6.20. Complete the proof of Lemma 6.18 by carrying out the cases where the right premise is proved by IR or CR.

Problem 6.21. Complete case (3) in the proof of Lemma 6.19.

Problem 6.22. Explicitly verify the cases in the proof of Lemma 6.19 where the left premise of the MIX ends in either ∧R or ∀R.

6.6 Reducing the rank: example

Let us now examine a proof containing CUT and let us show how it is possible to eliminate CUT on the basis of some of the cases taken into consideration in the proof of the *Hauptsatz*. The application of CUT will of course be considered as already reformulated into applications of MIX. We begin by focusing on the cases that reduce the rank.

$$C \supset A, A \supset B \Rightarrow C \supset B$$

Consider the proof of the above sequent using MIX which displays the occurrences of the MIX formula, in the left and right branches, that are relevant for computing the rank.

Since we are interested first in the steps that reduce the rank only, we will assume that A is atomic. Then the degree of A, and hence of the MIX, is 0, while the rank is $4 + 6 = 10$.

The MIX formula A does not occur in the antecedent of the left premise or in in the succedent of the right premise. Therefore, we cannot apply Lemma 6.15. Since the right rank is > 1, we are in the situation dealt with in Lemma 6.18, specifically the case where the right premise of the MIX is the conclusion of IL. Since A is a principal formula of the IL inference, the second sub-case applies: we can simply remove the IL inference to obtain:

$$
\cfrac{
\cfrac{
\cfrac{\cfrac{C \Rightarrow C}{C \Rightarrow C, \boxed{A}}\text{ WR}}{C \Rightarrow \boxed{A}, C}\text{ IR} \quad A \Rightarrow A
}{
\cfrac{C \supset A, C \Rightarrow A, \boxed{A}}{C \supset A, C \Rightarrow \boxed{A}}\text{ CR}
}\supset\!\text{L}
\quad
\cfrac{
\cfrac{
\cfrac{\cfrac{\boxed{A} \Rightarrow A}{\boxed{A} \Rightarrow A, B}\text{ WR}}{\boxed{A} \Rightarrow B, A}\text{ IR} \quad B \Rightarrow B
}{
\cfrac{A \supset B, \boxed{A} \Rightarrow B, B}{A \supset B, \boxed{A} \Rightarrow B}\text{ CR}
}\supset\!\text{L}
}{
\cfrac{\cfrac{C \supset A, C, A \supset B \Rightarrow B}{C, C \supset A, A \supset B \Rightarrow B}\text{ IL}}{C \supset A, A \supset B \Rightarrow C \supset B}\supset\!\text{R}
}\text{ MIX}
$$

Obviously the right rank has decreased by 1 and by the inductive hypothesis the new proof is reducible. We can actually prove this effectively. Indeed, the new proof also falls within the same case as the original proof. Only the rule changes as we now must consider CR. Since the principal formula of the CR inference is B, i.e., not the MIX formula, we can simply move the CR inference below the MIX. The proof becomes:

$$
\cfrac{
\cfrac{
\cfrac{\cfrac{C \Rightarrow C}{C \Rightarrow C, \boxed{A}}\text{ WR}}{C \Rightarrow \boxed{A}, C}\text{ IR} \quad A \Rightarrow A
}{
\cfrac{C \supset A, C \Rightarrow A, \boxed{A}}{C \supset A, C \Rightarrow \boxed{A}}\text{ CR}
}\supset\!\text{L}
\quad
\cfrac{
\cfrac{\cfrac{\boxed{A} \Rightarrow A}{\boxed{A} \Rightarrow A, B}\text{ WR}}{\boxed{A} \Rightarrow B, A}\text{ IR} \quad B \Rightarrow B
}{A \supset B, \boxed{A} \Rightarrow B, B}\supset\!\text{L}
}{
\cfrac{\cfrac{\cfrac{C \supset A, C, A \supset B \Rightarrow B, B}{C \supset A, C, A \supset B \Rightarrow B}\text{ CR}}{C, C \supset A, A \supset B \Rightarrow B}\text{ IL}}{C \supset A, A \supset B \Rightarrow C \supset B}\supset\!\text{R}
}\text{ MIX}
$$

The MIX has been pushed upward and in this way the rank is lowered. Here we can proceed to eliminate the application of CR in the left branch. (Note that the proof of the *Hauptsatz* would require that we keep lowering the right rank until it is 1 before proceeding to the left premise of the MIX. We'll ignore this here, as the purpose of the example is to illustrate how the reduction steps work.) We get:

$$
\cfrac{
  \cfrac{
    \cfrac{
      \cfrac{C \Rightarrow C}{C \Rightarrow C, \boxed{A}} \text{ WR}
    }{C \Rightarrow \boxed{A}, C} \text{ IR}
    \qquad A \Rightarrow A
  }{C \supset A, C \Rightarrow A, \boxed{A}} \supset\!\text{L}
  \qquad
  \cfrac{
    \cfrac{
      \cfrac{
        \cfrac{\boxed{A} \Rightarrow A}{\boxed{A} \Rightarrow A, B} \text{ WR}
      }{\boxed{A} \Rightarrow B, A} \text{ IR}
      \qquad B \Rightarrow B
    }{A \supset B, \boxed{A} \Rightarrow B, B} \supset\!\text{L}
  }{} \text{ MIX}
}{
  \cfrac{
    \cfrac{
      \cfrac{C \supset A, C, A \supset B \Rightarrow B, B}{C \supset A, C, A \supset B \Rightarrow B} \text{ CR}
    }{C, C \supset A, A \supset B \Rightarrow B} \text{ IL}
  }{C \supset A, A \supset B \Rightarrow C \supset B} \supset\!\text{R}
}
$$

The rank has decreased once again even though there is still one MIX occurring. However, the proof of the right premise of the MIX now depends on the application of an operational rule to a formula which is different from the MIX formula. We are thus facing the case considered in Lemma 6.18, sub-case C3, dealing with ⊃L. Let us consider first the proof of the right premise. In our example, the MIX formula A only occurs in the antecedent of the left premise and not in that of the right premise of ⊃L. Thus we must proceed according to sub-case C3.2. Let us construct the following piece of proof:

$$
\cfrac{
  \cfrac{
    \cfrac{
      \cfrac{C \Rightarrow C}{C \Rightarrow C, \boxed{A}} \text{ WR}
    }{C \Rightarrow \boxed{A}, C} \text{ IR}
    \qquad A \Rightarrow A
  }{C \supset A, C \Rightarrow A, \boxed{A}} \supset\!\text{L}
  \qquad
  \cfrac{
    \cfrac{\boxed{A} \Rightarrow A}{\boxed{A} \Rightarrow A, B} \text{ WR}
  }{\boxed{A} \Rightarrow B, A} \text{ IR}
}{C \supset A, C \Rightarrow B, A} \text{ MIX}
$$

To obtain the end-sequent, we do not need to apply another MIX. It suffices in fact to proceed as follows:

$$
\cfrac{
  \cfrac{
    \cfrac{
      \cfrac{
        \cfrac{C \Rightarrow C}{C \Rightarrow C, \boxed{A}} \text{ WR}
      }{C \Rightarrow \boxed{A}, C} \text{ IR}
      \qquad A \Rightarrow A
    }{C \supset A, C \Rightarrow A, \boxed{A}} \supset\!\text{L}
    \qquad
    \cfrac{
      \cfrac{\boxed{A} \Rightarrow A}{\boxed{A} \Rightarrow A, B} \text{ WR}
    }{\boxed{A} \Rightarrow B, A} \text{ IR}
  }{C \supset A, C \Rightarrow B, A} \text{ MIX}
  \qquad B \Rightarrow B
}{
  \cfrac{
    \cfrac{
      \cfrac{A \supset B, C \supset A, C \Rightarrow B, B}{A \supset B, C \supset A, C \Rightarrow B} \text{ CR}
    }{C, C \supset A, A \supset B \Rightarrow B} \text{ IL}
  }{C \supset A, A \supset B \Rightarrow C \supset B} \supset\!\text{R}
} \supset\!\text{L}
$$

As is easy to see, the sub-proof ending in the new MIX has smaller rank: it is now $3 + 3 = 6$, i.e., the rank decreased by one. Let us now try to push the MIX further and further upwards in order to eliminate it. This can be obtained by transforming the sub-proof ending in the MIX,

$$
\cfrac{
  \cfrac{
    \cfrac{C \Rightarrow C}{C \Rightarrow C, \boxed{A}}\text{ WR}
  }{C \Rightarrow \boxed{A}, C}\text{ IR} \qquad A \Rightarrow A
}{C \supset A, C \Rightarrow A, \boxed{A}}\text{ ⊃L}
\qquad
\cfrac{
  \cfrac{
    \cfrac{\boxed{A} \Rightarrow A}{\boxed{A} \Rightarrow A, B}\text{ WR}
  }{\boxed{A} \Rightarrow B, A}\text{ IR}
}{}
$$
$$
\overline{\qquad\qquad\qquad C \supset A, C \Rightarrow B, A \qquad\qquad\qquad}\text{ MIX}
$$

into the following new proof, by applying Lemma 6.18, case B2 and B1, to switch the MIX with the IR and WR inferences that yields the left premise of the MIX:

$$
\cfrac{
  \cfrac{
    \cfrac{
      \cfrac{C \Rightarrow C}{C \Rightarrow C, \boxed{A}}\text{ WR}
    }{C \Rightarrow \boxed{A}, C}\text{ IR} \qquad A \Rightarrow A
  }{C \supset A, C \Rightarrow A, \boxed{A}}\text{ ⊃L} \qquad \boxed{A} \Rightarrow A
}{
  \cfrac{
    \cfrac{C \supset A, C \Rightarrow A}{C \supset A, C \Rightarrow A, B}\text{ WR}
  }{C \supset A, C \Rightarrow B, A}\text{ IR}
}\text{ MIX}
$$

Now the right rank is 1, and Lemma 6.19 applies to the sub-proof ending in the MIX: we switch the MIX with the ⊃L inference, and obtain:

$$
\cfrac{
  \cfrac{
    \cfrac{
      \cfrac{C \Rightarrow C}{C \Rightarrow C, \boxed{A}}\text{ WR}
    }{C \Rightarrow \boxed{A}, C}\text{ IR} \qquad \boxed{A} \Rightarrow A
  }{
    \cfrac{C \Rightarrow C, A}{C \Rightarrow A, C}\text{ IR}
  }\text{ MIX} \qquad A \Rightarrow A
}{
  \cfrac{
    \cfrac{
      \cfrac{C \supset A, C \Rightarrow A, A}{C \supset A, C \Rightarrow A}\text{ CR}
    }{C \supset A, C \Rightarrow A, B}\text{ WR}
  }{C \supset A, C \Rightarrow B, A}\text{ IR}
}\text{ ⊃L}
$$

We have obtained a proof with rank = 3. The left premise of the new MIX is the conclusion of IR, which can be switched with the MIX. We get:

$$
\cfrac{
  \cfrac{
    \cfrac{C \Rightarrow C}{C \Rightarrow C, \boxed{A}}\text{ WR} \qquad \boxed{A} \Rightarrow A
  }{
    \cfrac{C \Rightarrow C, A}{C \Rightarrow A, C}\text{ IR}
  }\text{ MIX} \qquad A \Rightarrow A
}{
  \cfrac{
    \cfrac{
      \cfrac{C \supset A, C \Rightarrow A, A}{C \supset A, C \Rightarrow A}\text{ CR}
    }{C \supset A, C \Rightarrow A, B}\text{ WR}
  }{C \supset A, C \Rightarrow B, A}\text{ IR}
}\text{ ⊃L}
$$

Now the left rank is 1 and the left premise is the result of WR, so Lemma 6.13 applies. We can remove the MIX to obtain:

$$
\cfrac{
\cfrac{
\cfrac{
\cfrac{C \Rightarrow C}{C \Rightarrow C, A} \text{ WR}
}{C \Rightarrow A, C} \text{ IR}
\qquad A \Rightarrow A
}{
\cfrac{
\cfrac{
\cfrac{C \supset A, C \Rightarrow A, A}{C \supset A, C \Rightarrow A} \text{ CR}
}{C \supset A, C \Rightarrow A, B} \text{ WR}
}{C \supset A, C \Rightarrow B, A} \text{ IR}
} \supset \text{L}
}{}
$$

In conclusion, the final proof without MIXes is the following:

$$
\cfrac{
\cfrac{
\cfrac{
\cfrac{
\cfrac{
\cfrac{
\cfrac{\cfrac{C \Rightarrow C}{C \Rightarrow C, A}\text{ WR}}{C \Rightarrow A, C}\text{ IR} \quad A \Rightarrow A
}{C \supset A, C \Rightarrow A, A}\supset\text{L}
}{C \supset A, C \Rightarrow A}\text{ CR}
}{C \supset A, C \Rightarrow A, B}\text{ WR}
}{C \supset A, C \Rightarrow B, A}\text{ IR} \qquad B \Rightarrow B
}{
\cfrac{
\cfrac{A \supset B, C \supset A, C \Rightarrow B, B}{A \supset B, C \supset A, C \Rightarrow B}\text{ CR}
}{C, C \supset A, A \supset B \Rightarrow B}\text{ IL}
}\supset\text{L}
}{C \supset A, A \supset B \Rightarrow C \supset B}\supset\text{R}
$$

6.7   Reducing the degree: example

Consider the *Law of minimal refutation*

$$A \supset B, A \supset \neg B \Rightarrow \neg A$$

and the following proof of it:

$$
\cfrac{
\begin{array}{c} \vdots \text{ LNC} \\ \vdots \end{array}
\quad
\cfrac{
\cfrac{
\cfrac{\dfrac{A \Rightarrow A \quad B \Rightarrow B}{A \supset B, A \Rightarrow B}\supset\text{L}}{A \supset B, A \supset \neg B, A \Rightarrow B}\text{ WL, IL}
\qquad
\cfrac{\dfrac{A \Rightarrow A \quad \neg B \Rightarrow \neg B}{A \supset \neg B, A \Rightarrow \neg B}\supset\text{L}}{A \supset B, A \supset \neg B, A \Rightarrow \neg B}\text{ WL}
}{A \supset B, A \supset \neg B, A \Rightarrow B \wedge \neg B}\wedge\text{R}
}{\neg(B \wedge \neg B), A \supset B, A \supset \neg A, A \Rightarrow}\neg\text{L}
}{}\text{ MIX}
$$

$$
\cfrac{
\cfrac{
\cfrac{\Rightarrow \neg(B \wedge \neg B)}{A \supset B, A \supset \neg B, A \Rightarrow} \text{ MIX}
}{A, A \supset B, A \supset \neg B \Rightarrow}\text{ IL}
}{A \supset B, A \supset \neg B \Rightarrow \neg A}\neg\text{R}
$$

Let us isolate the part of the proof until MIX and let us complete it with the insertion of the proof of the law of non-contradiction on the left. Our proof π is:

$$
\pi_1 \left\{
\begin{array}{c}
\dfrac{\dfrac{\dfrac{\dfrac{\dfrac{\dfrac{B \Rightarrow B}{\neg B, B \Rightarrow} \neg L}{B, \neg B \Rightarrow} IL}{B \wedge \neg B, \neg B \Rightarrow} \wedge_1 L}{\neg B, B \wedge \neg B \Rightarrow} IL}{\dfrac{B \wedge \neg B, B \wedge \neg B \Rightarrow}{B \wedge \neg B \Rightarrow} CL}}{\Rightarrow \neg(B \wedge \neg B)} \neg R
\end{array}
\right.
$$

$$
\pi_2 \left\{
\begin{array}{cc}
\dfrac{\dfrac{A \Rightarrow A \quad B \Rightarrow B}{A \supset B, A \Rightarrow B} \supset L}{\begin{array}{c} A \supset B, \\ A \supset \neg B, A \Rightarrow B \end{array}} WL, IL
&
\dfrac{\dfrac{A \Rightarrow A \quad \neg B \Rightarrow \neg B}{A \supset \neg B, A \Rightarrow \neg B} \supset L}{\begin{array}{c} A \supset B, \\ A \supset \neg B, A \Rightarrow \neg B \end{array}} WL
\end{array}
\right\}
$$

$$
\dfrac{A \supset B, A \supset \neg B, A \Rightarrow B \wedge \neg B}{\dfrac{\neg(B \wedge \neg B),}{A \supset B, A \supset \neg B, A \Rightarrow}} \neg L
$$

$$
\dfrac{\Rightarrow \neg(B \wedge \neg B) \qquad \dfrac{A \supset B, A \supset \neg B, A \Rightarrow B \wedge \neg B}{\neg(B \wedge \neg B),\ A \supset B, A \supset \neg B, A \Rightarrow}}{A \supset B, A \supset \neg B, A \Rightarrow} \text{MIX}
$$

Let's assume that A and B are atomic. Then the degree $\mathrm{dg}(\pi)$ of the MIX is 3 (the degree of $\neg(B \wedge \neg B)$) and the rank $\mathrm{rk}(\pi) = \mathrm{rk}_l(\pi) + \mathrm{rk}_r(\pi) = 1 + 1 = 2$. We are therefore in the situation dealt with in Lemma 6.16, which is characterized by the fact that the two premises of the MIX are the result of the application of an operational rule in which M occurs as principal formula. The relevant rules in this case are $\neg L$ in the right premise and $\neg R$ in the left premise, so case D4 of Lemma 6.16 applies. The proof can accordingly be transformed into π':

$$
\pi_2 \left\{
\begin{array}{cc}
\dfrac{\dfrac{A \Rightarrow A \quad B \Rightarrow B}{A \supset B, A \Rightarrow B} \supset L}{\begin{array}{c} A \supset B, \\ A \supset \neg B, A \Rightarrow B \end{array}} WL, IL
&
\dfrac{\dfrac{A \Rightarrow A \quad \neg B \Rightarrow \neg B}{A \supset \neg B, A \Rightarrow \neg B} \supset L}{\begin{array}{c} A \supset B, \\ A \supset \neg B, A \Rightarrow \neg B \end{array}} WL
\end{array}
\right.
\qquad
\pi_1 \left\{
\dfrac{\dfrac{\dfrac{\dfrac{\dfrac{B \Rightarrow B}{\neg B, B \Rightarrow} \neg L}{B, \neg B \Rightarrow} IL}{B \wedge \neg B, \neg B \Rightarrow} \wedge_1 L}{\neg B, B \wedge \neg B \Rightarrow} IL}{\dfrac{B \wedge \neg B, B \wedge \neg B \Rightarrow}{B \wedge \neg B \Rightarrow} CL}
\right.
$$

$$
\dfrac{\dfrac{A \supset B, A \supset \neg B, A \Rightarrow B \wedge \neg B}{} \wedge R \qquad B \wedge \neg B \Rightarrow}{A \supset B, A \supset \neg B, A \Rightarrow} \text{MIX}
$$

The proof π' still presents one MIX but of lower degree. By contrast, the rank has increased significantly. Indeed, if we display the occurrence of the new MIX formula in the antecedents of the right branch and in the succedents of the left branch, we have:

$$
\dfrac{\dfrac{A \Rightarrow A \quad B \Rightarrow B}{A \supset B, A \Rightarrow B} \supset L}{\begin{array}{c} A \supset B, \\ A \supset \neg B, A \Rightarrow B \end{array}} WL, IL
\qquad
\dfrac{\dfrac{A \Rightarrow A \quad \neg B \Rightarrow \neg B}{A \supset \neg B, A \Rightarrow \neg B} \supset L}{\begin{array}{c} A \supset B, \\ A \supset \neg B, A \Rightarrow \neg B \end{array}} WL
\qquad
\dfrac{\dfrac{\dfrac{\dfrac{\dfrac{B \Rightarrow B}{\neg B, B \Rightarrow} \neg L}{B, \neg B \Rightarrow} IL}{\boxed{B \wedge \neg B}, \neg B \Rightarrow} \wedge_1 L}{\neg B, \boxed{B \wedge \neg B} \Rightarrow} IL}{\dfrac{\boxed{B \wedge \neg B}, B \wedge \neg B \Rightarrow}{\boxed{B \wedge \neg B} \Rightarrow} CL} \wedge_2 L
$$

$$
\dfrac{\dfrac{A \supset B, A \supset \neg B, A \Rightarrow \boxed{B \wedge \neg B}}{} \wedge R \qquad \boxed{B \wedge \neg B} \Rightarrow}{A \supset B, A \supset \neg B, A \Rightarrow} \text{MIX}
$$

From this we see that $\mathrm{rk}(\pi') = \mathrm{rk}_l(\pi') + \mathrm{rk}_r(\pi') = 1 + 4 = 5$. The MIX-formula of π' is $B \wedge \neg B$, so $\mathrm{dg}(\pi') = 2$. The rank has significantly increased. The increase in rank is however compensated by the lowering of the degree which, as we know, has more "weight" than the rank. By inductive hypothesis, the new proof is reducible and, as a consequence, the original proof is too. We'll carry out the reduction process in detail.

The proof displayed above is a proof of degree still > 0 and of rank > 2. This corresponds to the case dealt with in Lemma 6.18, and more precisely case A of the proof, since the rule ending with the right sequent of the MIX is a CL. The proof can be transformed as follows:

$$
\cfrac{
\cfrac{\cfrac{A \Rightarrow A \quad B \Rightarrow B}{A \supset B, A \Rightarrow B}\ \supset\!L}{\begin{array}{c} A \supset B, \\ A \supset \neg B, A \Rightarrow B \end{array}}\ \text{WL, IL}
\qquad
\cfrac{\cfrac{\cfrac{A \Rightarrow A \quad \neg B \Rightarrow \neg B}{A \supset \neg B, A \Rightarrow \neg B}\ \supset\!L}{\begin{array}{c} A \supset B, \\ A \supset \neg B, A \Rightarrow \neg B \end{array}}\ \text{WL}}
{\begin{array}{c}\end{array}}
\qquad
\cfrac{
\cfrac{
\cfrac{
\cfrac{\cfrac{B \Rightarrow B}{\neg B, B \Rightarrow}\ \neg L}{B, \neg B \Rightarrow}\ \text{IL}
}{\boxed{B \wedge \neg B}, \neg B \Rightarrow}\ \wedge_1 L
}{\neg B, \boxed{B \wedge \neg B} \Rightarrow}\ \text{IL}
}{\boxed{B \wedge \neg B}, \boxed{B \wedge \neg B} \Rightarrow}\ \wedge_2 L
}{\cfrac{A \supset B, A \supset \neg B, A \Rightarrow \boxed{B \wedge \neg B} \qquad \qquad}{A \supset B, A \supset \neg B, A \Rightarrow}\ \text{MIX}}
$$

(with $\wedge$R combining the first two pieces into $A \supset B, A \supset \neg B, A \Rightarrow \boxed{B \wedge \neg B}$)

The rank of the proof has decreased by one. Let us keep going. We are still in the general situation contemplated by the preceding case with the only difference that the rule that now takes us to the right premise of the MIX is an operational rule. We are thus in the situation contemplated in Lemma 6.18, sub-case B4(b) where the MIX-formula $M = B \wedge \neg B$ is the principal formula of a $\wedge$L inference. Let us construct the first piece of the proof:

$$
\cfrac{
\cfrac{\cfrac{A \Rightarrow A \quad B \Rightarrow B}{A \supset B, A \Rightarrow B}\ \supset\!L}{A \supset B,\ A \supset \neg B, A \Rightarrow B}\ \text{WL, IL}
\quad
\cfrac{\cfrac{A \Rightarrow A \quad \neg B \Rightarrow \neg B}{A \supset \neg B, A \Rightarrow \neg B}\ \supset\!L}{A \supset B,\ A \supset \neg B, A \Rightarrow \neg B}\ \text{WL}
\quad \Big\}\ \pi_3
}{
\cfrac{\cfrac{A \supset B, A \supset \neg B, A \Rightarrow \boxed{B \wedge \neg B}}{\cfrac{A \supset B, A \supset \neg B, A, \neg B \Rightarrow}{\cfrac{\neg B, A \supset B, A \supset \neg B, A \Rightarrow}{B \wedge \neg B, A \supset B, A \supset \neg B, A \Rightarrow}\ \wedge_2 L}\ \text{IL}}}{}
}
$$

where the $\wedge$R and MIX are as in the first block, with
$$\cfrac{\cfrac{\cfrac{\cfrac{B \Rightarrow B}{\neg B, B \Rightarrow}\ \neg L}{B, \neg B \Rightarrow}\ \text{IL}}{\boxed{B \wedge \neg B}, \neg B \Rightarrow}\ \wedge_1 L}{\neg B, \boxed{B \wedge \neg B} \Rightarrow}\ \text{IL} \qquad \text{MIX}$$

To that piece we add the following second piece:

$$
\cfrac{
\cfrac{\cfrac{A \Rightarrow A \quad B \Rightarrow B}{A \supset B, A \Rightarrow B}\ \supset\!L}{\begin{array}{c}A \supset B,\\ A \supset \neg B, A \Rightarrow B\end{array}}\ \text{WL, IL}
\quad
\cfrac{\cfrac{A \Rightarrow A \quad \neg B \Rightarrow \neg B}{A \supset \neg B, A \Rightarrow \neg B}\ \supset\!L}{\begin{array}{c}A \supset B,\\ A \supset \neg B, A \Rightarrow \neg B\end{array}}\ \text{WL}
\quad \vdots\ \pi_3
}{
\cfrac{A \supset B, A \supset \neg B, A \Rightarrow \boxed{B \wedge \neg B} \qquad \begin{array}{c}\boxed{B \wedge \neg B},\\ A \supset B,\\ A \supset \neg B, A \Rightarrow\end{array}}{\cfrac{A \supset B, A \supset \neg B, A, A \supset B, A \supset \neg B, A \Rightarrow}{A \supset B, A \supset \neg B, A \Rightarrow}\ \text{IL, CL}}\ \text{MIX}
}
$$

(with $\wedge$R yielding $A \supset B, A \supset \neg B, A \Rightarrow \boxed{B \wedge \neg B}$)

By assembling the two pieces together we get the proof displayed in Section 6.7.

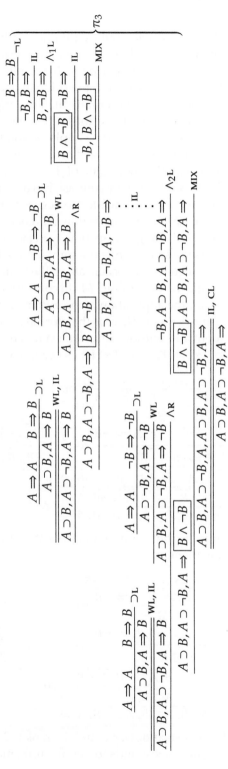

Figure 6.1: Reducing the rank

The exercise we have just now carried out exemplifies with great clarity the features of the transformation presented in the general discussion of the case in question. In the general discussion we mentioned that the resulting transformation presented one additional MIX in comparison to the original one. Here the same thing occurs. However, both MIXes are of lower rank than the original one for two different reasons. Let us consider the first MIX (the higher one). Its rank is equal to the rank of the original one minus 1 for the simple reason that the left rank remains the same and the right rank is now affected by the removal of one sequent, namely the sequent that makes up the conclusion of the rule ∧L. Admittedly, the rule is still applied but after the MIX which, by contrast, has been pushed upwards. As a consequence, the rank decreases by 1 while the degree remains unchanged. The second MIX (the added one, i.e. the one lower down) is characterized by a rank which is equal to the left rank of the original MIX plus 1. In fact, the left rank is the same, since the left sequent is the same, while the right rank is 1, for M does not occur in the antecedent of the sequent that occurs above the right premise. Indeed, earlier on the first MIX had removed the formula M. In conclusion, the constructed proof satisfies the condition for the application of the inductive hypothesis on the rank (lower rank and the same degree). There is therefore also in this case a proof having the same end-sequent and without MIX. Yet, even here we want to show effectively that such a proof exists. To this effect let us begin removing the first MIX.

$$
\cfrac{
\cfrac{
\cfrac{\cfrac{A \Rightarrow A \quad B \Rightarrow B}{A \supset B, A \Rightarrow B}\supset\!\text{L}}{\genfrac{}{}{0pt}{}{A \supset B,}{A \supset \neg B, A \Rightarrow B}}\text{WL, IL}
\quad
\cfrac{\cfrac{\cfrac{A \Rightarrow A \quad \neg B \Rightarrow \neg B}{A \supset \neg B, A \Rightarrow \neg B}\supset\!\text{L}}{\genfrac{}{}{0pt}{}{A \supset B,}{A \supset \neg B, A \Rightarrow \neg B}}\text{WL}}{A \supset B, A \supset \neg B, A \Rightarrow \boxed{B \wedge \neg B}}\wedge\text{R}
\quad
\cfrac{\cfrac{\cfrac{\cfrac{B \Rightarrow B}{\neg B, B \Rightarrow}\neg\text{L}}{B, \neg B \Rightarrow}\text{IL}}{\boxed{B \wedge \neg B}, \neg B \Rightarrow}\wedge_1\text{L}}{\neg B, \boxed{B \wedge \neg B} \Rightarrow}\text{IL}
}{
\cfrac{\cfrac{A \supset B, A \supset \neg B, A, \neg B \Rightarrow}{\neg B, A \supset B, A \supset \neg B, A \Rightarrow}\text{IL}}{B \wedge \neg B, A \supset B, A \supset \neg B, A \Rightarrow}\wedge_2\text{L}
}\text{MIX}
$$

In this case the right premise is the result of an application of IL. We can thus transform it in the following manner:

$$
\cfrac{
\cfrac{
\cfrac{\cfrac{A \Rightarrow A \quad B \Rightarrow B}{A \supset B, A \Rightarrow B}\supset\!\text{L}}{\genfrac{}{}{0pt}{}{A \supset B,}{A \supset \neg B, A \Rightarrow B}}\text{WL, IL}
\quad
\cfrac{\cfrac{\cfrac{A \Rightarrow A \quad \neg B \Rightarrow \neg B}{A \supset \neg B, A \Rightarrow \neg B}\supset\!\text{L}}{\genfrac{}{}{0pt}{}{A \supset B,}{A \supset \neg B, A \Rightarrow \neg B}}\text{WL}}{A \supset B, A \supset \neg B, A \Rightarrow \boxed{B \wedge \neg B}}\wedge\text{R}
\quad
\cfrac{\cfrac{\cfrac{B \Rightarrow B}{\neg B, B \Rightarrow}\neg\text{L}}{B, \neg B \Rightarrow}\text{IL}}{\boxed{B \wedge \neg B}, \neg B \Rightarrow}\wedge_1\text{L}
}{
\cfrac{\cfrac{A \supset B, A \supset \neg B, A, \neg B \Rightarrow}{\neg B, A \supset B, A \supset \neg B, A \Rightarrow}\text{IL}}{B \wedge \neg B, A \supset B, A \supset \neg B, A \Rightarrow}\wedge_2\text{L}
}\text{MIX}
$$

At this point the MIX has been pushed upwards and the rank is 2 while the degree remains 2 > 0. This now again corresponds to Lemma 6.16; more precisely, we find ourselves in case D1 of the proof. Its transformation is:

$$
\cfrac{
  \cfrac{
    \cfrac{
      \cfrac{A \Rightarrow A \qquad B \Rightarrow \boxed{B}}{A \supset B, A \Rightarrow \boxed{B}} \; \supset\text{L}
    }{A \supset B, A \supset \neg B, A \Rightarrow \boxed{B}} \; \text{WL, IL}
    \qquad
    \cfrac{
      \cfrac{
        \cfrac{\boxed{B} \Rightarrow B}{\neg B, \boxed{B} \Rightarrow} \; \neg\text{L}
      }{\boxed{B}, \neg B \Rightarrow} \; \text{IL}
    }{}
  }{A \supset B, A \supset \neg B, A, \neg B \Rightarrow} \; \text{MIX}
}{
  \cfrac{
    \cfrac{\neg B, A \supset B, A \supset \neg B, A \Rightarrow}{B \wedge \neg B, A \supset B, A \supset \neg B, A \Rightarrow} \; \wedge_2\text{L}
  }{}
} \; \text{IL}
$$

The degree of the MIX is now 0 while the rank has increased. To reduce the rank, we again consult the proof of Lemma 6.18, since the right rank is > 1. Since the right premise of the MIX is the conclusion of IL, we are dealing with Case (A). Since the MIX-formula B is a principal formula of the IL rule ending in the right premise of the mix, we are dealing with the second sub-case. The IL rule can simply be removed, and we obtain:

$$
\cfrac{
  \cfrac{
    \cfrac{A \Rightarrow A \qquad B \Rightarrow \boxed{B}}{A \supset B, A \Rightarrow \boxed{B}} \; \supset\text{L}
  }{A \supset B, A \supset \neg B, A \Rightarrow \boxed{B}} \; \text{WL, IL}
  \qquad
  \cfrac{\boxed{B} \Rightarrow B}{\neg B, \boxed{B} \Rightarrow} \; \neg\text{L}
}{
  \cfrac{
    \cfrac{A \supset B, A \supset \neg B, A, \neg B \Rightarrow}{\neg B, A \supset B, A \supset \neg B, A \Rightarrow} \; \text{IL}
  }{B \wedge \neg B, A \supset B, A \supset \neg B, A \Rightarrow} \; \wedge_2\text{L}
} \; \text{MIX}
$$

This decreases the right rank, but the right rank is still > 1. So Lemma 6.18 still applies; now it is sub-case B7(a), since the principal formula of the ¬L rule leading to the right premise of the MIX is not the MIX-formula. The proof can accordingly be transformed as follows:

$$
\cfrac{
  \cfrac{
    \cfrac{A \Rightarrow A \qquad B \Rightarrow \boxed{B}}{A \supset B, A \Rightarrow \boxed{B}} \; \supset\text{L}
  }{A \supset B, A \supset \neg B, A \Rightarrow \boxed{B}} \; \text{WL, IL}
  \qquad
  \boxed{B} \Rightarrow B
}{
  \cfrac{
    \cfrac{A \supset B, A \supset \neg B, A \Rightarrow B}{\neg B, A \supset B, A \supset \neg B, A \Rightarrow} \; \neg\text{L}
  }{B \wedge \neg B, \neg A \supset B, \neg A \supset \neg B, \neg A \Rightarrow} \; \wedge_2\text{L}
} \; \text{MIX}
$$

Now the right premise of the MIX is an axiom, and Lemma 6.11 applies. We obtain:

$$
\cfrac{
  \cfrac{
    \cfrac{A \Rightarrow A \qquad B \Rightarrow B}{A \supset B, A \Rightarrow B} \; \supset\text{L}
  }{A \supset B, A \supset \neg B, A \Rightarrow B} \; \text{WL, IL}
}{
  \cfrac{
    \cfrac{\neg B, A \supset B, A \supset \neg B, A \Rightarrow}{B \wedge \neg B, A \supset B, A \supset \neg B, A \Rightarrow} \; \wedge_2\text{L}
  }{} \; \neg\text{L}
}
$$

We have thus managed to eliminate the first MIX of the proof in Figure 6.7, which had been constructed above during the reduction of the original proof. Replacing the proof ending in the right premise of the lower MIX by the proof we just obtained results in:

$$
\cfrac{
  \cfrac{
    \cfrac{A \Rightarrow A \quad B \Rightarrow B}{A \supset B, A \Rightarrow B} \supset L
  }{
    \cfrac{\begin{array}{c} A \supset B, \\ A \supset \neg B, A \Rightarrow B \end{array}}{\begin{array}{c} A \supset B, \\ A \supset \neg B, A \Rightarrow B \end{array} } \text{WL, IL}
  }
  \quad
  \cfrac{
    \cfrac{A \Rightarrow A \quad \neg B \Rightarrow \neg B}{A \supset \neg B, A \Rightarrow \neg B} \supset L
  }{
    \begin{array}{c} A \supset B, \\ A \supset \neg B, A \Rightarrow \neg B \end{array}
  } \text{WL}
}{
  A \supset B, A \supset \neg B, A \Rightarrow \boxed{B \wedge \neg B}
} \wedge R
$$

combined with the right branch:

$$
\cfrac{
  \cfrac{
    \cfrac{A \Rightarrow A \quad B \Rightarrow B}{A \supset B, A \Rightarrow B} \supset L
  }{
    \cfrac{A \supset B,}{A \supset \neg B, A \Rightarrow B} \text{WL, IL}
  }
}{
  \cfrac{
    \cfrac{\neg B, A \supset B,}{A \supset \neg B, A \Rightarrow} \neg L
  }{
    \boxed{B \wedge \neg B}, A \supset B, A \supset \neg B, A \Rightarrow
  } \wedge 2 L
}
$$

$$
\cfrac{A \supset B, A \supset \neg B, A, A \supset B, A \supset \neg B, A \Rightarrow}{A \supset B, A \supset \neg B, A \Rightarrow} \text{MIX, IL, CL}
$$

The MIX in this proof is on $B \wedge \neg B$, i.e., has degree 2. The rank is 2; so Lemma 6.16 applies. The proof can be transformed into:

$$
\cfrac{
  \cfrac{
    \cfrac{A \Rightarrow A \quad \neg B \Rightarrow \boxed{\neg B}}{A \supset \neg B, A \Rightarrow \boxed{\neg B}} \supset L
  }{
    A \supset B, A \supset \neg B, A \Rightarrow \boxed{\neg B}
  } \text{WL}
  \quad
  \cfrac{
    \cfrac{
      \cfrac{A \Rightarrow A \quad B \Rightarrow B}{A \supset B, A \Rightarrow B} \supset L
    }{
      A \supset B, A \supset \neg B, A \Rightarrow B
    } \text{WL, IL}
  }{
    \boxed{\neg B}, A \supset B, A \supset \neg B, A \Rightarrow
  } \neg L
}{
  \cfrac{A \supset B, A \supset \neg B, A, A \supset B, A \supset \neg B, A \Rightarrow}{A \supset B, A \supset \neg B, A \Rightarrow} \text{IL, CL}
} \text{MIX}
$$

It still contains a MIX but of lower degree. Now the MIX formula is $\neg B$ and the rule that introduces it in the right branch is $\neg L$. The right rank is 1 while the left rank is 3; thus the rank is 4. The right rank is already minimal. We must therefore focus on the left branch in order to lower the rank; this is done in the proof of Lemma 6.19. In order to reduce the left rank we need to take into consideration the last sequent on the left branch that is obtained by means of WL. The WL can be moved below the MIX (as in case A of Lemma 6.18). We thus have:

$$
\cfrac{
  \cfrac{
    \cfrac{A \Rightarrow A \quad \neg B \Rightarrow \boxed{\neg B}}{A \supset \neg B, A \Rightarrow \boxed{\neg B}} \supset L
    \quad
    \cfrac{
      \cfrac{
        \cfrac{A \Rightarrow A \quad B \Rightarrow B}{A \supset B, A \Rightarrow B} \supset L
      }{
        A \supset B, A \supset \neg B, A \Rightarrow B
      } \text{WL, IL}
    }{
      \boxed{\neg B}, A \supset B, A \supset \neg B, A \Rightarrow
    } \neg L
  }{
    A \supset \neg B, A, A \supset B, A \supset \neg B, A \Rightarrow
  } \text{MIX}
}{
  \cfrac{A \supset B, A \supset \neg B, A, A \supset B, A \supset \neg B, A \Rightarrow}{A \supset B, A \supset \neg B, A \Rightarrow} \text{IL, CL}
} \text{WL}
$$

Once again the rank has decreased; it is now 3. However, at this point we are in the case dealt with in Lemma 6.19: $rk_r(\pi) = 1$ and $rk_l(\pi) > 1$. The relevant rule is $\supset L$. Since there are two cases for the $\supset L$ rule, it is important to determine whether we are in case (1), (2) or in case (3). In case (1), the formula M occurs in both Θ and Λ; in cases (2) and (3), M occurs in one but not the other. In order to determine whether we are in the first or in the second case, it is useful to examine the proof in light of the general schema. The outcome of this examination is the following figure, where the relevant components of the proof are labeled according to the general schema. We'll indicate empty components by $\emptyset$:

$$
\cfrac{
\cfrac{
\underbrace{A \Rightarrow}_{\Gamma} \ \underbrace{\emptyset}_{\Theta} \ , A \qquad \underbrace{\neg B, \ \emptyset}_{\Delta} \Rightarrow \underbrace{\boxed{\neg B}}_{\Lambda}
}{
\underbrace{A \supset \neg B, \ A}_{\Gamma} \ , \ \underbrace{\emptyset}_{\Delta} \Rightarrow \underbrace{\emptyset}_{\Theta} \ , \underbrace{\boxed{\neg B}}_{\Lambda}
} \supset\text{L}
\qquad
\cfrac{
\cfrac{
\cfrac{
\cfrac{A \Rightarrow A \qquad B \Rightarrow B}{A \supset B, A \Rightarrow B} \supset\text{L}
}{A \supset B, A \supset \neg B, A \Rightarrow B} \text{WL, IL}
}{\underbrace{\boxed{\neg B}, A \supset B,}_{\Sigma} \ A \supset \neg B, A \Rightarrow \underbrace{\emptyset}_{\Pi}} \neg\text{L}
}
}{
\cfrac{
\underbrace{A \supset \neg B, \ A}_{\Gamma} \ , \ \underbrace{\emptyset}_{\Delta} \ , \underbrace{A \supset B, A \supset \neg B, A}_{\Sigma^*} \Rightarrow \underbrace{\emptyset}_{\Theta^*} \ , \underbrace{\emptyset}_{\Lambda^*} \ , \underbrace{\emptyset}_{\Pi} \ ,
}{
\cfrac{A \supset B, A \supset \neg B, A, A \supset B, A \supset \neg B, A \Rightarrow}{A \supset B, A \supset \neg B, A \Rightarrow} \text{IL, CL}
} \text{WL}
} \text{MIX}
$$

Now we see that the formula M, which here coincides with $\neg B$, does not occur in Θ—which is empty—but only in Λ—which coincides with M. Hence, we are in case (2) and therefore we can transform the proof as follows:

$$
\cfrac{
\cfrac{
\underbrace{A \Rightarrow}_{\Gamma} \ \underbrace{\emptyset}_{\Theta} \ , A
\qquad
\cfrac{
\underbrace{\neg B, \ \emptyset}_{\Delta} \Rightarrow \underbrace{\boxed{\neg B}}_{\Lambda}
\qquad
\cfrac{
\cfrac{
\cfrac{\cfrac{A \Rightarrow A \qquad B \Rightarrow B}{A \supset B, A \Rightarrow B} \supset\text{L}}{A \supset B, A \supset \neg B, A \Rightarrow B} \text{WL,IL}
}{\underbrace{\boxed{\neg B}, A \supset B, A \supset \neg B, A \Rightarrow}_{\Sigma} \ \underbrace{\emptyset}_{\Pi}} \neg\text{L}
}{\underbrace{\neg B, \ \emptyset}_{\Delta} \ , \underbrace{A \supset B, A \supset \neg B, A}_{\Sigma^*} \Rightarrow \underbrace{\emptyset}_{\Lambda^*} \ , \underbrace{\emptyset}_{\Pi}} \text{MIX}
}{
\underbrace{A \supset \neg B, \ A}_{\Gamma} \ , \ \underbrace{\emptyset}_{\Delta} \ , \underbrace{A \supset B, A \supset \neg B, A}_{\Sigma^*} \Rightarrow \underbrace{\emptyset}_{\Theta^*} \ , \underbrace{\emptyset}_{\Lambda^*} \ , \underbrace{\emptyset}_{\Pi}
} \supset\text{L}
}{
\cfrac{A \supset B, A \supset \neg B, A, A \supset B, A \supset \neg B, A \Rightarrow}{A \supset B, A \supset \neg B, A \Rightarrow} \text{IL, CL}
} \text{WL}
$$

We have further reduced the rank of the MIX. Not only is the rank now minimal (i.e., 2), the left premise is an axiom. We can remove the MIX by Lemma 6.11. Indeed, we have:

$$\frac{\dfrac{\dfrac{\dfrac{A \Rightarrow A \quad B \Rightarrow B}{A \supset B, A \Rightarrow B} \supset L}{A \supset B, A \supset \neg B, A \Rightarrow B} \text{WL, IL}}{A \Rightarrow A \quad \dfrac{\neg B, A \supset B, A \supset \neg B, A \Rightarrow}{\neg B, A \supset B, A \supset \neg B, A \Rightarrow} \neg L}}{A \supset \neg B, A, A \supset B, A \supset \neg B, A \Rightarrow}$$

$$\frac{A \Rightarrow A \quad \neg B, A \supset B, A \supset \neg B, A \Rightarrow}{A \supset \neg B, A, A \supset B, A \supset \neg B, A \Rightarrow} \supset L$$

$$\frac{A \supset B, A \supset \neg B, A, A \supset B, A \supset \neg B, A \Rightarrow}{A \supset B, A \supset \neg B, A \Rightarrow} \text{WL}$$

$$\frac{}{A \supset B, A \supset \neg B, A \Rightarrow} \text{IL, CL}$$

We have thus reached the complete elimination of MIX from the original proof. The proof of the original sequent is now obtained by tacking on the last two lines of the original proof with MIX.

Problem 6.23. Eliminate the use of CUT from proofs c and d in Section 5.5.9 (laws of transformation of the quantifiers). As you do so, keep track of how the rank and degree change.

Problem 6.24. First label the proof below by justifying the inferential steps. Then test your understanding of MIX elimination by systematically eliminating all the MIXes in the derivation and providing a cut-free proof of the end-sequent. Mention explicitly which cases of which lemmas treat the cases you need to eliminate the MIXes. While working on each MIX also state explicitly the degree and rank of the relevant (sub)derivations and show how each intermediate reduction step lowers the rank or degree of the relevant (sub)derivation.

$$\frac{\dfrac{\dfrac{\dfrac{A \Rightarrow A}{\neg A, A \Rightarrow}}{\neg A, A \Rightarrow \neg B} \quad \dfrac{B \Rightarrow B}{\neg B, B \Rightarrow}}{\dfrac{\neg A, A, B \Rightarrow}{\neg A, A, B \Rightarrow A \wedge B}} \text{MIX} \quad \dfrac{\dfrac{\dfrac{A \Rightarrow A}{A \wedge B \Rightarrow A}}{A \wedge B \Rightarrow A \vee C} \quad \dfrac{\dfrac{B \Rightarrow B}{A \wedge B \Rightarrow B}}{A \wedge B \Rightarrow B \vee C}}{A \wedge B \Rightarrow (A \vee C) \wedge (B \vee C)}}{\neg A, A, B \Rightarrow (A \vee C) \wedge (B \vee C)} \text{MIX}$$

6.8   Intuitionistic sequent calculus **LJ**

The proof of cut-elimination for **LJ** consists in adapting the line of reasoning used for **LK** to the rules of **LJ**. On account of the fact that **LJ** is obtained by restricting the form of the sequents, the general structure of the argument does not change. The types of arguments are moreover exactly the same. The task of the reader is that of adapting the procedures used before to the novel form of the sequents. With the goal of making explicit the proof of the *Hauptsatz* for **LJ** in relation to the new form of the sequents, we present the calculus **LJ** in an explicit way. As we have already mentioned at the beginning of the previous chapter, the intuitionistic calculus **LJ** is obtained from **LK** by imposing the restriction that the succedent of the sequents is constituted by at most one formula and by adapting the rules of the calculus accordingly. When this is done, the rules take the form which we will presently give. Please note that the formulas displayed in square brackets need not occur.

6.8.1   Axioms

Axioms are the same as in **LK**:

Definition 6.25. Sequents of the type $A \Rightarrow A$, where A is any formula, are *axioms*.

6.8.2   Structural rules

Weakening

$$\frac{\Gamma \Rightarrow [C]}{A, \Gamma \Rightarrow [C]} \text{ WL} \qquad \frac{\Gamma \Rightarrow}{\Gamma \Rightarrow A} \text{ WR}$$

Contraction

$$\frac{A, A, \Gamma \Rightarrow [C]}{A, \Gamma \Rightarrow [C]} \text{ CL}$$

There is no CR rule.

Interchange

$$\frac{\Delta, A, B, \Gamma \Rightarrow [C]}{\Delta, B, A, \Gamma \Rightarrow [C]} \text{ IL}$$

There is no IR rule.

6.8.3   Cut

$$\frac{\Gamma \Rightarrow A \quad A, \Delta \Rightarrow [C]}{\Gamma, \Delta \Rightarrow [C]} \text{ CUT}$$

6.8.4   Operational rules

Conjunction

$$\frac{A, \Gamma \Rightarrow [C]}{A \wedge B, \Gamma \Rightarrow [C]} \wedge \text{L} \qquad \frac{B, \Gamma \Rightarrow [C]}{A \wedge B, \Gamma \Rightarrow [C]} \wedge \text{L} \qquad \frac{\Gamma \Rightarrow A \quad \Gamma \Rightarrow B}{\Gamma \Rightarrow A \wedge B} \wedge \text{R}$$

Disjunction

$$\frac{A, \Gamma \Rightarrow [C] \quad B, \Gamma \Rightarrow [C]}{A \vee B, \Gamma \Rightarrow [C]} \vee \text{L} \qquad \frac{\Gamma \Rightarrow A}{\Gamma \Rightarrow A \vee B} \vee \text{R} \qquad \frac{\Gamma \Rightarrow B}{\Gamma \Rightarrow A \vee B} \vee \text{R}$$

Conditional

$$\frac{\Gamma \Rightarrow A \quad B, \Delta \Rightarrow [C]}{A \supset B, \Gamma, \Delta \Rightarrow [C]} \supset\text{L} \qquad \frac{A, \Gamma \Rightarrow B}{\Gamma \Rightarrow A \supset B} \supset\text{R}$$

Negation

$$\frac{\Gamma \Rightarrow A}{\neg A, \Gamma \Rightarrow} \neg\text{L} \qquad \frac{A, \Gamma \Rightarrow}{\Gamma \Rightarrow \neg A} \neg\text{R}$$

Universal quantifier

$$\frac{A(t), \Gamma \Rightarrow [C]}{\forall x\, A(x), \Gamma \Rightarrow [C]} \forall\text{L} \qquad !\,\frac{\Gamma \Rightarrow A(a)}{\Gamma \Rightarrow \forall x\, A(x)} \forall\text{R}$$

Existential quantifier

$$!\,\frac{A(a), \Gamma \Rightarrow [C]}{\exists x\, A(x), \Gamma \Rightarrow [C]} \exists\text{L} \qquad \frac{\Gamma \Rightarrow A(t)}{\Gamma \Rightarrow \exists x\, A(x)} \exists\text{R}$$

The ∀R and ∃L rules qualified by the exclamation mark ! are again critical rules: the eigenvariable a must not appear in the conclusion sequent.

Problem 6.26. State the MIX rule for **LJ**. Show that **LJ** − CUT + MIX proves CUT and that **LJ** (which includes CUT) proves MIX, all with the restriction to at most one formula in the succedent.

Theorem 6.27. *Cut-elimination holds for* **LJ**.

Problem 6.28. **LJ** is like **LK** except that the succedent of any sequent is restricted to at most one formula. This makes IR and CR redundant. If we also remove WR, we obtain **LM**, the sequent calculus for minimal logic. Formulate and give a proof of the cut-elimination theorem for **LM** with the rules restricted to those for ¬, ∧ and ⊃. See the comments in n. 2 on p. 210.

6.9   Why MIX?

Why prove that we can eliminate MIX instead of CUT directly? The reason is that using the strategy we have employed, namely either to reduce the degree of a MIX (if the rank is minimal, i.e., 2) or else to reduce the rank, relies on switching the order of the MIX with an inference ending in one of the premises. Reducing the degree of a CUT would work in the same way; but it is the strategy for reducing the rank where we run into difficulties. There is no clear way to apply the strategy if we want to permute a CUT with a contraction. For instance, suppose our proof contains a single CUT as its last inference, and the right premise is the conclusion of CL:

$$
\cfrac{\Pi \Rightarrow A \qquad \cfrac{\vdots\ \pi_2 \\ A, A, \Psi \Rightarrow B}{A, \Psi \Rightarrow B}\ \text{CL}}{\Pi, \Psi \Rightarrow B}\ \text{CUT}
$$

Simply permuting CUT and CL doesn't work, since the antecendent of the CL contains two occurrences of A. So we would have to use CUT twice, e.g., as follows:

$$
\cfrac{\Pi \Rightarrow A \qquad \cfrac{\left.\cfrac{\Pi \Rightarrow A \qquad A, A, \Psi \Rightarrow B}{\cfrac{\Pi, A, \Psi \Rightarrow B}{A, \Pi, \Psi \Rightarrow B}\ \text{IL}}\ \text{CUT}\right\}\pi_3}{}}{\cfrac{\Pi, \Pi, \Psi \Rightarrow B}{\Pi, \Psi \Rightarrow B}\ \text{IL, CL}}\ \text{CUT}
$$

Now we have two CUTS with the same cut-formula A. Take the sub-proof π_3 ending in the conclusion of the higher of the two CUTS. This would be a proof ending in a single CUT of lower rank than our original proof, so the inductive hypothesis applies. This would give us a cut-free proof π'_3 of the conclusion of the CUT inference, i.e., of $\Pi, A, \Psi \Rightarrow B$, which we can replace in the transformed proof to get:

$$
\cfrac{\Pi \Rightarrow A \qquad \left.\cfrac{\cfrac{\vdots\ \pi'_3 \\ \Pi, A, \Psi \Rightarrow B}{A, \Pi, \Psi \Rightarrow B}\ \text{IL}}{}\right\}\pi_4}{\cfrac{\Pi, \Pi, \Psi \Rightarrow B}{\Pi, \Psi \Rightarrow B}\ \text{IL, CL}}\ \text{CUT}
$$

However, we have no guarantee that the derivation π_4 that ends with $\Pi, \Pi, \Psi \Rightarrow B$ has lower rank than the original proof. This is mainly because we don't know how many sequents in the new proof π'_3 resulting from eliminating the topmost CUT contain A in the antecedent. Thus we cannot apply the inductive hypothesis to the resulting proof. (The additional IL inferences required to place A on the far left of the antecedent could be taken care of by simply not counting interchanges when computing the rank.)

Of course, this is not just a problem in **LJ**. In fact, in **LK** the situation is even more difficult. In the following proof, for instance, both premises of the CUT are conclusions of contractions with the cut-formula principal:

$$
\cfrac{
  \cfrac{\vdots}{\Pi \Rightarrow \Theta, A, A} \text{ CR}
  \qquad
  \cfrac{\vdots}{A, A, \Psi \Rightarrow \Lambda} \text{ CL}
}{
  \begin{array}{c} \Pi \Rightarrow \Theta, A \qquad A, \Psi \Rightarrow \Lambda \end{array}
} \text{ CUT}
$$

$$
\dfrac{\Pi \Rightarrow \Theta, A \qquad A, \Psi \Rightarrow \Lambda}{\Pi, \Psi \Rightarrow \Theta, \Lambda} \text{ CUT}
$$

The situation is not hopeless, however. A direct way to eliminate CUTs using a similar strategy as Gentzen is given by Borisavljević (2003). It is also possible to avoid the need for MIX by pursuing entirely different strategies; see, e.g., the presentation in Troelstra and Schwichtenberg (2000).

6.10   Consequences of the *Hauptsatz*

The cut-elimination theorem owes its extraordinary relevance in proof theory to the results that follow from it. Quite rightly, one often emphasizes the general philosophical significance of the theorem, but one cannot have access to it without taking into account some of the most important corollaries that flow from it. Gentzen conceived of the *Hauptsatz* as a tool for proving the consistency of mathematical theories, especially arithmetic. Such a consistency proof requires a more involved proof transformation (see Chapter 7). In this chapter, we will restrict ourselves to a treatment of some of the immediate consequences of the *Hauptsatz* in accordance with the following plan. After some preliminaries, centered on the notion of sub-formula, we will present the sub-formula theorem. Then we will show some easy consequences of the theorem concerning the non-derivability of certain sequents either in **LK** or in **LJ** and, consequently, the consistency of both calculi. We will then show two constructive features of intuitionistic logic: the disjunction property and the existence property. Finally, we will delve a little deeper into the mid-sequent theorem and Herbrand's theorem as its corollary.

Let us begin with the concept of sub-formula. We repeat Definition 4.13, but now no longer have to consider $\bot$ a sub-formula of $\neg A$.

Definition 6.29. The sub-formulas of a formula A are defined inductively as follows:

1. *Basis clause:* If A is atomic, A is the only sub-formula of A.

2. *Inductive clauses:* If A is of the form $\neg B$, the sub-formulas of A are A itself and the sub-formulas of B.

3. If A is of the form $(B \wedge C)$, $(B \vee C)$, or $(B \supset C)$, then the sub-formulas of A are A itself and the sub-formulas of B and C.

4. If A is of the form $\exists x\, B(x)$ or $\forall x\, B(x)$, and $B(x)$ is $B[x/a]$, then the sub-formulas of A are A itself and the sub-formulas of all formulas $B[t/a]$, t any term.

5. *Extremal clause:* Nothing else is a sub-formula of A.

Theorem 6.30 (Sub-formula property). *Let $\Theta \Rightarrow \Lambda$ be a sequent occurring in a cut-free proof π of $\Gamma \Rightarrow \Delta$. Then any formula in $\Theta \cup \Lambda$ is a sub-formula of some formula in $\Gamma \cup \Delta$.*

Proof. The proof follows by induction on the size of the proof. The key fact is that the proof is cut-free. Indeed, with the exception of CUT, all other rules are such that a formula that occurs in a premise also occurs in the conclusion (by itself or as a sub-formula of a formula introduced by an operational rule). The formulas occurring in the end-sequent are therefore the result of a finite number of introductions of some logical sign during the course of the proof starting from their sub-formulas (in axioms). □

The theorem holds for all three systems, **LM**, **LJ** and **LK**.

Let us recall the fact that to derive in **LK** (or in **LJ**, **LM**) a sequent with empty succedent, such as $\Gamma \Rightarrow$, is to show that the formulas constituting the antecedent Γ imply in **LK** (or in **LJ**, **LM**) a contradiction. Generalizing: to derive the empty sequent, namely $\Rightarrow$, means to derive the contradiction (unconditionally). Thus, to prove that a sequent calculus is consistent means to show that one cannot derive the empty sequent $\Rightarrow$. In axiomatic systems J_0 and K_0, if we can prove $A \wedge \neg A$, we can prove anything (since we can prove $(A \wedge \neg A) \supset B$). In **NJ** and **NK**, if we can prove $\bot$, we can prove anything (by $\bot_J$). And in **LJ** and **LK**, if we can derive $\Rightarrow$, we can prove anything (using WR and WL). So in each system we have a convenient single formula or sequent the provability of which indicates that the system becomes trivial in the sense that anything becomes provable.

Theorem 6.31 (Consistency). *LK, LJ, and LM are consistent, i.e., they do not prove the empty sequent.*

Proof. If one could derive the empty sequent $\Rightarrow$ then, by the *Hauptsatz*, the same sequent could be derived in a cut-free proof. But this is impossible. Indeed, reasoning by induction on the size of the proof we can show that the empty sequent has no cut-free proof.

Induction basis: If the size of the proof is 0, it consists only of an axiom. But the empty sequent is not an axiom: axioms are sequents of the form $A \Rightarrow A$, where A is an arbitrary formula.

Inductive step: Suppose the size of the proof is > 0. In this case the sequent is the result of an application of a structural rule or of an operational rule. In the first case, it cannot be the result of a weakening W, nor of an interchange I, nor of a contraction C, since none of these rules can yield the empty sequent. In the second case, all the operational rules introduce a complex formula with the corresponding operator and thus, according to the inductive hypothesis according to which the premises cannot be empty sequents, the end-sequent cannot be itself empty. □

Theorem 6.32 (Disjunction property of LJ). *Let $\Rightarrow A \vee B$ be derivable in **LJ**. Then in **LJ** one can derive either $\Rightarrow A$ or $\Rightarrow B$.*

Proof. By the *Hauptsatz*, if $\Rightarrow A \vee B$ has a proof in **LJ**, it has a proof in **LJ** without CUT. The last inference in a cut-free proof of $\Rightarrow A \vee B$ can only be $\vee$R or WR. It clearly cannot be WR, since then the premise would also have a proof, but the premise would be the empty sequent $\Rightarrow$. So, the last rule applied must be either $\vee_1$R or $\vee_2$R. Thus either $\Rightarrow A$ or $\Rightarrow B$ must be derivable. $\square$

In Section 2.11, we showed that the axiomatic proof system J_0 for intuitionistic propositional logic does not prove either $A \vee \neg A$ nor $\neg\neg A \supset A$. We can use the disjunction property to prove independence results proof-theoretically.

Proposition 6.33. *LJ does not prove* $\Rightarrow A \vee \neg A$ *for all formulas A.*

Proof. Suppose it did. Then by Theorem 6.32, **LJ** proves either $\Rightarrow A$ or $\Rightarrow \neg A$. If it proves the latter, it must also prove $A \Rightarrow$, since the only possible last inference of a cut-free proof of $\Rightarrow \neg A$ in **LJ** is $\neg$R.

Now A may be any formula: so if **LJ** proves $\Rightarrow A$ it also proves $\Rightarrow B \wedge \neg B$, and if it proves $A \Rightarrow$ it also proves $B \supset B \Rightarrow$. In the former case, since **LJ** also proves $B \wedge \neg B \Rightarrow$, using a CUT, **LJ** proves the empty sequent. But this contradicts Theorem 6.31. Similarly in the second case, as $\Rightarrow B \supset B$ is provable in **LJ**. $\square$

Corollary 6.34. *LJ does not prove* $\Rightarrow \neg\neg A \supset A$ *for all formulas A.*

Proof. It is an easy exercise to find a proof in **LJ** of

$$\neg\neg A \supset A \Rightarrow A \vee \neg A.$$

If there were a derivation of $\neg\neg A \supset A$, using a CUT we would obtain a proof of $\Rightarrow A \vee \neg A$, contradicting Proposition 6.33. $\square$

Corollary 6.35. *LM does not prove* $\Rightarrow (A \wedge \neg A) \supset B$ *for all formulas A and B.*

Proof. It is easily seen that **LM** proves $\Rightarrow (A \wedge \neg A) \supset B$ if and only if it proves $A, \neg A \Rightarrow B$. If A and B are atomic and different, the only possible rules applied in a cut-free proof ending in $A, \neg A \Rightarrow B$ are CL, IL, or WL. $\neg$L is not an option because the premise would have both A and B in the succedent, and WR is not an option because it is not a rule of **LM**. By induction, we can show that if the sequent $A, \neg A \Rightarrow B$ can be obtained by only these rules from some sequent $\Gamma \Rightarrow \Delta$, then Δ must be B and Γ must contain only A and $\neg A$. None of these is an axiom, if A and B are different. $\square$

We obtained these results also using the normalization theorem for **NJ**; compare Corollaries 4.46 and 4.47.

6.11   The mid-sequent theorem

A standard result of elementary logic shows that every formula of predicate logic is provably equivalent (within **LK**) to one in prenex normal form, defined as follows:

Definition 6.36. A formula is in *prenex normal form* if it has the form

$$Q_1 x_1, \ldots Q_n x_n \, A(x_1, \ldots, x_n),$$

where Q_i is a universal or an existential quantifier and $A(a_1, \ldots, a_n)$ is a formula without quantifiers.

The mid-sequent theorem concerns the form of proofs of sequents consisting only in formulas in prenex normal form. It is related to a classical result known as Herbrand's theorem, and can be used to prove other useful results.

Theorem 6.37 (Mid-sequent theorem). *Suppose the sequent $\Delta \Rightarrow \Lambda$ only contains formulas in prenex normal form. If $\Delta \Rightarrow \Lambda$ has an **LK**-proof π, then there exists a cut-free proof π' of $\Delta \Rightarrow \Lambda$ in which there occurs a sequent, called* mid-sequent, *that only contains formulas without quantifiers.*

The mid-sequent divides the proof into two parts. The first part ends with the mid-sequent and uses only axioms, propositional rules, and structural rules (weakenings, interchanges, and contractions). In the second part, which begins with the mid-sequent, there are only applications of quantifier rules and structural rules.

Remark. Note that the part of the proof π' below the mid-sequent constitutes a non-branching path since all the rules that are applied are one premise rules. Moreover, the formulas that occur in the mid-sequent result from the elimination of the quantifiers, if any such occur, from formulas occurring in the end-sequent.

Proof. We may assume, by the cut-elimination theorem, that the proof π is cut-free. Since π is cut-free, the sub-formula property holds. Since the end-sequent contains only formulas in prenex normal form, if follows that *all* formulas that occur in π are in prenex normal form (or contain no quantifiers at all). Indeed, let us suppose that a formula that occurs in some sequent belonging to the proof π contains a quantifier but has as a principal logical symbol a propositional operator. From this it would follow, in virtue of the sub-formula theorem, that this formula should occur in the end-sequent of the proof. But this is impossible, for by assumption all the formulas occurring in the end-sequent are in prenex normal form. Consequently, from this fact we can infer that if there are axioms containing quantifiers then they must have one of the following forms: $\forall x \, B(x) \Rightarrow \forall x \, B(x)$ or $\exists x \, B(x) \Rightarrow \exists x \, B(x)$, where the quantified formulas are in prenex normal form.

It is easy to show that π can be transformed into a proof whose initial axioms do not contain quantifiers. Consider the first form of axiom:

$$\forall x \, B(x) \Rightarrow \forall x \, B(x).$$

Such an axiom can be proved as follows, with a a suitable eigenvariable:

$$\cfrac{\cfrac{B(a) \Rightarrow B(a)}{\forall x\, B(x) \Rightarrow B(a)} \text{ } \forall \text{L}}{\forall x\, B(x) \Rightarrow \forall x\, B(x)} \text{ } \forall \text{R}$$

We proceed analogously in the case of the second type of axiom:

$$\cfrac{\cfrac{B(a) \Rightarrow B(a)}{B(a) \Rightarrow \exists x\, B(x)} \text{ } \exists \text{R}}{\exists x\, B(x) \Rightarrow \exists x\, B(x)} \text{ } \exists \text{L}$$

The proof π can thus be transformed into a new proof in which the formulas containing quantifiers in the original axioms are replaced by new formulas in which the occurrence of the external quantifier is removed. Obviously, the transformation can be iterated until all the quantified formulas are eliminated from the axioms of the original proof. With the aim of avoiding useless complexity let us denote with π the outcome of the transformation of the original proof.

The proof of the mid-sequent theorem now proceeds by induction on the "order" of a proof.

Definition 6.38. Let I be a quantificational inference, i.e., the application of a quantificational rule in π. Then the *order* of I, written $o(I)$, is the number of propositional rules applied below the inference I in the proof π.

The order of a proof π, written $o(\pi)$, is the sum of the orders of all the quantificational inferences occurring in π.

For instance, consider the proof:

$$\cfrac{\cfrac{\cfrac{\cfrac{A(a,b) \Rightarrow A(a,b)}{A(a,b) \wedge B \Rightarrow A(a,b)} \wedge \text{L}}{\forall x(A(x,b) \wedge B) \Rightarrow A(a,b)} \forall \text{L} = I_1 \quad \cfrac{\cfrac{B \Rightarrow B}{A(a,b) \wedge B \Rightarrow B} \wedge \text{L}}{\forall x(A(x,b) \wedge B) \Rightarrow B} \forall \text{L} = I_2}{\cfrac{\cfrac{\forall x(A(x,b) \wedge B) \Rightarrow A(a,b) \wedge B}{\forall x(A(x,b) \wedge B) \Rightarrow \exists y(A(a,y) \wedge B)} \exists \text{R} = I_3}{\exists y \forall x(A(x,y) \wedge B) \Rightarrow \exists y(A(a,y) \wedge B)} \exists \text{L} = I_4}}{\exists y \forall x(A(x,y) \wedge B) \Rightarrow \forall x \exists y(A(x,y) \wedge B)} \forall \text{R} = I_5$$

There are no propositional inferences below I_3, I_4, or I_5, so $o(I_3) = o(I_4) = o(I_5) = 0$. The only propositional inference below I_1 and I_2 is $\wedge \text{R}$, so $o(I_1) = o(I_2) = 1$. The order of π is the sum of all these, i.e., $o(\pi) = 1 + 1 + 0 + 0 + 0 = 2$.

A cut-free proof π has order $o(\pi) = 0$, iff π contains a sequent such that all the inferences above it are propositional or structural, and all the inferences below it are structural or quantificational. For when $o(\pi) = 0$, any topmost quantifier inference I must have order 0, and so I has no propositional inferences below it. In other words, all inferences below I are either structural or quantifier inferences.

Since these are all inferences with a single premise, there can be only one topmost quantifier inference I. Since the axioms contain no quantifiers, the premise of I contains no quantifiers. No inference above I is a quantifier inference, otherwise I would not be the topmost quantifier inference—so all the inferences above I are propositional or structural. No inference below I is propositional, since $o(I) = 0$. Consequently, the premise of I has the required property. This argument does not establish, however, that the premise of I contains no quantifiers: a weakening above I may have introduced them.

To establish the theorem, we show that every cut-free proof from quantifier-free axioms can be transformed into one of order 0 *and* no quantified formulas are introduced by weakening above the mid-sequent. We show this by induction on $o(\pi)$.

Induction basis: $o(\pi) = 0$. First, suppose that there are no propositional inferences in π. In this case we have a linear demonstration without any branching. The mid-sequent can then be unambiguously determined as the upper sequent of the first application of a quantificational rule or of a weakening rule introducing a quantifier. Since the axioms do not contain quantified formulas, it is clear that the sequent in question is quantifier-free.

Inductive step: Now assume that there is at least one propositional inference. In this case we look for the lowest application of a propositional rule. Let its lower sequent be $\Delta_0 \Rightarrow \Lambda_0$.[3] Since $o(\pi) = 0$, there does not exist any application of a quantifier rule above this one. Thus, if any quantified formula occurs in $\Delta_0 \Rightarrow \Lambda_0$, it is the result of a weakening. On account of the sub-formula property, this quantified formula will occur as a sub-formula of a formula in $\Delta \Rightarrow \Lambda$ (the end-sequent of the proof). Hence, it must be in prenex normal form. This means that it cannot be the principal formula of a propositional rule. So if it occurs in $\Delta_0 \Rightarrow \Lambda_0$, it does so because it was introduced by a weakening somewhere above $\Delta_0 \Rightarrow \Lambda_0$. Any such weakening can be moved below $\Delta_0 \Rightarrow \Lambda_0$, resulting in the lower sequent of the lowermost propositional inference being quantifier-free.

Take the proof ending in $\Delta_0 \Rightarrow \Lambda_0$, and remove from it and all the sequents above it all the formulas in which quantifiers occur and inferences acting on them. This results in a correct proof since

(a) the axioms do not contain quantifiers, so axioms are unchanged;

(b) there are no quantifier inferences; and

(c) no quantified formula is an auxiliary formula of a propositional inference (since that would result in a formula not in prenex normal form).

If we remove formulas from the premises and conclusion of a structural rule, the inference is either still correct, or it results in identical lower and upper sequents,

---

[3] Gentzen denotes it as $\mathfrak{S}$q.

in which case we remove the inference. The result is a new sub-proof of a sequent $\Delta_0^* \Rightarrow \Lambda_0^*$, where $\Delta_0^* \subseteq \Delta_0$ and $\Lambda_0^* \subseteq \Lambda_0$.[4] We obtain a proof of $\Delta_0 \Rightarrow \Lambda_0$ by adding weakenings and interchanges. In the new proof, $\Delta_0^* \Rightarrow \Lambda_0^*$ is the mid-sequent: it contains no quantifiers, no rule above it is a quantificational rule, and no rule below it is a propositional rule.

Let's consider an example:

$$
\dfrac{
  \dfrac{
    \dfrac{
      \dfrac{
        \dfrac{A(a) \Rightarrow A(a)}{\forall x\, B(x), A(a) \Rightarrow A(a)}\ \text{WL}
      }{A(a), \forall x\, B(x) \Rightarrow A(a)}\ \text{IL}
    }{\underbrace{A(a) \wedge B(c), \forall x\, B(x)}_{\Delta_0} \Rightarrow \underbrace{A(a)}_{\Lambda_0}}\ \wedge\text{L}
  }{
    \dfrac{\forall x(A(x) \wedge B(c)), \forall x\, B(x) \Rightarrow A(a)}{\forall y \forall x(A(x) \wedge B(y)), \forall x\, B(x) \Rightarrow A(a)}\ \forall\text{L}
  }\ \forall\text{L}
}{\forall y \forall x(A(x) \wedge B(y)), \forall x\, B(x) \Rightarrow \exists z\, A(z)}\ \exists\text{R}
$$

This proof has order 0, and the lower sequent of the $\wedge$L inference is the only sequent which has no quantifier inferences above it, and also no propositional inferences below it. However, it contains $\forall x\, B(x)$. If we remove all occurrences of $\forall x\, B(x)$ in the sub-proof leading to it, we obtain:

$$
\dfrac{A(a) \Rightarrow A(a)}{\underbrace{A(a) \wedge B(c)}_{\Delta_0^*} \Rightarrow \underbrace{A(a)}_{\Lambda_0^*}}\ \wedge\text{L}
$$

We recover $\Delta_0 \Rightarrow \Lambda_0$ by adding WL and IL.

$$
\dfrac{
  \dfrac{A(a) \Rightarrow A(a)}{\underbrace{A(a) \wedge B(c)}_{\Delta_0^*} \Rightarrow \underbrace{A(a)}_{\Lambda_0^*}}\ \wedge\text{L}
}{
  \dfrac{\forall x\, B(x), A(a) \wedge B(c) \Rightarrow A(a)}{\underbrace{A(a) \wedge B(c), \forall x\, B(x)}_{\Delta_0} \Rightarrow \underbrace{A(a)}_{\Lambda_0}}\ \text{IL}
}\ \text{WL}
$$

[4] Gentzen calls it $\mathfrak{S}\mathfrak{q}^*$.

The resulting proof is:

$$
\cfrac{
  \cfrac{
    \cfrac{
      \cfrac{
        \cfrac{
          \cfrac{
            \cfrac{A(a) \Rightarrow A(a)}{A(a) \wedge B(c) \Rightarrow A(a)} \; \wedge\text{L}
            \;\;\; \underbrace{}_{\Delta_0^*} \;\;\; \underbrace{}_{\Lambda_0^*}
          }{\forall x\, B(x), A(a) \wedge B(c) \Rightarrow A(a)} \; \text{WL}
        }{A(a) \wedge B(c), \forall x\, B(x) \Rightarrow A(a)} \; \text{IL}
      }{\forall x (A(x) \wedge B(c)), \forall x\, B(x) \Rightarrow A(a)} \; \forall\text{L}
    }{\forall y \forall x (A(x) \wedge B(y)), \forall x\, B(x) \Rightarrow A(a)} \; \forall\text{L}
  }{\forall y \forall x (A(x) \wedge B(y)), \forall x\, B(x) \Rightarrow \exists z\, A(z)} \; \exists\text{R}
}{}
$$

The lower sequent of the $\wedge$L inference is now the mid-sequent.

Inductive step: $o(\pi) > 0$. Let $o(\pi) = n + 1$ and assume the inductive hypothesis: that the result holds for all proofs of order n. The first step[5] consists in renaming the free variables of π in such a way that all the eigenvariables of quantificational rules are distinct and that they occur only above the application of the respective quantificational rules, i.e., the proof is regular. We proved this in Proposition 5.23.

Then we move to the key step of the transformation.[6] Because we are in the inductive step we know that $o(\pi) \geq 1$. This means that there must be at least one application of a propositional rule that follows a quantificational inference. Among these propositional inferences that follow a quantificational inference, let I be a topmost one, and let J be a bottommost quantifier inference above I. In other words, pick a quantifier inference J followed by a propositional inference I with no intervening quantifier or propositional inferences. The general strategy consists in pushing the application of the propositional rule above the application of the quantificational rule. We have to take into consideration all the different cases generated by the application of each quantificational rule.

First consider the case where J is a $\forall$R inference followed by a propositional inference I with one premise. Then the sub-proof ending in I has the form:

$$
\begin{array}{c}
\vdots \\
\cfrac{\Gamma \Rightarrow \Theta, B(a)}{\Gamma \Rightarrow \Theta, \forall x\, B(x)} \; \forall\text{R} = J \\
\vdots \;\; \pi' \\
\vdots \\
\cfrac{\Delta' \Rightarrow \Lambda'}{\Delta \Rightarrow \Lambda} \; I
\end{array}
$$

---

[5] This first step corresponds to 2.232.1 in Gentzen.
[6] Case 2.232.2 in Gentzen.

Here, π' may contain structural inferences, but no propositional or quantifier inferences. We can replace the sub-proof by the following one:

$$
\begin{array}{c}
\vdots \\
\dfrac{\Gamma \Rightarrow \Theta, B(a)}{\Gamma \Rightarrow B(a), \Theta} \ \text{IR} \\[4pt]
\hline
\Gamma \Rightarrow B(a), \Theta, \forall x\, B(x) \ \text{WR} \\
\vdots \ \pi' \\
\vdots \\
\dfrac{\Delta' \Rightarrow B(a), \Lambda'}{\Delta \Rightarrow B(a), \Lambda} \ I \\[4pt]
\hline
\Delta \Rightarrow B(a), \Pi, \forall x\, B(x) \ \text{IR} \\
\hline
\Delta \Rightarrow \Pi, \forall x\, B(x), B(a) \ \text{IR} \\
\hline
\Delta \Rightarrow \Pi, \forall x\, B(x), \forall x\, B(x) \ \ \forall\text{R} = J \\
\hline
\Delta \Rightarrow \Pi, \forall x\, B(x) \ \text{CR} \\
\hline
\Delta \Rightarrow \Lambda \ \text{IR}
\end{array}
$$

Here, Π is Λ with one occurrence of $\forall x\, B(x)$ removed. Note that in the original as well as in the new proof, one of the formulas in Λ must be $\forall x\, B(x)$ since π' contains only structural inferences and $\forall x\, B(x)$ cannot be an auxiliary formula of I. If it were, the formula resulting from the application of I would not be in prenex normal form and, on account of the sub-formula property, it would occur in the end-sequent. But this contradicts the fact that all formulas in the end-sequent are in prenex normal form. Hence, a sequence of IR can transform Λ into $\Pi, \forall x\, B(x)$ and vice versa.

The assumption that all eigenvariables are distinct and occur only above the quantifier inference they are eigenvariables for is important here. For we must be able to assume that the application of $\forall$R satisfies the eigenvariable condition. It would be violated if Δ or $\Pi, \forall x\, B(x)$ contained a. Of course, $\forall x\, B(x)$ cannot contain a, otherwise the eigenvariable condition of $\forall$R would have been violated in the original proof. But Δ or Π might contain a if it were introduced, say, by weakening in π' or if the principal formula of I contained it. This is ruled out by the condition that a only occurs above J in the original proof.

Every quantifier inference that lies above I or below J still has the same number of propositional inferences below it, so its order is unchanged. The order of J, however, is now reduced by one, since I is now above J. Therefore, the inductive hypothesis applies.

The case where I has two premises is only slightly more complicated. Assume that the $\forall$R inference occurs above the left premise. We have a proof of the form:

$$
\frac{\Gamma \Rightarrow \Theta, B(a)}{\Gamma \Rightarrow \Theta, \forall x\, B(x)}\ \forall\text{R} = J
$$

$$
\vdots\ \pi'
$$

$$
\frac{\Delta' \Rightarrow \Lambda' \qquad\qquad \Delta'' \Rightarrow \Lambda''}{\Delta \Rightarrow \Lambda}\ I
$$

We switch the order of I and J by replacing this with the proof:

$$
\frac{\dfrac{\dfrac{\Gamma \Rightarrow \Theta, B(a)}{\Gamma \Rightarrow B(a), \Theta}\ \text{IR}}{\Gamma \Rightarrow B(a), \Theta, \forall x\, B(x)}\ \text{WR}}
$$

$$
\vdots\ \pi'
$$

$$
\Delta' \Rightarrow B(a), \Lambda' \qquad \frac{\dfrac{\dfrac{\Delta'' \Rightarrow \Lambda''}{\Delta'' \Rightarrow \Lambda'', B(a)}\ \text{WR}}{\Delta'' \Rightarrow B(a), \Lambda''}\ \text{IR}}
$$

$$
\frac{\Delta \Rightarrow B(a), \Lambda}{\frac{\Delta \Rightarrow B(a), \Pi, \forall x\, B(x)}{\frac{\Delta \Rightarrow \Pi, \forall x\, B(x), B(a)}{\frac{\Delta \Rightarrow \Pi, \forall x\, B(x), \forall x\, B(x)}{\frac{\Delta \Rightarrow \Pi, \forall x\, B(x)}{\Delta \Rightarrow \Lambda}\ \text{IR}}\ \text{CR}}\ \forall\text{R} = J}\ \text{IR}}\ \text{IR}}\ I
$$

The main difference to the case where I has a single premise consists in the addition of $B(a)$ also to the succedent of the second premise, as the side formulas in the premises of propositional rules must match. (Note that this is not the case, and the addition of $B(a)$ not necessary, in the case of $\supset$L.)

For instance, consider:

$$
\frac{\dfrac{\dfrac{\dfrac{A \Rightarrow C, B(a)}{A \Rightarrow C, \forall x\, B(x)}\ \forall\text{R}}{\dfrac{A \Rightarrow \forall x\, B(x), C}{D, A \Rightarrow \forall x\, B(x), C}\ \text{WL}}\ \text{IR} \qquad \dfrac{D \Rightarrow D}{D, A \Rightarrow \forall x\, B(x), D}\ \text{W, I}}{D, A \Rightarrow \forall x\, B(x), C \wedge D}\ \wedge\text{R}
$$

The procedure would switch the order of ∧R and ∀R as follows:

$$
\cfrac{
\cfrac{
\cfrac{
\cfrac{
\cfrac{
\cfrac{A \Rightarrow C, B(a)}{A \Rightarrow B(a), C}\ \text{IR}
}{A \Rightarrow B(a), C, \forall x\, B(x)}\ \text{WR}
}{A \Rightarrow B(a), \forall x\, B(x), C}\ \text{IR}
}{D, A \Rightarrow B(a), \forall x\, B(x), C}\ \text{WL}
\qquad
\cfrac{
\cfrac{
\cfrac{
\cfrac{D \Rightarrow D}{D, A \Rightarrow \forall x\, B(x), D}\ \text{W, I}
}{D, A \Rightarrow \forall x\, B(x), D, B(a)}\ \text{WR}
}{D, A \Rightarrow B(a), \forall x\, B(x), D}\ \text{IR}
}{\ }
}{D, A \Rightarrow B(a), \forall x\, B(x), C \wedge D}\ \text{∧R}
}{D, A \Rightarrow C \wedge D, \forall x\, B(x), B(a)}\ \text{IR}
$$

Continuing:

$$
\cfrac{
\cfrac{
\cfrac{D, A \Rightarrow C \wedge D, \forall x\, B(x), B(a)}{D, A \Rightarrow C \wedge D, \forall x\, B(x), \forall x\, B(x)}\ \text{∀R}
}{D, A \Rightarrow C \wedge D, \forall x\, B(x)}\ \text{CR}
}{D, A \Rightarrow \forall x\, B(x), C \wedge D}\ \text{IR}
$$

The case where J is ∃R is treated exactly the same. In the case of ∀L and ∃L, the procedure is similar, except that we work on the left side of the sequents. E.g., for the case of ∀L and a single-premise propositional rule I we replace:

$$
\cfrac{
\cfrac{B(t), \Gamma \Rightarrow \Theta}{\forall x\, B(x), \Gamma \Rightarrow \Theta}\ \forall\text{L} = J
}{\quad\vdots\ \pi' \quad}
\quad
\cfrac{\Delta' \Rightarrow \Lambda'}{\Delta \Rightarrow \Lambda}\ I
$$

by

$$
\cfrac{
\cfrac{
\cfrac{
\cfrac{
\cfrac{
\cfrac{
\cfrac{
\cfrac{B(t), \Gamma \Rightarrow \Theta}{\Gamma, B(t) \Rightarrow \Theta}\ \text{IL}
}{\forall x\, B(x), \Gamma, B(t) \Rightarrow \Theta}\ \text{WL}
}{\vdots\ \pi'}
}{\Delta', B(t) \Rightarrow \Lambda'}
}{\Delta, B(t) \Rightarrow \Lambda}\ I
}{\forall x\, B(x), \Sigma, B(t) \Rightarrow \Lambda}\ \text{IL}
}{B(t), \forall x\, B(x), \Sigma \Rightarrow \Lambda}\ \text{IL}
}{\forall x\, B(x), \forall x\, B(x), \Sigma \Rightarrow \Lambda}\ \forall\text{L} = J
}{\forall x\, B(x), \Sigma \Rightarrow \Lambda}\ \text{CL}
$$

$$
\cfrac{\forall x\, B(x), \Sigma \Rightarrow \Lambda}{\Delta \Rightarrow \Lambda}\ \text{IL}
$$

Here, Σ is Δ with one occurrence of $\forall x\, B(x)$ removed. Again, Δ must contain $\forall x\, B(x)$, and so we can obtain $\forall x\, B(x), \Sigma$ from Δ by interchanges alone, and vice versa. □

Problem 6.39. Describe the inductive step of the proof of the mid-sequent theorem in detail for the case where ∃L is followed by ⊃L.

Let us consider again the example we used to explain the notion of order.

$$
\cfrac{
  \cfrac{
    \cfrac{
      \cfrac{A(a,b) \Rightarrow A(a,b)}{A(a,b) \wedge B \Rightarrow A(a,b)} \wedge\text{\tiny L}
    }{\forall x(A(x,b) \wedge B) \Rightarrow A(a,b)} \forall\text{\tiny L} = J
    \qquad
    \cfrac{
      \cfrac{B \Rightarrow B}{A(a,b) \wedge B \Rightarrow B} \wedge\text{\tiny L}
    }{\forall x(A(x,b) \wedge B) \Rightarrow B} \forall\text{\tiny L}
  }{\forall x(A(a,b) \wedge B) \Rightarrow A(a,b) \wedge B} \wedge\text{\tiny R} = I
}{
  \cfrac{
    \cfrac{\forall x(A(x,b) \wedge B) \Rightarrow \exists y(A(a,y) \wedge B)}{\exists y \forall x(A(x,y) \wedge B) \Rightarrow \exists y(A(a,y) \wedge B)} \exists\text{\tiny L}
  }{\exists y \forall x(A(x,y) \wedge B) \Rightarrow \forall x \exists y(A(x,y) \wedge B)} \forall\text{\tiny R}
}
$$

In the part below $\wedge$R, we only have quantifier inferences, and none of them contribute to the order of π. So, only the part above the conclusion of $\wedge$R is relevant. We have two candidates for the quantificational inference that's followed by $\wedge$R: either one of the topmost $\forall$L inferences. Let's start with the one on the left, modifying, at the same time, the right branch accordingly.

$$
\cfrac{
  \cfrac{
    \cfrac{
      \cfrac{A(a,b) \Rightarrow A(a,b)}{A(a,b) \wedge B \Rightarrow A(a,b)} \wedge\text{\tiny L}
    }{\forall x(A(x,b) \wedge B), A(a,b) \wedge B \Rightarrow A(a,b)} \text{\tiny WL}
    \quad
    \cfrac{
      \cfrac{
        \cfrac{B \Rightarrow B}{A(a,b) \wedge B \Rightarrow B} \wedge\text{\tiny L}
      }{\forall x(A(x,b) \wedge B) \Rightarrow B} \forall\text{\tiny L} = J
    }{\forall x(A(x,b) \wedge B), A(a,b) \wedge B \Rightarrow B} \text{\tiny WL, IL}
  }{
    \cfrac{
      \cfrac{
        \cfrac{
          \cfrac{\forall x(A(x,b) \wedge B), A(a,b) \wedge B \Rightarrow A(a,b) \wedge B}{A(a,b) \wedge B, \forall x(A(x,b) \wedge B) \Rightarrow A(a,b) \wedge B} \text{\tiny IL}
        }{\forall x(A(x,b) \wedge B), \forall x(A(x,b) \wedge B) \Rightarrow A(a,b) \wedge B} \forall\text{\tiny L}
      }{\forall x(A(x,b) \wedge B) \Rightarrow A(a,b) \wedge B} \text{\tiny CL}
    }{
      \cfrac{
        \cfrac{\forall x(A(x,b) \wedge B) \Rightarrow \exists y(A(a,y) \wedge B)}{\exists y \forall x(A(x,y) \wedge B) \Rightarrow \exists y(A(a,y) \wedge B)} \exists\text{\tiny L}
      }{\exists y \forall x(A(x,y) \wedge B) \Rightarrow \forall x \exists y(A(x,y) \wedge B)} \forall\text{\tiny R}
    }
  } \wedge\text{\tiny R} = I
}{}
$$

We have one remaining pair of inferences I, J where J is a quantifier inference followed by a propositional inference. Removing it results in:

$$
\cfrac{
  \cfrac{
    \cfrac{
      \cfrac{A(a,b) \Rightarrow A(a,b)}{A(a,b) \wedge B \Rightarrow A(a,b)} \wedge\text{\tiny L}
    }{\forall x(A(x,b) \wedge B), A(a,b) \wedge B \Rightarrow A(a,b)} \text{\tiny WL}
    \qquad
    \cfrac{
      \cfrac{B \Rightarrow B}{A(a,b) \wedge B \Rightarrow B} \wedge\text{\tiny L}
    }{\forall x(A(x,b) \wedge B), A(a,b) \wedge B \Rightarrow B} \text{\tiny WL}
  }{
    \cfrac{
      \cfrac{
        \cfrac{\forall x(A(x,b) \wedge B), A(a,b) \wedge B \Rightarrow A(a,b) \wedge B}{A(a,b) \wedge B, \forall x(A(x,b) \wedge B) \Rightarrow A(a,b) \wedge B} \text{\tiny IL}
      }{\forall x(A(x,b) \wedge B), \forall x(A(x,b) \wedge B) \Rightarrow A(a,b) \wedge B} \forall\text{\tiny L}
    }{
      \cfrac{
        \cfrac{
          \cfrac{\forall x(A(x,b) \wedge B) \Rightarrow A(a,b) \wedge B}{\forall x(A(x,b) \wedge B) \Rightarrow \exists y(A(a,y) \wedge B)} \exists\text{\tiny R}
        }{\exists y \forall x(A(x,y) \wedge B) \Rightarrow \exists y(A(a,y) \wedge B)} \exists\text{\tiny L}
      }{\exists y \forall x(A(x,y) \wedge B) \Rightarrow \forall x \exists y(A(x,y) \wedge B)} \forall\text{\tiny R}
    } \text{\tiny CL}
  } \wedge\text{\tiny R}
}{}
$$

Note that we've cheated a little bit here. According to the official procedure, the right sub-proof

$$\cfrac{\cfrac{\cfrac{\cfrac{B \Rightarrow B}{A(a,b) \wedge B \Rightarrow B} \text{ ∧L}}{\forall x(A(x,b) \wedge B) \Rightarrow B} \text{ ∀L} = J}{\forall x(A(x,b) \wedge B), A(a,b) \wedge B \Rightarrow B} \text{ WL, IL}}{}$$

should actually be replaced by

$$\cfrac{\cfrac{\cfrac{\cfrac{B \Rightarrow B}{A(a,b) \wedge B \Rightarrow B} \text{ ∧L}}{\forall x(A(x,b) \wedge B), A(a,b) \wedge B \Rightarrow B} \text{ WL}}{\forall x(A(x,b) \wedge B), \forall x(A(x,b) \wedge B), A(a,b) \wedge B \Rightarrow B} \text{ WL, IL}}{}$$

since the new WL should be followed by all the structural inferences between J and I. To match the side formula of the ∧R, we'd also have to replace the left sub-proof with

$$\cfrac{\cfrac{\cfrac{A(a,b) \Rightarrow A(a,b)}{A(a,b) \wedge B \Rightarrow A(a,b)} \text{ ∧L}}{\forall x(A(x,b) \wedge B), A(a,b) \wedge B \Rightarrow A(a,b)} \text{ WL}}{\forall x(A(x,b) \wedge B), \forall x(A(x,b) \wedge B), A(a,b) \wedge B \Rightarrow A(a,b)} \text{ WL}$$

and follow the ∧R rule by additional IL and CL inferences.

We have a proof of order 0, but not yet a mid-sequent. For this it is still necessary to apply the considerations in the induction basis to the proof ending in the bottommost propositional inference:

$$\cfrac{\cfrac{\cfrac{A(a,b) \Rightarrow A(a,b)}{A(a,b) \wedge B \Rightarrow A(a,b)} \text{ ∧L}}{\forall x(A(x,b) \wedge B), A(a,b) \wedge B \Rightarrow A(a,b)} \text{ WL} \qquad \cfrac{\cfrac{B \Rightarrow B}{A(a,b) \wedge B \Rightarrow B} \text{ ∧L}}{\forall x(A(x,b) \wedge B), A(a,b) \wedge B \Rightarrow B} \text{ WL}}{\forall x(A(x,b) \wedge B), A(a,b) \wedge B \Rightarrow A(a,b) \wedge B} \text{ ∧R}$$

We remove all quantified formulas:

$$\cfrac{\cfrac{A(a,b) \Rightarrow A(a,b)}{A(a,b) \wedge B \Rightarrow A(a,b)} \text{ ∧L} \qquad \cfrac{B \Rightarrow B}{A(a,b) \wedge B \Rightarrow B} \text{ ∧L}}{A(a,b) \wedge B \Rightarrow A(a,b) \wedge B} \text{ ∧R}$$

Now we add a wl to obtain a proof of the original conclusion of ∧R. The completed proof is:

$$
\cfrac{
\cfrac{
\cfrac{
\cfrac{
\cfrac{
\cfrac{
\cfrac{
\cfrac{
\cfrac{A(a,b) \Rightarrow A(a,b)}{A(a,b) \wedge B \Rightarrow A(a,b)} \wedge\text{L}
\qquad
\cfrac{B \Rightarrow B}{A(a,b) \wedge B \Rightarrow B} \wedge\text{L}
}{A(a,b) \wedge B \Rightarrow A(a,b) \wedge B} \wedge\text{R}
}{\forall x(A(x,b) \wedge B), A(a,b) \wedge B \Rightarrow A(a,b) \wedge B} \text{WL}
}{A(a,b) \wedge B, \forall x(A(x,b) \wedge B) \Rightarrow A(a,b) \wedge B} \text{IL}
}{\forall x(A(x,b) \wedge B), \forall x(A(x,b) \wedge B) \Rightarrow A(a,b) \wedge B} \forall\text{L}
}{\forall x(A(x,b) \wedge B) \Rightarrow A(a,b) \wedge B} \text{CL}
}{\forall x(A(x,b) \wedge B) \Rightarrow \exists y(A(a,y) \wedge B)} \exists\text{R}
}{\exists y \forall x(A(x,y) \wedge B) \Rightarrow \exists y(A(a,y) \wedge B)} \exists\text{L}
}{\exists y \forall x(A(x,y) \wedge B) \Rightarrow \forall x \exists y(A(x,y) \wedge B)} \forall\text{R}
$$

Theorem 6.40. *The mid-sequent theorem holds for* **LJ**.

Problem 6.41. Verify that the proof of the mid-sequent theorem indeed goes through for **LJ**. Does the mid-sequent theorem hold for **LM**?

Corollary 6.42 (Herbrand's theorem). *Let $B(a_1, \ldots, a_n)$ be a formula without quantifiers and let the sequent $\Rightarrow \exists x_1 \ldots \exists x_n \, B(x_1, \ldots, x_n)$ be provable in* **LK**. *Then there exist m sequences of n terms t_i^j (for $i = 1, \ldots, n$ and $j = 1, \ldots, m$), all of which belong to the set of terms occurring in the proof, such that:*

$$
\Rightarrow B(t_1^1, \ldots, t_n^1), B(t_1^2, \ldots, t_n^2), \ldots, B(t_1^m, \ldots, t_n^m)
$$

is provable in **LK**.

Proof. The result follows from the mid-sequent theorem. Let a proof of

$$
\Rightarrow \exists x_1 \ldots \exists x_n \, B(x_1, \ldots, x_n)
$$

be given. On account of the mid-sequent theorem there is a proof in which there exists a sequent from which the final sequent in the proof is obtained by applications of ∃R, IR, WR, and CR. This means that the mid-sequent must have empty antecedent.

The succedent of the mid-sequent consists of several instances of $B(x_1, \ldots, x_n)$; that is, of formulas that are obtained from $B(x_1, \ldots, x_n)$ by replacement of the n bound variables x_i by an n-tuple of terms. Due to the finiteness of the proof, the number of these instances cannot be infinite. Therefore, there exist a number m of n-tuples of terms such that a given permutation of them occurs in the formulas constituting the mid-sequent. From these one can obtain several occurrences of $\exists x_1 \ldots \exists x_n \, B(x_1, \ldots, x_n)$ by iterating applications of ∃R. Finally, these can be contracted to a single occurrence by successive applications of IR and CR. □

Example 6.43. Let a proof of $\Rightarrow \exists x_1 \exists x_2\, B(x_1, x_2)$ be given. Let us suppose that the terms occurring in the proof are t_1, t_2, t_3. Then, there are nine possible closures of $B(x_1, x_2)$:

$$
\begin{array}{ccc}
B(t_1, t_1) & B(t_1, t_2) & B(t_1, t_3) \\
B(t_2, t_1) & B(t_2, t_2) & B(t_2, t_3) \\
B(t_3, t_1) & B(t_3, t_2) & B(t_3, t_3)
\end{array}
$$

Now, the mid-sequent is constituted of a given subset of these closures. Clearly, one obtains $\Rightarrow \exists x_1 \exists x_2\, B(x_1, x_2)$ from this mid-sequent by successive applications of $\exists$R, IR, WR, and CR.

Corollary 6.44 (Existence property). *Let an **LM**- or **LJ**-proof of*

$$
\Rightarrow \exists x_1 \ldots \exists x_n\, B(x_1, \ldots, x_n)
$$

*be given, with $B(x_1, \ldots, x_n)$ quantifier-free. Then there exist terms $t_1, \ldots, t_n$ such that $\Rightarrow B(t_1, \ldots, t_n)$ is derivable in **LM** or **LJ**, respectively. Moreover, the terms t_i can be effectively determined from the proof.*

Proof. If the proof of $\Rightarrow \exists x_1 \ldots \exists x_n\, B(x_1, \ldots, x_n)$ were an **LK**-proof, then, on account of the previous restricted Herbrand theorem, there are m (for some m) such sequences of terms, such that

$$
\Rightarrow B(t_1^1, \ldots, t_n^1), B(t_1^2, \ldots, t_n^2), \ldots, B(t_1^m, \ldots, t_n^m)
$$

is provable in **LK**. However, the proof of $\Rightarrow \exists x_1 \ldots \exists x_n\, B(x_1, \ldots, x_n)$ is a proof in **LM** or **LJ**. This means that in this proof sequents with succedents containing more than one formula cannot occur. On the other hand, the mid-sequent theorem applies to **LM** and **LJ** as well, as this is just a corollary of the *Hauptsatz*. Therefore the proof of $\Rightarrow \exists x_1 \ldots \exists x_n\, B(x_1, \ldots, x_n)$ can be transformed into a proof whose mid-sequent contains only one instance of $B(x_1, \ldots, x_n)$. Thus, this sequent will have the form $B(t_1, \ldots, t_n)$, for some terms t_i. In addition, the mid-sequent can be effectively determined from the original proof: the transformation (by virtue of the mid-sequent theorem) of the proof of $\Rightarrow \exists x_1 \ldots \exists x_n\, B(x_1, \ldots, x_n)$ into a proof with a mid-sequent is effective. As a consequence, the construction of the terms t_i is effective, as well. $\square$

7

The consistency of arithmetic

7.1 Introduction

In discussing the sequent calculus and natural deduction, we saw that the cut-elimination and normalization theorems have an important consequence: they show that these systems are consistent. Consistency means that there are no proofs of $\perp$ (in the case of natural deduction) or the empty sequent (in the case of the sequent calculus). For if there were, these theorems would imply that any such proof could be transformed into a normal, or cut-free, proof. Normal and cut-free proofs have the sub-formula property, and so can't be proofs of $\perp$ or the empty sequent, since these don't have any proper sub-formulas.

The consistency of mathematical theories is of course also a crucial question. In fact, it is the question that was at the center of Hilbert's program. This question motivated Gentzen's development of natural deduction and the sequent calculus in the first place. One theory is of particular interest: arithmetic. It is formulated in a language with a constant symbol 0 for the number zero, a one-place function symbol ′ (prime) for the successor operation $x \mapsto x+1$, two binary function symbols, $+$ and $\cdot$, and a binary relation symbol, $=$. In this language, every natural number has a canonical term—called a *numeral*—representing it: zero is represented by 0, one by 0′, two by 0″, and so on. We'll abbreviate the term consisting of n ′ symbols following 0 by $\bar{n}$. Thus $\bar{1}$ abbreviates 0′, $\bar{2}$ abbreviates 0″, etc.[1]

We have seen how Peano arithmetic can be characterized in an axiomatic system in Section 2.15. This system is given in the language of arithmetic. Its terms are inductively defined as follows:

1. The constant symbol 0 and the free variables $a_1, a_2, \dots$, are terms.

2. If t_1, t_2 are terms, so are t_1', $(t_1 + t_2)$, and $(t_1 \cdot t_2)$.

---

[1] We start the natural numbers here with zero. Gentzen started with one. His language of arithmetic then includes a constant 1 for one, and $\bar{n}$ would be the term consisting of 1 followed by $n - 1$ primes. For the consistency proof, nothing complex hinges on whether one starts with 0 or 1.

An Introduction to Proof Theory: Normalization, Cut-Elimination, and Consistency Proofs.
Paolo Mancosu, Sergio Galvan, and Richard Zach, Oxford University Press. © Paolo Mancosu,
Sergio Galvan and Richard Zach 2021. DOI: 10.1093/oso/9780192895936.003.0007

For readability, we will now leave out the outermost parentheses in terms of the form $(t_1 + t_2)$ and $(t_1 \cdot t_2)$. For instance, $a + 0$ is short for $(a + 0)$.

The well-formed formulas in this language are exactly as given in Section 2.15. Here are the axioms of $\mathbf{PA}_K$ again:

PA1. $a = a$

PA2. $a = b \supset b = a$

PA3. $(a = b \wedge b = c) \supset a = c$

PA4. $\neg a' = 0$

PA5. $a = b \supset a' = b'$

PA6. $a' = b' \supset a = b$

PA7. $[F(0) \wedge \forall x(F(x) \supset F(x'))] \supset \forall x\, F(x)$ for any formula $F(a)$

PA9. $a + 0 = a$

PA10. $a + b' = (a + b)'$

PA11. $a \cdot 0 = 0$

PA12. $a \cdot b' = (a \cdot b) + a$

We have left out PA8, $\neg\neg a = b \supset a = b$. It is only needed for intuitionistic arithmetic. We can prove $\Rightarrow \neg\neg a = b \supset a = b$ in **LK** since **LK** (but not **LJ**) proves $\neg\neg A \Rightarrow A$ for all A.

We can turn the axioms other than PA7 into sequents consisting only of atomic formulas, viz., all substitution instances of:

PA$_S$1. $\Rightarrow a = a$

PA$_S$2. $a = b \Rightarrow b = a$

PA$_S$3. $a = b, b = c \Rightarrow a = c$

PA$_S$4. $a' = 0 \Rightarrow$

PA$_S$5. $a = b \Rightarrow a' = b'$

PA$_S$6. $a' = b' \Rightarrow a = b$

PA$_S$7. $\Rightarrow a + 0 = a$

PA$_S$8. $\Rightarrow a + b' = (a + b)'$

PA$_S$9. $\Rightarrow a \cdot 0 = 0$

PA$_S$10. $\Rightarrow a \cdot b' = (a \cdot b) + a$

Just like in the axiomatic presentation of $\mathbf{PA}_K$, all substitution instances of these sequents count as initial sequents. For instance, for any terms t and s, $t = s \Rightarrow s = t$ is a substitution instance of PA$_S$2 and counts as an initial sequent. Note that t

and s may contain free variables or be closed terms. We call them "mathematical" initial sequents to distinguish them from the logical initial sequents $A \Rightarrow A$.[2]

The list of axioms of $\mathbf{PA}_K$ includes the induction schema PA7. A straightforward translation of it into a sequent would suggest we use

$$F(0), \forall x \big(F(x) \supset F(x') \big) \Rightarrow \forall x \, F(x).$$

as another initial sequent. In contrast to all other mathematical initial sequents, this would be an initial sequent that is not composed of only atomic formulas. For this reason, we do not add the induction principle as an initial sequent, but instead formulate it as an inference rule. The induction rule is:

$$! \; \frac{F(a), \Gamma \Rightarrow \Theta, F(a')}{F(0), \Gamma \Rightarrow \Theta, F(t)} \; \text{CJ}$$

$F(0)$ and $F(t)$ are the principal formulas of CJ.[3] The variable a in the premise is the eigenvariable of the inference, and like eigenvariables of $\forall$R and $\exists$L, may not occur in the lower sequent.

The induction rule CJ is needed to prove even basic arithmetical laws such as $\forall x \forall y \, (x + y) = (y + x)$, even though every variable-free substitution instance of it can be proved without it. We'll simply call this sequent calculus system for classical arithmetic $\mathbf{PA}$.[4]

Example 7.1. Here is a simple example of a proof using CJ. Take $F(a)$ to be

$$a = 0 \vee \exists x \, a = x'.$$

This formula says that every number is either zero or the successor of some number. Every natural number clearly has this property, and in fact, for each n, it is not hard to prove $\Rightarrow F(\overline{n})$ without CJ. To prove the general claim $\forall y \, F(y)$, though, we need to use CJ. To apply CJ, we have to prove the corresponding premise, $F(a) \Rightarrow F(a')$. Here's a proof π_1 of it:

---

[2] Gentzen treated = as a non-logical predicate, although = is now more commonly considered a logical primitive. So axiom sequents expressing basic properties of = such as $\Rightarrow a = a$ count as "mathematical" in Gentzen's writings.

[3] CJ translates Gentzen's "VJ" for "vollständige Induktion," i.e., complete induction, where again the standard German way of writing a capital I was as J.

[4] We'll simply use $\mathbf{PA}$ instead of complicating notation by, e.g., introducing a subscript $\mathbf{PA}_S$ for "sequent version of Peano arithmetic."

$$\dfrac{\dfrac{\dfrac{\Rightarrow a' = a'}{a = 0 \Rightarrow a' = 0, a' = a'}\ \text{W, I}}{a = 0 \Rightarrow a' = 0, \exists x\, a' = x'}\ \exists\text{R} \qquad \dfrac{\dfrac{\dfrac{\dfrac{a = c' \Rightarrow a' = c''}{a = c' \Rightarrow a' = 0, a' = c''}\ \text{WR, IR}}{a = c' \Rightarrow a' = 0, \exists x\, a' = x'}\ \exists\text{R}}{\exists x\, a = x' \Rightarrow a' = 0, \exists x\, a' = x'}\ \exists\text{L}}{}}{\ } $$

$$\dfrac{\dfrac{\dfrac{\dfrac{\dfrac{a = 0 \vee \exists x\, a = x' \Rightarrow a' = 0, \exists x\, a' = x'}{a = 0 \vee \exists x\, a = x' \Rightarrow a' = 0, a' = 0 \vee \exists x\, a' = x'}\ \text{VR}}{a = 0 \vee \exists x\, a = x' \Rightarrow a' = 0 \vee \exists x\, a' = x', a' = 0}\ \text{IR}}{a = 0 \vee \exists x\, a = x' \Rightarrow a' = 0 \vee \exists x\, a' = x', a' = 0 \vee \exists x\, a' = x'}\ \text{VR}}{a = 0 \vee \exists x\, a = x' \Rightarrow a' = 0 \vee \exists x\, a' = x'}\ \text{CR}}{}\ \text{VL}$$

The left branch uses an instance of axiom $PA_S 1$, the right branch an instance of axiom $PA_S 5$. The conclusion of the CJ-rule applied to the end-sequent yields any sequent of the form $F(0) \Rightarrow F(t)$, e.g., $F(0) \Rightarrow F(b)$ (taking t to be the free variable b). We can derive $\Rightarrow F(0)$ by a proof π_2:

$$\dfrac{\Rightarrow 0 = 0}{\Rightarrow 0 = 0 \vee \exists x\, 0 = x'}\ \text{VR}$$

This again uses an instance of axiom $PA_S 1$.

A proof of $\Rightarrow \forall y\, F(y)$ would now be

$$\dfrac{\dfrac{\vdots \pi_2 \qquad \dfrac{\dfrac{\vdots \pi_1}{F(a) \Rightarrow F(a')}}{F(0) \Rightarrow F(b)}\ \text{CJ}}{\Rightarrow F(0) \qquad\qquad}\ \ \dfrac{}{\Rightarrow F(b)}\ \text{CUT}}{\Rightarrow \forall y\, F(y)}\ \text{VR}$$

or, with $F(a)$ spelled out:

$$\dfrac{\dfrac{\vdots \pi_2 \qquad\qquad \dfrac{\dfrac{\vdots \pi_1}{a = 0 \vee \exists x\, a = x' \Rightarrow a' = 0 \vee \exists x\, a' = x'}}{0 = 0 \vee \exists x\, 0 = x' \Rightarrow b = 0 \vee \exists x\, b = x'}\ \text{CJ}}{\Rightarrow 0 = 0 \vee \exists x\, 0 = x' \qquad\qquad}\ \ \dfrac{}{\Rightarrow b = 0 \vee \exists x\, b = x'}\ \text{CUT}}{\Rightarrow \forall y\, (y = 0 \vee \exists x\, y = x')}\ \text{VR}$$

The previous example used CJ on a complex formula, but CJ is even useful—and sometimes necessary—to derive atomic sequents. For instance, note that our axioms of **PA** do not include what are usually called equality axioms, e.g., $a = b \Rightarrow (c + a) = (c + b)$. We can use CJ to prove this.

Our proof will use two inference patterns, corresponding to the transitivity and symmetry of identity.

$$\dfrac{\Gamma \Rightarrow s = t \qquad \Delta \Rightarrow t = u}{\Gamma, \Delta \Rightarrow s = u}\ \text{TR} \qquad\qquad \dfrac{\Gamma \Rightarrow s = t}{\Gamma \Rightarrow t = s}\ \text{SYM}$$

These are not new inference rules, but instead abbreviate inference patterns. Specifically, sym (for symmetry) abbreviates the following inference:

$$\frac{\Gamma \Rightarrow s = t \qquad s = t \Rightarrow t = s}{\Gamma \Rightarrow t = s} \text{ CUT}$$

Note that the right premise of the cut is an instance of axiom PA_S2.

The "rule" TR allows us to "string together" equalities, i.e., from $\Gamma \Rightarrow s = t$ and $\Delta \Rightarrow t = u$ we can derive $\Gamma, \Delta \Rightarrow s = u$. It abbreviates the following inference pattern, involving instances of axiom PA_S3 and cut:

$$\frac{\Delta \Rightarrow t = u \qquad \dfrac{\dfrac{\Gamma \Rightarrow s = t \qquad s = t, t = u \Rightarrow s = u}{\Gamma, t = u \Rightarrow s = u} \text{ CUT}}{\dfrac{t = u, \Gamma \Rightarrow s = u}{} \text{ IL}}}{\dfrac{\Delta, \Gamma \Rightarrow s = u}{\Gamma, \Delta \Rightarrow s = u} \text{ IL}} \text{ CUT}$$

Example 7.2. We now proceed with the proof of $a = b \Rightarrow (c + a) = (c + b)$. First we give a proof $\eta_1(c, d)$ of $\Rightarrow (c' + d) = (c + d')$, using induction on d. For this, we'll need a proof of the induction basis $d = 0$, i.e., of the sequent $\Rightarrow (c' + 0) = (c + 0')$. Let $\eta_1'(c)$ be:

$$\frac{\Rightarrow c + 0 = c \qquad \dfrac{\Rightarrow c' + 0 = c' \qquad \dfrac{\dfrac{\Rightarrow c + 0' = (c + 0)' \qquad c + 0 = c \Rightarrow (c + 0)' = c'}{c + 0 = c \Rightarrow c + 0' = c'} \text{ TR}}{\dfrac{c + 0 = c \Rightarrow c' = c + 0'}{c + 0 = c \Rightarrow c' + 0 = c + 0'} \text{ TR}} \text{ SYM}}{\Rightarrow c' + 0 = c + 0'} \text{ CUT}}{}$$

In the two uses of TR in $\eta_1'(c)$, the roles of s, t, and u are played by $c + 0'$, $(c + 0)'$, and c', respectively, in the upper use, and by $c' + 0$, c', and $c + 0'$, respectively, in the lower use.

Problem 7.3. Write out the complete proof without using the abbreviated proof segments sym and TR. For each initial sequent, indicate which axiom of **PA** it is an instance of.

For the inductive step, we have to prove the sequent

$$c' + e = c + e' \Rightarrow c' + e' = c + e''.$$

Let $\eta_1''(c, e)$ be the following proof of it:

$$\frac{\dfrac{\Rightarrow c' + e' = (c' + e)' \qquad \dfrac{c' + e = c + e' \Rightarrow \quad (c' + e)' = (c + e')'}{c' + e = c + e' \Rightarrow c' + e' = (c + e')'} \text{ TR}}{} \qquad \dfrac{\dfrac{\Rightarrow c + e'' = (c + e')'}{\Rightarrow (c + e')' = c + e''} \text{ SYM}}{}}{c' + e = c + e' \Rightarrow c' + e' = c + e''} \text{ TR}$$

The axioms used are instances of PA_S8, PA_S5, and PA_S8, respectively (from left to right).

Then $\eta_1(c, d)$ is:

$$
\cfrac{
\begin{array}{cc}
\vdots\,\eta_1'(c) & \cfrac{c' + e = c + e' \Rightarrow c' + e' = c + e''}{\cfrac{c' + 0 = c + 0' \Rightarrow c' + d = c + d'}{}} \text{ CJ} \\
\Rightarrow c' + 0 = c + 0' &
\end{array}
}{\Rightarrow c' + d = c + d'} \text{ CUT}
$$

with $\vdots\,\eta_1''(c, e)$ above.

Let $\eta_2(a)$ be the following proof of $\Rightarrow 0 + a = a$:

$$
\cfrac{
\begin{array}{cc}
& \cfrac{\cfrac{\Rightarrow 0 + f' = (0 + f)' \quad 0 + f = f \Rightarrow (0 + f)' = f'}{0 + f = f \Rightarrow 0 + f' = f'} \text{ TR}}{0 + 0 = 0 \Rightarrow 0 + a = a} \text{ CJ} \\
\Rightarrow 0 + 0 = 0 &
\end{array}
}{\Rightarrow 0 + a = a} \text{ CUT}
$$

Using proofs $\eta_2(a)$ of $\Rightarrow (0 + a) = a$ and $\eta_2(b)$ of $\Rightarrow (0 + b) = b$ plus SYM and TR gives us a proof of $a = b \Rightarrow (0 + a) = (0 + b)$. That's the induction basis of a proof of $a = b \Rightarrow (c + a) = (c + b)$ using induction on c.

For the inductive step, we have to prove $(d + a) = (d + b) \Rightarrow (d' + a) = (d' + b)$. Use the axioms $\Rightarrow (d' + a') = (d' + a)'$, $(d + a) = (d + b) \Rightarrow (d + a)' = (d + b)'$ plus proofs $\eta_1(d, a)$ of $\Rightarrow (d' + a) = (d + a')$ and $\eta_1(d, b)$ of $\Rightarrow (d' + b) = (d + b')$. We leave the details as an exercise.

Problem 7.4. To prove the sequent $a = b \Rightarrow (c + a) = (c + b)$ using induction on c we need a proof of $(d + a) = (d + b), a = b \Rightarrow (d' + a) = (d' + b)$, to which we apply CJ, followed by a CUT with the induction basis $a = b \Rightarrow (0 + a) = (0 + b)$. Write out the proof of $a = b \Rightarrow (c + a) = (c + b)$ using the proof segments we constructed above.

Example 7.5. We can use the above proof fragments (specifically, $\eta_1(g, a)$ and $\eta_2(a)$) to prove $\Rightarrow (a + b) = (b + a)$, too. We use induction on b. The inductive step is $(a + g) = (g + a) \Rightarrow (a + g') = (g' + a)$. Here is the proof $\eta_3(a, g)$ of it:

$$
\cfrac{
\begin{array}{cc}
\vdots\,\eta_3' & \cfrac{\cfrac{\vdots\,\eta_1(g, a)}{\Rightarrow (g' + a) = (g + a')}}{\cfrac{\Rightarrow (g + a') = (g' + a)}{}} \text{ SYM} \\
(a + g) = (g + a) \Rightarrow (a + g') = (g + a') &
\end{array}
}{(a + g) = (g + a) \Rightarrow (a + g') = (g' + a)} \text{ TR}
$$

where η_3' is:

$$
\cfrac{
\begin{array}{cc}
\cfrac{
\begin{array}{cc}
& (a + g) = (g + a) \Rightarrow \\
\Rightarrow (a + g') = (a + g)' & (a + g)' = (g + a)'
\end{array}
}{(a + g) = (g + a) \Rightarrow (a + g') = (g + a)'} \text{ TR}
&
\cfrac{\Rightarrow (g + a') = (g + a)'}{\Rightarrow (g + a)' = (g + a')} \text{ SYM}
\end{array}
}{(a + g) = (g + a) \Rightarrow (a + g') = (g + a')} \text{ TR}
$$

The induction basis is $\Rightarrow (a + 0) = (0 + a)$. Here is a proof $\eta_4(a)$ of that:

$$
\frac{\Rightarrow (a + 0) = a \qquad \dfrac{\begin{array}{c} \vdots\, \eta_2(a) \\ \Rightarrow (0 + a) = a \end{array}}{\Rightarrow a = (0 + a)}\ \text{SYM}}{\Rightarrow (a + 0) = (0 + a)}\ \text{TR}
$$

Together we get a proof $\eta_5(a, b)$:

$$
\frac{\begin{array}{c} \vdots\, \eta_4(a) \\ \Rightarrow (a + 0) = (0 + a) \end{array} \qquad \dfrac{(a + g) = (g + a) \Rightarrow (a + g') = (g' + a)}{(a + 0) = (0 + a) \Rightarrow (a + b) = (b + a)}\ \text{CJ}\quad \begin{array}{c} \vdots\, \eta_3(a, g) \\ \ \end{array}}{\Rightarrow (a + b) = (b + a)}\ \text{CUT}
$$

Problem 7.6. Give a proof in **PA** of $a = b \Rightarrow (a + c) = (b + c)$ directly using CJ, or using the previously constructed proof segments and the derived rule TR.

Problem 7.7. Give a proof in **PA** of $a = b \Rightarrow a \cdot c = b \cdot c$. You may use previously constructed proofs such as Example 7.2 and Problem 7.6.

Problem 7.8. Let S be the sequent containing only the induction schema PA7 on the right:
$$\Rightarrow \big[F(0) \wedge \forall x\big(F(x) \supset F(x')\big)\big] \supset \forall x\, F(x).$$

1. Give a proof of S in **PA**.
2. Show that **LK** with the axioms of **PA** plus S, but without CJ, can derive CJ.

Any proof in **LK** using the induction rule CJ and making use of mathematical initial sequents counts as a proof in **PA**. The system **PA** is consistent if, and only if, the empty sequent has no such proof.

Gentzen's strategy to establish the consistency of **PA** is similar to the strategy for proving consistency of **LK** using the cut-elimination theorem. There we showed that every proof can be transformed into one of a specific simple form, in that case, one that avoids the CUT rule. Proofs in this simple form have a nice property: every formula occurring in such a proof is a sub-formula of one of the formulas in the end-sequent. Every initial sequent, including the mathematical initial sequents, contains at least one formula. Every rule other than CUT leads from premises containing at least one formula to a conclusion containing at least one formula. So every proof in **PA** not using CUT contains at least one formula in every sequent, including the end-sequent. Since the empty sequent contains no formulas, there cannot be a proof of the empty end-sequent.

Unfortunately, the presence of the CJ rule prevents a straightforward application of the cut-elimination theorem to arithmetic. For if a CJ inference is followed by a MIX, we would have to be able to exchange the order of the CJ and MIX inferences for the inductive step to go through. Consider this example:

$$\frac{\dfrac{F(a), \Gamma \Rightarrow F(a')}{F(0), \Gamma \Rightarrow F(t)} \text{ CJ} \qquad F(t) \Rightarrow \Lambda}{F(0), \Gamma \Rightarrow \Lambda} \text{ MIX}$$

In order to apply the MIX rule first, we would have to get the right side of $F(a), \Gamma \Rightarrow F(a')$ to match up with the left side of $F(t) \Rightarrow \Lambda$. Since t may be any term, there is no guarantee that $F(a') \Rightarrow \Lambda$ can be derived if $F(t) \Rightarrow \Lambda$ can. The eigenvariable condition on a allows us to substitute any term for a, but nothing guarantees that t is such a substitution instance. Even if it were, say, t is s' for some s, we would have

$$\frac{F(s), \Gamma \Rightarrow F(s') \qquad F(s') \Rightarrow \Lambda}{F(s), \Gamma \Rightarrow \Lambda} \text{ MIX}$$

The conclusion of this MIX is not of the right form to apply CJ to it. And these considerations automatically extend to CUT.

Even though the cut-elimination theorem does not apply to **PA**, Gentzen still managed to find a procedure for "simplifying" proofs in **PA**. For Gentzen's proof of the consistency of **PA**, the strategy is similar to the proof of the cut-elimination theorem, but the procedure is more involved and the notion of "simple" proof is different. While the cut-elimination theorem required a double induction on rank and degree of cut-formulas, we will now have to use a more complicated induction. But before we go there, let's consider the notion of "simple proof" in play in Gentzen's proof and why simple proofs cannot be proofs of the empty sequent.[5]

7.2   Consistency of simple proofs

Let's call a CUT *atomic* if the cut-formula is atomic, and *complex* otherwise.

Definition 7.9. A proof in **PA** is *simple* if (a) it contains no free variables, (b) it consists only of atomic formulas, (c) the only rules applied are weakening, contraction, interchange, and atomic CUTs. In particular, in a simple proof, operational rules and CJ are not allowed, and all formulas in it are sentences.

---

[5] It would be possible to give a consistency proof also for natural deduction. In fact, Gentzen's first consistency proofs (Gentzen, 1935a, 1936) were formulated for a system that is in some ways closer to natural deduction than the sequent calculus. However, the consistency proof for the sequent calculus formulation of **PA** in (Gentzen, 1938) is much easier to follow, and so we only treat that proof.

In a simple proof, we essentially only use atomic CUTs to carry out basic operations with atomic formulas which contain no free variables. For instance, we can show that $(0 + 0)' = 0'$, i.e., $(\overline{0} + \overline{0})' = \overline{1}$, using mathematical initial sequents and an atomic CUT:

$$\frac{\Rightarrow 0 + 0 = 0 \qquad 0 + 0 = 0 \Rightarrow (0 + 0)' = 0'}{\Rightarrow (0 + 0)' = 0'} \text{ CUT}$$

If we disregard the axioms for addition and multiplication for a moment, and allow only numerals as variable-free terms, then clearly every simple proof contains as formulas only equations of the form $\overline{n} = \overline{m}$. We follow Gentzen and call such a formula *"true"* if the numerals flanking the $=$ sign are identical, i.e., $n = m$, and *"false"* otherwise. By extension, a sequent consisting only of atomic formulas without free variables is "true" if it contains a "false" formula in the antecedent or a "true" formula in the succedent. Simple proofs have the following property: every sequent in it is "true." This can be easily established using induction on the length of the proof.[6]

Proposition 7.10. *Every sequent in a simple proof is true.*

Proof. Induction basis. Since a simple proof contains no variables, all initial sequents must be substitution instances of $A \Rightarrow A$ (with A an atomic sentence) or of axioms PA_S1–PA_S6 (recall that we have excluded addition and multiplication for the time being), where the variables have been replaced by numerals. First, we verify that any such initial sequent is true:

1. $A \Rightarrow A$, where A is an atomic sentence, is true: if A is true, $A \Rightarrow A$ contains a true formula in the succedent; if it is false, it contains a false formula in the antecedent.

2. $\Rightarrow \overline{n} = \overline{n}$ is obviously true, since $\overline{n} = \overline{n}$ is a true formula by definition, and occurs in the succedent.

3. $\overline{n} = \overline{m} \Rightarrow \overline{m} = \overline{n}$ is true. If $n = m$, then the formula in the succedent is identical to $\overline{n} = \overline{n}$ and so is true. If $n \neq m$, then $\overline{n}$ and $\overline{m}$ are not identical, so the formula in the antecedent is false.

4. $\overline{n} = \overline{k}, \overline{k} = \overline{m} \Rightarrow \overline{n} = \overline{m}$ is true, since if both formulas in the antecedent are true, then $k = n$ and $k = m$, and hence $n = m$, and so the formula in the succedent is $\overline{n} = \overline{n}$, which is true. So either one of the formulas in the antecendent is false or, if both are true, the formula in the succedent is true.

---

[6] Note that our definition of "true" and "false" here is purely syntactical, decidable, and applies only to atomic identities without free variables. In particular, "true" and "false" are not defined for formulas containing connectives or quantifiers. (We could extend the notion to connectives, but not to quantifiers.) We'll leave out the scare quotes for readability in what follows, but keep in mind that this finitary notion of "true" does not presuppose any metaphysically or epistemically questionable notion of truth, such as the standard Tarskian notion of truth (which is not intelligible according to Hilbert-style finitism).

5. $\overline{n}' = 0 \Rightarrow$ is true, since $\overline{n}'$ is the same as $\overline{n+1}$ and $n+1 \neq 0$ if $n \geq 0$. So, $\overline{n}' = 0$ is always false.

6. $\overline{n} = \overline{m} \Rightarrow \overline{n}' = \overline{m}'$ is true, as whenever $\overline{n} = \overline{m}$ is true, then $n = m$ and so $\overline{n}' = \overline{m}'$ is true.

7. Similarly, $\overline{n}' = \overline{m}' \Rightarrow \overline{n} = \overline{m}$ is true.

Thus, every initial sequent in a simple proof is true.

Inductive step. We now show that all inferences in a simple proof preserve truth. The only allowed inferences in simple proofs are weakening, contraction, interchange, and atomic CUTs. Clearly, if the premise of a weakening, contraction, or interchange are true, so is the conclusion. If a false formula occurs in the antecendent of the premise of such a rule, it also occurs in the antecendent of the conclusion; and a true formula in the succedent of the premise also occurs in the succedent of the conclusion. Now suppose we have an atomic CUT:

$$\frac{\Gamma \Rightarrow \Theta, A \qquad A, \Delta \Rightarrow \Lambda}{\Gamma, \Delta \Rightarrow \Theta, \Lambda} \text{ CUT}$$

Suppose both premises are true. If the left premise is true because Γ contains a false formula or Θ contains a true formula, then the conclusion is true. Similarly, if the right premise is true because Δ contains a false formula or Λ a true formula. The only case not considered is when the left premise is true because A is true and the right premise because A is false. But since A cannot be both true and false, this case cannot arise. ☐

If we allow addition and multiplication, and add the remaining axioms PA$_S$7–PA$_S$10, we can extend the notion of "true" and "false" to formulas and simple proofs including such terms. Call the *value* of a term without free variables the number it denotes. The value val(t) of such a term can be defined by induction on the complexity of t:

$$\text{val}(0) = 0$$
$$\text{val}(t') = \text{val}(t) + 1$$
$$\text{val}(s + t) = \text{val}(s) + \text{val}(t)$$
$$\text{val}(s \cdot t) = \text{val}(s) \cdot \text{val}(t)$$

While these equations might seem trivial, the symbols 0, $+$, and $\cdot$ are being used to refer to different (albeit related) things on the two sides of the equations. The 0, $+$, and $\cdot$ on the left inside val($\dots$) are *constant* and *function symbols* of the language of **PA**, while on the right they are the ordinary number zero and the ordinary operations of addition and multiplication. As you'd expect, e.g., val($0'''$) = 3 and val($0'' \cdot (0' + 0''')$) = $2 \cdot (1 + 3) = 8$. Then we can say that an atomic formula $s = t$ is true if, and only if, val(s) = val(t), and false otherwise.

As noted in n. 6 on p. 277, this notion of "true" formula is finitarily meaningful, since the definition of the value of a term allows us to compute it, and so whether an atomic formula is "true" or not remains decidable. Likewise as before, we can verify that all initial sequents are true under this expanded definition of "true formula." Weakening, contraction, interchange, and atomic CUTs then let us prove only true sequents. Specifically, every variable-free substitution instance of any mathematical initial sequent is true. For instance, an instance of PA_S7 is $\Rightarrow t+0 = t$. It is true since $val(t + 0) = val(t) + val(0) = val(t) + 0 = val(t)$. An instance of PA_S8 is $\Rightarrow s + t' = (s + t)'$. It is true, as $val(s + t') = val(s) + val(t') = val(s) + val(t) + 1 = val(s + t) + 1 = val((s + t)')$. (Note here that $+$ inside an argument to val is the $+$ symbol of the language of **PA**, but, e.g., the $+$ between $val(t)$ and 0 in $val(t) + 0$ is addition between natural numbers.) In other words, Proposition 7.10 remains true for the language including $+$ and $\cdot$ and proofs including axioms PA_S7–PA_S10.

Problem 7.11. Show that any instance of PA_S9 and PA_S10 is true.

Note that "true" in the sense defined above only applies to sequents with variable-free atomic formulas, so this result is still quite limited. In particular, there are sequents which are clearly true, but have no simple proofs, such as $\Rightarrow 0 = 0 \land 0 = 0$ or $\neg(0 = 0) \Rightarrow$. In order to derive these, we need operational rules, which are not allowed in simple proofs.

Here is an example that shows that $\Rightarrow \bar{0} + \bar{1} = \bar{1} + \bar{0}$, i.e., $\Rightarrow 0 + 0' = 0' + 0$ is derivable via a simple proof.

First, observe that

$$\Rightarrow 0 + 0 = 0$$
$$\Rightarrow 0' + 0 = 0'$$

are instances of axiom PA_S7, and

$$\Rightarrow 0 + 0' = (0 + 0)'$$

is an instance of axiom PA_S8. The following is an instance of axiom PA_S5:

$$0 + 0 = 0 \Rightarrow (0 + 0)' = 0'$$

and finally we have two instances of axiom PA_S3:

$$(0 + 0)' = 0'', 0'' = 0' + 0 \Rightarrow (0 + 0)' = 0' + 0$$
$$0 + 0' = (0 + 0)', (0 + 0)' = 0' + 0 \Rightarrow 0 + 0' = 0' + 0$$

Instances of axiom PA_S2 allow us to switch the order of equalities, e.g.,

$$\frac{\Rightarrow 0' + 0 = 0' \qquad 0' + 0 = 0' \Rightarrow 0' = 0' + 0}{\Rightarrow 0' = 0' + 0} \text{ CUT}$$

(This is the inference pattern SYM.)

Finally, note that instances of axiom PA$_S$5 allow us to pass from $\Rightarrow s = t$ to $\Rightarrow s' = t'$ using CUT:

$$\frac{\Rightarrow s = t \qquad s = t \Rightarrow s' = t'}{\Rightarrow s' = t'} \text{ CUT}$$

We'll abbreviate this as FNC (for functionality).

Using these abbreviations, the simple proof of $0 + 0' = 0' + 0$ can be given as follows:

$$\frac{\Rightarrow 0 + 0' = (0 + 0)' \qquad \dfrac{\dfrac{\Rightarrow 0 + 0 = 0}{\Rightarrow (0 + 0)' = 0'} \text{ FNC} \qquad \dfrac{\Rightarrow 0' + 0 = 0'}{\Rightarrow 0' = 0' + 0} \text{ SYM}}{\Rightarrow (0 + 0)' = 0' + 0} \text{ TR}}{\Rightarrow 0 + 0' = 0' + 0} \text{ TR}$$

Recall that TR abbreviates an inference pattern that allows us to derive $\Rightarrow s = u$ from $\Rightarrow t = s$ and $\Rightarrow s = u$ using an instance of axiom PA$_S$3 and atomic CUTs. E.g, the last use of TR in the above proof abbreviates

$$\frac{\Rightarrow (0+0)' = 0' + 0 \quad \dfrac{\Rightarrow 0 + 0' = (0+0)' \quad \dfrac{0 + 0' = (0+0)', (0+0)' = 0' +0 \Rightarrow}{0 + 0' = 0' + 0}}{(0+0)' = 0' + 0 \Rightarrow 0 + 0' = 0' + 0} \text{ CUT}}{\Rightarrow 0 + 0' = 0' + 0} \text{ CUT}$$

Again, this example only illustrates that atomic, variable-free sequents have simple proofs. In particular, no sequent containing free variables or logical operators has a simple proof, because simple proofs must have atomic, variable-free end-sequents.

Gentzen's consistency proof developed a method for transforming any proof in **PA** whose end-sequent contains only variable-free atomic formulas into a simple proof of the same end-sequent. This yields a consistency proof of **PA**. Indeed, if there were a proof of the empty sequent in **PA**, the procedure of the consistency proof would produce a simple proof of the empty sequent. As we have shown, the end-sequent of any simple proof is true. On the other hand, the empty sequent is not true, since it contains neither a false formula on the left nor a true formula on the right.

Before we go on to describe the proof, we'll mention a detail that will become important later. In point of fact, Gentzen (1938) does not assume a particular collection of mathematical initial sequents, such as those we have obtained from the axiomatic system **PA** presented at the beginning of this chapter. Rather, he only required that all their variable-free mathematical initial sequents are "true" in the special syntactic sense defined above. The mathematical initial sequents of our sequent calculus version of **PA** have this property, as we showed in the proof of Proposition 7.10. Gentzen's definition allows us to add further initial sequents, however. For instance, the sequents

$$s = t, u = v \Rightarrow s + u = t + v \text{ and } s = t, u = v \Rightarrow (s \cdot u) = (t \cdot v)$$

can be added as standard axioms for identity. They have the required property: if both $s = t$ and $u = v$ are "true," the value of s is the same as the value of t, and likewise the value of u is the same as the value of v. Hence, by definition of the value of $(s + u)$ and $(t + v)$, these are the same as well, and hence $s + u = t + v$ is "true."

We proved a special case of these identity axioms in Section 7.1, using CJ. If instead we add them as initial sequents, we can obtain the following:

Proposition 7.12. *Let t be a term without free variables, and $n = \text{val}(t)$ its value. Then $\Rightarrow t = \bar{n}$ has a simple proof.*

Proof. We first establish two lemmas, namely (1) if $k = n + m$, then $\Rightarrow \bar{n} + \bar{m} = \bar{k}$ has a simple proof; and (2) if $k = nm$, then $\Rightarrow \bar{n} \cdot \bar{m} = \bar{k}$ has a simple proof.

Here is a proof of the first lemma. We use induction on m.

Induction basis. For the base case, we have to show that if $k = n + 0$, then $\Rightarrow (\bar{n} + 0) = \bar{k}$. But note that in this case $n = k$ and so $\bar{k}$ and $\bar{n}$ are the same term. Thus, this sequent is a case of axiom PA$_S$7.

Inductive step. Now suppose the claim is true for m and n; we want to show it for $m + 1$ and n. Let $k = n + m$. By inductive hypothesis there is a simple proof π_1 of $\Rightarrow \bar{n} + \bar{m} = \bar{k}$. What we want to show is that $\bar{n} + \bar{m}' = \bar{k}'$ has a simple proof. Here it is:

$$
\frac{
\begin{array}{cc}
\Rightarrow \bar{n} + \bar{m}' = (\bar{n} + \bar{m})' &
\dfrac{\vdots\, \pi_1 \quad \dfrac{\Rightarrow \bar{n} + \bar{m} = \bar{k} \quad \bar{n} + \bar{m} = \bar{k} \Rightarrow (\bar{n} + \bar{m})' = \bar{k}'}{\Rightarrow (\bar{n} + \bar{m})' = \bar{k}'}\ \text{CUT}}{\Rightarrow (\bar{n} + \bar{m})' = \bar{k}'}
\end{array}
}{\Rightarrow \bar{n} + \bar{m}' = \bar{k}'}\ \text{TR}
$$

We leave the other lemma as an exercise.

Now we can prove the main claim that $\Rightarrow t = \bar{n}$ has a simple proof. We use induction on the number of occurrences of $'$, $+$, and $\cdot$ in t. If that number is 0, then t is just the symbol 0, its value $n = 0$, and $\bar{n}$ is also just the symbol 0. Hence, t and $\bar{n}$ are the same term, namely 0. The sequent $\Rightarrow \bar{n} = t$, i.e., $\Rightarrow 0 = 0$, is an instance of axiom PA$_S$1. If the number of occurrences of $'$, $+$, and $\cdot$ in t is greater than 0, we have three cases: (1) t is of the form s'; (2) t is of the form $(s_1 + s_2)$; or (3) t is of the form $(s_1 \cdot s_2)$. We prove case (2) and leave the others as exercises. The terms s_1 and s_2 contain fewer occurrences of $'$, $+$, and $\cdot$ than t and so the inductive hypothesis applies to them. Let $n_1 = \text{val}(s_1)$ and $n_2 = \text{val}(s_2)$; clearly $n = n_1 + n_2$. We have simple proofs π_1 and π_2 of $\Rightarrow s_1 = \bar{n_1}$ and $\Rightarrow s_2 = \bar{n_2}$ by inductive hypothesis. By the first lemma, we also have a simple proof π of $\Rightarrow \bar{n_1} + \bar{n_2} = \bar{n}$. Here's the simple proof of $\Rightarrow t = \bar{n}$:

$$
\begin{array}{c}
\cfrac{\displaystyle \cfrac{\vdots\ \pi_1 \quad\quad s_1 = \overline{n_1}, s_2 = \overline{n_2} \Rightarrow}{\Rightarrow s_1 = \overline{n_1} \quad\quad s_1 + s_2 = \overline{n_1} + \overline{n_2}}\ \text{CUT}}{\cfrac{s_2 = \overline{n_2} \Rightarrow s_1 + s_2 = \overline{n_1} + \overline{n_2}}{\Rightarrow s_1 + s_2 = \overline{n_1} + \overline{n_2}}\ \text{CUT} \quad\quad \cfrac{\vdots\ \pi}{\Rightarrow \overline{n_1} + \overline{n_2} = \overline{n}}\ \text{TR}}{\Rightarrow s_1 + s_2 = \overline{n}}
\end{array}
$$

The right premise of the topmost CUT is an identity axiom. $\qquad\square$

Problem 7.13. Show that if $k = nm$, then $\Rightarrow \overline{n} \cdot \overline{m} = \overline{k}$ has a simple proof.

Problem 7.14. Carry out the remaining cases in the proof of Proposition 7.12, namely (1) that if $\mathrm{val}(s) = n$ and $\Rightarrow s = \overline{n}$ has a simple proof then so does $\Rightarrow s' = \overline{n+1}$ and (3) if $\mathrm{val}(s_1) = n_1$ and $\mathrm{val}(s_2) = n_2$ and $\Rightarrow s_1 = \overline{n_1}$ and $\Rightarrow s_2 = \overline{n_2}$ have simple proofs, then so does $\Rightarrow s_1 \cdot s_2 = \overline{n_1 n_2}$.

7.3 Preliminary details

In order to describe the consistency proof, we must recall a few definitions and results from Chapter 5 and introduce a bit of additional terminology.

First, recall that by Proposition 5.17, we can replace logical axiom sequents $A \Rightarrow A$ by derivations from atomic logical axioms. So, we may assume that all proofs from now on contain only atomic logical axioms.

Let's also remember the notion of a regular proof in **LK**: a proof is regular if every eigenvariable is used for only one application of $\forall$R or $\exists$L, and appears only above that inference (Definition 5.22). We can extend this definition to cover proofs in **PA**.

Definition 7.15. A proof π in **PA** is *regular* if every eigenvariable occurring in π is the eigenvariable of exactly one $\forall$R, $\exists$L, or cj inference and only occurs above it.

Proposition 7.16. *Any proof can be turned into a regular proof.*

Proof. The proof of Proposition 5.23 goes through, if we cover the case of cj in the proofs of Lemma 5.19 and Corollary 5.21 as well. We leave this as an exercise. $\square$

Problem 7.17. Carry out the inductive step of Lemma 5.19 for the case where $\pi(a)$ ends in a cj inference.

Problem 7.18. Corollary 5.21 is used repeatedly in the proof of Proposition 5.23 for **LK**. Reformulate the corollary so that it can be used to prove that every proof in **PA** can be transformed into a regular proof, and prove it using Lemma 5.19 expanded to cover proofs in **PA** as in the preceding exercise.

Proposition 7.19. *If $\pi(a)$ is a regular proof of a sequent $S(a)$ containing the free variable a, t is a term not containing any eigenvariables of $\pi(a)$, then replacing a everywhere in π by t results in a proof $\pi(t)$ of the sequent $S(t)$.*

Proof. Since in a regular proof, any eigenvariable c is used for exactly one inference in π and does not occur below it, c cannot occur in the end-sequent. Hence the free variable a is not used as an eigenvariable. Each inference remains correct if the same free variable is replaced by the same term in both the conclusion and the premises as long as it isn't an eigenvariable of that inference.   □

Proposition 7.20. *If π is a regular proof and the end-sequent contains no free variables, then π' resulting from π by replacing every free variable that isn't used as an eigenvariable by 0 is a proof of the same end-sequent.*

Proof. By induction on the size of the proof. Uniform replacement of free variables by terms preserves correctness of inferences, except possibly if substituted into eigenvariables in applications of $\forall$R, $\exists$L, and CJ, which are excluded. If the end-sequent contains no free variables, the resulting proof is a proof of the same end-sequent.   □

Of course, the result also holds if the end-sequent does contain free variables, but then the resulting proof has a different sequent as end-sequent. For instance, the proof on the left below can be transformed into the proof on the right, and all axioms remain axioms and all inferences remain correct:

$$
\cfrac{\cfrac{a = b \Rightarrow b = a}{a = b \land \neg b = a \Rightarrow b = a}\ \land\text{L}}{\neg b = a, a = b \land \neg b = a \Rightarrow}\ \neg\text{L}
\qquad\qquad
\cfrac{\cfrac{0 = 0 \Rightarrow 0 = 0}{0 = 0 \land \neg 0 = 0 \Rightarrow 0 = 0}\ \land\text{L}}{\neg 0 = 0, 0 = 0 \land \neg 0 = 0 \Rightarrow}\ \neg\text{L}
$$

However, in cases where a variable is used as an eigenvariable, it is not replaced by the constant 0, e.g.,

$$
\cfrac{\cfrac{\cfrac{\cfrac{\cfrac{a = b \Rightarrow b = a}{a = b \land \neg b = a \Rightarrow b = a}\ \land\text{L}}{\neg b = a, a = b \land \neg b = a \Rightarrow}\ \neg\text{L}}{a = b \land \neg b = a, a = b \land \neg b = a \Rightarrow}\ \land\text{L}}{a = b \land \neg b = a \Rightarrow}\ \text{CL}}{\cfrac{\Rightarrow \neg(a = b \land \neg b = a)}{\Rightarrow \forall y\, \neg(a = y \land \neg y = a)}\ \forall\text{R}}\ \neg\text{R}
$$

turns into:

$$
\cfrac{\cfrac{\cfrac{\cfrac{\cfrac{0 = b \Rightarrow b = 0}{0 = b \land \neg b = 0 \Rightarrow b = 0}\ \land\text{L}}{\neg b = 0, 0 = b \land \neg b = 0 \Rightarrow}\ \neg\text{L}}{0 = b \land \neg b = 0, 0 = b \land \neg b = 0 \Rightarrow}\ \land\text{L}}{0 = b \land \neg b = 0 \Rightarrow}\ \text{CL}}{\cfrac{\Rightarrow \neg(0 = b \land \neg b = 0)}{\Rightarrow \forall y\, \neg(0 = y \land \neg y = 0)}\ \forall\text{R}}\ \neg\text{R}
$$

The previous propositions establish that when dealing with proofs whose end-sequents do not contain free variables, we may assume that the proof is regular and that all free variables occurring in it are eigenvariables. This property is crucial for the elimination of cj inferences.

In order to describe the method used in Gentzen's consistency proof, we'll have to be able to talk more precisely about the relationship between occurrences of sequents and formulas in a proof. In particular, we are interested in which formula occurrences in a proof arise out of which previous formula occurrences.

Definition 7.21. A sequent S occurs *below* (or *above*) another sequent S' in a proof if the sub-proof ending with S contains S' (or vice versa).

Definition 7.22. In any inference I occurring in a proof, we say that a formula occurrence F' in the conclusion is the *successor* of a formula occurrence F in the premise if, and only if, one of the following holds:

1. I is an operational inference, F is an auxiliary formula, and F' is the principal formula (e.g., the $A \wedge B$ in the conclusion of an $\wedge$R rule is the successor of the occurrence of A in the left premise and of the occurrence of B in the right premise).

2. I is a contraction, F' is the occurrence of the contraction formula A in the conclusion, and F is one of the two indicated occurrences of the contraction formula A in the premise.

3. I is an interchange, F' and F are the indicated occurrences of one of the interchanged formulas in the conclusion and premise, respectively.

4. I is an induction rule, and F' is the occurrence of $A(0)$ on the left side of the conclusion, and F is the occurrence of $A(a)$ on the left side of the premise.

5. I is an induction rule, F' is the occurrence of $A(t)$ on the right side of the conclusion, and F is the occurrence of $A(a')$ on the right side of the premise.

6. I is any rule, and F' and F are corresponding occurrences of the same formula in the side formulas in conclusion and premise, respectively.

If F' is the successor of F, then F is a *predecessor* of F'.

Problem 7.23. Write out the proof π_1 from Example 7.1, circling every formula occurrence in it. Then draw an arrow from each formula to its successor in each inference.

Note that cut-formulas and formulas in the end-sequent have no successors, and that neither weakening formulas nor formulas in an initial sequent have predecessors. Next we'll define a bundle of formula occurrences linked by the successor relation.

Definition 7.24. A *bundle* is a sequence of formula occurrences $F_1, \ldots, F_n$ in a proof such that F_{i+1} is the successor of F_i, F_1 has no predecessor (i.e., is a weakening formula or occurs in an initial sequent), and F_n has no successor (i.e., is a cut-formula or occurs in the end-sequent). If $i < j$, then F_i is called an *ancestor* of F_j, and F_j a *descendant* of F_i.

We have to introduce the notion of a bundle and the notion of ancestors and descendant formulas in a proof because—in contrast to the cut-elimination theorem—Gentzen's reduction operations are not strictly local: they don't just operate on sub-proofs, but on parts of the proof between occurrences of the same formula in a bundle.[7] Moreover, for the reduction procedure it is important to know which formulas will "disappear" in a CUT (and so do not make a necessary contribution to the end-sequent), and which formula occurrences can be traced to the end-sequent. For this purpose, we'll make the following definition:

Definition 7.25. A bundle is *implicit* if its last formula is the cut-formula in a CUT, and *explicit* otherwise. Furthermore, an inference in a proof is implicit if its principal formula belongs to an implicit bundle, and explicit otherwise.

Clearly, the last formula in an explicit bundle must belong to the end-sequent. Note that if the end-sequent contains only atomic sentences, or if it is empty (the special case considered by Gentzen for proving consistency), every operational inference is implicit. If it were not, the principal formula of the operational inference would not be atomic, but also be an ancestor of a formula in the end-sequent.

You may want to think of a bundle as the entire history of a formula in a proof, consisting of the first appearance of a formula in an initial sequent or as a weakening formula, through all operations on that formula, until the final occurrence of the formula in the premise of a CUT or in the end-sequent. An implicit bundle tracks the history of a formula that doesn't appear in the end-sequent because it is a cut-formula (i.e., it is eliminated from the proof because of a CUT). An explicit bundle is the history of a formula that still appears in the end-sequent.[8]

In the following example, the two sequences of formula occurrences in dashed boxes constitute implicit bundles, since their last formulas ($\neg B$ and $A \wedge B$) are cut-formulas. (There are actually five implicit bundles in the example but for the moment we only highlight two of them.) The two sequences of formulas in

---

[7] By contrast, the cut-elimination theorem uses local operations. For instance, the sub-proofs ending in premises of inferences immediately preceding a MIX are combined using a simpler MIX, but without changing the sub-proofs themselves.

[8] In Gentzen's original paper, he defines the notion of a *cluster*. A cluster is a set of occurrences of a formula any two of which are either (a) the two occurrences of the contracted formula in the premise of a contraction inference, (b) the two occurrences of a cut-formula in a CUT, or (c) one is a descendant of the other. We'll use Takeuti's notion of a bundle instead since it serves the same purpose but is a bit more precise and versatile.

solid boxes form explicit bundles. The inferences with dashed lines are implicit inferences, those with solid lines, explicit.

$$
\cfrac{
  \cfrac{
    \cfrac{
      \cfrac{A \Rightarrow \boxed{A}}{\boxed{\neg A}, A \Rightarrow} \ \neg\text{L}
    }{\boxed{\neg A}, A \Rightarrow \neg B} \ \text{WR} \qquad
    \cfrac{B \Rightarrow \boxed{B} \qquad \neg B , B \Rightarrow \ \neg\text{L}}{} \ \text{CUT}
  }{\boxed{\neg A}, A, B \Rightarrow}
}{
  \cfrac{\boxed{\neg A}, A, B \Rightarrow A \wedge B}{\ } \ \text{WR}
}
$$

$$
\neg A , A, B \Rightarrow (A \vee C) \wedge (B \vee C)
$$

Of particular interest in the reduction procedure is that part of the proof which lies between the lowermost implicit operational inferences and the end-sequent. Since these inferences are implicit, their principal formulas "disappear," i.e., the bundles they belong to terminate in CUTs and do not reach all the way down to the end-sequent. It's this part of the proof that Gentzen's reduction procedure focuses on. We'll call it the *end-part* of the proof (Gentzen uses "ending") and it is defined as follows:

Definition 7.26. The *end-part* of a proof is the smallest part of the proof that satisfies:

1. The end-sequent belongs to the end-part.

2. If the conclusion of an inference belongs to the end-part, so do the premises, unless the inference is an implicit operational inference.

In other words, the end-part consists of all sequents lying on threads between the end-sequent and the conclusions of the lowermost implicit operational inferences (inclusive), or between the end-sequent and initial sequents if that thread contains no implicit operational inferences at all.[9] These lowermost implicit inferences are called *boundary* inferences. For instance, in the preceding example, there are five implicit bundles, two ending in the premises of the CUT on $\neg B$ (of length 1 and 2), and three ending in the two occurrences of the cut-formula $A \wedge B$ (one of length 1, two of length 4). We indicate them below by putting dashed boxes around the formulas in them. The three operational inferences with dashed inference lines are the implicit operational inferences, since the principal formula belongs to an implicit bundle. They are also boundary inferences, since they are *lowermost*

---

[9] A thread is a sequence of occurrences of sequents in the proof where each sequent is a premise of an inference, the conclusion of which is the following sequent in the thread.

implicit operational inferences, and anything below them belongs to the end part.

$$
\cfrac{
\cfrac{
\cfrac{\cfrac{A \Rightarrow A}{\neg A, A \Rightarrow} \, \neg\text{L}}{\neg A, A \Rightarrow \boxed{\neg B}} \, \text{WR} \qquad
\cfrac{B \Rightarrow \boxed{B}}{\boxed{\neg B}, B \Rightarrow} \, \neg\text{L}
}{
\cfrac{\neg A, A, B \Rightarrow}{\neg A, A, B \Rightarrow \boxed{A \wedge B}} \, \text{WR}
} \, \text{CUT}
\qquad
\cfrac{
\cfrac{
\cfrac{\boxed{A} \Rightarrow A}{\boxed{A \wedge B} \Rightarrow A} \, \wedge\text{L}}{\boxed{A \wedge B} \Rightarrow A \vee C} \, \vee\text{R} \qquad
\cfrac{\cfrac{\boxed{B} \Rightarrow B}{\boxed{A \wedge B} \Rightarrow B} \, \wedge\text{L}}{\boxed{A \wedge B} \Rightarrow B \vee C} \, \vee\text{R}
}{\boxed{A \wedge B} \Rightarrow (A \vee C) \wedge (B \vee C)} \, \wedge\text{R}
}{\neg A, A, B \Rightarrow (A \vee C) \wedge (B \vee C)} \, \text{CUT}
$$

We see that the topmost sequents in the end-part are all either conclusions of (implicit) operational inferences or they are initial sequents. In this example, the end-sequent contains complex formulas and so the end-part contains operational inferences in addition to structural inferences. However, notice that all the operational inferences below the three boundary inferences with dashed lines are explicit since their main formulas are not eliminated by a CUT.

In the consistency proof, we will be focusing on proofs of atomic end-sequents. Since the end-sequent is atomic, any operational inference in such a proof must be implicit. If it were not, again, the principal formula of an operational inference would then be an ancestor of a non-atomic formula in the end-sequent, which is impossible since the end-sequent consists of atomic formulas only. Thus, in such a proof, the end-part consists of all sequents between the end-sequent and either an initial sequent or a lowermost operational inference.

7.4   Overview of the consistency proof

Suppose π is a regular proof from atomic axioms with an end-sequent consisting only of atomic sentences, i.e., atomic formulas without variables, and in which the only free variables are eigenvariables. (This includes the special case where the end-sequent is empty, i.e., we are dealing with a putative proof of a contradiction.) We want to find a way of transforming π into a simple proof, i.e., one in which no variables at all occur, and which contains no operational inferences, induction inferences, or complex CUTs. If we manage to do this, we will have shown that **PA** is consistent, since no false sequent (including the empty sequent) can have a simple proof by Proposition 7.10.

The overall strategy of Gentzen's consistency proof is to apply, successively, a number of reduction steps to the proof. They are:

1. replacing "suitable" induction inferences by CUTs (with possibly complex cut-formulas),

2. removing weakenings, and

3. removing "suitable" complex CUT inferences.

These steps all operate only on the end-part of the proof, since the existence of "suitable" induction inferences and complex CUTS is only guaranteed in the end-part.

Step 1 can be accomplished directly for a suitable induction inference: the first crucial step of the consistency proof is to show how a single induction inference can be replaced by a sequence of CUT inferences. We will describe it in detail in Section 7.5. By applying this step until all CJ inferences are removed from the end-part, we obtain a proof π^*.

After we remove all CJ inferences from the end-part, we remove weakening inferences from the end-part (step 2) to obtain a proof π^{**}. This is a relatively straightforward but necessary step in preparation for the crucial step of removing CUTS. We will postpone describing it in detail to Section 7.8.

Step 3 is not accomplished directly, but—like in the cut-elimination theorem—a suitably chosen CUT is "reduced" in the sense that a sub-proof ending with it is replaced by a modified sub-proof. This sub-proof is generally larger, and may contain multiple copies of existing CUTS (and induction inferences) as well as newly introduced CUTS. We consider sub-proofs of π^{**} ending in a complex CUT on a formula C which is introduced on the left and the right by operational inferences. We replace it by a sub-proof ending in a CUT on a formula of lower degree, in which the operational inferences introducing the cut-formula are removed. As in the cut-elimination theorem, this requires that we can find sequents where C is the principal formula on the right (in the proof of the left premise of the CUT) and one where it is principal on the left (in the proof of the right premise). These sequents are the conclusions of boundary inferences (they are the boundary between the end-part of the proof and the rest). In "reducing" the CUT on C to CUTS on a sub-formula or sub-formulas of C, these boundary inferences are replaced by weakenings. Consequently, in the new proof, on the threads containing these inferences the boundary is then raised—parts of the new proof that were above the boundary originally are now below it. This means in particular that in the new, "reduced" proof, additional induction inferences and complex CUTS may now be part of the end-part of the proof. We describe the reduction of CUTS in detail in Section 7.6.

By repeating 1–3 along each thread of the proof, the boundary is successively raised this way until there are no boundary inferences on it at all. When the entire proof is its own end-part, and all induction inferences and complex CUTS in it are removed, the reduction is complete.

7.5 Replacing inductions

Let us now address step 1 in more detail. Recall that a CJ inference is of the following form:

$$\frac{F(a), \Gamma \Rightarrow \Theta, F(a')}{F(0), \Gamma \Rightarrow \Theta, F(t)} \text{ CJ}$$

$F(0)$ and $F(t)$ are the principal formulas of cj. Note that t may in general be any term, including one that contains—or is identical to—a free variable. The "suitable" induction inferences are those in which t does not contain any free variables, i.e., t is a closed term. In general, t could be constructed using 0, $'$, $+$ and $\cdot$. In order to keep things simple, we will assume that $+$ and $\cdot$ do not occur in the proof, i.e., that t is a closed term consisting only of 0 and $'$. In other words, t is of the form $\overline{n}$. So, a suitable induction inference is of the form

$$\frac{F(a), \Gamma \Rightarrow \Theta, F(a')}{F(0), \Gamma \Rightarrow \Theta, F(\overline{n})} \ \text{cj}$$

The removal of complex CUTS, to be carried out in step 3, will sometimes transform an induction inference into a new, but suitable, induction inference. As in the cut-elimination theorem, these reduction steps do not work on all inductions and CUTS simultaneously, but on a single, suitably chosen induction or CUT at a time.

First, let's convince ourselves that there is always a suitable cj inference in the end-part, if there is a cj inference in the end-part at all.

Proposition 7.27. *Suppose π is a regular proof without free variables other than eigenvariables. If I is a lowermost induction inference in the end-part of π, then the principal formula $F(t)$ of I contains no free variables.*

Proof. There can be no eigenvariable inferences below I. This is because, first of all, the end-part contains no operational inferences such as $\forall$R and $\exists$L. Second, because I is a lowermost cj inference in the end-part, there are no cj inferences below I. As π contains no free variables other than eigenvariables, this means that $F(t)$ contains no free variables. □

Proposition 7.28. *Suppose π is a regular proof of an atomic sequent with no free variables, in which all free variables are eigenvariables and all axioms are atomic. There is a proof π^* of the same end-sequent containing no cj inferences in its end-part.*

Proof. First we show that a lowermost cj inference in the end-part of π can be removed. By the preceding proposition, its principal formula $F(t)$ contains no free variables, i.e., t is a closed term $\overline{n}$. Now consider a sub-proof that ends in such a cj:

$$\vdots \ \pi'(a)$$
$$\vdots$$
$$\frac{F(a), \Gamma \Rightarrow \Theta, F(a')}{F(0), \Gamma \Rightarrow \Theta, F(\overline{n})} \ \text{cj}$$

By Proposition 7.19, we can substitute any term s for a in the proof $\pi'(a)$ to obtain a proof $\pi'(s)$ of $F(s), \Gamma \Rightarrow \Theta, F(s')$. For instance, if we take s to be $\overline{k}$, then the end-sequent is $F(\overline{k}), \Gamma \Rightarrow \Theta, F(\overline{k+1})$, since $\overline{k}'$ is just $\overline{k+1}$. If we choose s

successively to be $0, 0', \ldots, \overline{n-1}$, we obtain proofs of

$$F(0), \Gamma \Rightarrow \Theta, F(0')$$
$$F(0'), \Gamma \Rightarrow \Theta, F(0'')$$
$$F(0''), \Gamma \Rightarrow \Theta, F(0''')$$

$$\ddots$$

$$F(\overline{n-1}), \Gamma \Rightarrow \Theta, F(\overline{n})$$

Since the formula $F(\overline{k}')$ (not necessarily atomic) in the succedent of one of these sequents is the same as the formula $F(\overline{k+1})$ in the antecendent of the next sequent, we can apply the CUT rule to any two adjacent sequents in the series. In the end, we have a proof of $F(0), \Gamma \Rightarrow \Theta, F(\overline{n})$:

In the case where $n = 0$, the conclusion of the CJ inference is $F(0), \Gamma \Rightarrow \Theta, F(0)$. In that case, we can replace the entire sub-proof ending in the CJ inference by

$$\frac{F(0) \Rightarrow F(0)}{F(0), \Gamma \Rightarrow \Theta, F(0)} \text{ W, I}$$

Since we are now considering CJ inferences in the end-part of the proof, there are no quantifier inferences below it. So we can guarantee that the t in the principal formula $F(t)$ of a lowermost CJ inference is a numeral $\overline{n}$, as long as the end-sequent contains no free variables, by applying Proposition 7.20 and replacing free variables by 0.[10] In general, the induction rule allows conclusion sequents

---

[10] This is true as it stands if our proof does not contain $+$ and $\cdot$, as we have assumed. In the more general case, we need to also know that if $\mathrm{val}(t) = n$ then $F(\overline{n}) \Rightarrow F(t)$ is provable without CJ or complex CUTs. This follows by a simple induction on the degree of $F(a)$ from Proposition 7.12, but requires additional identity axioms as initial sequents, namely, $a = b \Rightarrow a + c = b + c$, $a = b \Rightarrow c + a = c + b$, $a = b \Rightarrow a \cdot c = b \cdot c$, and $a = b \Rightarrow c \cdot a = c \cdot b$.

with any substitution instance $F(t)$ on the right, including where t contains free variables. In that case, replacing the induction inferences with a sequence of CUTS is not possible, and the presence of instances $F(t)$ with free (eigen)variables is unavoidable if below that induction a $\forall$L or $\exists$R inference operates on $F(t)$. But in that case, the induction inference wouldn't be in the end-part of the proof.

We now apply induction to obtain the result. The induction is not on the number of CJ inferences in the end-part, since the result of removing the lowermost CJ inference results in a new proof in which the end-part may contain more CJ inferences in the proof: $\pi'(a)$ itself may contain CJ inferences, so the new proof contains n times as many in the end-part, since it contains n copies of $\pi'(a)$: $\pi'(0)$, ..., $\pi'(\overline{n-1})$. But, each CJ inference in it still contains as many CJ inferences above it as it does in $\pi'(a)$. So, induction is on the maximum number of CJ inferences along a thread between a boundary inference and the end-sequent. This measure is reduced if we remove a lowermost CJ inference.

A bit more precisely: Call $I_1, \ldots, I_k$ an *induction chain* in the end-part of π if, and only if, each I_j is a CJ inference in the end-part, I_{j+1} occurs below I_j in π, and there are no CJ inferences in the end-part above I_1, between I_j and I_{j+1}, or below I_k. Let $m(\pi)$ be the maximum length of an induction chain in the end-part of π, and let $o(\pi)$ be the number of induction chains in the end-part of π of length $m(\pi)$. Clearly, $m(\pi) = o(\pi) = 0$ if, and only if, there are no CJ inferences in the end-part. Now consider an induction chain of maximal length $m = m(\pi)$. The sub-proof $\pi'(a)$ ending in the premise of I_m cannot contain an induction chain of length m, since then adding I_m to it would produce a longer induction chain and so the induction chain would not be of maximal length. There are no CJ inferences below the last inference I_m in that chain by definition. Since there are no CJ inferences below I_m, the right principal formula of I_m is of the form $F(\overline{n})$. We replace the sub-proof ending in I_m with a sequence of n CUT inferences as described above. Each one of the sub-proofs $\pi'(\overline{k})$ leading to the premises of the new CUT inferences contains induction chains only of length $< m$. So, in the new proof, the maximal length of induction chains is $\leq m$, and the number of induction chains of length m has been decreased by at least 1. The result follows by double induction on $m(\pi)$ and $o(\pi)$. $\square$

Example 7.29. Let us illustrate this part of the proof with an example. Consider the following derivation π:

$$
\pi_1(a)
\begin{cases}
\dfrac{\Rightarrow 0 = 0}{a = a \Rightarrow 0 = 0}\ \text{WL}
\end{cases}
\qquad
\pi_2(b)
\begin{cases}
\dfrac{\dfrac{\dfrac{\Rightarrow b' = b'}{b = b \Rightarrow b' = b'}\ \text{WL}}{0 = 0 \Rightarrow a' = a'}\ I_1 = \text{CJ}}{}
\end{cases}
$$

$$
\dfrac{\dfrac{}{a = a \Rightarrow a' = a'}\ \text{CUT}}{\dfrac{a = a \Rightarrow a' = a'}{0 = 0 \Rightarrow \overline{2} = \overline{2}}\ I_2 = \text{CJ}}
$$

This derivation has an atomic end-sequent, its only free variables are eigenvariables of the two CJ inferences. As predicted by Proposition 7.27, the principal formula $F(t)$ of the lowermost CJ inference (I_2) contains no free variables, i.e., the term t is of the form $\overline{n}$ (in our case, $n = 2$). The derivation is regular and is its own end-part.

So, Proposition 7.28 applies. Since there are two CJ inferences, we must make use of the inductive step. There is an induction chain of length 2, given by I_1 and I_2. Since it is the only induction chain, it is of maximal length, so $o(\pi) = 1$. We replace the lowermost CJ inference in the chain of maximal length, viz., I_2 by CUTs. In our case, $F(a)$ is $a = a$. So we take the premise of I_2, $a = a \Rightarrow a' = a'$, and substitute $\bar{k}$ for a, for $k = 0, \ldots, n - 1$. This yields the two sequents $0 = 0 \Rightarrow 0' = 0'$ and $0' = 0' \Rightarrow 0'' = 0''$ with derivations $\pi_1(0)$ and $\pi_1(0')$. They look as follows:

$$\pi_1(0)\left\{ \begin{array}{c} \dfrac{\dfrac{\Rightarrow 0 = 0}{0 = 0 \Rightarrow 0 = 0}\ \text{WL} \qquad \dfrac{\dfrac{\Rightarrow b' = b'}{b = b \Rightarrow b' = b'}\ \text{WL}}{0 = 0 \Rightarrow 0' = 0'}\ I_1 = \text{CJ}}{0 = 0 \Rightarrow 0' = 0'}\ \text{CUT} \end{array} \right.$$

and

$$\pi_1(0')\left\{ \begin{array}{c} \dfrac{\dfrac{\Rightarrow 0 = 0}{0' = 0' \Rightarrow 0 = 0}\ \text{WL} \qquad \dfrac{\dfrac{\Rightarrow b' = b'}{b = b \Rightarrow b' = b'}\ \text{WL}}{0 = 0 \Rightarrow 0'' = 0''}\ I_1 = \text{CJ}}{0' = 0' \Rightarrow 0'' = 0''}\ \text{CUT} \end{array} \right.$$

The combined proof π' is:

$$\dfrac{\vdots\ \pi_1(0) \qquad\qquad \vdots\ \pi_1(0')}{\dfrac{0 = 0 \Rightarrow 0' = 0' \qquad 0' = 0' \Rightarrow 0'' = 0''}{0 = 0 \Rightarrow \bar{2} = \bar{2}}\ \text{CUT}}$$

If we replace $t = t$ everywhere by $F(t)$ we can write π' out in full:

$$\dfrac{\dfrac{\dfrac{\Rightarrow F(0)}{F(0) \Rightarrow F(0)}\ \text{WL} \quad \dfrac{\dfrac{\Rightarrow F(b')}{F(b) \Rightarrow F(b')}\ \text{WL}}{F(0) \Rightarrow F(0')}\ I_1 = \text{CJ}}{F(0) \Rightarrow F(0')}\ \text{CUT} \quad \dfrac{\dfrac{\Rightarrow F(0)}{F(0') \Rightarrow F(0)}\ \text{WL} \quad \dfrac{\dfrac{\Rightarrow F(b')}{F(b) \Rightarrow F(b')}\ \text{WL}}{F(0) \Rightarrow F(0'')}\ I_1' = \text{CJ}}{F(0'') \Rightarrow F(0'')}\ \text{CUT}}{F(0) \Rightarrow F(0'')}\ \text{CUT}$$

The new derivation π' still has two induction inferences, as did the original derivation π. It has two induction chains (I_1, and I_1'), rather than one, but each is of length 1. So although $o(\pi') = 2$ is larger than $o(\pi) = 1$, the maximum length $m(\pi')$ of induction chains decreased from 2 to 1. In fact, both remaining CJ inferences (I_1, I_1') are now lowermost and hence, by Proposition 7.27, the principal formula is of the form $F(\bar{n})$. In I_1, we have $n = 1$, and in I_1', $n = 2$. Consider the two sub-proofs ending in I_1 and I_1':

$$\pi_2(b)\left\{ \dfrac{\dfrac{\Rightarrow F(b')}{F(b) \Rightarrow F(b')}\ \text{WL}}{F(0) \Rightarrow F(0')}\ I_1 = \text{CJ} \right. \qquad\qquad \pi_2(b)\left\{ \dfrac{\dfrac{\Rightarrow F(b')}{F(b) \Rightarrow F(b')}\ \text{WL}}{F(0) \Rightarrow F(0'')}\ I_1' = \text{CJ} \right.$$

We can replace the sub-proof ending in I_1 by

$$\pi_2(0)\left\{ \frac{\Rightarrow F(0')}{F(0) \Rightarrow F(0')} \text{ WL} \right.$$

and the sub-proof ending in I_1' by:

$$\pi_2(0)\left\{ \frac{\Rightarrow F(0')}{F(0) \Rightarrow F(0')} \text{ WL} \right. \quad \pi_2(0')\left\{ \frac{\Rightarrow F(0'')}{F(0') \Rightarrow F(0'')} \text{ WL} \right.$$
$$\frac{}{F(0) \Rightarrow F(0'')} \text{ CUT}$$

The resulting proof π^* contain no CJ inferences.

7.6   Reducing suitable CUTs

The reduction of complex CUTs is the most involved part of the reduction process. To describe it completely, we must first consider how the complexity of proofs in **PA** is measured using ordinal notations, which we won't get to until Chapter 9. But we can describe the simplest case, where the complex CUT we replace has no CUTs of equal or higher degree below it. Here's how the reduction works in the case where the cut-formula C is $\forall x\, F(x)$. We start with a proof π of the form:

$$
\begin{array}{c}
\vdots\, \pi_1(a) \qquad\qquad\qquad\qquad\qquad\qquad \vdots\, \pi_2 \\
\frac{\Gamma_1 \Rightarrow \Theta_1, F(a)}{\Gamma_1 \Rightarrow \Theta_1, \forall x\, F(x)}\, \forall\text{R} \qquad\qquad \frac{F(\overline{n}), \Delta_1 \Rightarrow \Lambda_1}{\forall x\, F(x), \Delta_1 \Rightarrow \Lambda_1}\, \forall\text{L} \\
\ddots\, \pi_1' \qquad\qquad\qquad\qquad\qquad\qquad \pi_2' \\
\frac{\Gamma \Rightarrow \Theta, \forall x\, F(x) \qquad \forall x\, F(x), \Delta \Rightarrow \Lambda}{\Gamma, \Delta \Rightarrow \Theta, \Lambda}\, \text{CUT} \\
\vdots\, \pi_4 \\
\Pi \Rightarrow \Xi
\end{array}
$$

Boundary inferences are indicated by dashed lines. The occurrences of the cut-formula $\forall x\, F(x)$ are descendants of principal formulas of boundary inferences.

We'll replace the sub-proof π' of π ending with $\Gamma, \Delta \Rightarrow \Theta, \Lambda$ by:[11]

$$
\begin{array}{c}
\vdots\ \pi_1(\overline{n}) \\
\dfrac{\Gamma_1 \Rightarrow \Theta_1, F(\overline{n})}{\Gamma_1 \Rightarrow F(\overline{n}), \Theta_1}\ \text{I} \\
\dfrac{\ }{\Gamma_1 \Rightarrow F(\overline{n}), \Theta_1,}\ \text{WR}
\end{array}
$$

The sub-proofs π_1'' and π_2'' are just like π_1' and π_2' from the original proof, respectively, except that each sequent in them contains the formula $F(\overline{n})$ immediately to the right (left) of the sequent arrow.

Clearly, the resulting proof is much larger. Both operational boundary inferences are still present (at the end of the left copy of π_2 and at the end of $\pi_1(a)$), plus a copy each of the proofs ending in their premises with the operational inference replaced by a weakening ($\pi_1(\overline{n})$ and the right copy of π_2). Because we replaced the operational inferences in the latter two proofs by weakenings, they no longer end in boundary inferences—the boundary now lies somewhere within $\pi_1(\overline{n})$ and in the right copy of π_2. Instead of one CUT, we now have three, only one of which is on a proper sub-formula of the original cut-formula. So it is not at all clear that repeatedly applying this reduction constitutes a procedure that must always terminate. In contrast to the cut-elimination procedure, the order in which the reduction steps 1–3 are applied is not determined by some measure on the complexity of complex CUTS and induction inferences, but by a more complicated measure on the entire proof. We will show in Chapter 9 that the entire reduction procedure in fact terminates.

Problem 7.30. Convince yourself that the procedure for reducing a CUT on $\exists x\, F(x)$ is the same as for $\forall x\, F(x)$.

---

[11] Here and below we will sometimes display the formulas in a sequent across multiple lines to save space.

Problem 7.31. Suppose you have a suitable CUT on the formula $A \supset B$ in the end-part, e.g.:

$$
\begin{array}{ccc}
\vdots \ \pi_1 & \vdots \ \pi_2 & \vdots \ \pi_3 \\[4pt]
\dfrac{A, \Gamma_1 \Rightarrow \Theta_1, B}{\Gamma_1 \Rightarrow \Theta_1, A \supset B} \supset\!R &
\dfrac{\Delta_1 \Rightarrow \Lambda_1, A \quad B, \Delta_2 \Rightarrow \Lambda_2}{A \supset B, \Delta_1, \Delta_2 \Rightarrow \Lambda_1, \Lambda_2} \supset\!L & \\[10pt]
\vdots \ \pi_1' & \vdots \ \pi_{23}' & \\[4pt]
\Gamma \Rightarrow \Theta, A \supset B & A \supset B, \Delta \Rightarrow \Lambda &
\end{array}
$$

$$
\dfrac{\Gamma \Rightarrow \Theta, A \supset B \qquad A \supset B, \Delta \Rightarrow \Lambda}{\Gamma, \Delta \Rightarrow \Theta, \Lambda} \ \text{CUT}
$$

$$
\vdots
$$

$$
\Pi \Rightarrow \Xi
$$

What is the reduced proof?

Problem 7.32. Give the form of the original proof and the new proof for the case of the reduction of a complex CUT where the cut-formula is $A \wedge B$. How does the reduction change in the case of $A \vee B$?

7.7   A first example

We will give a first example showing the removal of CJ and complex CUTs in the end-part. We do this by slightly modifying Example 7.29:

$$
\dfrac{
\dfrac{\dfrac{\Rightarrow 0 = 0}{a = a \Rightarrow 0 = 0}\,\text{WL} \quad \dfrac{\Rightarrow \bar{1} = \bar{1}}{a = a \Rightarrow \bar{1} = \bar{1}}\,\text{WL}}{a = a \Rightarrow 0 \wedge \bar{1} = \bar{1}}\,\wedge R \quad
\dfrac{\dfrac{\dfrac{\Rightarrow b' = b'}{b = b \Rightarrow b' = b'}\,\text{WL}}{0 = 0 \Rightarrow a' = a'}\,I_1 = \text{CJ}}{0 = 0 \wedge \bar{1} = \bar{1} \Rightarrow a' = a'}\,\wedge L
}{
\dfrac{a = a \Rightarrow a' = a'}{0 = 0 \Rightarrow \bar{2} = \bar{2}}\,I_2 = \text{CJ}
}\,\text{CUT}
$$

As before, we indicate boundary inferences by dashed inference lines. To increase readability, again, we will abbreviate $a = a$ by $F(a)$:

$$
\pi_1(a)\left\{
\dfrac{
\dfrac{\dfrac{\Rightarrow F(0)}{F(a) \Rightarrow F(0)}\,\text{WL} \quad \dfrac{\Rightarrow F(\bar{1})}{F(a) \Rightarrow F(\bar{1})}\,\text{WL}}{F(a) \Rightarrow F(0) \wedge F(\bar{1})}\,\wedge R \quad
\dfrac{\dfrac{\dfrac{\Rightarrow F(b')}{F(b) \Rightarrow F(b')}\,\text{WL}}{F(0) \Rightarrow F(a')}\,I_1 = \text{CJ}}{F(0) \wedge F(\bar{1}) \Rightarrow F(a')}\,\wedge L
}{
\dfrac{F(a) \Rightarrow F(a')}{F(0) \Rightarrow F(\bar{2})}\,I_2 = \text{CJ}
}\,\text{CUT}
\right\}\pi_2(a)
$$

In step 1 we have to replace the CJ inferences in the end-part by CUTS. In this case, there is only one CJ inference in the end-part, I_2. Let $\pi_3(a)$ be the sub-proof ending in its premise, $F(a) \Rightarrow F(a')$. We replace it by a CUT as before, between $F(0) \Rightarrow F(0')$ and $F(\overline{1}) \Rightarrow F(\overline{1}')$, the results of $\pi_3(0)$ and $\pi_3(\overline{1})$:

$$
\frac{
\frac{
\begin{array}{cc}
\vdots\; \pi_1(0) & \vdots\; \pi_2(0) \\[4pt]
F(0) \Rightarrow & F(0) \wedge F(\overline{1}) \\
F(0) \wedge F(\overline{1}) & \Rightarrow F(0')
\end{array}
}{F(0) \Rightarrow F(0')}\text{CUT}
\qquad
\frac{
\begin{array}{cc}
\vdots\; \pi_1(\overline{1}) & \vdots\; \pi_2(\overline{1}) \\[4pt]
F(\overline{1}) \Rightarrow & F(0) \wedge F(\overline{1}) \\
F(0) \wedge F(\overline{1}) & \Rightarrow F(\overline{1}')
\end{array}
}{F(\overline{1}) \Rightarrow F(\overline{1}')}\text{CUT}
}{F(0) \Rightarrow F(\overline{2})}\text{CUT}
$$

The sub-proofs $\pi_1(0)$, $\pi_1(\overline{1})$, $\pi_2(0)$, and $\pi_2(\overline{1})$ all end in operational inferences, so the displayed sequents are all the sequents in the end-part of the new proof. We see that the end-part contains no weakenings, so we can skip step 2 for now. Step 3 requires that we reduce a lowermost complex CUT. Both complex CUTS are lowermost complex CUTS; we can choose on or the other to operate on. We will pick the one on the left, with conclusion $F(0) \Rightarrow F(0')$. The previous Problem 7.32 asked for the formulation of the reduction of such CUTS in the general case. Here is the result in this specific case. The sub-proof ending in the left CUT is:

$$
\frac{
\dfrac{\dfrac{\Rightarrow F(0)}{F(0) \Rightarrow F(0)}\text{WL}\quad \dfrac{\Rightarrow F(\overline{1})}{F(0) \Rightarrow F(\overline{1})}\text{WL}}{F(0) \Rightarrow F(0) \wedge F(\overline{1})}\text{$\wedge$R}
\qquad
\dfrac{\dfrac{\dfrac{\Rightarrow F(b')}{F(b) \Rightarrow F(b')}\text{WL}}{F(0) \Rightarrow F(0')}\text{CJ}}{F(0) \wedge F(\overline{1}) \Rightarrow F(0')}\text{$\wedge$L}
}{F(0) \Rightarrow F(0')}\text{CUT}
$$

It is replaced by:

$$
\frac{
\dfrac{\begin{array}{cc}\vdots\; \pi_4 & \vdots\; \pi_4' \\[4pt] F(0) \Rightarrow F(0'), F(0) & F(0), F(0) \Rightarrow F(0')\end{array}}{F(0), F(0) \Rightarrow F(0'), F(0')}\text{CUT}
}{F(0) \Rightarrow F(0')}\text{CL, CR}
$$

where π_4 is:

$$
\frac{
\dfrac{\dfrac{\Rightarrow F(0)}{F(0) \Rightarrow F(0)}\text{WL}}{F(0) \Rightarrow F(0), F(0) \wedge F(\overline{1})}\text{WR}
\qquad
\dfrac{\dfrac{\dfrac{\Rightarrow F(b')}{F(b) \Rightarrow F(b')}\text{WL}}{F(0) \Rightarrow F(0')}\text{CJ}}{F(0) \wedge F(\overline{1}) \Rightarrow F(0')}\text{$\wedge$L}
}{\dfrac{F(0) \Rightarrow F(0), F(0')}{F(0) \Rightarrow F(0'), F(0)}\text{IR}}\text{CUT}
$$

and π'_4 is:

$$
\frac{\dfrac{\Rightarrow F(0)}{F(0) \Rightarrow F(0)}\ \text{WL} \qquad \dfrac{\Rightarrow F(\bar{1})}{F(0) \Rightarrow F(\bar{1})}\ \text{WL}}{F(0) \Rightarrow F(0) \wedge F(\bar{1})}\ \wedge\text{R} \qquad \dfrac{\dfrac{\dfrac{\Rightarrow F(b')}{F(b) \Rightarrow F(b')}\ \text{WL}}{F(0) \Rightarrow F(0')}\ \text{CJ}}{F(0) \wedge F(\bar{1}), F(0) \Rightarrow F(0')}\ \text{WL}}{F(0), F(0) \Rightarrow F(0')}\ \text{CUT}
$$

Note that the ∧R inference has turned into a WR in π_4, and the ∧L has turned into a WL in π'_4. In the resulting proof, therefore, inferences above these new weakenings now belong to the end-part. This includes a CJ inference in π'_4. Going back to step 1, it has to be removed. In this case that is simply a replacement of

$$
\frac{\dfrac{\Rightarrow F(b')}{F(b) \Rightarrow F(b')}\ \text{WL}}{F(0) \Rightarrow F(0')}\ \text{CJ} \qquad \text{by} \qquad \frac{\Rightarrow F(0')}{F(0) \Rightarrow F(0')}\ \text{WL}
$$

The left half of our new proof looks like this:

$$
\frac{\dfrac{\dfrac{\dfrac{\Rightarrow F(0)}{F(0) \Rightarrow F(0)}\ \text{WL}}{F(0) \Rightarrow F(0),}\ \text{WR} \qquad \begin{array}{c}\vdots\ \pi_2(0) \\ \vdots \\ F(0) \wedge F(\bar{1})\end{array}}{\dfrac{F(0) \Rightarrow F(0), F(0')}{F(0) \Rightarrow F(0'), F(0)}\ \text{IR}}\ \text{CUT} \qquad \dfrac{\begin{array}{c}\vdots\ \pi_1(0)\\ \vdots \\ F(0) \Rightarrow \\ F(0) \wedge F(\bar{1})\end{array} \qquad \dfrac{\dfrac{\Rightarrow F(0')}{F(0) \Rightarrow F(0')}\ \text{WL}}{F(0) \wedge F(\bar{1}), F(0) \Rightarrow F(0')}\ \text{WL}}{F(0), F(0) \Rightarrow F(0')}\ \text{CUT}}{F(0), F(0) \Rightarrow F(0'), F(0')}\ \text{CL, CR}
$$
$$
F(0) \Rightarrow F(0')
$$

Recall that $\pi_1(0)$ ends in ∧R and $\pi_2(0)$ in ∧L, so all sequents displayed here belong to the end-part. The proof is not complete, of course. The proof displayed ends in the left premise of a CUT, with right premise $F(\bar{1}) \Rightarrow F(\bar{2})$. We have to apply step 2 to this proof by removing weakenings. We will discuss the process of doing this in the following section. In fact, this process will turn out to replace the entire left half of the proof simply by the axiom $\Rightarrow F(0')$. Including displaying the missing right part, we end up with the following proof:

$$
\frac{\Rightarrow F(\bar{1}) \qquad \dfrac{F(\bar{1}) \Rightarrow F(0) \wedge F(\bar{1}) \qquad F(0) \wedge F(\bar{1}) \Rightarrow F(\bar{1}')}{F(\bar{1}) \Rightarrow F(\bar{2})}\ \text{CUT}}{\Rightarrow F(\bar{2})}\ \text{CUT}
$$

with $\begin{array}{c}\vdots\ \pi_1(\bar{1})\\ \vdots\end{array}$ and $\begin{array}{c}\vdots\ \pi_2(\bar{1})\\ \vdots\end{array}$ above.

Problem 7.33. Complete the example by applying step 3 to the remaining CUT on $F(0) \wedge F(\bar{1})$; replace the CJ inference in the end-part of the resulting derivation.

7.8   Elimination of weakenings

It is a bit more complicated to ensure the presence of a suitable complex CUT to which step 3 can be applied. Step 3 requires a complex CUT in which the cut-formula is descended on the left and the right from principal formulas of boundary inferences. So for instance, if the CUT is on $\forall x\, F(x)$, then on the left it must be descended from some $F(a)$ in the succedent of a premise to $\forall$R, and on the right side of the CUT it must be descended from some $F(t)$ in the antecedent of a premise to $\forall$L. In general, however, not every complex formula need be descended from a principal formula of a corresponding logical inference—it might also be the descendant of a formula introduced by a weakening. So we first have to remove weakenings from the end-part, resulting in a simplified proof. The new proof may not be a proof of the same end-sequent: the new end-sequent is, however, a sub-sequent of the original one. (In the case of a putative proof of the empty sequent, the end-sequent would be the same since the empty sequent has no proper sub-sequents.)

Proposition 7.34. *Suppose π is a proof of an atomic end-sequent $\Gamma \Rightarrow \Theta$ in which the end-part contains no free variables and no induction inferences. There is a proof π^* in which the end-part contains no weakenings, of a sequent $\Gamma^* \Rightarrow \Theta^*$, where $\Gamma^* \subseteq \Gamma$ and $\Theta^* \subseteq \Theta$.*

Proof. First, if the end-part of π contains no weakenings, we can obviously just take $\pi^* = \pi$.

We now proceed by induction on the number of inferences in the end-part of π.

Induction basis. If that number is 0, the end-part consists only of the end-sequent. Then of course the end-part contains no weakening inferences, and we can take $\pi^* = \pi$.

Inductive step. Assume that the end-part contains at least one inference, and consider cases according to the last inference in π. That last inference cannot be a boundary inference (otherwise the end-part would consist just of the end-sequent and so would not contain any weakenings). So it can only be a structural inference (weakening, contraction, interchange, or CUT). We distinguish cases according to that inference.

If the last inference is WL, π is:

$$\vdots\, \pi_1$$
$$\frac{\Gamma \Rightarrow \Theta}{A, \Gamma \Rightarrow \Theta}\ \text{WL}$$

By the inductive hypothesis, there is a proof π_1^* of $\Gamma^* \Rightarrow \Theta^*$ in which the end-part contains no weakenings. Let $\pi^* = \pi_1^*$. (When the number of inferences in the end-part is 1 this amounts to simply deleting the formula introduced by weakening.) The case where the last inference is a WR is handled the same way.

If the last inference is a CL, π is:

$$
\begin{array}{c}
\vdots\ \pi_1 \\[6pt]
\dfrac{A, A, \Gamma \Rightarrow \Theta}{A, \Gamma \Rightarrow \Theta}\ \text{CL}
\end{array}
$$

By the inductive hypothesis, there is a proof π_1^* in which the end-part contains no weakening. Its end-sequent is $\Gamma^* \Rightarrow \Theta^*$, or $A, \Gamma^* \Rightarrow \Theta^*$, or $A, A, \Gamma^* \Rightarrow \Theta^*$. In the first two cases, let $\pi^* = \pi_1^*$. In the last case, let π^* be:

$$
\begin{array}{c}
\vdots\ \pi_1^* \\[6pt]
\dfrac{A, A, \Gamma^* \Rightarrow \Theta^*}{A, \Gamma^* \Rightarrow \Theta^*}\ \text{CL}
\end{array}
$$

The case where the last inference is CR is treated the same.

If the last inference is IL, π is:

$$
\begin{array}{c}
\vdots\ \pi_1 \\[6pt]
\dfrac{\Gamma, A, B, \Delta \Rightarrow \Theta}{\Gamma, B, A, \Delta \Rightarrow \Theta}\ \text{IL}
\end{array}
$$

Again, the inductive hypothesis yields a proof π_1^* with the end-sequent being $\Gamma^*, A, \Delta^* \Rightarrow \Theta^*$, or $\Gamma^*, B, \Delta^* \Rightarrow \Theta^*$, or $\Gamma^*, A, B, \Delta^* \Rightarrow \Theta^*$. In the first two cases, let $\pi^* = \pi_1^*$; in the third, π^* is

$$
\begin{array}{c}
\vdots\ \pi_1^* \\[6pt]
\dfrac{\Gamma^*, A, B, \Delta^* \Rightarrow \Theta^*}{\Gamma^*, B, A, \Delta^* \Rightarrow \Theta^*}\ \text{IL}
\end{array}
$$

Similarly if the last inference is IR.

If the last inference is a CUT, π is:

$$
\begin{array}{c}
\vdots\ \pi_1 \qquad\qquad \vdots\ \pi_2 \\[6pt]
\dfrac{\Gamma \Rightarrow \Theta, A \qquad A, \Delta \Rightarrow \Lambda}{\Gamma, \Delta \Rightarrow \Theta, \Lambda}\ \text{CUT}
\end{array}
$$

By the inductive hypothesis, there are proofs π_1^* and π_2^* of sub-sequents of $\Gamma^* \Rightarrow \Theta^*, A^*$ and $A^*, \Delta^* \Rightarrow \Lambda^*$, respectively, in which the end-parts contain no weakenings. On either side, the sub-sequent may but need not contain the formula A, i.e., A^* is either A or empty. Suppose first A^* is empty in the end-sequent of π_1^*, i.e., the end-sequent of π_1^* is just $\Gamma^* \Rightarrow \Theta^*$. Since $\Gamma^* \subseteq \Gamma, \Delta$ and

$\Theta^* \subseteq \Theta, \Lambda, \pi^* = \pi_1^*$ serves our purpose: it is a proof of a sub-sequent of the end-sequent of π in which the end-part contains no weakenings. Similarly, if A^* is empty in the end-sequent of π_2^*, then we let $\pi^* = \pi_2^*$.

If A is present in the end-sequents of both π_1^* and π_2^* we take π^* to be:

$$
\frac{
\begin{array}{cc}
\vdots\ \pi_1^* & \vdots\ \pi_2^* \\
\Gamma^* \Rightarrow \Theta^*, A \quad & A, \Delta^* \Rightarrow \Lambda^*
\end{array}
}{\Gamma^*, \Delta^* \Rightarrow \Theta^*, \Lambda^*}\ \text{CUT}
$$

Note that the removal of weakenings may remove parts of the proof and so change the end-part. However, since we halt the removal of weakenings at operational (i.e., boundary) inferences, any boundary inference in the new proof was also a boundary inference in the old proof. No inferences that were above the boundary in the old proof are part of the end-part of the new proof. □

Here is an example to illustrate the removal of weakenings from the end-part.

$$
\pi_4\left\{
\frac{
\left.\frac{\dfrac{C \Rightarrow C}{\dfrac{B, C \Rightarrow C}{C \Rightarrow C, \neg B}\ \neg \text{R}}\ \text{WL}}\right\}\pi_1\ \ \left.\frac{\dfrac{B \Rightarrow B}{\neg B, B \Rightarrow}\ \neg \text{L}}{\neg B, B \Rightarrow D}\ \text{WR}\right\}\pi_2\ \ \Bigg\}\pi_3
}{
\dfrac{\dfrac{C, B \Rightarrow C, D}{C, B \Rightarrow C, D, A \wedge B}\ \text{WR}\ \ \ \ \left.\dfrac{B \Rightarrow B}{A \wedge B \Rightarrow B}\ \wedge\text{L}\right\}\pi_5}{C, B \Rightarrow C, D, B}\ \text{CUT}
}\ \text{CUT}
\right.
$$

We begin at the boundary inferences and work our way down, removing weakenings. The sub-proofs π_1 and π_5 contain no inferences in their end-parts, in particular, no weakenings. (Although π_1 contains a weakening, that weakening does not occur in the end-part.) The sub-proof π_2 ending in $\neg B, B \Rightarrow D$ contains a single inference in its end-part, WR. If we remove it, we obtain a proof π_2^* of the sub-sequent $\neg B, B \Rightarrow$:

$$
\frac{B \Rightarrow B}{\neg B, B \Rightarrow}\ \neg \text{L}
$$

Now consider the proof π_3, ending in the CUT on the left. By the inductive hypothesis, we have proofs of sub-sequents of the premises of these CUTs in which the end-parts contain no weakenings. These are π_1 and π_2^*. Since the end-sequents of both proofs still contain the cut-formula $\neg B$, we combine them using a CUT to obtain π_3^*:

$$
\left.\frac{\left.\dfrac{\dfrac{C \Rightarrow C}{B, C \Rightarrow C}\ \text{WL}}{C \Rightarrow C, \neg B}\ \neg \text{R}\right\}\pi_1\ \ \left.\dfrac{B \Rightarrow B}{\neg B, B \Rightarrow}\ \neg\text{L}\right\}\pi_2^*}{C, B \Rightarrow C}\ \text{CUT}\right\}\pi_3^*
$$

The proof π_4 ends in a weakening. The proof of a sub-sequent of its premise $C, B \Rightarrow C, D$ we have just obtained is π_3^*. We leave out the weakening, and take $\pi_4^* = \pi_3^*$ to be the proof of a sub-sequent of the end-sequent of π_4 without weakenings in its end-part.

Finally, we consider the last CUT. The inductive hypothesis delivered proofs of sub-sequents of the premises: $\pi_4^* = \pi_3^*$ of $C, B \Rightarrow C$ and π_5 of $A \wedge B \Rightarrow B$. Here, the new left premise no longer contains the cut-formula $A \wedge B$ and so is a sub-sequent of the end-sequent. Thus, $\pi_4^* = \pi_3^*$ already is the desired proof, and $\Gamma^* \Rightarrow \Theta^*$ is $C, B \Rightarrow C$.

Problem 7.35. Apply the procedure for removing weakenings to the result of Problem 7.33.

7.9   Existence of suitable CUTS

The final step 3 in the reduction procedure applies to "suitable" CUTS in the end-part of a proof. A CUT is suitable if it is complex (i.e., the cut-formula is not atomic) and both occurrences of the cut-formula in the premises are descendants of operational inferences at the boundary. For instance, if the CUT is on $\neg A$, the proof might look like this:

$$
\cfrac{
\cfrac{A, \Gamma_1 \Rightarrow \Theta_1}{\Gamma_1 \Rightarrow \Theta_1, \neg A}\ \neg R
\qquad
\cfrac{\Delta_1 \Rightarrow \Pi_1, A}{\neg A, \Delta_1 \Rightarrow \Pi_1}\ \neg L
}{
\cfrac{\Gamma \Rightarrow \Theta, \neg A \qquad \neg A, \Delta \Rightarrow \Pi}{\Gamma, \Delta \Rightarrow \Theta, \Pi}\ \text{CUT}
}
$$

$$\Pi \Rightarrow \Xi$$

where the dashed inferences are operational inferences at the boundary with the occurrences of $\neg A$ in their conclusions principal.

In other words, these CUTS remove complex formulas from the proof which were (a) introduced on both the left and the right side by an operational inference, and where (b) between these operational inferences and the CUT only structural inferences (including possibly other CUTS) occur. The reduction then replaces the CUT with new CUTS in such a way as to keep the reduction procedure going. So it is essential that we verify that a suitable CUT always exists.

Not every complex CUT in the end-part of the proof necessarily is a suitable CUT. How and why is this so? Recall that a suitable CUT is one in which the occurrences of the cut-formula (both in the left and in the right premise of the CUT) are descendants of principal formulas of operational inferences at the boundary.

Here is a case where the lowermost complex CUT in the end-part is not suitable.

$$
\cfrac{
  \cfrac{A \Rightarrow A}{A, B \Rightarrow A} \text{WL, IL}
  \qquad
  \cfrac{B \Rightarrow B}{A, B \Rightarrow A} \text{WL}
}{A, B \Rightarrow A \wedge B} \text{$\wedge$R}
\qquad
\cfrac{
  \cfrac{
    \cfrac{A \Rightarrow \boxed{A}}{\boxed{\neg A}, A \Rightarrow} \text{$\neg$L}
  }{A, \boxed{\neg A} \Rightarrow A} \text{WR, IL}
}{A \wedge B, \boxed{\neg A} \Rightarrow A} \text{$\wedge$L}
$$

$$
\cfrac{A \Rightarrow A}{\Rightarrow A, \neg A} \text{$\neg$R}
\qquad
\cfrac{A \wedge B, \boxed{\neg A} \Rightarrow A}{
  \cfrac{A, B, \boxed{\neg A} \Rightarrow A}{\boxed{\neg A}, A, B \Rightarrow A} \text{IL}
} \text{CUT}
$$

$$
\cfrac{A, B \Rightarrow A, A}{A, B \Rightarrow A} \text{CR}
$$

On the right hand side $\neg A$ is not a descendant of a principal formula of an implicit operational inference on the boundary—it is a descendant only of the principal formula of the $\neg$L inference, which is implicit, but not at the boundary.

Incidentally, notice the following small variation of the proof showing also that a cut-formula in the end-part can be the descendant of a weakening, even when all the weakenings have been eliminated form the end-part. As you can see, $\neg A$ on the right branch now originates from a weakening above the boundary.

$$
\cfrac{
  \cfrac{A \Rightarrow A}{A, B \Rightarrow A} \text{WL, IL}
  \qquad
  \cfrac{B \Rightarrow B}{A, B \Rightarrow A} \text{WL}
}{A, B \Rightarrow A \wedge B} \text{$\wedge$R}
\qquad
\cfrac{
  \cfrac{
    \cfrac{A \Rightarrow A}{\boxed{\neg A}, A \Rightarrow A} \text{WL}
  }{A, \boxed{\neg A} \Rightarrow A} \text{IL}
}{A \wedge B, \boxed{\neg A} \Rightarrow A} \text{$\wedge$L}
$$

$$
\cfrac{A \Rightarrow A}{\Rightarrow A, \neg A} \text{$\neg$R}
\qquad
\cfrac{A \wedge B, \boxed{\neg A} \Rightarrow A}{
  \cfrac{A, B, \boxed{\neg A} \Rightarrow A}{\boxed{\neg A}, A, B \Rightarrow A} \text{IL}
} \text{CUT}
$$

$$
\cfrac{A, B \Rightarrow A, A}{A, B \Rightarrow A} \text{CR}
$$

In both proofs, the CUT on $A \wedge B$ is a suitable CUT. This illustrates a general phenomenon, namely that if there is a complex CUT in the end-part then there is at least one suitable CUT. (The suitable CUT need not be the lowermost CUT, although it could be.) This is established in the following proposition.

Proposition 7.36. *Suppose π is a proof in which the end-part contains no induction inferences, no non-atomic logical initial sequents, and no weakenings. Then π is either its own end-part, or the end-part contains a complex CUT in which both cut-formulas are descendants of principal formulas of boundary inferences.*

Proof. Note that we do not require here that the end-sequent of π is atomic. Consider again Definition 7.26 of the end-part of π: if the conclusion of an inference is in the end-part, and that inference is not an *implicit* operational inference, its premises also belong to the end part. In a proof of an atomic sequent, every operational inference is implicit: its principal formula is non-atomic, and so must belong to a bundle ending in a cut-formula (an implicit bundle). If the end-sequent is not atomic, the end-part may contain explicit operational inferences. However, if every operational inference in π is explicit, then π is its own end-part. So, if π is not its own end-part, it must contain an implicit operational inference, and hence also contain a complex CUT (the cut-formula of which is the last formula in the implicit bundle that starts with the principal formula of the implicit operational inference).

We can now prove the claim by induction on the number of complex CUTs in the end-part of π. If π is not its own end-part, there must be a complex CUT in the end-part of π (as discussed above).

Consider a lowermost complex CUT in the end-part of π:

$$\frac{\Gamma \Rightarrow \Theta, A \quad A, \Delta \Rightarrow \Lambda}{\Gamma, \Delta \Rightarrow \Theta, \Lambda} \text{ CUT}$$

$$\vdots \pi_1 \qquad \vdots \pi_2$$

$$\Pi \Rightarrow \Xi$$

Suppose this is not a suitable CUT, i.e., one of the two occurrences of the cut-formula A is not the descendant of the principal formula of a boundary inference. Let's assume that the occurrence of A in the left premise $\Gamma \Rightarrow \Theta, A$ is not a descendant of principal formula of a boundary inference. (The case where the occurrence is in the right premise can be dealt with the same way.)

First of all, there must be a boundary inference of π in π_1. For if not, the entirety of π_1 is part of the end-part of π. That means that all operational inferences in π_1 are explicit, and so their principal formulas cannot be ancestors of the cut-formula A. Since the end-part of π contains no weakenings, the occurrence of A in the left premise of the CUT must be a descendant of an occurrence of the same formula in an initial sequent. But π also contains no non-atomic logical initial sequents and all non-logical initial sequents are atomic, so A must be atomic. However, the CUT we're considering is complex, i.e., A is not atomic.

Now consider the proof π_1 of the left premise $\Gamma \Rightarrow \Theta, A$. The end-part of π_1 considered on its own is the same as that part of π_1 which is in the end-part of π. On the one hand, any operational inference which counts as implicit in π_1 clearly also counts as implicit in π. That is to say, if the principal formula of an operational inference has as a descendant a cut-formula of a CUT in π_1, that CUT is also a CUT in π and so the inference is also implicit in π.

On the other hand, there can be no operational inferences which count as implicit in π but not in π_1. In order to do so, their principal formula would have to have a descendant which is the cut-formula in a CUT inference in π which does not belong to π_1. But we assumed that the CUT we're considering is a lowermost complex CUT in π, i.e., there are no complex CUTs below it. Its own cut-formula A, furthermore, is not the descendant of a boundary inference by assumption.

Together we now have that π_1 is a proof which is not its own end-part (since it contains a boundary inference, as shown above). Since its end-part is contained in the end-part of π, its end-part contains no logical initial sequents and no weakenings. So its end-part contains fewer complex CUTs than the end-part of π. By inductive hypothesis, its end-part must contain a suitable CUT. But that CUT is also a suitable CUT in the end-part of π. (Note that the end-sequent of π_1 need not be atomic. That is why we must formulate the statement of the result for proofs of non-atomic end-sequents, otherwise we would not be able to apply the inductive hypothesis to π_1.) □

7.10 A simple example

Let's give an example of the kind of proof to which the consistency procedure applies. The end-sequent will be similar to the simple proof we already saw: $\Rightarrow 0 + 0''' = 0''' + 0$. But the proof uses a CUT to derive it from the general law $\Rightarrow \forall x\, 0 + x = x + 0$. To derive this general law, we have to use induction. First, we prove the inductive step, which also serves as the premise of the CJ inference: $0 + a = a + 0 \Rightarrow 0 + a' = a' + 0$, i.e., our formula $F(a)$ is $0 + a = a + 0$.

$$
\frac{
\dfrac{
\dfrac{0 + a = a + 0 \Rightarrow}{
\dfrac{\Rightarrow 0 + a' = (0 + a)' \qquad (0 + a)' = (a + 0)'}{0 + a = a + 0 \Rightarrow 0 + a' = (a + 0)'}\ \text{TR}
}
\qquad
\dfrac{\dfrac{\Rightarrow a + 0 = a}{\Rightarrow (a + 0)' = a'}\ \text{FNC}}{}
}{0 + a = a + 0 \Rightarrow 0 + a' = a'}\ \text{TR}
\qquad
\dfrac{\dfrac{\Rightarrow a' + 0 = a'}{\Rightarrow a' = a' + 0}\ \text{SYM}}{}\ \text{TR}
}{0 + a = a + 0 \Rightarrow 0 + a' = a' + 0}
$$

Let's call this proof $\pi_1(a)$. Note that it uses no logical axioms and only atomic CUTs. The full proof now is:

$$
\vdots\, \pi_1(a)
$$

$$
\frac{
\dfrac{
\dfrac{\dfrac{0 + a = a + 0 \Rightarrow}{0 + a' = a' + 0}}{\dfrac{0 + 0 = 0 + 0 \Rightarrow}{0 + b = b + 0}}\ \text{CJ}
\qquad \Rightarrow 0 + 0 = 0 + 0}{
\dfrac{\Rightarrow 0 + b = b + 0}{\Rightarrow \forall x\, 0 + x = x + 0}\ \text{VR}
}\ \text{CUT}
\qquad
\dfrac{
\dfrac{0 + 0''' = 0''' + 0 \Rightarrow}{0 + 0''' = 0''' + 0}
}{
\dfrac{\forall x\, 0 + x = x + 0 \Rightarrow}{0 + 0''' = 0''' + 0}
}\ \text{VL}
}{\Rightarrow 0 + 0''' = 0''' + 0}\ \text{CUT}
$$

The dashed inferences are boundary inferences. So the end-part consists only of the lowermost three sequents, i.e., the conclusions of the VL and VR rules as well as the end-sequent. In order to transform this proof into a simple one, we must get rid of the CJ inference and the last CUT, as all other CUTs are atomic.

Note that the CJ inference is not in the end-part, so we skip step 1. It also contains no weakenings, so we can skip step 2. Since the end-part of the proof contains no CJ inference, weakenings, and no non-atomic initial sequents, it must contain a suitable CUT, as we proved in Proposition 7.36. How do we remove such a suitable CUT? The idea is this: if there is a complex CUT in the end-part, its cut-formula must be complex, i.e., not atomic. And the two occurrences of this complex cut-formula must have ancestors higher up in the proof. If these ancestors arise from a weakening, we show that these weakenings can be removed, and the part of the proof below them simplified and the CUT removed. If an ancestor of the cut-formula is a formula in a logical axiom, we derive the axiom from atomic axioms first. Thus, we can assume that the formula is introduced by a right operational rule in the proof leading to the left premise of the CUT, and by a left operational rule in the proof leading to the right premise of the CUT. In the simplest case, the premises of the CUT are introduced directly above the CUT, as in the previous example: $\forall x\, 0 + x = x + 0$ is the formula introduced by a $\forall$R above the left premise, and by a $\forall$L above the right premise. However, in more complicated cases, the cut-formula may be the descendant of multiple logical inferences on the left or right (which were contracted into one cut-formula on the left and the right).

When defining the reduction, we will not assume that there is only one operational inference in the left sub-proof that yields the cut-formula in the left premise, and one such inference that yields the right cut-formula. In order to take this into account, the reduction will be structured in such a way that a CUT on the original cut-formula will be retained and thus will appear in the new proof. At the same time, there will also be a CUT of lower complexity on the premises of the operational inferences. After the reduction, the CUT on the original formula will depend on one fewer pair of operational inferences.

Let's consider our example again, but abbreviate $0 + a = a + 0$ by $F(a)$. Then the proof is:

$$
\pi_2(b)\left\{
\begin{array}{c}
\begin{array}{cc}
 & \vdots\;\pi_1(a) \\
 & \vdots \\
\Rightarrow F(0) & \dfrac{F(a) \Rightarrow F(a')}{F(0) \Rightarrow F(b)}\;\text{CJ} \\
\end{array}
\end{array}
\right.
$$

$$
\dfrac{\dfrac{\dfrac{\Rightarrow F(0) \quad F(0) \Rightarrow F(b)}{\Rightarrow F(b)}\;\text{CUT}}{\Rightarrow \forall x\, F(x)}\;\forall\text{R} \qquad \dfrac{F(0''') \Rightarrow F(0''')}{\forall x\, F(x) \Rightarrow F(0''')}\;\forall\text{L}}{\Rightarrow F(0''')}\;\text{CUT}
$$

Let the sub-proof ending in $\Rightarrow F(b)$ be called $\pi_2(b)$. If we replace b in it everywhere by some term t, the resulting proof $\pi_2(t)$ has end-sequent $\Rightarrow F(t)$.

We now construct two proofs, one in which the $\forall$R inference is removed, and one in which the $\forall$L inference is removed. Note that the term that is universally quantified in the $\forall$L inference is $0'''$, which we'll label t to make comparison with the general case later easier. The first proof is:

$$
\frac{\dfrac{\vdots\ \pi_2(t)}{\Rightarrow F(t)}}{\Rightarrow F(t), \forall x\, F(x)}\ \text{WR} \qquad \frac{F(t) \Rightarrow F(0''')}{\forall x\, F(x) \Rightarrow F(0''')}\ \forall\text{L}
$$
$$
\frac{\Rightarrow F(t), F(0''')}{\Rightarrow F(0'''), F(t)}\ \text{IR}\ \ \text{CUT}
$$

The second proof:

$$
\frac{\dfrac{\vdots\ \pi_2(b)}{\Rightarrow F(b)}}{\Rightarrow \forall x\, F(x)}\ \forall\text{R} \qquad \frac{F(t) \Rightarrow F(0''')}{\forall x\, F(x), F(t) \Rightarrow F(0''')}\ \text{WL}
$$
$$
\frac{}{F(t) \Rightarrow F(0''')}\ \text{CUT}
$$

We can now combine these using a CUT to obtain a proof of $\Rightarrow F(0''')$.

$$
\frac{\dfrac{\Rightarrow F(t)}{F(t),\ \Rightarrow \forall x\, F(x)}\ \text{WR}\quad \dfrac{F(t) \Rightarrow F(0''')}{\forall x\, F(x) \Rightarrow F(0''')}\ \forall\text{L}}{\dfrac{\Rightarrow F(t), F(0''')}{\Rightarrow F(0'''), F(t)}\ \text{IR}}\ \text{CUT} \qquad \frac{\dfrac{\Rightarrow F(b)}{\Rightarrow \forall x\, F(x)}\ \forall\text{R}\quad \dfrac{F(t) \Rightarrow F(0''')}{\forall x\, F(x),\ F(t) \Rightarrow F(0''')}\ \text{WL}}{F(t) \Rightarrow F(0''')}\ \text{CUT}
$$
$$
\frac{\Rightarrow F(0'''), F(0''')}{\Rightarrow F(0''')}\ \text{CR}
$$

Notice that we introduce $\forall x\, F(x)$ by weakening in both proofs. This is only to make clear that the introduced formula will remain in the proof until the CUT. If there were additional inferences between the $\forall$-inference and the CUT in the original proof, or now between the weakening and the CUT, it might happen that in the intervening steps the same formula $\forall x\, F(x)$ would be introduced using another operational inference, and then contracted.

This does not happen in this case, so we can apply step 2: removal of weakenings from the end part.[12] In this step, we simply remove all weakening inferences and the formulas they introduce. This results in simplification of the proof as

---

[12] Notice that as we transform the proof the end-part of the new proof is different from the one we started with and, as we remarked before, new weakenings and inductions might appear.

inferences become redundant. In this case, for instance, if the weakened $\forall x\, F(x)$ is removed, the two CUTs on it become redundant, as do the sub-proofs ending in the premises of the two CUTs which don't involve the weakened $\forall x\, F(x)$. We are left with the following proof:

$$\frac{\begin{array}{c} \vdots\ \pi_2(t) \\ \vdots \\ \Rightarrow F(t) \end{array} \qquad F(t) \Rightarrow F(0''')}{\Rightarrow F(0''')}\ \text{CUT}$$

Now note that the right premise of the CUT is in fact a logical axiom: t is $0'''$, so that premise is actually $F(0''') \Rightarrow F(0''')$. Furthermore, a CUT where one premise is a logical axiom always has the property that the conclusion is the same as the premise that is not the axiom: in general, it looks like this:

$$\frac{\Gamma \Rightarrow \Theta, A \qquad A \Rightarrow A}{\Gamma \Rightarrow \Theta, A}\ \text{CUT}$$

So such CUT inferences can be simply removed. In fact it is clear that *any* logical axiom in the end-part of the proof from which all weakenings have been removed is the premise of a CUT: it can't be the premise of a CJ inference, since that must contain $F(a)$ on the left and $F(a')$ on the right. It can't be the premise of a contraction or an interchange, since those both require at least two formulas on one side of the sequent. And since weakenings have been removed, the end-part otherwise only contains CUTs.[13]

If we remove the final CUT with the logical axiom, we are left with the following proof:

$$\frac{\Rightarrow F(0) \qquad \dfrac{\begin{array}{c}\vdots\ \pi_1(a) \\ \vdots \\ F(a) \Rightarrow F(a')\end{array}}{F(0) \Rightarrow F(t)}\ \text{CJ}}{\Rightarrow F(t)}\ \text{CUT}$$

What has happened is that by removing the "boundary inferences" $\forall$L and $\forall$R, the parts of the proof previously above them now no longer lie above the boundary but in the end-part. Recall that the sub-proof π_1 contains only mathematical axioms and atomic CUTs and no logical inferences. That means that the new proof is its own end-part. Since there are no weakenings, step 2 does not apply. Hence, we apply step 1, removal of the CJ inference. Note that the conclusion a lowermost CJ inference in the end-part never contains a free variable. Indeed, in our proof the term t in the induction formula $F(t)$ on the right hand side of the conclusion of the

CJ inference now is a variable-free term; in fact it is a numeral $\overline{n}$, viz., $0'''$. Any CJ inference with this property can be replaced by a sequence of CUTs on premises resulting from substituting, successively, the numerals $\overline{k}$ (where $k = 0, \ldots, n-1$) for the eigenvariable. In our case, $n = 3$, so we consider the proof:

$$
\cfrac{\Rightarrow F(0) \qquad \cfrac{\cfrac{\vdots\,\pi_1(\overline{0}) \qquad\qquad \vdots\,\pi_1(\overline{1})}{\cfrac{F(0) \Rightarrow F(0') \quad F(0') \Rightarrow F(0'')}{F(0) \Rightarrow F(0'')}\;\text{CUT} \qquad \cfrac{\vdots\,\pi_1(\overline{2})}{F(0'') \Rightarrow F(0''')}}{\cfrac{F(0) \Rightarrow F(0''')}{}\;\text{CUT}}\;\text{CUT}}{\Rightarrow F(t) \quad [\text{i.e., } F(\overline{3})]}
$$

We are now done: the proof is simple, since $\pi_1(\overline{n})$ only contains mathematical axioms, atomic CUTs (recall that $F(a)$ is the atomic formula $0 + a = a + 0$), and the remaining inferences are also all atomic CUTs.

In a more general situation, the CUTs with which we replace CJ may be complex, viz., if the induction formula $F(a)$ is not atomic. In such a situation we would have to continue by removing those complex CUTs. Another situation that may occur is when the variable free term t in the conclusion of CJ is not a numeral: then we have to add a final CUT of the form $F(\overline{n}) \Rightarrow F(t)$. Since there is always a simple proof of $\Rightarrow t = \overline{n}$ where $n = \text{val}(t)$ for a variable-free term t (if we add equality axioms to the axioms of **PA**), such a sequent is always provable (without complex CUTs or CJs).

7.11 Summary

Let us recap what we have done. In the sequent calculus, we can accommodate theories such as Peano arithmetic by adding to the logical initial sequents $A \Rightarrow A$ also non-logical initial sequents. These play the role of the axioms of $\mathbf{PA}_K$ in the axiomatic development of arithmetic. All the non-logical initial sequents we consider are actually atomic, since the syntax of sequents gives us a bit more flexibility. For instance, we can replace the axiom $(a = b \wedge b = c) \supset a = c$ by the instances of $a = b, b = c \Rightarrow a = c$. The only axiom of $\mathbf{PA}_K$ that cannot be translated into an atomic initial sequent is the induction axiom scheme, which we replace by the induction rule,

$$
\cfrac{F(a), \Gamma \Rightarrow \Theta, F(a')}{F(0), \Gamma \Rightarrow \Theta, F(t)}\;\text{CJ}
$$

Any proof in $\mathbf{PA}_K$ can be transformed into a proof in the sequent calculus version of **PA**, and vice versa.

To prove that **PA** is consistent is to show that no proof of the empty sequent can exist in **PA**, i.e., in **LK** plus the induction rule CJ, from logical and mathematical initial sequents. Our strategy for proving this is to show that any proof in **PA** of an atomic, variable-free sequent can be transformed into a simple proof of the same sequent (or a sub-sequent thereof). (A simple proof is one in which every sequent is atomic and which does not use CJ.) The empty sequent is atomic and variable-free, and we showed (in Section 7.2) that simple proofs cannot be proofs of the empty sequent. So if any proof whatsoever of the empty sequent (if such a proof existed) could be transformed into a simple proof, but no simple proof of the empty sequent is possible, then no proof of the empty sequent is possible, simple or otherwise.

The transformation of proofs of atomic sequents into simple proofs is complicated. It involves repeating the following three steps. In step 1 we remove induction inferences from the end-part. In step 2, we remove weakenings from the end-part. In step 3, we reduce suitable CUTS in the end-part. The "end-part" of a proof is the part of the proof that lies between the end-sequent and any operational inferences the principal formulas of which "disappear" below it using CUTS. Such operational inferences are called implicit, and the lowermost ones are boundary inferences; the end-part of a proof consists of the sequents between a boundary inference and the end-sequent.

How are CJ-inferences removed in step 1? Consider a lowermost CJ inference in the end-part. Since the end-sequent contains no free variables, we can guarantee that in the conclusion of that inference $F(0), \Gamma \Rightarrow \Theta, F(t)$, the term t also contains no variable, and so is a numeral $\overline{n}$.[14] The premise of the CJ inference is $F(a), \Gamma \Rightarrow \Theta, F(a')$. By replacing the eigenvariable a in the proof of this premise by $\overline{0}, \overline{1}, \ldots, \overline{n-1}$, we obtain proofs of $F(\overline{k}), \Gamma \Rightarrow \Theta, F(\overline{k+1})$ for all $k < n$. We can replace the proof ending in the CJ inference by a larger proof that replaces the CJ inference by a series of CUT inferences. Although this may multiply the overall number of CJ inferences (and blows up the size of the proof), the number of CJ inferences between the boundary and the end-sequent is reduced. So we can see by induction on the maximum number of CJ inferences along a thread between the end-sequent and the boundary of π that repeating this procedure eventually produces a proof in which the end-part contains no CJ inferences (Proposition 7.28).

Step 2 is merely a preparatory step for step 3: it guarantees that the end-part of the proof contains no weakenings (Section 7.8). It does not increase the size of the proof. Once all weakenings are removed from the end-part, the end-part is guaranteed to contain a "suitable CUT," i.e., a CUT inference with a non-atomic cut-formula, such that the two occurrences of the cut-formula are descendants of principal formulas of operational inferences on the boundary of the end-part (Section 7.9). Here is a very simplified, schematic example where the cut-formula

---

[14] The term t is a numeral $\overline{n}$ if $+, \cdot$ do not occur; otherwise we can replace it with $\overline{\mathrm{val}(t)}$ and add a CUT with the sequent $F(\overline{\mathrm{val}(t)}) \Rightarrow F(t)$, which has a simple proof.

is $\neg A$, and we leave out all side formulas:

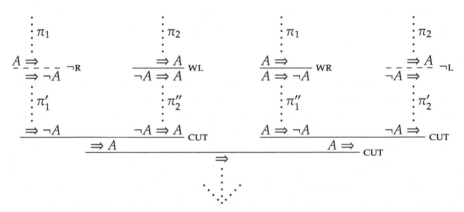

We replace the proof by

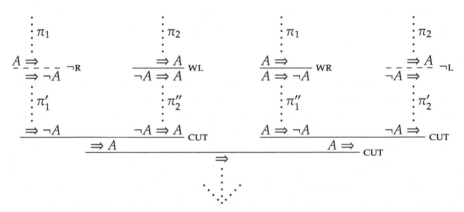

What has happened? We have turned one CUT into three, and only one of the three CUTS in the new proof is of lower degree. The size of the new proof is about double the size of the old proof. It's hard to see how this gets us closer to a simple proof!

But notice two things. First, in the new proof, the copies of π_1 and π_2 in the middle now end in weakening inferences instead of operational inferences. Thus, more of these sub-proofs belong to the end-part of the new proof. We have "raised the boundary," and brought more of the proof into the scope of the reduction steps 1–3. The end-part of the new proof may now contain CJ inferences (inside these two copies of π_1 and π_2) which step 1 can remove and replace by CUT inferences. Eventually, the procedure raises the boundary on all threads until all boundary inferences are removed.

The second thing to notice is that where we raised the boundary, we replaced operational inferences by weakenings. These weakenings now occur in the end-part of the new proof (and also in the end-part of the proof that has replaced any CJ-inferences in the copies of π_1 and π_2 by CUTS). In step 2, we remove those weakening inferences. What does that do to the part of the proof below these weakenings? For instance, if we remove the WL of $\neg A$ at the top of π_2'', the last sequent of π_2'' will also no longer contain $\neg A$ on the left. But if it no longer contains $\neg A$ on the left, the cut which removes it is superfluous! We have a proof of $\Rightarrow A$ resulting from just π_2 and π_2'' with $\neg A$ removed from the antecedents of relevant sequents. The same goes for the wr of $\neg A$ following the copy of π_1 on the right. So in effect, after we raise the boundary, remove cj inferences that now fall into the end-part, and remove weakenings from the end-part, we have replaced the one cut on $\neg A$ with a cut on just A, and so in effect have reduced the degree of the cut.

This summary of course waves a lot of hands. In both steps 1 and 3, we replace a proof by one that can obviously be very much larger, and in step 3 we replace one cut by three, only one of which is of lower degree. That the two cuts of same degree eventually are removed in step 2 is far from obvious. Nor is it clear that "raising the boundary" in step 3 will eventually raise it all the way to the initial sequents. (Once that happens, we are dealing with a proof that is its own end-part, from which we can remove all cj inferences, and so finally are left with a simple proof.) Note that we have not mentioned which suitable cut we should apply step 3 to. One might think that since we are in effect reducing the degree of the cut we should always pick one of maximal degree—but in fact we must pick a topmost suitable cut in the end-part.

In order to prove that applying steps 1–3 repeatedly and in that order eventually results in a simple proof, we have to find a way of measuring the complexity of proofs in **PA** in such a way that the complexity of proofs produced by successive steps decreases. We can't do this with a single or even a pair of numerical measures the way we did in the case of the cut-elimination theorem. More complex measures are needed, and we must show that these measures share with numbers (and pairs of numbers) a crucial property: we can't keep going to smaller and smaller ones without eventually coming to a stop. The measures Gentzen developed for this purpose are so-called *ordinal notations*.

8

Ordinal notations and induction

8.1   Orders, well-orders, and induction

Recall the principle of strong induction from Definition 2.13. To prove that $P(n)$ holds for all n, we show that,

(a) $P(0)$ holds and

(b) for arbitrary $n > 0$, if $P(m)$ holds for all $m < n$, then it holds for $P(n)$.

One way to see that this works is by reflecting on what would happen if $P(n)$ were false for some n. Suppose that did happen: (a) and (b) hold for the property P, but there is an m_1 such that $P(m_1)$ does not hold. Now the contrapositive of (b) is this: for any $n > 0$, if $P(n)$ fails, then there is an $m < n$ such that $P(m)$ fails. Since $P(m_1)$ is false, then by the contrapositive of (b), there is an $m_2 < m_1$ such that $P(m_2)$ is false. Of course, (b) also applies to m_2: There is a further $m_3 < m_2$ such that $P(m_3)$ is false, and so on. We get a sequence $m_1 > m_2 > m_3 > \dots$ where each $P(m_i)$ is false. Since these m_i are natural numbers, the sequence must eventually terminate, i.e., for some k, $m_k = 0$ and $P(m_k)$ is false. So we have both that $P(0)$ is true (by (a)) and that $P(0)$ is false (since $P(m_k)$ is false and $m_k = 0$). That is a contradiction. So there cannot be an m_1 where $P(m_1)$ is false.

What's important here is that this process of moving to smaller and smaller numbers eventually comes to an end—that is what guarantees that induction for natural numbers works as intended. This suggests that a similar proof method should work not just for numbers, but for any kind of objects, as long as these objects are ordered in such a way that any process of moving to smaller and smaller objects (in that ordering) eventually comes to an end. In other words, induction is possible along any ordering as long as that ordering has no infinitely decreasing sequences. Orderings with this property are called *well-orderings*.

Recall also the principle of successor induction from Definition 2.12: if it is both true that $P(0)$, and also that if $P(n)$ is true then so is $P(n+1)$, then $P(n)$ is true for all n. As a moment of reflection shows, the second part of the antecedent is equivalent to "$P(n)$ holds if $P(n-1)$ holds (for $n \geq 1$)." So this is a special case of

An Introduction to Proof Theory: Normalization, Cut-Elimination, and Consistency Proofs.
Paolo Mancosu, Sergio Galvan, and Richard Zach, Oxford University Press. © Paolo Mancosu,
Sergio Galvan and Richard Zach 2021. DOI: 10.1093/oso/9780192895936.003.0008

the application of the principle of strong induction, where the proof always uses a particular $m < n$, namely $m = n - 1$. Strong induction is equivalent to successor induction *on the natural numbers*. However, as a general principle, successor induction is weaker: it only works for the natural numbers because every n can be reached from 0 by repeatedly adding 1. There are other well-orders—and we'll meet some below—for which proof by strong induction is a valid principle, but where not every member of the order can be reached "from below" by a successor operation. So from here on in, we'll focus on strong induction, but consider sets other than the natural numbers, which are nevertheless ordered in such a way that every decreasing sequence always comes to an end.

We begin by reviewing some definitions.

Definition 8.1. If X is a set, a *relation* R on X is a set of pairs from X, i.e., $R \subseteq X \times X = \{\langle x, y \rangle : x, y \in X\}$. We write xRy for $\langle x, y \rangle \in R$.

Definition 8.2. A relation R on X is called a *strict linear order* (on X) if, and only if,

1. R is transitive: for all $x, y, z \in X$, if xRy and yRz then xRz.

2. R is asymmetric: for all x and $y \in X$, if xRy then not yRx.

3. R is total: for any x and $y \in X$, if $x \neq y$ then either xRy or yRx.

By convention, we use variations of the symbol $<$ for strict linear orders, such as $\prec$. We write $x \leq y$ for "$x < y$ or $x = y$," (and similarly for $\preceq$). The relation $\leq$ is called the *partial linear order* associated with $<$. We write $y > x$ for $x < y$ and $y \geq x$ for $x \leq y$. The set X together with a strict linear order $<$ is called an *ordered set* and written $\langle X, < \rangle$. We'll also use "ordering" as short for "strict linear order."

Orderings and ordered sets are familiar from mathematics, especially the so-called natural ordering "less than" on sets of numbers. It will be important for us to consider (a) ordered sets other than sets of numbers and (b) orderings other than "less than," so it'll be important to make clear which sets we have in mind and which orderings should go with them.

Among the linear orders there are some that are closely related to induction, as we will see. These are the so-called well-orderings:

Definition 8.3. A *well-ordering* $<$ of X is a relation which is a strict linear ordering of X and which has the property that every non-empty subset $Y \subseteq X$ has a least element under $<$.

A least element of Y is some $y \in Y$ such that $y \leq z$ for every $z \in Y$. The least element of a linearly ordered set is unique. For suppose y_1 and y_2 both satisfy $y_1 \leq z$ and $y_2 \leq z$ for *all* $z \in Y$. Then in particular $y_1 \leq y_2$ and $y_2 \leq y_1$, i.e., either $y_1 = y_2$ or both $y_1 < y_2$ and $y_2 < y_1$. The latter case is ruled out because $<$ is asymmetric. So, we have $y_1 = y_2$.

Here's an example of how to prove something about well-orders from this definition:

Proposition 8.4. *Suppose* $\langle X, < \rangle$ *is a well-ordered set. For every* $x \in X$*, if there is a* $y > x$ *in X at all, then there is a unique* $x' > x$ *such that there is no z with* $x < z < x'$ *(the* successor *of x in* $\langle X, < \rangle$*).*

Proof. Let $Y = \{y \in X : y > x\}$, i.e., the set of all elements y of X such that $x < y$. If $<$ is a well-ordering of X, then by definition Y has a least element x'. We show that this x' is the unique successor of x.

Because $x' \in Y$, $x < x'$. Because x' is least in Y, there can't be any $z \in X$ with $x < z < x'$. For suppose there were a z such that both $x < z$ and $z < x'$. Then because $x < z$, we have $z \in Y$. Since $z \in Y$, we have $x' \leq z$, since x' is the least element of Y, i.e., $x' = z$ or $x' < z$. Since $z < x'$, this rules out $x' = z$. So we must have $x' < z$. But we also have that $z < x'$, by the assumption that $x < z < x'$. Since $<$ is asymmetric, this is a contradiction. So, there is no z with $x < z < x'$.

Lastly, x' is unique: Suppose z is some element of X with $x < z$ and no u such that $x < u < z$. We have to show that $z = x'$. Since $x < z$, $z \in Y$. Since x' was least in Y, $x' \leq z$, i.e., $x' = z$ or $x' < z$. We can't have $x' < z$, since otherwise $x < x' < z$, in which case z is not a successor of x. So $x' = z$, i.e., x' is unique.   □

Proposition 8.5. *Every well-ordering has the property that there is no infinite sequence* $x_1 > x_2 > x_3 > \ldots$.

Proof. Suppose X is well-ordered by $<$ (i.e., every non-empty subset of X has a least element) but that there is such an infinite descending sequence $x_1 > x_2 > \ldots$. Let $Y = \{x_i : i \in \mathbb{N} \text{ and } i \geq 1\}$. This is a non-empty subset of X. It cannot have a least element. For suppose $y \in Y$ is a least element, i.e., for all $z \in Y$, if $z \neq y$ then $y < z$. The elements of Y are exactly the elements of the infinite sequence of x_i, so there must be some i such that $y = x_i$. But $x_i > x_{i+1}$ by hypothesis, and $x_{i+1} \in Y$ by the definition of Y, contradicting the assumption that y is a least element of Y.□

Corollary 8.6. *If* $\langle X, < \rangle$ *is a well-ordering, and* $x_1 \geq x_2 \geq x_3 \geq \ldots$ *is an infinite sequence, then the sequence eventually stabilizes: for some k,* $x_k = x_i$ *for all* $i \geq k$.

Proof. If this weren't the case, then for every k there would be an $i_k > k$ such that $x_k \neq x_{i_k}$. Since $x_k \geq \cdots \geq x_{i_k}$ and $\geq$ is transitive, also $x_k \geq x_{i_k}$. But since $x_k \neq x_{i_k}$, it would follow that $x_k > x_{i_k}$. This would mean that there is an infinite descending subsequence of elements of X, contrary to the previous proposition.   □

Problem 8.7. Show that every total ordering $\langle X, < \rangle$ in which there is no infinite descending sequence $x_1 > x_2 > x_3 > \ldots$ is a well-ordering. (Hint: suppose there is a set $Y \subseteq X$ without a least element. Construct an infinite descending sequence $x_1 > x_2 > \ldots$ from Y.)

The importance of well-orderings lies in the following principle:

Theorem 8.8 (Induction along well-orderings). *Suppose $\langle X, < \rangle$ is a well-ordering, and P is a property that may be true of elements of X. If, for every $x \in X$, it is true that:*

$$P(x) \text{ if for all } y < x, P(y) \qquad\qquad (I(P, x))$$

then $P(x)$ is true for all $x \in X$.

Proof. Suppose $I(P, x)$ holds for all x but for some z, $P(z)$ is false. So $\{y : P(y) \text{ is false}\}$ is non-empty. Since $<$ is a well-ordering, it has a least element x_1, i.e., for all $y < x_1$, $P(y)$ is true but $P(x_1)$ is false. This contradicts $I(P, x_1)$. □

Definition 8.9. An *order isomorphism* between two ordered sets $\langle X, < \rangle$ and $\langle Y, <' \rangle$ is a mapping $f : X \to Y$ such that

1. f is total, i.e., defined for all $x \in X$.

2. f is injective, i.e., if $x \neq y$ then $f(x) \neq f(y)$.

3. f is surjective onto Y, i.e., for every $y \in Y$, there is an $x \in X$ such that $f(x) = y$.

4. f is order-preserving, i.e., $x < y$ if, and only if, $f(x) <' f(y)$.

If there is an order isomorphism between two orderings, we say the two orderings are *order isomorphic*, or that they have the same *order type*. For instance, any two finite ordered sets A and B are order isomorphic if, and only if, A and B have the same number of elements. The set $\mathbb{Z}$ of integers and the set $2\mathbb{Z}$ of (positive and negative) even numbers, each with their usual order, are order isomorphic since $f(x) = 2x$ is an order isomorphism.

Crucially, if two ordered sets are order isomorphic, then either both are well-ordered or neither is.

Proposition 8.10. *If $\langle X, < \rangle$ and $\langle Y, <' \rangle$ are order isomorphic and $\langle X, < \rangle$ is well-ordered, so is $\langle Y, <' \rangle$.*

Proof. Let f be an order isomorphism between the two ordered sets and suppose $\langle X, < \rangle$ is a well-ordering. We show that $\langle Y, <' \rangle$ is as well. Suppose it weren't. Then there is an infinite descending sequence $y_1 >' y_2 >' \dots$. Because f is onto, for each i there is an $x_i \in X$ such that $f(x_i) = y_i$. Since $y_i >' y_{i+1}$ then $f(x_i) >' f(x_{i+1})$. By clause 4, $x_i > x_{i+1}$ for all i. But then $x_1 > x_2 > \dots$ is an infinite descending sequence in X. □

8.2   Lexicographical orderings

In Chapter 3 we first proved some proof-theoretic results by double induction. We explained (in Section 4.2), as we can now see, that double induction is induction on a certain well-ordering: the ordered set is the set of pairs of natural numbers $\mathbb{N}^2 = \{\langle n, m \rangle : n, m \in \mathbb{N}\}$. The ordering $<_{\text{lex}}$ is given by

$$\langle k, \ell \rangle <_{\text{lex}} \langle n, m \rangle \text{ if, and only if, (a) } k < n \text{ or (b) } k = n \text{ and } \ell < m.$$

The inductive step in a double induction proof was: Assume $P(k, l)$ holds for all $\langle k, l \rangle <_{\text{lex}} \langle n, m \rangle$, and show that $P(n, m)$ holds. This is how we proved normalization for natural deduction **NJ** in Chapter 4, and later the cut-elimination theorem in Chapter 6. By revisiting the discussion in Section 4.2, you should be able to prove that $\mathbb{N}^2$ ordered by $<_{\text{lex}}$ is well-ordered. The ordering $<_{\text{lex}}$ is called the *lexicographical ordering* on $\mathbb{N}^2$.

Problem 8.11. Show that $\langle \mathbb{N}^2, <_{\text{lex}} \rangle$ is a well-ordering.

Problem 8.12. Show that $\{i - (1/j) : i, j \in \mathbb{N}, j > 0\}$ with the natural ordering is order isomorphic to $\mathbb{N}^2$ with the lexicographical ordering as given in Problem 8.11.

The ordering $<_{\text{lex}}$ is called the lexicographical ordering because it orders pairs of numbers just like a dictionary sorts words: we compare the first digit (letter) and sort a pair (word) before all the pairs (word) where the first digit is greater. If two pairs (words) agree on the first digit (letter), we compare the second digit (letter). Of course this can be extended to sets of sequences of a fixed length. More precisely:

Definition 8.13. The *lexicographical ordering* on $\mathbb{N}^k$ is defined by:

$\langle n_1, \ldots, n_k \rangle <_{\text{lex}} \langle m_1, \ldots, m_k \rangle$ if, and only if,
for some $j < k$, both (a) $n_i = m_i$ for all i with $0 < i \leq j$ and (b) $n_{j+1} < m_{j+1}$

For instance, $\langle 1, 2, 3 \rangle <_{\text{lex}} \langle 2, 1, 3 \rangle$ because for $j = 0$ we have (a) that for all i with $0 < i \leq j$, $n_i = m_i$ (since there are no i with $0 < i \leq 0$, this holds vacuously) and (b) $n_1 < m_1$ since $1 < 2$ (note that $j + 1 = 0 + 1 = 1$). Similarly, $\langle 1, 2, 3 \rangle <_{\text{lex}} \langle 1, 2, 5 \rangle$ since we can take $j = 2$ and have that $n_i = m_i$ for all i with $0 < i \leq j$ (as $n_1 = m_1 = 1$ and $n_2 = m_2 = 2$) and also $n_{j+1} < m_{j+1}$ (since $j + 1 = 3$ and $n_3 = 3 < 5 = m_3$). On the other hand, $\langle 1, 2, 3 \rangle \not<_{\text{lex}} \langle 1, 2, 3 \rangle$ since for all possible values of $j < 3$ for which (a) holds (namely, $j = 0$, 1, or 2), (b) does not as $n_{j+1} \not< m_{j+1}$.

The ordering $<_{\text{lex}}$ on $\mathbb{N}^k$ is a total order and indeed a well-order.

Proposition 8.14. $\langle \mathbb{N}^k, <_{\text{lex}} \rangle$ *is a well-ordering.*

Proof. Suppose $Y \subseteq \mathbb{N}^k$ is a set of k-tuples. Let n_1 be the least n such that $\langle n, m_2, \ldots, m_k \rangle \in Y$. Then there is no $\langle m_1, \ldots, m_k \rangle \in Y$ such that $m_1 < n_1$. Now let n_2 be the least n such that $\langle n_1, n, m_3, \ldots, m_k \rangle \in Y$. Again, for no $\langle m_1, \ldots, m_k \rangle \in Y$ do we have $m_1 < n_1$ or $m_1 = n_1$ and $m_2 < n_2$. Continue in this way and finally define n_k as the least n such that $\langle n_1, \ldots, n_{k-1}, n \rangle \in Y$. Then $\langle n_1, \ldots, n_k \rangle \in Y$ and is the least element in the $<_{\text{lex}}$ ordering.

Alternatively, we can show that $\prec_{\text{lex}}$ is a well-ordering by showing that there cannot be an infinite descending sequence of k-tuples. For suppose there were:

$$\langle n_1^1, \ldots, n_k^1 \rangle \succ_{\text{lex}} \langle n_1^2, \ldots, n_k^2 \rangle \succ_{\text{lex}} \langle n_1^3, \ldots, n_k^3 \rangle \succ_{\text{lex}} \cdots$$

(Here we let the superscripts index the position of the tuple in the sequence, and the subscript the element of the tuple, i.e., n_j^i is the j-th element of the i-th k-tuple. In particular, the superscripts here are *not* meant to be read as exponents. We choose the notation for readability, but it could be replaced by double subscripts, e.g., $n_{i,j}$.)

Consider the sequence $n_1^1, n_1^2, n_1^3, \ldots$ of the first elements of these k-tuples. This sequence can never increase: $n_1^i \geq n_1^{i+1}$ for all i, because of the way $\prec_{\text{lex}}$ is defined (if $n_1^i < n_1^{i+1}$ then $\langle n_1^i, \ldots, n_k^i \rangle \prec_{\text{lex}} \langle n_1^{i+1}, \ldots, n_k^{i+1} \rangle$ contrary to assumption). Hence, from some index j_1 forward, $n_1^{j_1} = n_1^i$ for all $i > j_1$: the first components of a decreasing sequence of k-tuples must eventually "stabilize." Now consider the sequence of second components from j_1 onward: $n_2^{j_1}, n_2^{j_1+1}, \ldots$. For the same reason, this sequence must eventually stabilize: for some $j_2 > j_1$, we have $n_2^{j_2} = n_2^i$ for all $i > j_2$. Continue in this way until we have found a j_{k-1} after which all $(k-1)$-st components have stabilized. Consider the k-th components in the remainder of the sequence: $n_k^{j_{k-1}}, n_k^{j_{k-1}+1}, \ldots$. This sequence eventually stabilizes as well, and we must have reached the end of the sequence.   □

Problem 8.15. Take the sequence $\langle n_1^i, n_2^i, n_3^i \rangle$ where $n_1^1 = n_2^1 = n_3^1 = p$ for some number p, defined as follows: Let j be the least number such that $n_j^i = 0$ or $j = 4$ if no element of $\langle n_1^i, n_2^i, n_3^i \rangle$ is equal to 0. Then let

$$n_\ell^{i+1} = \begin{cases} j - 1 & \text{if } \ell = j \\ n_\ell^i - 1 & \text{if } \ell = j - 1 \\ n_\ell^i & \text{if } \ell < j - 1 \text{ or or } \ell > j \end{cases}$$

The sequence ends if the next tuple $\langle n_1^{i+1}, n_2^{i+1}, n_3^{i+1} \rangle$ is the same as $\langle n_1^i, n_2^i, n_3^i \rangle$.

Write out the sequence so defined for $p = 2$. Check that it is a decreasing sequence of 3-tuples in the order $\prec_{\text{lex}}$. Then consider the proof that this sequence must be finite. What is the sequence of first elements n_1^1, n_1^2, etc., and at what index j_1 does it stabilize? What is the sequence of second elements from j_1 onward, and at what index j_2 does it stabilize? Now consider the elements of the sequence as a set $Y \subseteq \mathbb{N}^3$. Consider the preceding proof that Y must have a least element: What is n_1, i.e., the least n such that $\langle n, m_2, m_3 \rangle \in Y$? What are n_2 and n_3? Is the resulting tuple $\langle n_1, n_2, n_3 \rangle$ really the least element of Y?

We said that the lexicographical ordering is akin to the order of words in a dictionary. The words in a dictionary, of course, are not all of the same length. The same ordering principle, however, applies to words of arbitrary length. We can apply the same principle to define an ordering of sequences of arbitrary objects (which are themselves ordered). For instance, take $\mathbb{N}^*$, the set of finite sequences of natural numbers. These are:

$$\langle 0 \rangle, \langle 1 \rangle, \langle 2 \rangle, \ldots$$
$$\langle 0,0 \rangle, \langle 0,1 \rangle, \langle 1,0 \rangle, \langle 1,1 \rangle, \langle 0,2 \rangle, \ldots$$
$$\langle 0,0,0 \rangle, \langle 0,0,1 \rangle, \langle 0,1,0 \rangle, \langle 0,1,1 \rangle, \langle 1,0,0 \rangle, \ldots$$
$$\vdots$$

We have an ordering $<$ on the basis of which we can define a lexicographical ordering $\prec_{\text{lex}}$: Suppose

$$s = \langle n_1, n_2, \ldots, n_k \rangle$$
$$t = \langle m_1, m_2, \ldots, m_\ell \rangle$$

Then $s \prec_{\text{lex}} t$ if, and only if, either

1. $k < \ell$ and $n_1 = m_1, \ldots, n_k = m_k$ or

2. there is a $j < k$ such that $n_1 = m_1, \ldots, n_j = m_j$ and $n_{j+1} < m_{j+1}$

In other words, $s \prec_{\text{lex}} t$ if, and only if, either (a) s is a proper initial subsequence of t (s is shorter than t but agrees with t, positionwise, for all elements of s) or (b) at the first position where s and t disagree, the element of s is less than the corresponding element of t.

A moment's reflection shows that the definition is equivalent to the following one: $s \prec_{\text{lex}} t$ if, and only if, for some j, we have $n_i = m_i$ for all $i \leq j$ and either $j = k$ or $n_{j+1} < m_{j+1}$.

For instance, $\langle 1,3,5 \rangle \prec_{\text{lex}} \langle 1,3,5,7,9 \rangle$ since the first is a proper initial segment of the second, and $\langle 1,3,5,9,11 \rangle \prec_{\text{lex}} \langle 1,3,7,2 \rangle$ because at the first position where they disagree (the third), the element of the first sequence (5) is less than the corresponding element of the second sequence (7). In this case, $j = 2$, $n_1 = m_1$ and $n_2 = m_2$ but $n_3 < m_3$.

Although $\langle \mathbb{N}^*, \prec_{\text{lex}} \rangle$ is a linear order, it is not a well-order. Here is an infinite decreasing sequence of elements:

$$\langle 1 \rangle >_{\text{lex}} \langle 0,1 \rangle >_{\text{lex}} \langle 0,0,1 \rangle >_{\text{lex}} \langle 0,0,0,1 \rangle >_{\text{lex}} \cdots$$

It is of course possible to well-order $\mathbb{N}^*$ as well, but the ordering must—as we just saw—be a different one than the lexicographical order. One such ordering is the "short-lex" ordering: in the short-lex order, we first order sequences by length, and among sequences of equal length, we use the lexicographical order. In other words, $s \prec_{\text{slex}} t$ if, and only if, either

1. $k < \ell$ or

2. $k = \ell$ and for some $j < k$, $n_i = m_i$ for $i = 1, \ldots, j$ and $n_{j+1} < m_{j+1}$

Problem 8.16. Show that $\langle \mathbb{N}^*, <_{\text{slex}} \rangle$ is a well-ordering.

Although $<_{\text{lex}}$ is not a well-ordering of $\mathbb{N}^*$, it *is* a well-ordering of those sequences of numbers in which the elements themselves are decreasing. Let $\mathbb{N}^*_>$ be the set of all $\langle n_1, \ldots, n_k \rangle \in \mathbb{N}^*$ where $n_i > n_{i+1}$. Note in particular that the elements of the infinite decreasing sequence in $\mathbb{N}^*$ given above are *not* of this form! Then $\langle \mathbb{N}^*_>, <_{\text{lex}} \rangle$ is a well-ordering. For suppose $s_1 >_{\text{lex}} s_2 >_{\text{lex}} s_3 >_{\text{lex}} \ldots$ were an infinite descending sequence of elements of $\mathbb{N}^*_>$. Let $s_i = \langle n_1^i, \ldots, n_{k_i}^i \rangle$. We'll argue in a way similar to the proof that $\mathbb{N}^k$ is well-ordered by $<_{\text{lex}}$. We consider the first elements n_1^i and show that they must eventually stabilize. The sequence $n_1^1, n_1^2, \ldots$ is a non-increasing sequence of numbers. (Again, if it were increasing at some point, we'd have $n_1^i < n_1^{i+1}$, but then by the definition of $<_{\text{lex}}$ we'd have $s_i <_{\text{lex}} s_{i+1}$ contrary to the assumption that the s_i are decreasing.) Let m_1 be the least of these n_1^i and let j_1 be the least i such that $s_i = \langle m_1, n_2^i, \ldots, n_{k_i}^i \rangle$. If $i > j_1$ we have: (a) $s_i = \langle m_1, n_2^i, \ldots, n_{k_i}^i \rangle$ (the first component has stabilized by position j_1 in the sequence) and (b) the numbers n_2^i form a non-increasing sequence. As before, there must be a least number among these, let us call it m_2, and there must be a least position $j_2 > j_1$ where $s_{j_2} = \langle m_1, m_2, n_3^{j_2}, \ldots, n_{k_{j_2}}^{j_2} \rangle$. (In other words, at some place j_2 after j_1, the second components of the sequences must stabilize, and after j_2, every sequence must begin with m_1, m_2.) Now *because the elements of sequences in $\mathbb{N}^*_>$ are decreasing*, $m_1 > m_2$. Continuing in this way we obtain a sequence $m_1 > m_2 > m_3 > \ldots$ of natural numbers, which is of course impossible.

Note that we have assumed nothing about the elements of the sequences beyond the fact that they are well-ordered by $<$. So we obtain an immediate generalization:

Proposition 8.17. *Suppose $\langle X, < \rangle$ is a well-ordering. Then the set $X^*_>$ of $<$-decreasing sequences of elements of X is well-ordered by $<_{\text{lex}}$.*

We'll add one more wrinkle to the discussion of lexicographically well-ordered sets: Not just $X^*_>$, but even $X^*_\geq$, the set of *non-increasing* sequences of elements of X, is well-ordered by $<_{\text{lex}}$ (if X itself is well-ordered).

A non-increasing sequence of elements of X is a sequence $\langle x_1, x_2, \ldots, x_k \rangle$ where $x_i \geq x_{i+1}$ (rather than *decreasing*, i.e., $x_i > x_{i+1}$). Each such sequence can be thought of as a sequence $\langle y^1, \ldots, y^\ell \rangle$ where each y^j is a finite sequence where every element is one of the x_i. For instance, the non-increasing sequence $\langle 4, 4, 3, 2, 2, 2 \rangle$ can be thought of as $\langle \langle 4, 4 \rangle, \langle 3 \rangle, \langle 2, 2, 2 \rangle \rangle$. Let's call such sequences *constant* sequences: $X^*_=$ is the set of all sequences $\langle x, \ldots, x \rangle$ where $x \in X$. It is an easy exercise to see that $X^*_=$ is well-ordered by $<_{\text{lex}}$.

Proposition 8.18. *If $\langle X, < \rangle$ is a well-ordering, $\langle X^*_=, <_{\text{lex}} \rangle$ is a well-ordering.*

Proof. Exercise. □

Problem 8.19. Prove Proposition 8.18.

We'll use this fact to show that $X_{\geq}^{*}$ is well-ordered.

Proposition 8.20. *If $\langle X, < \rangle$ is a well-ordering, $\langle X_{\geq}^{*}, <_{\text{lex}} \rangle$ is a well-ordering.*

Proof. The set $X_{\geq}^{*}$ of *non-increasing* sequences of elements of X is order-isomorphic to the set $(X_{=}^{*})_{>_{\text{lex}}}^{*}$ of *decreasing* sequences of elements of $X_{=}^{*}$ (decreasing in the lexicographical ordering $<_{\text{lex}}$ on $X_{=}^{*}$). Here, we take both sets to be ordered lexicographically: $X_{\geq}^{*}$ is ordered by the lexicographical order $<_{\text{lex}}$ of sequences of elements of X based on $<$. $(X_{=}^{*})_{>_{\text{lex}}}^{*}$ is ordered by the lexicographical order $\ll$ based on $<_{\text{lex}}$ itself, i.e., the ordering of sequences of constant sequences of elements of X where

$$\underbrace{\langle x, \ldots, x \rangle}_{n\ x's} \ll \underbrace{\langle y, \ldots, y \rangle}_{m\ y's} \text{ if, and only if, either } x < y \text{ or } x = y \text{ and } n < m.$$

Here is the isomorphism:

$$f(\langle y_1, \ldots, y_{k'} \rangle) = \langle \underbrace{x_1, \ldots, x_1}_{y_1}, \ldots, \underbrace{x_{k'}, \ldots, x_{k'}}_{y_{k'}} \rangle$$

Since $X_{=}^{*}$ is well-ordered by $<_{\text{lex}}$, it follows from Proposition 8.17 that $(X_{=}^{*})_{>_{\text{lex}}}^{*}$, the set of decreasing sequences of elements of $X_{=}^{*}$, is well-ordered by $\ll$. Since this is order-isomorphic to $X_{\geq}^{*}$ ordered by $<_{\text{lex}}$, the latter is well-ordered as well by Proposition 8.10. $\square$

8.3 Ordinal notations up to ε_0

For Gentzen's consistency proof of number theory, a particular well-ordered set will play a special role. Its definition is a bit arcane, if you do not already know about set theory, transfinite ordinals numbers, arithmetical operations on ordinals, and the Cantor-Bachmann normal form of ordinals. Here, we will define ordinal notations without assuming such knowledge as background. First of all, such background knowledge is not necessary to understand ordinal notations. But more importantly, by avoiding appeal to set theory in the definition of ordinal notations, we can avoid a common misconception. Gentzen's proof of consistency of **PA** proceeds by induction along a well-ordering (namely along the well-ordering of ordinal notations which we'll presently introduce) and this well-ordering is the same as the well-ordering of infinite ordinal numbers less than a certain ordinal called ε_0.[1] The misconception then easily arises that Gentzen's consistency proof

---

[1] There are systems of ordinal notations for ordinals larger than ε_0 as well. Ordinal notations are also called "constructive ordinals."

somehow depends on the infinite ordinal ε_0, the theory of transfinite ordinals, or set theory generally. It does not.

We will later discuss the relevant theory of transfinite ordinals. However, the ordinal *notations* themselves are not transfinite objects, they are finite strings generated from the symbols **0** and ω using the symbol **+** and superscripts ("exponentiation"). For instance,

$$\mathbf{0}, \quad \omega^{\omega^0}, \quad \text{and} \quad \omega^{\omega^0 + \omega^0} + \omega^0$$

are ordinal notations.

The reason for the choice of these symbols will become clear later; what's important for now is just that these are certain strings of symbols, and that they can be well-ordered in a certain way. Not every string involving these symbols and operations is a legal ordinal notation (e.g., ω and $\omega^0 + \omega^{\omega^0}$ will not count as legal ordinal notations), but to say which ones do, we have to make reference to an ordering of ordinal notations. So we define both the set of ordinal notations and its ordering together.

We construct an ordered set O in stages. Once we have introduced a stage of ordinal notations, the next stage will include all "sums" of "powers" of ω of non-decreasing ordinal notations already defined, with the first "exponent" an ordinal notation added during the immediately preceding stage.

As the first stage, we include **0** in O. It is the smallest element of O. Since **0** is the only ordinal notation we have so far, in the second stage we get the ordinal notations of height 1, i.e.,

$$\omega^0, \omega^0 + \omega^0, \omega^0 + \omega^0 + \omega^0,$$

etc. They are all larger than **0**, and each one is smaller than the next (because shorter). In the third stage, we can already construct all "powers" of ω where each exponent is one of the infinitely many ordinal notations we've constructed so far, but the first one has to be an ordinal notation of height 1. These new ordinals are all greater than the ordinals already constructed in the first two stages.

More precisely, the first stage is the set of ordinal notations of *height* 0. This set contains only **0**. We stipulate that this set has the empty ordering: $\mathbf{0} \not\prec \mathbf{0}$.

Suppose we have already defined the sets of ordinal notations of heights $0, \ldots, k$, and the ordering $\prec$ on all of these. Then we let the ordinal notations of height $k + 1$ be all strings of the form

$$\omega^{\alpha_1} + \cdots + \omega^{\alpha_n}$$

where the α_i are ordinal notations of height $\leq k$ and non-increasing according to $\prec$ as already defined (i.e., if $i < n$, $\alpha_i \succeq \alpha_{i+1}$), and α_1 is of height k.

We now extend the ordering $\prec$ to cover the new ordinals of height $k + 1$: first of all, every ordinal notation of height $k + 1$ is greater (according to the ordering $\prec$) than any ordinal notation of height $\leq k$. If we have two ordinal notations of height

$k + 1$, we say that the first is less than the second if the leftmost exponent of ω where they differ is less in the first than in the second, or, if they do not differ, the first sum is shorter than the second. More precisely, suppose that

$$\alpha = \omega^{\alpha_1} + \cdots + \omega^{\alpha_n} \text{ and}$$
$$\beta = \omega^{\beta_1} + \cdots + \omega^{\beta_m}$$

are ordinal notations of height $k + 1$. Then $\alpha \prec \beta$ if, and only if, either $\alpha_i = \beta_i$ for $i = 1, \ldots, j$ and $\alpha_{j+1} \prec \beta_{j+1}$, or $n < m$ and $\alpha_i = \beta_i$ for all $i = 1, \ldots, n$.

For instance, ω^0, $\omega^0 + \omega^0$, and $\omega^0 + \omega^0 + \omega^0$ are ordinal notations of height 1, since 0 is (the only) ordinal notation of height 0. We have

$$\omega^0 \prec \omega^0 + \omega^0 \prec \omega^0 + \omega^0 + \omega^0,$$

just because of the second clause since the α_i and β_i here are all just 0. All of them are $\succ 0$. An ordinal notation of height 2 starts with ω^{α_1} with α_1 of height 1, and followed by a non-increasing sequence of ω^{α_i} with α_i either of height 0 or 1. So for instance, we have

$$\omega^{\omega^0} \prec \omega^{\omega^0 + \omega^0} \prec \omega^{\omega^0 + \omega^0} + \omega^{\omega^0} + \omega^0 + \omega^0.$$

Definition 8.21 (Ordinal notations $< \varepsilon_0$). We define inductively the set $O_{=k}$ of ordinal notations $< \varepsilon_0$ of height k, the set $O_{\leq k}$ of ordinal notations of height up to k, and an ordering $\prec_k$ on $O_{\leq k}$ as follows:

1. *Basis clause:* 0 is the only ordinal notation of height 0, i.e., $O_{=0} = O_{\leq 0} = \{0\}$ and $\prec_0 = \emptyset$, i.e., $0 \nprec_0 0$.

2. *Inductive clause:* If $O_{=k}$ and $O_{\leq k}$ have been defined, then the ordinal notations of height $k + 1$ is the set $O_{=k+1}$ of all expressions of the form

$$\omega^{\alpha_1} + \omega^{\alpha_2} + \cdots + \omega^{\alpha_n}$$

where

   (i) $\alpha_1 \in O_{=k}$,
   (ii) if $1 < i \leq n$, $\alpha_i \in O_{\leq k}$, and
   (iii) $\alpha_i \succeq_k \alpha_{i+1}$ for $i = 1, \ldots, n - 1$.

   $O_{\leq k+1}$ is $O_{\leq k}$ together with all ordinal notations of height $k + 1$, i.e., $O_{\leq k+1} = O_{\leq k} \cup O_{=k+1}$.

The ordering $\prec_{k+1}$ on $O_{\leq k+1}$ is defined by: $\alpha \prec_{k+1} \beta$ if, and only if,

(a) α and $\beta \in O_{\leq k}$ and $\alpha \prec_k \beta$, or

(b) $\alpha \in O_{\leq k}$ and $\beta \in O_{=k+1}$, or

(c) α and $\beta \in O_{=k+1}$ are of the forms

$$\alpha = \omega^{\alpha_1} + \cdots + \omega^{\alpha_n}$$
$$\beta = \omega^{\beta_1} + \cdots + \omega^{\beta_m}$$

and $\alpha_i = \beta_i$ for $i \leq j$, with either $j = n < m$ or $j < n$ and $\alpha_{j+1} \prec_k \beta_{j+1}$.

Finally, we let $O = O_{\leq 0} \cup O_{\leq 1} \cup \ldots$ and $\prec = \prec_0 \cup \prec_1 \cup \ldots$, i.e., $\alpha \prec \beta$ if, and only if, $\alpha \prec_j \beta$ for some j.

Let's abbreviate ω^0 as 1. Examples of ordinal notations are

$$0, 1, 1+1, \omega^{1+1}, \omega^{1+1} + 1, \omega^{\omega^{1+1}+1} + \omega^{1+1} + 1.$$

By contrast, $\omega^0 + 0$ and $\omega^1 + \omega^{1+1}$ are not ordinal notations! Clause 2 only allows "sums" of "powers" of ω, so while $\omega^0 + 1$ is short for $\omega^0 + \omega^0$ and hence is in the required form, 0 is not (short for) a power of ω, and so $\omega^0 + 0$ is not of the required form. Also, the exponents in a sum of powers of ω must be non-increasing from left to right. Since $1 \prec 1+1$, the expression $\omega^1 + \omega^{1+1}$ does not count as an ordinal notation.

The choice of the word "height" to describe the stage in which an ordinal notation is constructed is not accidental. Every ordinal notation (other than 0) is clearly of the form

$$\omega^{\omega^{\cdots^{\omega^0 + \cdots}} + \cdots} + \ldots$$

and the number of ω symbols on the far left of this expression is the height of the ordinal notation. So, any ordinal notation of the form $\omega^0 + \ldots$ is of height 1, any of the form $\omega^{\omega^0 + \cdots} + \ldots$ of height 2, and so on. In particular, it is clear that the height of an ordinal notation is uniquely determined; no ordinal notation is both an element of $O_{=k}$ and $O_{=\ell}$ if $k \neq \ell$. Furthermore, if β is of height k, a pair $\alpha \prec \beta$ is added at stage k only, i.e., $\alpha \prec \beta$ if, and only if, $\alpha \prec_k \beta$.

Problem 8.22. Which of the following expressions are ordinal notations? What is the height of the legal ordinal notations on the list? Put them in the right order according to $\prec$.

$\omega^{\omega^{\omega^0}}$

$\omega^{\omega^0 + \omega^0} + \omega^{\omega^0} + \omega^0$

$\omega^0 + \omega^0$

$\omega^0 + \omega^{\omega^0}$

$\omega^{\omega^{\omega^0} + \omega^{\omega^0} + \omega^0} + \omega^{\omega^0} + \omega^0$

$\omega^{\omega^{\omega^0} + \omega^0} + \omega^{\omega^0} + \omega^{\omega^0} + \omega^0$

We'll first verify some basic facts about ordinal notations, directly from the definitions.

Proposition 8.23. 0 is least in the ordering $\prec$, i.e., for all $\alpha \in O$ with $\alpha \neq 0$, $0 \prec \alpha$.

Proof. Since $O_{=0} = \{\mathbf{0}\}$, if $\alpha \neq \mathbf{0}$, then $\alpha \in O_{=k}$ for some $k > 0$. Then $\alpha \succ \mathbf{0}$ by clause b of Definition 8.21.   □

Proposition 8.24. *If β is of height k and $\alpha \prec \beta$ then $\alpha \prec_k \beta$.*

Proof. Since $\mathbf{0}$ is the only ordinal notation of height 0, $k > 0$ and $\beta \neq \mathbf{0}$. Since $\alpha \prec \beta$, for some j, $\alpha \prec_{j+1} \beta$. We may assume that j is the least number with this property, i.e., not: $\alpha \prec_j \beta$. Then by the definition of $\prec_{j+1}$, $\beta \in O_{=j+1}$, that is $k = j + 1$.   □

In Definition 8.21 we defined the orderings $\prec_k$ on $O_{\leq k}$ and then took O as the union of all $O_{\leq k}$ and $\prec$ the union of all $\prec_k$. We'll now prove that the ordering $\prec$ so defined is in fact the "lexicographical" ordering of ordinal notations other than $\mathbf{0}$, i.e., $\alpha \prec \beta$ if, and only if, either the sum of powers of ω that α represents is a proper initial segment of β, or, if not, at the first place where the two differ the exponent of ω in α is less than the exponent of ω in β. In other words, the defining property of $\prec_k$ on $O_{=k}$ is also the defining property of $\prec$ on O. Of course, $\mathbf{0}$ is not a sum of powers of ω, so it is excluded. (We've just shown that $\mathbf{0} \prec \alpha$ for all α which are sums of powers of ω.)

Proposition 8.25. *Suppose*

$$\alpha = \omega^{\alpha_1} + \cdots + \omega^{\alpha_n} \text{ and}$$

$$\beta = \omega^{\beta_1} + \cdots + \omega^{\beta_m}.$$

Then $\alpha \prec \beta$ if, and only if, for some j, $\alpha_i = \beta_i$ when $i \leq j$ and either $j = n < m$ (i.e., α is "shorter" than β but the exponents agree) or $j < n$ and $\alpha_{j+1} \prec \beta_{j+1}$.

Proof. First we prove the "if" part. Assume the condition on the right is satisfied. We distinguish cases according to whether α and β are of the same height or not. Suppose first that both α and β are in $O_{=k+1}$ for some k. Then the condition coincides with clause c, and so $\alpha \prec_{k+1} \beta$. By the definition of $\prec$, also $\alpha \prec \beta$. Suppose now that α and β are not of the same height. Then clearly α_1 and β_1 must be of different heights as well, because any ordinal notation is of height $k + 1$ if, and only if, the exponent of the first ω is of height k. Since ordinal notations of different heights are syntactically different, $\alpha_1 \neq \beta_1$. By the condition, we have $j = 0$ and $\alpha_1 \prec \beta_1$. So $\beta_1 \succ \mathbf{0}$, i.e., it is of some height $k + 1$, and therefore β is of height $k + 2$. Since α_1 and β_1 have different heights and $\alpha_1 \prec \beta_1$, the height of α_1 is less than or equal to k, and so the height of α is less than or equal to $k + 1$. By clause b, $\alpha \prec_{k+1} \beta$.

Now the "only if" part. Suppose $\alpha \prec \beta$. Then $\alpha \prec_{k+1} \beta$ for some k, by the definition of $\prec$. Let k be the least number satisfying this, i.e., $\alpha \prec_{k+1} \beta$ but not $\alpha \prec_k \beta$. Then clause a can't apply. If $\alpha \prec_{k+1} \beta$ because of clause b, then $\beta \in O_{=k+1}$ and $\alpha \in O_{\leq k}$. In other words, $\beta_1 \in O_{=k}$ (is of height k) and $\alpha_1 \in O_{\leq k-1}$ (is of height $< k$). Since the height of an ordinal notation is uniquely determined, $\alpha_1 \neq \beta_1$. By clause b, $\alpha_1 \prec_k \beta_1$ and so $\alpha_1 \prec \beta_1$. So, there is a j (namely 0) such that $\alpha_i = \beta_i$

for all $i \leq j$ (vacuously, since there are no α_i or β_i with $i \leq 0$) and $\alpha_{j+1} \prec \beta_{j+1}$. If clause b does not apply, clause c applies. In that case, α and β are of the same height $k + 1$. By clause c, for some j, for all $i \leq j$, $\alpha_i = \beta_i$ and either $j = n < m$ or $\alpha_{j+1} \prec_k \beta_{j+1}$. Since when $\alpha_{j+1} \prec_k \beta_{j+1}$ also $\alpha_{j+1} \prec \beta_{j+1}$, we are done. □

Proposition 8.26. *For any ordinal notation α, $\alpha \prec \omega^\alpha$.*

Proof. If α is an ordinal notation, it must have been constructed at some stage, i.e., α is of height k. But then ω^α is of height $k + 1$. By clause b of the definition of $\prec_{k+1}$, $\alpha \prec_{k+1} \omega^\alpha$. So $\alpha \prec \omega^\alpha$. □

8.4   Operations on ordinal notations

Definition 8.27 (Natural sum). If α and β are ordinal notations, then $\alpha \# \beta$ is defined as:

1. If $\alpha = 0$, then $\alpha \# \beta = \beta$.

2. If $\beta = 0$, then $\alpha \# \beta = \alpha$.

3. Otherwise, there are n, m, α_i and β_i such that

$$\alpha = \omega^{\alpha_1} + \cdots + \omega^{\alpha_n} \text{ and}$$
$$\beta = \omega^{\beta_1} + \cdots + \omega^{\beta_m}.$$

Let π be a permutation[2] of $\{1, \dots, n + m\}$ and

$$\gamma_i = \begin{cases} \alpha_{\pi(i)} & \text{if } \pi(i) \leq n \\ \beta_{\pi(i)-n} & \text{if } \pi(i) > n, \end{cases}$$

such that $\gamma_i \succeq \gamma_{i+1}$ for $i < n + m$.[3] Then

$$\alpha \# \beta = \omega^{\gamma_1} + \cdots + \omega^{\gamma_{n+m}}.$$

In less technical language: the γ_i are a rearrangement of α_i together with the β_i in non-increasing order. Although there may be multiple permutations π that satisfy the condition, the resulting ordinal notation is unique. For instance, if

$$\alpha = \omega^{\omega^{\omega^0} + \omega^0} + \omega^{\omega^0 + \omega^0} + \omega^0 \text{ and} \qquad \beta = \omega^{\omega^{\omega^0 + \omega^0}} + \omega^{\omega^{\omega^0}} + \omega^0,$$

---

[2] A permutation of X is a bijection of X onto X, i.e., a mapping π such that (a) for every $x \in X$, $x = \pi(y)$ for some $y \in X$, and also such that whenever $x \neq y$, $\pi(x) \neq \pi(y)$.

[3] We know that the α_i are sorted in non-increasing order, and so are the β_i. But to sort them into a non-decreasing order together, i.e., to guarantee that there is such a permutation with $\gamma_i \succeq \gamma_{i+1}$, it's necessary that for any α_i and β_j, one of the following holds: $\alpha_i \prec \beta_j$ or $\beta_j \prec \alpha_i$ or $\alpha_i = \beta_j$. In other words, $\prec$ must be a linear ordering. This is the content of Proposition 8.43 below.

then the α_i and β_i together, in non-increasing order, are:

$$\underbrace{\omega^{\omega^0+\omega^0}}_{\beta_1}, \underbrace{\omega^{\omega^0}+\omega^0}_{\alpha_1}, \underbrace{\omega^{\omega^0}}_{\beta_2}, \underbrace{\omega^0+\omega^0}_{\alpha_2}, \underbrace{0}_{\alpha_3}, \underbrace{0}_{\beta_3}.$$

Thus $\alpha \# \beta$ is

$$\omega^{\omega^{\omega^0+\omega^0}} + \omega^{\omega^{\omega^0}+\omega^0} + \omega^{\omega^{\omega^0}} + \omega^{\omega^0+\omega^0} + \omega^0 + \omega^0.$$

Recall the definition of ordinal notations of height $h + 1$: They are of the form

$$\omega^{\alpha_1} + \cdots + \omega^{\alpha_n}$$

for some α_i of height less than or equal to h, α_1 of height h, and with $\alpha_i \geq \alpha_{i+1}$. If we group identical exponents α_i together, we can write such an ordinal notation also as

$$\underbrace{\omega^{\gamma_1} + \cdots + \omega^{\gamma_1}}_{c_1 \text{ copies}} + \cdots + \underbrace{\omega^{\gamma_k} + \cdots + \omega^{\gamma_k}}_{c_k \text{ copies}}$$

where there are c_j copies of ω^{γ_j} and each $\gamma_j \succ \gamma_{j+1}$. Let's abbreviate an ordinal notation of this form by $\omega^{\gamma_1} \cdot c_1 + \cdots + \omega^{\gamma_k} \cdot c_k$. Slightly generalizing, we may allow expressions $\omega^{\gamma} \cdot 0$ as well, if γ is not among the α_i, and $c_j = 0$ for all j if $\alpha = 0$. We'll record this in a definition.

Definition 8.28. Suppose $\gamma_1 \succ \cdots \succ \gamma_k$ are distinct ordinal notations and $c_1, \ldots, c_k \geq 0$. Then

$$\omega^{\gamma_1} \cdot c_1 + \cdots + \omega^{\gamma_k} \cdot c_k$$

abbreviates

$$\underbrace{\omega^{\gamma_1} + \cdots + \omega^{\gamma_1}}_{c_1 \text{ copies}} + \cdots + \underbrace{\omega^{\gamma_k} + \cdots + \omega^{\gamma_k}}_{c_k \text{ copies}}.$$

if at least one $c_i > 0$, and abbreviates 0 otherwise.

With this notation, we can describe the ordinal sum more succinctly. Suppose

$$\alpha = \omega^{\alpha_1} + \cdots + \omega^{\alpha_n} \text{ and}$$
$$\beta = \omega^{\beta_1} + \cdots + \omega^{\beta_m}.$$

Now let $\gamma_1 \succ \cdots \succ \gamma_k$ and $\{\gamma_1, \ldots, \gamma_k\} = \{\alpha_1, \ldots, \alpha_n, \beta_1, \ldots, \beta_m\}$. In other words, $\gamma_1, \ldots, \gamma_k$ are the ordinal notations $\alpha_1, \ldots, \alpha_n, \beta_1, \ldots, \beta_m$ ordered according to strictly decreasing size and without multiples. Then we can write, using the above notation,

$$\alpha = \omega^{\gamma_1} \cdot a_1 + \cdots + \omega^{\gamma_k} \cdot a_k \text{ and}$$
$$\beta = \omega^{\gamma_1} \cdot b_1 + \cdots + \omega^{\gamma_k} \cdot b_k$$

(where $a_j = 0$ if γ_j is not among the α_i and $b_j = 0$ if γ_j is not among the β_i).[4] The natural sum $\alpha \# \beta$ then is

$$\alpha \# \beta = \omega^{\gamma_1} \cdot (a_1 + b_1) + \cdots + \omega^{\gamma_k} \cdot (a_k + b_k).$$

For instance, if

$$\alpha = \qquad\quad \omega^{\omega^0 + \omega^0} + \omega^{\omega^0 + \omega^0} + \omega^{\omega^0} + \omega^0 + \omega^0$$

$$\beta = \omega^{\omega^{\omega^0}} + \omega^{\omega^0 + \omega^0} + \qquad\qquad \omega^0 + \omega^0 + \omega^0$$

we can write this as

$$\alpha = \omega^{\omega^{\omega^0}} \cdot 0 + \omega^{\omega^0 + \omega^0} \cdot 2 + \omega^{\omega^0} \cdot 1 + \omega^0 \cdot 2$$

$$\beta = \omega^{\omega^{\omega^0}} \cdot 1 + \omega^{\omega^0 + \omega^0} \cdot 1 + \omega^{\omega^0} \cdot 0 + \omega^0 \cdot 3$$

and consequently we have

$$\alpha \# \beta = \omega^{\omega^{\omega^0}} \cdot 1 + \omega^{\omega^0 + \omega^0} \cdot 3 + \omega^{\omega^0} \cdot 1 + \omega^0 \cdot 5 =$$

$$= \omega^{\omega^{\omega^0}} + \omega^{\omega^0 + \omega^0} + \omega^{\omega^0 + \omega^0} + \omega^{\omega^0 + \omega^0} + \omega^{\omega^0} + \omega^0 + \omega^0 + \omega^0 + \omega^0 + \omega^0.$$

We'll record this as a proposition.

Proposition 8.29. *If $\gamma_1 \succ \cdots \succ \gamma_k$ and*

$$\alpha = \omega^{\gamma_1} \cdot a_1 + \cdots + \omega^{\gamma_k} \cdot a_k$$
$$\beta = \omega^{\gamma_1} \cdot b_1 + \cdots + \omega^{\gamma_k} \cdot b_k$$

Then

$$\alpha \# \beta = \omega^{\gamma_1} \cdot c_1 + \cdots + \omega^{\gamma_k} \cdot c_k$$

where $c_i = a_i + b_i$.

Proposition 8.30. *For any α and β, $\alpha \# \beta = \beta \# \alpha$.*

Proof. It is clear from the introduced notation that

$$\alpha \# \beta = \omega^{\gamma_1} \cdot (a_1 + b_1) + \cdots + \omega^{\gamma_k} \cdot (a_k + b_k) =$$
$$\beta \# \alpha = \omega^{\gamma_1} \cdot (b_1 + a_1) + \cdots + \omega^{\gamma_k} \cdot (b_k + a_k)$$

since $a_i + b_i = b_i + a_i$. $\qquad\qquad\qquad\qquad\qquad\qquad\qquad\square$

---

[4] It is important to keep in mind that writing α and β in this way is only an abbreviation. It is, however, a very useful abbreviation as it allows us to "normalize" the two ordinal notations and to compare them term by term.

With the introduced notation, it is also easier to compare two ordinal notations α and β. Let's write α and β again as

$$\alpha = \omega^{\gamma_1} \cdot a_1 + \cdots + \omega^{\gamma_k} \cdot a_k \text{ and}$$
$$\beta = \omega^{\gamma_1} \cdot b_1 + \cdots + \omega^{\gamma_k} \cdot b_k$$

with γ_j, a_j, and b_j as before.

Proposition 8.31. $\alpha \prec \beta$ if for the least ℓ such that $a_\ell \neq b_\ell$, $a_\ell < b_\ell$.

Proof. Suppose ℓ is least such that $a_\ell \neq b_\ell$ and $a_\ell < b_\ell$.

If $a_\ell < b_\ell$ for some ℓ then we must have $\beta \neq \mathbf{0}$. If $a_i = 0$ for all i, then $\alpha = \mathbf{0}$ and so $\alpha \prec \beta$. Otherwise, let j be least such that $a_j > 0$. We have two cases.

The first case is where γ_j is of smaller height than the height of γ_1. In that case, clearly $\gamma_j \neq \gamma_1$ since otherwise the height of γ_j would be the same as the height of γ_1. Hence, $j \neq 1$ and $a_1 = 0$. In fact, for the same reason, $\gamma_j \neq \gamma_i$ for all i where γ_i is of the same height as γ_1, and $a_i = 0$ for all these i. On the other hand, $b_i > 0$ for all these i since otherwise γ_i would not be among the α_i and β_i at all. Thus, in that case, α is of lower height than β and hence $\alpha \prec \beta$.

The second case is where γ_j is of the same height as γ_1. Then α and β are of the same height. But ordinal notations of the same height are ordered lexicographically, and since $a_\ell < b_\ell$ and $a_i = b_i$ for all $i < \ell$, $\alpha \prec \beta$. $\square$

Problem 8.32. Show that $\omega^\mu \# \omega^\nu$ is either $\omega^\mu + \omega^\nu$ (if $\mu \geq \nu$) or $\omega^\nu + \omega^\mu$ (otherwise).

Proposition 8.33. For any α and β, $\alpha \preceq \alpha \# \beta$. Specifically, if $\beta = \mathbf{0}$ then $\alpha = \alpha \# \beta$, and otherwise $\alpha \prec \alpha \# \beta$.

Proof. If we have

$$\alpha = \omega^{\gamma_1} \cdot a_1 + \cdots + \omega^{\gamma_k} \cdot a_k \text{ and}$$
$$\beta = \omega^{\gamma_1} \cdot b_1 + \cdots + \omega^{\gamma_k} \cdot b_k$$

then

$$\alpha \# \beta = \omega^{\gamma_1} \cdot (a_1 + b_1) + \cdots + \omega^{\gamma_k} \cdot (a_k + b_k).$$

Clearly $a_j \leq a_j + b_j$ for all $j \leq k$. If $a_j = a_j + b_j$ for all $j \leq k$, then $b_j = 0$ for all $j \leq k$, i.e., $\beta = \mathbf{0}$, and $\alpha \# \mathbf{0} = \alpha$. Otherwise, for the least j such that $a_j \neq a_j + b_j$, $a_j < a_j + b_j$. By Proposition 8.31, $\alpha \prec \alpha \# \beta$. $\square$

Proposition 8.34. If $\alpha \preceq \delta$ and $\beta \preceq \eta$ then $\alpha \# \beta \preceq \delta \# \eta$.

Proof. Exercise. $\square$

Problem 8.35. Prove Proposition 8.34. Write all four ordinal notations $\alpha, \beta, \delta, \eta$ as sums of $\omega^{\gamma_i} \cdot a_i$ (or b_i, d_i, e_i) and use the fact that $a_i \leq d_i$ and $b_i \leq e_i$ for all i (why?).

Proposition 8.36. *If $\beta \prec \delta$, then $\alpha \# \beta \prec \alpha \# \delta$.*

Proof. We can again write

$$\alpha = \omega^{\gamma_1} \cdot a_1 + \cdots + \omega^{\gamma_k} \cdot a_k,$$
$$\beta = \omega^{\gamma_1} \cdot b_1 + \cdots + \omega^{\gamma_k} \cdot b_k, \text{ and}$$
$$\delta = \omega^{\gamma_1} \cdot d_1 + \cdots + \omega^{\gamma_k} \cdot d_k$$

for some $\gamma_i, a_i, b_i,$ and d_i. Then

$$\alpha \# \beta = \omega^{\gamma_1} \cdot (a_1 + b_1) + \cdots + \omega^{\gamma_k} \cdot (a_k + b_k) \text{ and}$$
$$\alpha \# \delta = \omega^{\gamma_1} \cdot (a_1 + d_1) + \cdots + \omega^{\gamma_k} \cdot (a_k + d_k).$$

If $\beta \prec \delta$, for the smallest j such that $b_j \neq d_j$ we have $b_j < d_j$. Since j is smallest with that property, for all $i < j$ we have $b_i = d_i$ and hence $a_i + b_i = a_i + d_i$. So j is smallest such that $a_j + b_j \neq a_j + d_j$. But since $b_j < d_j$ also $a_j + b_j < a_j + d_j$. So, $\alpha \# \beta \prec \alpha \# \delta$ by Proposition 8.31. $\square$

By Proposition 8.30, $\alpha \# \beta = \beta \# \alpha$ and $\alpha \# \delta = \delta \# \alpha$. So it follows that if $\beta \prec \delta$, then $\beta \# \alpha \prec \delta \# \alpha$.

For the following, recall that we abbreviate ω^0 as **1**.

Corollary 8.37. *For all α, $\alpha \prec \alpha \# \mathbf{1}$.*

Proof. Take $\beta = 0$ and $\delta = \mathbf{1}$ in Proposition 8.36: $\alpha \# 0 \prec \alpha \# \mathbf{1}$. But $\alpha \# 0 = \alpha$. $\square$

Proposition 8.38. *If $\alpha = \omega^{\alpha_1} + \cdots + \omega^{\alpha_n}$ then $\alpha \prec \omega^{\alpha_1 \# \mathbf{1}}$*

Proof. First suppose $\alpha_1 = 0$, i.e., $\alpha = \omega^0 \cdot n$ for some n. Then α is of height 1. Since $0 \# \beta = \beta$ for any β and $\alpha_1 = 0$, we have $\alpha_1 \# \mathbf{1} = \mathbf{1}$. Thus the ordinal notation $\omega^{\alpha_1 \# \mathbf{1}}$ is identical to the ordinal notation ω^{ω^0}, which is of height 2. Hence, $\alpha \prec \omega^{\alpha_1 \# \mathbf{1}}$.

Now assume $\alpha_1 \neq 0$. Then α_1 and $\alpha_1 \# \mathbf{1}$ are of the same height, and so

$$\alpha = \omega^{\alpha_1} + \cdots + \omega^{\alpha_n} \text{ and}$$
$$\omega^{\alpha_1 \# \mathbf{1}}$$

are of the same height. By Corollary 8.37, $\alpha_1 \prec \alpha_1 \# \mathbf{1}$, and so $\alpha \prec \omega^{\alpha_1 \# \mathbf{1}}$. $\square$

Definition 8.39. We define the function $\omega_n(\alpha)$ for ordinal notations α by

$$\omega_0(\alpha) = \alpha$$
$$\omega_{n+1}(\alpha) = \omega^{\omega_n(\alpha)}$$

In other words,

$$\omega_n(\alpha) = \omega^{\cdot^{\cdot^{\cdot^{\omega^{\alpha}}}}},$$

a "tower" of n ω's with the topmost ω having α as its "exponent."

Proposition 8.40. $\alpha \preceq \omega_n(\alpha)$.

Proof. By induction on n. If $n = 0$, then $\omega_n(\alpha) = \alpha$. For the induction step, assume that $\alpha \preceq \omega_n(\alpha)$. By definition, $\omega_{n+1}(\alpha) = \omega^{\omega_n(\alpha)}$. By Proposition 8.26, $\omega_n(\alpha) \prec \omega_{n+1}(\alpha)$ so also $\omega_n(\alpha) \preceq \omega_{n+1}(\alpha)$. By transitivity of $\preceq$ we have $\alpha \preceq \omega_{n+1}(\alpha)$. $\square$

Proposition 8.41. *If $\alpha \prec \beta$, then $\omega_n(\alpha) \prec \omega_n(\beta)$ for all n.*

Proof. Exercise (use induction on n). $\square$

Problem 8.42. Prove Proposition 8.41, using induction on n.

8.5   Ordinal notations are well-ordered

We have introduced the relation $\prec$ in Definition 8.21 and so far called it an ordering, but strictly speaking, we have not demonstrated that it has all the properties of an ordering. Let's prove this.

Proposition 8.43. *The relation $\prec$ is a strict linear order (an ordering).*

Proof.   1. $\prec$ is transitive: Suppose $\alpha \prec \beta$ and $\beta \prec \delta$. If $\alpha = 0$, i.e., α is of height 0, then β and δ must be of height greater than 0 and so $\alpha \prec \delta$. Otherwise, all of α, β, δ are of height $k > 0$. We can write all three as sums of multiples of powers of ω with exponents γ_i (where $\gamma_i \in O_{\leq k}$):

$$\alpha = \omega^{\gamma_1} \cdot a_1 + \cdots + \omega^{\gamma_n} \cdot a_n$$
$$\beta = \omega^{\gamma_1} \cdot b_1 + \cdots + \omega^{\gamma_n} \cdot b_n$$
$$\delta = \omega^{\gamma_1} \cdot d_1 + \cdots + \omega^{\gamma_n} \cdot d_n$$

For the least j where $a_j \neq b_j$ we have $a_j < b_j$ and the least ℓ where $b_\ell \neq d_\ell$ we have $b_\ell < d_\ell$. If $j < \ell$, then $a_i = b_i = d_i$ for all $i < j$ and $a_j < b_j = d_j$, so $\alpha \prec \delta$. If $\ell \leq j$ then $a_i = b_i = d_i$ for all $i < \ell$ and $a_\ell \leq b_\ell < d_\ell$, and again $\alpha \prec \delta$.

2. $\prec$ is asymmetric. Suppose $\alpha \prec \beta$. By 1, if $\beta \prec \alpha$, then $\alpha \prec \alpha$. So it is enough to show that for all α, $\alpha \not\prec \alpha$. Suppose for a contradiction that $\alpha \prec \alpha$. By Proposition 8.24, $\alpha \prec_k \alpha$, where k is the height of α. But by definition of $\prec_k$, this can happen only by clause b or clause c. The height of α cannot be less than the height of α, since the height of α is uniquely determined. Thus, clause b does not apply. On the other hand, the length of α and the exponents of ω in it are also uniquely determined since α is a string of symbols. So α is neither a proper initial segment of itself, nor do its exponents at some place disagree. So, clause c also does not apply.

3. $\prec$ is total: exercise. $\square$

Problem 8.44. Show that $\prec$ is total, i.e., if $\alpha \neq \beta$ then $\alpha \prec \beta$ or $\beta \prec \alpha$.

Finally, since our consistency proof will depend on induction along the ordering $\langle O, \prec \rangle$, we have to show that it is actually a well-ordering. The best way to see this is to make use of the results on lexicographical orderings from the previous section. There we showed that if $\langle X, < \rangle$ is a well-ordering, then the set $X^*_{\geq}$ of non-increasing sequences of elements of X is well-ordered by the lexicographical ordering $<_{\text{lex}}$. When constructing the ordinal notations of height $k + 1$, we take ordinal notations of height $\leq k$ and form non-increasing sequences of them and order them lexicographically. (We just write them in a complicated way: instead of $\langle \alpha_1, \ldots, \alpha_k \rangle$ with $\alpha_1 \geq_k \alpha_{i+1}$ we write $\omega^{\alpha_1} + \cdots + \omega^{\alpha_k}$.) So we can use Proposition 8.20 to show that each set $\langle O_{\leq k}, \prec_k \rangle$ is a well-ordering.

Proposition 8.45. $\langle O_{\leq k}, \prec \rangle$ is a well-ordering.

Proof. By induction on k. The set $O_{\leq 0}$ is trivially well-ordered, since $\mathbf{0}$ by itself is the only possible decreasing sequence of elements—there are no elements of $O_{=0}$ less than $\mathbf{0}$—and it is finite. Now assume as induction hypothesis that $\langle O_{\leq k}, \prec_k \rangle$ is well-ordered. The set $\langle O_{\leq k+1}, \prec_{k+1} \rangle$ is order-isomorphic to the set of non-increasing sequences of ordinals notations from $O_{\leq k}$, ordered by the lexicographical order based on $\prec_k$. By Proposition 8.20, the latter is well-ordered. $\square$

Corollary 8.46. $\langle O, \prec \rangle$ is a well-ordering.

Proof. Suppose $\alpha_1 \succ \alpha_2 \succ \ldots$ is a decreasing sequence of ordinal notations. The first ordinal α_1 is of some height k. Since any ordinal of height $> k$ is greater than any ordinal of height k, the height of every α_i is $\leq k$. So $\alpha_1 \succ \alpha_2 \succ \ldots$ is a descending sequence also in $\langle O_{\leq k}, \prec_k \rangle$, which is well-ordered by Proposition 8.45.$\square$

We stressed above that it is a common misconception that Gentzen's consistency proof relies on transfinite set theory. We now see a bit more clearly why this is not so. Ordinal notations use symbols reminiscent of the theory of transfinite ordinals, such as the symbol ω also used for the first transfinite ordinal. Nevertheless, they really are just complicated sequences of sequences of . . . sequences of symbols, ordered lexicographically. Think of the ordinal notation **0** of height 0 as just a single symbol—perhaps the digit 0 itself. Then the ordinal notations of height 1 are just sequences of 0s. For instance, the ordinal notation $\omega^0 + \cdots + \omega^0$ (with ℓ ω^0's) corresponds to the sequence $\langle 0, \ldots, 0 \rangle$ (with ℓ 0's). If we add 0 to the set of these sequences (ordered lexicographically, with 0 less than any sequence of 0's), we get a well-ordering order isomorphic to $O_{\leq 1}$. Now we take the non-decreasing sequences, each element of which is either 0 or a sequence of 0's, and well-order them by the lexicographical ordering. This gives us a set order-isomorphic to the ordinal notations of height 2, and so on. For instance, the ordinal notation $\omega^{\omega^0 + \omega^0} + \omega^0$ can be thought of simply as the sequence $\langle \langle 0, 0 \rangle, 0 \rangle$. Ordinal notations are not themselves transfinite sets, and the fact that they are well-ordered does not require facts about transfinite ordinals—just the elementary facts about lexicographical orderings of sequences from Section 8.2.

Problem 8.47. Consider the expressions from Problem 8.22. Write them out as sequences of . . . sequences of 0s. Identify those sequences not corresponding to ordinal notations: the requirement that constituent sequences are non-increasing is violated. Identify which sequences violate it and why.

8.6   Set-theoretic definitions of the ordinals

The reason the expressions we have introduced above are called "ordinal notations" is that they have a close relationship with certain sets called *ordinals*. The ordinal numbers are a generalization of the natural numbers into the transfinite. In set theory, we can define the natural numbers as specific sets in such a way that every number n is represented by a set X_n. X_n has exactly n elements, and so X_n can be put into a one-to-one correspondence with any set that has n elements. Thus these sets X_n can be used as "yardsticks" for the size, or *cardinality*, of finite sets: the cardinality of a finite set is whichever X_n can be put into a one-to-one correspondence with it. This can be extended into the infinite. For instance, the set $\mathbb{N}$ of natural numbers can be put into a one-to-one correspondence with any countably infinite set. (In fact, this is the definition of "countably infinite.")

The same idea can be applied also to well-ordered sets, if we replace "one-to-one correspondence" with the more fine-grained notion of being order isomorphic. For finite well-ordered sets, the two notions coincide (any two finite well-orderings are order isomorphic if, and only if, they have the same number of elements). But for infinite sets, having the same cardinality and being order-isomorphic are not the same. For instance, $\langle \mathbb{N}, < \rangle$ and $\langle \mathbb{N}^2, <_{\text{lex}} \rangle$ are not order-isomorphic even though $\mathbb{N}$ and $\mathbb{N}^2$ have the same cardinality. However, just like we can pick certain sets that represent the cardinality of arbitrary sets, we can pick certain sets to represent the "ordinality," or *order type* of arbitrary well-ordered sets.

In fact, every well-ordered set is isomorphic to an ordinal number.[5] The ordinal notations described above can be viewed as names for ordinals in much the same way that, say, sequences of strokes or a zero followed by a number of primes can be seen as names for the natural numbers. Of course, there are many more ordinals than can be named by our ordinal notations. The first ordinal that cannot be named by them is called ε_0—hence "ordinal notations $< \varepsilon_0$." We'll explain this in more detail after we've taken a closer look at ordinals.

In set theory, the *ordinals* play an important role. Ordinals are the hereditarily transitive sets—intuitively, an ordinal is a set which contains all the sets which are members of members of . . . members of it. Each ordinal is well-ordered by the membership relation $\in$ (which we write as $<$ when we talk about ordinals), and so is any set of ordinals.[6] They are interesting because *every well-ordered set is order-isomorphic to an ordinal*. We can therefore use ordinals as representative for all possible well-orders, and sometimes say that an ordinal represents an *order type*, i.e., all the well-orders that are order-isomorphic to it. It's thus no surprise that ordinals play a special role when talking about well-orderings and induction.

The von Neumann definition of an ordinal is the following:[7]

Definition 8.48. A set α is an *ordinal* if, and only if, it is

1. transitive, i.e., for all $\beta \in \alpha$, $\beta \subseteq \alpha$, and

2. $\langle \alpha, \in \rangle$ is a well-ordering.

If a set α is an ordinal, all its members are ordinals as well. To see this, suppose $\beta \in \alpha$. We first verify that β is transitive, i.e., if $\delta \in \beta$ then $\delta \subseteq \beta$. Let $\gamma \in \delta$; we have to show that $\gamma \in \beta$. Since α is transitive, we have $\beta \subseteq \alpha$, i.e., $\delta \in \alpha$. Again because α is transitive, we have $\delta \subseteq \alpha$ and so $\gamma \in \alpha$. So we have that γ, δ, and β are all members of α. We also have that $\gamma \in \delta$ and $\delta \in \beta$. Since α is an ordinal, $\in$ is a well ordering on α, in particular, a strict linear order, and so is transitive (as an ordering, not in the sense of being a transitive set as defined above). So $\gamma \in \beta$. Since $\beta \subseteq \alpha$, $\langle \beta, \in \rangle$ is a sub-ordering of $\langle \alpha, \in \rangle$, and so also a well-ordering.

The first finite ordinal is $\emptyset$. It is transitive: since it has no elements, all its elements are trivially also subsets of it. It is well-ordered by $\in$, also trivially. The next ordinal is the set containing only $\emptyset$, etc. So the finite ordinals are the following:

$$0 = \emptyset$$
$$1 = \{0\} = \{\emptyset\}$$

---

[5] This relies on a set-theoretic assumption called the Axiom of Choice.

[6] Note that there is no set of all ordinals. This is the Burali-Forti Paradox: Suppose there is a set Ω of all ordinals. As we'll see below, every element of an ordinal is an ordinal. So, every element of an element of Ω is an element of Ω. As just pointed out, every set of ordinals is well-ordered by $\in$. This means that Ω itself is an ordinal: it is transitive and well-ordered by $\in$. So $\Omega \in \Omega$. This, however, contradicts the assumption made in set theory that no set is an element of itself.

[7] Refer to Appendix B for set-theoretic notation and definitions.

$$2 = \{0, 1\} = \{\emptyset, \{\emptyset\}\}$$
$$3 = \{0, 1, 2\} = \{\emptyset, \{\emptyset\}, \{\emptyset, \{\emptyset\}\}\}$$
$$4 = \{0, 1, 2, 3\} = \{\emptyset, \{\emptyset\}, \{\emptyset, \{\emptyset\}\}, \{\emptyset, \{\emptyset\}, \{\emptyset, \{\emptyset\}\}\}\}$$
$$\vdots$$

Not every transitive set is an ordinal. For instance, $\{0, 1, 2, \ldots, \{0, 2, 4, 6, \ldots\}\}$ is transitive, but it is not well-ordered by $\in$. To see this, note that $1 \in 2$ and $2 \in \{0, 2, 4, 6, \ldots\}$, but $1 \notin \{0, 2, 4, 6, \ldots\}$, so $\in$ as a relation on this set is not transitive. Since it is not transitive, it is not a linear order, and so also not a well-order.

Problem 8.49. Verify that 3 is an ordinal.

As you might guess from the abbreviations we've given to the finite ordinals, they are identified in set theory with the natural numbers. In addition to this definition, the ordinals—like the natural numbers—come with a natural ordering and with basic operations. We can order the ordinals by the membership relation $\in$, and write $\alpha < \beta$ for $\alpha \in \beta$. This is a well-ordering, i.e., every set of ordinals has a least element.

You can see in the list of finite ordinals that each one is the set of all ordinals preceding it. This is a general fact: every ordinal is identical to the set of ordinals that precede it.

Proposition 8.50. *If α is an ordinal then $\alpha = \{\beta : \beta$ is an ordinal and $\beta < \alpha\}$.*

Proof. Suppose α is an ordinal. Since $<$ is just $\in$, if $\beta < \alpha$ then $\beta \in \alpha$, so the right-to-left inclusion is trivial. To show that $\alpha \subseteq \{\beta : \beta$ is an ordinal and $\beta < \alpha\}$, recall that we've shown above that every member of an ordinal is itself an ordinal.□

Problem 8.51. Show that we could alternatively define $\alpha < \beta$ as $\alpha \subsetneq \beta$, i.e., show for ordinals α, β that $\alpha \in \beta$ if, and only if, $\alpha \subsetneq \beta$.

Problem 8.52. Show that for any two ordinals α and β, $\alpha \cap \beta$ is an ordinal.

Problem 8.53. Show that for any two ordinals α and β, either $\alpha \in \beta$, $\alpha = \beta$, or $\beta \in \alpha$.

Problem 8.54. Show that for any two ordinals α and β, if α and β are order isomorphic, then $\alpha = \beta$.

The flip side of the last proposition is that the set consisting of a given ordinal α together with all the ordinals that precede α, i.e., $\{\alpha\} \cup \alpha$ is itself an ordinal.

Definition 8.55 (Successor ordinal). If α is an ordinal, then $S(\alpha) = \alpha \cup \{\alpha\}$.

Proposition 8.56. *If α is an ordinal, $\alpha < S(\alpha)$ and no ordinal β is such that $\alpha < \beta < S(\alpha)$.*

Proof. That $\alpha < S(\alpha)$ is obvious, since $\alpha \in \alpha \cup \{\alpha\}$. Now suppose $\alpha < \beta$. Then $\alpha \in \beta$ (by definition) and $\alpha \subsetneq \beta$ (by the previous problem). But that means that $\alpha \cup \{\alpha\} \subseteq \beta$, i.e., $S(\alpha) \le \beta$. $\qquad\qquad\square$

We'll define addition for ordinals in Section 8.8, and it will turn out that $\alpha + 1$ and $S(\alpha)$ are the same. So from now on, we'll also use $\alpha + 1$ as the successor of α.

The successor operation gives us all the finite ordinals other than 0, but there are infinite ordinals as well. For instance, the set of *all* finite ordinals,

$$\omega = \{0, 1, 2, 3, \dots\}$$

is an ordinal, as each of its members is also a subset (note that $n = \{0, \dots, n-1\}$ and $0, \dots, n-1 \in \omega$), so ω is transitive, and $\in$ is a well-ordering of ω, since it is order-isomorphic to $<$ on $\mathbb{N}$.

Proposition 8.57. *If $\langle \alpha_i : i = 1, 2, \dots \rangle$ is a sequence of ordinals that is increasing (i.e., $\alpha_i < \alpha_{i+1}$) then*

$$\bigcup_i \alpha_i = \{x : x \in \alpha_i \text{ for some } i\}$$

is an ordinal and is the least ordinal $\beta > \alpha_i$ for all $i = 1, 2, \dots$.[8]

Proof. Suppose $\alpha \in \bigcup_i \alpha_i$. Then for some i, $\alpha \in \alpha_i$. Since α_i is an ordinal, so are all its elements, i.e., α is an ordinal. This means that $\bigcup_i \alpha_i$ is a set of ordinals.

A set of ordinals is *downward closed* if it contains all ordinals less than any member of it. Just as in the case of ω discussed above, any such set is itself an ordinal, since every member of it is an ordinal, and an ordinal is equal to the set of ordinals less than it. If X is downward closed, all these ordinals are members of X. Since α_i is an ordinal, all ordinals $\beta < \alpha \in \alpha_i$ are in α_i and hence in $\bigcup_i \alpha_i$, so $\bigcup_i \alpha_i$ is downward closed. That makes $\bigcup_i \alpha_i$ an ordinal.

Obviously, for every i, $\alpha_i < \bigcup_i \alpha_i$, since $\alpha_i \in \alpha_{i+1} \subseteq \bigcup_i \alpha_i$. Suppose β is an ordinal that also has this property, i.e., $\alpha_i \in \beta$ for all i. Then, because β is transitive, $\alpha_i \subseteq \beta$, i.e., $\bigcup_i \alpha_i \subseteq \beta$. And that means that $\bigcup_i \alpha_i \le \beta$. $\qquad\square$

The ordinal $\bigcup_i \alpha_i$ is called the *limit* or *supremum* of the sequence α_i. ω is the limit of the sequence $\langle 1, 2, 3, \dots \rangle$ of finite ordinals. It is the least ordinal which is greater than all finite ordinals.

Every ordinal that is the successor of some other ordinal is called a *successor ordinal*. All finite ordinals except 0 are successor ordinals. Ordinals other than 0 that are not successor ordinals are called *limit ordinals*.

---

[8] The axioms of set theory guarantee the existence of this set.

8.7   Constructing ε_0 from below

We can always generate a new ordinal from an increasing sequence of ordinals. For instance, we can consider the limit (union) of

$$\omega < \omega + 1 < (\omega + 1) + 1 < \ldots,$$

i.e., the limit of the sequence ω, $S(\omega)$, $S(S(\omega))$, etc. That ordinal is called $\omega + \omega$. In the same way we can consider the limit of

$$(\omega + \omega) < (\omega + \omega) + 1 < ((\omega + \omega) + 1) + 1 < \ldots$$

which is called $\omega + \omega + \omega$. Obviously we can keep going, and then consider the sequence of all these ordinals:

$$\omega < \omega + \omega < \omega + \omega + \omega < \ldots$$

and that gives us an ordinal called $\omega \cdot \omega$, or ω^2. We can repeat the entire process starting from ω^2 to obtain

$$\omega^2 + \omega = \omega^2 \cup (\omega^2) + 1 \cup ((\omega^2) + 1) + 1 \cup \ldots$$
$$\omega^2 + \omega + \omega = (\omega^2 + \omega) \cup (\omega^2 + \omega) + 1 \cup ((\omega^2 + \omega) + 1) + 1 \cup \ldots$$
$$\omega^2 + \omega + \omega + \omega = (\omega^2 + \omega + \omega) \cup (\omega^2 + \omega + \omega) + 1 \cup ((\omega^2 + \omega + \omega) + 1) + 1 \cup \ldots$$

$$\vdots$$

The union of all of these is $\omega^2 + \omega^2$. In the same way we can obtain $\omega^2 + \omega^2 + \omega^2$ and so on; the union of all of these is ω^3. From this we can get in the same way by iterating the successor operation and then taking unions, first $\omega^3 + \omega$, then iterating this entire operation and taking unions $\omega^3 + \omega^2$, then iterating *that* entire operation and taking the union $\omega^3 + \omega^3$. If we iterate this entire operation yet again, we get the sequence

$$\omega^3 < \omega^3 + \omega^3 < \omega^3 + \omega^3 + \omega^3 < \ldots$$

and the union of these is ω^4. Repeat this process to get ω^5, ω^6, etc., and call the union ω^ω.

   If we go back to the beginning and start with ω^ω instead of 0 and run through this entire process we'll end with $\omega^\omega + \omega^\omega$. Starting with this instead of 0, we get $\omega^\omega + \omega^\omega + \omega^\omega$. Taking the union of the resulting increasing sequence, we obtain an ordinal called $\omega^{\omega+1}$. The entire process thus gives us an "exponential successor," a process for going from an ordinal ω^α to $\omega^{\alpha+1}$. If we now go through the entire exercise again but use this process of "exponential successor" where we originally used +1, we'd construct the ordinals

$$\omega^\omega < \omega^{\omega+1} < \omega^{(\omega+1)+1} < \ldots \omega^{\omega+\omega} < \omega^{\omega+\omega+\omega} < \cdots < \omega^{\omega^2} < \omega^{\omega^3} < \cdots < \omega^{\omega^\omega}.$$

Now let's keep iterating this to get an increasing sequence $\omega < \omega^\omega < \omega^{\omega^\omega} < \ldots$. The union of this sequence is ε_0.

8.8 Ordinal arithmetic

In the preceding section we looked at a way of constructing new, larger ordinals from old ones by using the successor operation and taking infinite unions. The names we've used for the new ordinals are suggestive: there are indeed ways to define addition, multiplication, and exponentiation of ordinals, and the ordinals we've defined can also be obtained using these operations, e.g., ω^2 is indeed $\omega \cdot \omega$.

The definition of arithmetical operations on ordinals, however, requires the following important fact, which we'll just state here without proof.

Theorem 8.58. *If $\langle X, < \rangle$ is a well-ordered set then $\langle X, < \rangle$ is order isomorphic to an ordinal.*

Definition 8.59. If α and β are ordinals, then $\alpha + \beta$ is the ordinal isomorphic to the set $\{\langle 0, \alpha' \rangle : \alpha' < \alpha\} \cup \{\langle 1, \beta' \rangle : \beta' < \beta\}$ with the ordering $<$ defined by

$$\langle i, \gamma \rangle < \langle j, \gamma' \rangle \text{ if, and only if, } i < j \text{ or } i = j \text{ and } \gamma < \gamma'.$$

Problem 8.60. Show that the ordinal sum of α and $1 = \{\emptyset\}$ is the successor $S(\alpha) = \alpha \cup \{\alpha\}$. In other words, it's ok to use $\alpha + 1$ as an alternative notation for the successor of α.

The well-ordering corresponding to $\alpha + 1$ would be α "followed by" a single point which is greater than all the ordinals $< \alpha$. The ordinal isomorphic to this well-order is obviously $\alpha \cup \{\alpha\}$, i.e., $\alpha + 1$ as defined in Definition 8.55. In general, the well-ordering defined for $+$ is the well-ordering of the ordinals $< \alpha$ "followed by" the ordinals $< \beta$. Note that $+$ therefore is not commutative. For instance, $2 + \omega$ is the finite well-ordering 2 followed by the infinite well-ordering ω, i.e.,

$$\mathsf{\cdot\,\mathsf{I}\quad\cdot\mathsf{I}\mathsf{I}\mathsf{I}\mathsf{I}\mid\cdots}$$

But that is clearly isomorphic to

$$\mathsf{\cdot\mathsf{I}\mathsf{I}\mathsf{I}\mathsf{I}\mathsf{I}\mid\cdots}\,,$$

i.e., the well-ordering ω. By contrast, $\omega + 2$ is

$$\mathsf{\cdot\mathsf{I}\mathsf{I}\mathsf{I}\mathsf{I}\mathsf{I}\mathsf{I}\mid\cdots\quad\cdot\mathsf{I}\mathsf{,}}$$

which is not isomorphic to ω, but to $(\omega + 1) + 1$.

Definition 8.61. If α and β are ordinals, then $\alpha \cdot \beta$ is the ordinal isomorphic to the set $\alpha \times \beta = \{\langle \alpha_1, \beta_1 \rangle : \alpha_1 < \alpha \text{ and } \beta_1 < \beta\}$ with the ordering $<$ defined by

$$\langle \alpha_1, \beta_1 \rangle < \langle \alpha_1', \beta_1' \rangle \text{ if, and only if, } \beta_1 < \beta_1' \text{ or } \beta_1 = \beta_1' \text{ and } \alpha_1 < \alpha_1'.$$

The well-ordering so defined on $\alpha \times \beta$ is essentially the well-ordering β, where each element is replaced with a copy of α. Multiplication is also not commutative, as e.g., $2 \cdot \omega$ is the ordinal isomorphic to ω-many copies of $\shortmid\shortmid$, i.e.,

$$\shortmid\shortmid \quad \shortmid\shortmid \quad \shortmid\shortmid \quad \cdots$$

which is ω. On the other hand, $\omega \cdot 2$ is the ordinal isomorphic to two copies of ω, one after another, i.e.,

$$\shortmid\shortmid\shortmid\shortmid\shortmid\shortmid \Big| \cdots \shortmid\shortmid\shortmid\shortmid\shortmid\shortmid \Big| \cdots$$

And as you might expect, $\omega \cdot 2$ is the limit (union) of

$$\omega < \omega + 1 < \omega + 2 < \omega + 3 < \ldots$$

and $\omega \cdot \omega$ the limit (union) of

$$\omega < \omega \cdot 2 < \omega \cdot 3 < \omega \cdot 4 < \ldots.$$

Once we have ordinal multiplication, we can use iteration to define finite exponentiation. So $\omega^2 = \omega \cdot \omega$, $\omega^3 = \omega^2 \cdot \omega$, etc. Each one of these ordinals ω^n is order isomorphic to the set of n-tuples of natural numbers, ordered lexicographically. Equivalently, we can consider the anti-lexicographical order:

$$\langle k_1, \ldots, k_n \rangle \leq \langle k'_1, \ldots, k'_n \rangle \text{ if, and only if, } k_j < k'_j \text{ and, for some } j \leq n,$$
$$k_i = k'_i \text{ for } i = j + 1, \ldots, n.$$

The union of all ordinals ω^n for $n = 1, 2, \ldots$ is itself an ordinal of course, and it is order-isomorphic to the set of finite sequences of natural numbers, ordered by the anti-shortlex order: shorter sequences are smaller, and for sequences of the same length, sequences are ordered anti-lexicographically:

$$\langle k_1, \ldots, k_n \rangle \leq \langle k'_1, \ldots, k'_{n'} \rangle \text{ if, and only if, } n < n', \text{ or } n = n' \text{ and for some } j \leq n,$$
$$k_j < k'_j \text{ and } k_i = k'_i \text{ for } i = j + 1, \ldots, n.$$

This suggests the following definition for exponentiation:

Definition 8.62. If α and β are ordinals, then α^β is the ordinal isomorphic to the set

$$\{f : \beta \to \alpha : f(\gamma) > 0 \text{ for finitely many } \gamma < \beta\}$$

with the ordering $<$ defined by

$$f < f' \text{ if, and only if, } f(\gamma) < f'(\gamma) \text{ and } f(v) = f'(v) \text{ for all } v > \gamma.$$

For $\alpha = \beta = \omega$, the functions $f : \omega \to \omega$ which yield 0 for all but finitely many arguments correspond to the finite sequences of numbers $\langle f(0), f(1), \ldots, f(n) \rangle$ where $f(n) \neq 0$ but $f(i) = 0$ for all $i > n$. If $\beta > \omega$, then we're no longer dealing with finite sequences of natural numbers, but with sequences of ordinals less than α of "length" less than β.

The limit of the sequence $\omega, \omega^{\omega}, \omega^{\omega^{\omega}}$ is ε_0. It satisfies $\varepsilon_0 = \omega^{\varepsilon_0}$, and obviously is the smallest ordinal that does. If you think of ε_0 as an exponential stack of infinitely many ω's this is clear: if

$$\varepsilon_0 = \omega^{\omega^{\omega^{\cdot^{\cdot^{\cdot}}}}}, \text{ then } \omega^{\varepsilon_0} = \omega^{\left(\omega^{\omega^{\omega^{\cdot^{\cdot^{\cdot}}}}}\right)} = \omega^{\omega^{\omega^{\cdot^{\cdot^{\cdot}}}}} = \varepsilon_0.$$

To see that the ordinal notations $< \varepsilon_0$ introduced earlier deserve their name, we show that they are order-isomorphic to ε_0. We make use of the following fundamental theorem, which we won't prove.

Theorem 8.63 (Cantor normal form). *For every ordinal α there is a unique sequence of ordinals $\beta_1 \geq \beta_2 \geq \cdots \geq \beta_n$ such that*

$$\alpha = \omega^{\beta_1} + \omega^{\beta_2} + \cdots + \omega^{\beta_n}.$$

If $\alpha < \varepsilon_0$, then each $\beta_i < \alpha$. Since the exponents β_i can itself be written in this form, we can keep doing this until $\beta_i < \omega$, then $\beta_i = \omega^0 + \omega^0 + \cdots + \omega^0$ (sum of n ω^0's, for some n), or $\beta_i = 0$. Hence, the Cantor normal form of an ordinal $< \varepsilon_0$ corresponds exactly to an expression in our ordinal notation system.

Definition 8.64. The ordinal notation $o(\alpha)$ for an ordinal $\alpha < \varepsilon_0$ is:

$$o(\alpha) = \begin{cases} 0 & \text{if } \alpha = 0 \\ \omega^{o(\beta_1)} + \cdots + \omega^{o(\beta_n)} & \text{if the Cantor normal form of } \alpha \text{ is } \omega^{\beta_1} + \cdots + \omega^{\beta_n} \end{cases}$$

This is well-defined since the Cantor normal form is unique, and $\beta_i < \alpha$ if $\alpha < \varepsilon_0$.

The ordering defined on ordinal notations is the same as the ordering of the corresponding ordinals, i.e., $\alpha < \alpha'$ if, and only if, $o(\alpha) \prec o(\alpha')$.

8.9   Trees and Goodstein sequences

We now have two order-isomorphic examples of well-orders of order type ε_0: the ordinals $< \varepsilon_0$ ordered by $<$, and the ordinal notations $< \varepsilon_0$ ordered by $\prec$.

An additional, and useful, example of a well-ordering of type ε_0 is provided by finite, finitely branching trees.

Definition 8.65. The set $\mathbb{T}$ of finite, finitely branching trees can be inductively defined as follows:

1. *Basis clause:* The tree consisting of a single node is a tree (of height 1).

2. *Inductive clause:* If $T_1, \ldots, T_n$ are trees, then the following is a tree of height $k+1$ (where k is the maximum among the heights of $T_1, \ldots, T_n$):

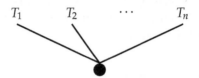

3. *Extremal clause:* Nothing else is a tree.

When we draw trees, the order in which the subtrees $T_1, \ldots, T_n$ are attached to a node does not matter, i.e.,

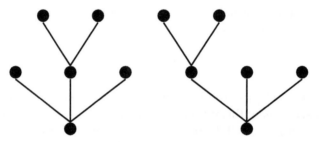

are the same tree. The bottommost node is called the *root*, the topmost nodes the *leaves*. If a node is connected to another above it, the latter is called a *descendant* of the former, and the former the *parent* of the latter.

Since trees are constructed inductively, we can define an ordering $<$ on them inductively (according to height) as well.

Definition 8.66.

1. *Basis clause:* The tree of height 1 (i.e., consisting only of its root) is $=$ only to itself, and $<$ all trees of height > 1.

2. *Inductive clause:* Suppose T, T' are trees of height > 1, and let h be the maximum of their heights. We may assume that $\leq$ has been defined for trees of height $< h$. Then we can write T and T' as

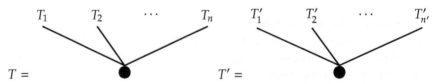

where each T_i, T'_i has height $< h$, and where $T_i \geq T_{i+1}$ and $T'_i \geq T'_{i+1}$. (In other words, we order the subtrees attached to the root according to $\geq$, i.e., in non-increasing order, just as we do the "exponents" in an ordinal notation $\omega^{\alpha_1 + \cdots + \alpha_n}$.)

$T = T'$ if, and only if, $n = n'$ and for every $i = 1, \ldots, n$, $T_i = T_i'$.

$T < T'$ if, and only if, either

(a) $n < n'$ and $T_i = T_i'$ for all $i \le n$, or

(b) there is a k such that for all $i < k$, $T_i = T_i'$ and $T_k < T_k'$

3. *Extremal clause:* No trees T, T' stand in the relation $T < T'$ unless they do so by clause 1 or 2.

Problem 8.67. Verify that $\langle \mathbb{T}, < \rangle$ is actually an ordering (i.e., a strict linear order; see Definition 8.2).

Problem 8.68. Show that $\langle \mathbb{T}, < \rangle$ is order isomorphic to ε_0.

Induction up to ε_0 can be used to prove some interesting results. Two that can be described relatively easily are the termination of the Hydra game and Goodstein's theorem.

Imagine a tree (as defined in the previous section) as a many-headed monster, with the leaves of the tree being the heads. In the ancient Greek myth about Hercules and the Hydra, for each head that Hercules cut off, the Hydra grew two new ones. In the Hydra game, the leaf nodes represent the "heads." Instead of every head removed being replaced by two new ones, when a leaf node is removed, the removed leaf's parent node and its remaining descendants are multiplied n times. For instance, if you cut off the indicated head below (white), the remaining subtree (dashed) of height 2 will be multiplied n times (any n will do). Here's what this looks like for $n = 4$. The tree:

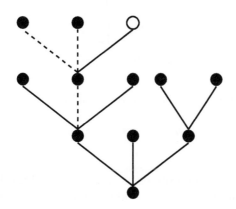

turns into the tree:

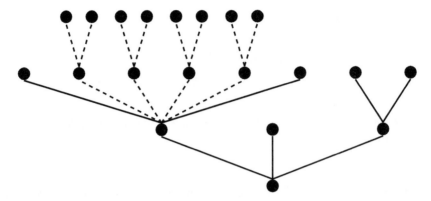

Theorem 8.69 (Kirby and Paris, 1982). *If you keep cutting off heads (in any manner), the Hydra eventually dies.*

Proof. We assign ordinal notations $< \varepsilon_0$ to nodes in trees as follows:

1. Heads (i.e., leaf nodes) are assigned 0.

2. If the ordinal notations $\alpha_1 \succeq \cdots \succeq \alpha_n$ are assigned to the descendants of a node, the node itself is assigned $\omega^{\alpha_1} + \cdots + \omega^{\alpha_n}$.

For instance, in the example above we have

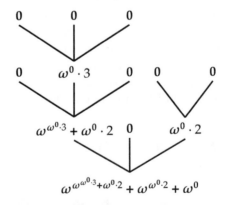

Under this assignment, a node underneath a head is assigned

$$\underbrace{\omega^{\alpha_1} + \cdots + \omega^{\alpha_n}}_{\alpha} + \omega^0,$$

where $\alpha_1 \succeq \cdots \succeq \alpha_n$ are the ordinals assigned to the other descendants of that node, in non-increasing order. The node two levels below a head is therefore assigned the ordinal

$$\underbrace{\omega^{\beta_1} \# \cdots \# \omega^{\beta_m}}_{\beta} \# \omega^{\alpha + \omega^0},$$

where the ordinals β_i are the ordinals assigned to the descendants of that node other than the one leading to the head in question.

If the head is removed, the ordinal notation assigned to its immediate predecessor is now $\alpha < \alpha + 1 = \alpha + \omega^0$. If the remaining part is copied n times, the ordinal assigned to its descendant is now

$$\beta \# \underbrace{\omega^{\alpha} + \cdots + \omega^{\alpha}}_{\gamma}$$

But $\gamma < \omega^{\alpha+1}$. So removing a head always decreases the ordinal assigned to the root. Eventually, therefore, the root must be assigned 0. □

Problem 8.70. Compute the ordinal notation assigned to the result of removing the head from the example tree and verify that the ordinal notation is less than that assigned to the original tree, $\omega^{\omega^{\omega^{0.3}+\omega^{0.2}}} + \omega^{\omega^{0.2}} + \omega^0$.

Another application of ordinal notations $< \varepsilon_0$ is Goodstein's theorem (Goodstein, 1944). To state it, recall that every number n can be written in base b notation, for any number $b \geq 2$, as

$$n = a_k b^k + a_{k-1} b^{k-1} + \cdots + a_1 b + a_0$$

with $a_i < b$ and $a_k > 0$. For instance, the base $b = 10$ notation is simply ordinary decimal notation where the a_i are the digits of the number, e.g.,

$$25{,}739 = 2 \cdot 10^4 + 5 \cdot 10^3 + 7 \cdot 10^2 + 3 \cdot 10 + 9.$$

In *hereditary* base b notation, the exponents themselves are expressed in hereditary base b notation. The above way of expressing 25,739 is already in hereditary base 10 notation, since each exponent of 10 in it is less than 10. If we write it in base 3 notation, however, we get

$$25{,}739 = 3^9 + 2 \cdot 3^7 + 2 \cdot 3^4 + 2 \cdot 3^3 + 2 \cdot 3 + 2.$$

Here, the exponents have non-trivial base 3 representations: $9 = 3^2$, $7 = 2 \cdot 3 + 1$, and $4 = 3 + 1$. The hereditary base 3 representation then is

$$25{,}739 = 3^{3^2} + 2 \cdot 3^{2 \cdot 3 + 1} + 2 \cdot 3^{3+1} + 2 \cdot 3^3 + 2 \cdot 3 + 2.$$

The *Goodstein sequence* for some number n is the sequence $n = n_2, n_3, \ldots$ where the number n_{i+1} is obtained by writing n_i in hereditary base-i notation, changing the i's into $(i+1)$'s and subtracting 1. If n_3 were the preceding number, n_4 would be

$$4{,}297{,}067{,}018 = (4^{4^2} + 2 \cdot 4^{2 \cdot 4 + 1} + 2 \cdot 4^{4+1} + 2 \cdot 4^4 + 2 \cdot 4 + 2) - 1.$$

So, elements of a Goodstein sequence can become very large very quickly. Let's compute the first few elements of the Goodstein sequence starting with $n_2 = 10$. First we write 10 in hereditary base 2 notation to obtain

$$n_2 = 10 = 2^3 + 2 = 2^{2+1} + 2.$$

Then n_3 is

$$n_3 = 3^{3+1} + 3 - 1 = 83.$$

For n_4, n_5, n_6, n_7, we get the values:

$$n_4 = 4^{4+1} + 2 - 1 = 4^{4+1} + 1 = 1{,}025$$
$$n_5 = 5^{5+1} + 1 - 1 = 5^{5+1} = 15{,}625$$
$$n_6 = 6^{6+1} - 1 = 279{,}935 =$$
$$= 5 \cdot 6^6 + 5 \cdot 6^5 + 5 \cdot 6^4 + 5 \cdot 6^3 + 5 \cdot 6^2 + 5 \cdot 6 + 5$$
$$n_7 = 5 \cdot 7^7 + 5 \cdot 7^5 + 5 \cdot 7^4 + 5 \cdot 7^3 + 5 \cdot 7^2 + 5 \cdot 7 + 5 - 1 =$$
$$= 4{,}215{,}754$$

Problem 8.71. If $n_2 = 100$, what are n_3 and n_4?

Despite appearances, any Goodstein sequence eventually terminates at 0! The idea to prove this is actually quite simple. Imagine writing a number n_b given in hereditary base b notation instead in "hereditary base-ω" notation $\alpha(n_b, b)$ by replacing each b by ω: then it turns into one of our ordinal notations $< \varepsilon_0$. And it is not hard to see that as a Goodstein sequence progresses from n_i to n_{i+1}, the corresponding ordinal notation decreases. For instance, the ordinal notations corresponding to n_4, n_5, and n_6 in the preceding example would be:

$$\alpha(n_4, 5) = \omega^{\omega+1} + 1$$
$$\alpha(n_5, 5) = \omega^{\omega+1}$$
$$\alpha(n_6, 6) = \omega^{\omega} \cdot 5 + \omega^5 \cdot 5 + \omega^4 \cdot 5 + \omega^3 \cdot 5 + \omega^2 \cdot 5 + \omega^1 \cdot 5 + 5$$

Theorem 8.72. *The Goodstein sequence for n eventually reaches 0, for all n.*

Proof. Given the base b representation of a number n, let $\alpha(n, b)$ be the ordinal notation resulting from it by replacing every b by ω. Obviously, $\alpha(n, b)$ is an ordinal notation $< \varepsilon_0$. Then let the ordinal assigned to $n_i = \alpha(n_i, i)$. We show that $\alpha(n_i, i) \succ \alpha(n_{i+1}, i+1)$.

Suppose the base-i representation of n_i is $a_k i^k + \cdots + a_j i^j + a_0$ with $a_i < i$ and $a_j > 0$. Then $\alpha(n_i, i) = \omega^{\alpha_k} \cdot a_k + \cdots + \omega^{\alpha_j} \cdot a_j + a_0$. The number n_{i+1} was defined as $n_i' - 1$ where n_i' is the result of replacing all i's in the hereditary base-i representation of n_i by $(i+1)$'s. Obviously, $\alpha(n_i', i+1) = \alpha(n_i, i)$.

Now if $a_0 \geq 1$, then

$$n_{i+1} = n_i' - 1 = a_k i^k + \cdots + a_j i^j + (a_0 - 1) \text{ and}$$
$$\alpha(n_{i+1}, i+1) = \omega^{\alpha_k} \cdot a_k + \cdots + \omega^{\alpha_j} \cdot a_j + \ell \prec \alpha(n_i, i),$$

where $\ell = (a_0 - 1)$. If $a_0 = 0$, then

$$n_{i+1} = n_i' - 1 = a_k(i+1)^k + \cdots + (a_j - 1)(i+1)^j + \sum_{s=0}^{j-1} i(i+1)^s \text{ hence}$$

$$\alpha(n_{i+1}, i+1) = \omega^{\alpha_k} \cdot a_k + \cdots + \omega^{\alpha_j} \cdot (a_j - 1) + \sum_{s=0}^{j-1} \omega^{\alpha(s,i+1)} \cdot i.$$

It's easy to see that if $s < r$ then $\alpha(s, b) \prec \alpha(r, b)$, so the additional $\alpha(s, i+1) \prec \alpha_j$. The highest power of ω where $\alpha(n_i, i)$ and $\alpha(n_{i+1}, i+1)$ differ is ω^{α_j}. In $\alpha(n_i, i)$, this occurs as $\omega^{\alpha_j} \cdot a_j$ and in $\alpha(n_{i+1}, i+1)$ as $\omega^{\alpha_j} \cdot (a_j - 1)$. So we again have $\alpha(n_i, i) \succ \alpha(n_{i+1}, i+1)$. $\qquad\square$

Problem 8.73. Compute the first 5 entries of the Goodstein sequence for $n_2 = 5$. What is the corresponding sequence of $\alpha(n_i, i)$?

9

The consistency of arithmetic, continued

9.1 Assigning ordinal notations $< \varepsilon_0$ to proofs

Having considered an example in Section 7.10, let's now consider the general cases of the reduction steps. We are given a proof in **PA** of a sequent consisting of only atomic sentences. Axioms are also atomic (Proposition 5.17), the proof has already been made regular (Proposition 7.16), and all free variables in it are eigenvariables (Proposition 7.20).

Recall that there are three transformations we have to apply to the end-part of the proof.

1. Replace suitable induction inferences (those where the formula $F(\overline{n})$ in the succedent of the conclusion contains no free variables) in the end-part by successive CUTS.

2. Eliminate weakenings from the end-part.

3. Reduce suitable complex CUTS in the end-part.

The reduction of a complex CUT may introduce new weakenings, and since it also removes operational inferences, the boundary may change and so weakenings, logical axioms, and induction inferences that were above the boundary inferences before may now fall in the end-part.

We have already shown (Proposition 7.36) that the procedure does not get stuck: there is always a suitable induction inference in the end-part if there are induction inferences at all, and once steps 1 and 2 have been applied enough times and all induction inferences and weakenings have been removed from the end-part, there must exist a suitable complex CUT to which step 3 applies. What we have not shown is that this procedure eventually terminates, and when it does, it has produced a simple proof.

This is where ordinal notations come in. We assign to each proof an ordinal notation $< \varepsilon_0$. Simple proofs are assigned ordinal notations $\prec \omega^1$, and non-simple proofs get notations $\succ \omega^1$. We show that in the reduction during step 2, the ordinal notation assigned to the proof does not increase. However, in this case,

An Introduction to Proof Theory: Normalization, Cut-Elimination, and Consistency Proofs.
Paolo Mancosu, Sergio Galvan, and Richard Zach, Oxford University Press. © Paolo Mancosu,
Sergio Galvan and Richard Zach 2021. DOI: 10.1093/oso/9780192895936.003.0009

the proof becomes simpler and these steps can only be applied a finite number of times. In the reduction of CJ inferences and CUTS, the ordinal notation assigned to the resulting proof is smaller than that assigned to the original. Since the ordinal notations $< \varepsilon_0$ are well-ordered (Corollary 8.46), the procedure must eventually reach a proof with ordinal notation $< \omega^1$. A proof with an ordinal notation $< \omega^1$ assigned to it cannot contain complex CUTS, so if it is a proof whose end-sequent contains only atomic sentences, it must be simple.

It should be stressed here that the assignment of ordinal notations to proofs is to some degree arbitrary: it is determined by the particular reduction method we have defined. Other reduction methods are in principle possible which might require a different assignment of ordinal notations.[1] However, the order type of the ordinal notations is not quite arbitrary. No natural ordinal notation system of order type $< \varepsilon_0$ would work. In fact, Gentzen (1943) showed that for each $\alpha < \varepsilon_0$, induction up to α is provable in **PA**. So if we could prove the consistency of **PA** using only ordinal notations less than some $\alpha < \varepsilon_0$, **PA** would prove its own consistency, contrary to Gödel's second incompleteness theorem. Gentzen also showed, without appealing to Gödel's incompleteness theorem, that induction up to ε_0 itself is not provable in **PA**. The proof of this is beyond the scope of this book, however.

In order to describe how the ordinal notations are assigned to sequents and inferences in a proof, we must first define the *level* of a sequent in a proof. Let the degree of a CUT or of a CJ inference be the degree of the cut-formula, or of the induction formula $A(a)$ in a CJ inference, respectively.

Definition 9.1. The *level* of a sequent S in a proof π is the maximum degree of all CUTS and induction inferences below S in π. If there are no such inferences below S in π, the level is 0.[2]

To find the level of a sequent, trace the path between it and the end-sequent, note the degree of every CUT and every induction along that path, and take the largest of these. Thus, since the level of each sequent depends on the degrees of all CUTS and induction inferences below it, as we go down that path the level of sequents along the path stays the same or it decreases; it can never increase. We'll call the topmost inference below S where the premise has the same level as S and the conclusion has a lower level, the *level transition* for S (Gentzen uses "level line" in a similar way). This inference must be a CUT or induction, since in other inferences the level cannot change. Obviously if a CUT or CJ inference is a

---

[1] In fact, Gentzen's original method is slightly different from ours: he allowed non-atomic logical axioms in proofs, but included a separate step that removed them from the end-part.

[2] Here we diverge from Gentzen slightly as well: In Gentzen's definition, the degree of atomic formulas is 1, so the level of a sequent with a CUT or CJ inference below it is always greater than 0. This means that the ordinal notations his proof assigns are different from those we will assign, and his procedure eliminates *all* CUTS, including atomic CUTS. But this is only possible if, as he did, we assume that we start with a proof of the empty sequent.

level transition, the level of its premise(s) is equal to its degree. So, the degree of the CUT or induction formula in the level transition for S is equal to the level of S. All inductions and CUTs below the level transition have a lower degree. If a CUT inference is a level transition, both of its premises have the same level, namely the degree of the cut-formula.

For instance, consider the example from Section 7.10 (see p. 305), with levels of the sequents marked explicitly:

$$
\cfrac{1 \ \Rightarrow F(0) \quad \cfrac{\cfrac{\vdots \ \pi_1(a)}{1 \ F(a) \Rightarrow F(a')}\ \text{CJ}}{1 \ F(0) \Rightarrow F(b)}}{\cfrac{\cfrac{1 \ \Rightarrow F(b)}{1 \ \Rightarrow \forall x\, F(x)}\ \forall\text{R}}{\quad} \quad \cfrac{\cfrac{1 \ F(\overline{3}) \Rightarrow F(\overline{3})}{1 \ \forall x\, F(x) \Rightarrow F(\overline{3})}\ \forall\text{L}}{\quad}}{0 \ \Rightarrow F(\overline{3})}\ \text{CUT}
$$

Only the last CUT has a cut-formula of degree 1, the CUT inferences in $\pi_1(a)$ as well as the CJ inference and the CUT immediately following it have degree 0 (recall that $F(a)$ is $0 + a = a + 0$, so it is atomic and has degree 0). Consider for instance the sequent $F(a) \Rightarrow F(a')$. Along the path between it and the end-sequent there is one CJ inference with principal formula of degree 0, a CUT of degree 0, and a CUT of degree 1. So the level of that sequent is the maximum of these numbers, namely 1. In fact, the level of every sequent along that path other than the end-sequent is 1.

Problem 9.2. Assign levels to each sequent in the derivation on p. 286.

We are now ready to define the assignment of ordinal notations to sequents and inferences. If S is a sequent in π or I is an inference in the proof, we write $o(S; \pi)$ or $o(I; \pi)$ for the ordinal notation assigned. If S is the end-sequent of π, the ordinal notation $o(\pi)$ assigned to the proof is $o(S; \pi)$.

The assignment of ordinal notations to sequents in a proof π is defined inductively, similarly to how we have defined measures of complexity of sequents in proofs before (recall, e.g., the definition of degree and rank of a mix). However, here there is a wrinkle. The ordinal notation assigned to a sequent will not just depend on the sub-proof ending in that sequent, but also on the position of the sequent in π. Specifically, the ordinal notation assigned to S depends on the level of S, and the level of S in turn depends on what CUT and CJ inferences appear in π *below S.*

Definition 9.3. We inductively define the ordinal notation $o(I; \pi)$ assigned to an inference I in π on the basis of the type of inference and the ordinal notation assigned to the premise(s) of I. The ordinal notation assigned to the conclusion of I is then determined by the ordinal notation assigned to I itself and the levels of the premise(s) and conclusion of I. Specifically, we define $o(I; \pi)$ and $o(S; \pi)$ as follows.

1. *Basis clause:* If S is an initial sequent, then $o(S;\pi) = 1 = \omega^0$.

2. *Inductive clauses:* If I is a weakening, contraction, or interchange with premise S', then $o(I;\pi) = o(S';\pi)$.

3. If I is an operational inference with one premise S', then $o(I;\pi) = o(S';\pi) + 1$.

4. If I is an operational inference with two premises S' and S'', then $o(I;\pi) = o(S';\pi) \# o(S'';\pi)$.[3]

5. If I is a CUT with premises S' and S'', then $o(I;\pi) = o(S';\pi) \# o(S'';\pi)$.

6. If I is an induction inference with premise S', and $o(S';\pi) = \omega^{\alpha_1} + \cdots + \omega^{\alpha_n}$, then $o(I;\pi) = \omega^{\alpha_1 \# 1}$.

7. If S is the conclusion of an inference I, the level of the premise(s) is k and the level of the conclusion is ℓ, then $o(S;\pi) = \omega_{k-\ell}(o(I;\pi))$.

Note that if I is not a level transition, then $k = \ell$ and therefore $k - \ell = 0$, and $o(S;\pi) = o(I;\pi)$. Only CUT and CJ inferences can be level transitions. So for structural inferences other than CUT and operational inferences, $o(S;\pi) = o(I;\pi)$.[4]

It is difficult to see at this point why the ordinal assignments are given as they are. This will become clear later when we show that the ordinal notation assigned to the end-sequent of proofs decreases as we apply the reduction steps. Note at this point that the level of a sequent is not the ordinal notation assigned to it. The level is just a natural number determined by the degrees of CUT and CJ inferences below it. It is, however, used in the computation of the ordinal notations assigned to sequents.

Let's see how the assignment of ordinal notations to proofs works in our example. We first have to consider the proof $\pi_1(a)$ from p. 304:

$$\frac{\dfrac{0 + a = a + 0 \Rightarrow}{\Rightarrow 0 + a' = (0 + a)' \quad (0 + a)' = (a + 0)'}{\dfrac{0 + a = a + 0 \Rightarrow 0 + a' = (a + 0)'}{0 + a = a + 0 \Rightarrow 0 + a' = a'} \text{TR}} \quad \dfrac{\dfrac{\Rightarrow a + 0 = a}{\Rightarrow (a + 0)' = a'} \text{FNC} \quad \dfrac{\Rightarrow a' + 0 = a'}{\Rightarrow a' = a' + 0} \text{SYM}}{\Rightarrow a' = a' + 0} \text{TR}}{0 + a = a + 0 \Rightarrow 0 + a' = a' + 0}$$

The inferences SYM, FNC, and TR, remember, abbreviate short proof patterns which involve some non-logical axioms and otherwise only CUT inferences. For instance, TR is short for the following derivation of $\Gamma, \Delta \Rightarrow s = u$ from $\Gamma \Rightarrow s = t$ and $\Delta \Rightarrow t = u$:

---

[3] Gentzen originally assigned the maximum of the ordinal notations assigned to S' and S'', plus 1. We use Takeuti's assignment here. The difference is inconsequential, since both assignments guarantee that $o(I;\pi) \succ o(S';\pi)$ and $\succ o(S'';\pi)$. Takeuti's assignment allows us to avoid having to define the maximum of two ordinal notations.

[4] Of course we could have defined the ordinal notation assigned to conclusions of such inference without the detour via the ordinal notation assigned to the inference itself. We'll need the more general definition later on in Section 9.3.

$$\frac{\Gamma \Rightarrow s = t \quad s = t, t = u \Rightarrow s = u}{\Delta \Rightarrow t = u \quad \cfrac{\cfrac{\Gamma, t = u \Rightarrow s = u}{t = u, \Gamma \Rightarrow s = u} \text{ IL}}{\cfrac{\Delta, \Gamma \Rightarrow s = u}{\Gamma, \Delta \Rightarrow s = u} \text{ IL}} \text{ CUT}} \text{ CUT}$$

In the top left inference in $\pi_1(a)$,

$$\frac{\Rightarrow 0 + a' = (0 + a)' \quad 0 + a = a + 0 \Rightarrow (0 + a)' = (a + 0)'}{0 + a = a + 0 \Rightarrow 0 + a' = (a + 0)'} \text{ TR}$$

the role of s is played by $0 + a'$, t by $(0 + a)'$, and u by $(a + 0)'$, while Γ is empty and Δ is $0 + a = a + 0$, i.e., $F(a)$:

$$\frac{\mathbf{1} \; F(a) \Rightarrow (0+a)' = (a+0)' \quad \cfrac{\mathbf{1} \Rightarrow 0 + a' = (0 + a)' \quad \cfrac{\begin{array}{c}0 + a' = (0 + a)', \\ (0 + a)' = (a + 0)' \Rightarrow \\ \mathbf{1} \qquad 0 + a' = (a + 0)'\end{array}}{\mathbf{1} \# \mathbf{1} \; (0 + a)' = (a + 0)' \Rightarrow 0 + a' = (a + 0)'} \text{ CUT}}{\mathbf{1} \# (\mathbf{1} \# \mathbf{1}) \; F(a) \Rightarrow 0 + a' = (a + 0)'} \text{ CUT}}$$

The ordinal notations assigned to the axioms are all **1**. Thus the ordinal notation assigned to the first CUT inference is **1 # 1**, i.e., **1 + 1**. The levels of all sequents of our proof that fall into the sub-proof $\pi_1(a)$ is 1. Therefore, none of the CUT inferences that occur as part of $\pi_1(a)$ is a level transition. So, the ordinal notation assigned to the conclusion of the right CUT is the same as the ordinal notation assigned to the CUT itself, i.e., **1 + 1**. The ordinal notation assigned to the axiom on the left is also **1**, so the ordinal notation assigned to the lower CUT and its conclusion is **1 # (1 + 1)**, i.e., **1 + 1 + 1**.

In general, if α and β are assigned to the "premises" of TR, then $\alpha \# (\beta \# \mathbf{1})$ is assigned to the conclusion. Similarly, since SYM and FNC abbreviate derivations consisting of a single CUT with a mathematical axiom sequent, the ordinal notation assigned to their "conclusions" is always $\alpha + \mathbf{1}$.

Problem 9.4. Verify that if α and β are assigned to the "premises" of TR in a proof π, then $\alpha \# (\beta \# \mathbf{1})$ is assigned to the "conclusion," and that if α is assigned to the "premise" of SYM or FNC, the ordinal notation $\alpha + \mathbf{1}$ is assigned to the "conclusion."

For simplicity, let's abbreviate $\mathbf{1} + \cdots + \mathbf{1}$ (with n **1**'s) as n. With the notation of Definition 8.28, $n = \omega^0 \cdot n$. By Proposition 8.29, $n \# m = k$ where $k = n + m$.

Then the ordinal notations assigned to the sequents in $\pi_1(a)$ are:

$$\frac{\cfrac{\begin{array}{cc} \mathbf{1} \Rightarrow 0 + a' = & \mathbf{1} \; F(a) \Rightarrow (0 + a)' = \\ (0 + a)' & (a + 0)' \end{array}}{\mathbf{3} \; F(a) \Rightarrow 0 + a' = (a + 0)'} \text{ TR}}{\mathbf{6} \; F(a) \Rightarrow 0 + a' = a'} \qquad \cfrac{\cfrac{\mathbf{1} \Rightarrow a + 0 = a}{\mathbf{2} \Rightarrow (a + 0)' = a'} \text{ FNC}}{} \text{ TR} \qquad \cfrac{\mathbf{1} \Rightarrow a' + 0 = a'}{\mathbf{2} \Rightarrow a' = a' + 0} \text{ SYM}$$

$$\frac{\mathbf{6} \; F(a) \Rightarrow 0 + a' = a'}{\mathbf{9} \; F(a) \Rightarrow F(a')} \text{ TR}$$

Let's now consider the rest of the proof on p. 305, which was:

$$\vdots\ \pi_1(a)$$

$$
\begin{array}{c}
\mathbf{9}\ \dfrac{F(a) \Rightarrow F(a')}{F(0) \Rightarrow F(b)}\ \text{CJ} \\
\mathbf{1}\ \Rightarrow F(0) \qquad\qquad\qquad\qquad \\
\end{array}
$$

We know that every initial sequent is assigned **1**, and the conclusion of $\pi_1(a)$, as we've just calculated, is assigned **9**.

Consider the CJ inference from $F(a) \Rightarrow F(a')$ (assigned **9**) to $F(0) \Rightarrow F(b)$. According to the definition, if the premise of CJ is assigned $\omega^{\alpha_1} + \cdots + \omega^{\alpha_n}$, then the ordinal notation assigned to the inference is $\omega^{\alpha_1 \# 1}$. In our case, the premise is assigned **9** which abbreviates

$$\mathbf{1+1+1+1+1+1+1+1+1}, \text{ i.e.,}$$
$$\omega^0 + \omega^0 + \omega^0 + \omega^0 + \omega^0 + \omega^0 + \omega^0 + \omega^0 + \omega^0.$$

So we have $\alpha_1 = 0$, and the ordinal notation assigned to the CJ inference is $\omega^{0\#1}$, i.e., ω^1. By our discussion on p. 348, the level of all sequents except the end-sequent is 1, in other words only the final CUT is a level transition. In particular, the CJ inference is not a level transition: the level k of its premise is 1 and the level ℓ of its conclusion is also 1. Thus the ordinal notation assigned to the conclusion of the CJ inference is $\omega_{k-\ell}(o(I;\pi)) = \omega_0(\omega^1) = \omega^1$.

The following CUT is assigned $\mathbf{1} \# \omega^1$, i.e., $\omega^1 + 1$. The $\forall$R inference adds **1** to the ordinal notation assigned to its premise, so its conclusion is assigned $\omega^1 + 1 + 1$. On the right, we have an axiom (assigned **1**) followed by $\forall$L, so the conclusion is assigned **2**.

The premises of the final CUT inference are assigned $\omega^1 + 1 + 1$ and $1 + 1$. Thus the inference itself is assigned $\omega^1 + 1 + 1 + 1 + 1$, or $\omega^1 + 4$ for short. The last CUT now *is* a level transition. The level k of its premises (like every sequent above it) is 1, the level ℓ of its conclusion is 0. So the ordinal notation assigned to the conclusion, and hence the end-sequent, and hence to π, is: ω^{ω^1+4}. Overall, we have:

$$\vdots\ \pi_1(a)$$

$$
\mathbf{9}\ \dfrac{F(a) \Rightarrow F(a')}{\omega^1\ F(0) \Rightarrow F(b)}\ \text{CJ}
$$

Problem 9.5. Compute the ordinal notation assigned to the sequents in the proof given in Example 7.1. (Although this is not a proof of a sequent with only atomic formulas, the assignment of ordinal notations works for this case as well.)

Before we go on to consider the way ordinal notations assigned to proofs decrease when the proof is reduced, we'll record some facts about ordinal notations assigned to sequents which we'll need later on.

First, when assigning ordinal notations to sequents, the ordinal notations can never decrease as we proceed downwards in a proof.

Proposition 9.6. *If S occurs below S' in π, then $o(S;\pi) \succeq o(S';\pi)$.*

Proof. If S occurs below S' in π, then there is a sequence $S = S_1, \ldots, S_n = S'$ such that S_i is the conclusion of an inference of which S_{i+1} is a premise. So it is enough to show that the conclusion of an inference is always assigned an ordinal notation $\succeq$ that assigned to any of the premises. For instance, if S_i is the conclusion of a weakening, contraction, or interchange, then $o(S_i;\pi) = o(S_{i+1};\pi)$ and so $o(S_i;\pi) \succeq o(S_{i+1};\pi)$. If S_i is the conclusion of an operational inference of which S_{i+1} is the only premise, then $o(S_i;\pi) = o(S_{i+1};\pi) + 1$, and $\alpha + 1 \succeq \alpha$. If S_i is the conclusion of a CUT, with one premise being S_{i+1} assigned α and the other a sequent assigned the ordinal notation β, then $o(S_i;\pi) = \omega_{k-\ell}(\alpha \# \beta))$. But obviously $\alpha \# \beta \succeq \alpha$. So $\omega_{k-\ell}(\alpha \# \beta) \succeq \alpha$, by Propositions 8.33 and 8.40. We leave the other cases as exercises. $\square$

Problem 9.7. Complete the proof of Proposition 9.6 by verifying the remaining cases.

We've seen in our example that sequents in proofs containing only atomic CUTs and no CJ inferences (such as $\pi_1(a)$) are assigned ordinal notations of the form $1 + \cdots + 1 \prec \omega^1$. This is true in general. Furthermore, any proof of an atomic end-sequent with an ordinal notation $\prec \omega^1$ cannot contain complex CUTs, CJ inferences, or operational inferences. Together this gives us the following result.

Proposition 9.8. *Suppose π is a proof of an atomic, variable-free end-sequent. Then $o(\pi) \prec \omega^1$ if, and only if, π is simple.*

Proof. Suppose π is simple. Recall that a simple proof is one that contains only atomic CUTs and structural inferences. Since there are no CJ inferences and all CUTs are atomic, the level of all sequents is 0, i.e., there are no level transitions. Ordinal notations assigned to initial sequents are always 1, structural inferences other than CUT do not change the ordinal notation assigned, and the ordinal notation assigned to the conclusion of a CUT which is not a level transition is $\alpha \# \beta$ where α and β are the ordinal notations assigned to the premises. Thus, the ordinal notations assigned to sequents in a simple proof are all of the form $1 + \cdots + 1 \prec \omega^1$.

Now suppose $o(\pi) \prec \omega^1$ and the end-sequent of π is atomic and contains no free variables. We show that π is simple.

Observe that by Proposition 9.6, since the end-sequent occurs below any other sequent in π, for every sequent S in π, $o(S;\pi) \preceq o(\pi)$.

First, we show that π cannot contain a CJ inference I. For suppose it did. Let α be the ordinal notation assigned to its premise. Since $\alpha \preceq o(\pi) \prec \omega^1$ by the preceding observation, we have that $\alpha = \omega^0 + \cdots + \omega^0$. Then $o(I;\pi) = \omega^{0\#1} = \omega^1$. By the way ordinal notations are assigned, we have that $o(I;\pi) \preceq o(S;\pi)$, where S is the conclusion of I, and $o(S;\pi) \preceq o(\pi)$ by the observation above. By transitivity of $\preceq$, $o(I;\pi) \preceq o(\pi)$. But $o(I;\pi) = \omega^1$ and $o(\pi) \prec \omega^1$, which is impossible as it would imply $\omega^1 \prec \omega^1$.

There can also be no complex CUTS in π. A complex CUT is one where the cut-formula is not atomic. Suppose π contained a complex CUT. Clearly, there must be a lowermost one; let's call it I and let $n > 0$ be its degree. Since there are no complex cuts below I and, as we have shown, also no CJ inferences, the level of the conclusion of I is 0. The level of the premises is n. If the premises are assigned ordinal notations α and β, then the ordinal notation assigned to the conclusion is $\omega_{n-0}(\alpha \# \beta)$. But α and β are $\succ 0$, so $\alpha \# \beta \succ 1$. Since $n > 0$, $\omega_{n-0}(\alpha \# \beta) \succ \omega^1$, contradicting the assumption that $o(\pi) \prec \omega^1$.

Finally, there can be no operational inferences in π since the end-sequent is atomic and π contains no complex CUTS. $\qquad\square$

The consistency proof proceeds by applying reduction steps repeatedly. In each reduction step, a sub-proof of the proof is replaced by a new sub-proof. We will show that in each step, the ordinal notation assigned to the new sub-proof is smaller than that assigned to the original sub-proof. The ordinal notations assigned to the bigger proofs inside of which this replacement is carried out, of course, depend on the ordinal notations assigned to the sub-proofs. However, they also depend on the level transitions *below* these sub-proofs, and it is not immediately obvious that by decreasing the ordinal notation assigned to a sub-proof we decrease the ordinal notation of the entire proof. The next proposition establishes that the ordinal notations assigned to the bigger proofs do decrease as well in that case, provided there are no CJ inferences below the sub-proof being replaced.

Proposition 9.9. *Suppose that π_1 and π_2 are the proofs*

$$
\begin{array}{cc}
\vdots\, \pi_1' & \vdots\, \pi_2' \\
\beta_1\ S' & \beta_2\ S' \\
\vdots\, \pi_3 & \vdots\, \pi_3 \\
\alpha_1\ S & \alpha_2\ S
\end{array}
$$

and the sequent S' is assigned the ordinal notation β_1 in π_1 and β_2 in π_2, and the end-sequent S is assigned α_1 and α_2, respectively. If $\beta_1 \succ \beta_2$ and there are no CJ inferences below S' in π_3, then $\alpha_1 \succ \alpha_2$.

Proof. By induction on the number n of inferences between S' and S in π_3.

Induction basis. If $n = 0$, then $S' = S$, $\alpha_1 = \beta_1$, $\alpha_2 = \beta_2$, and the claim holds by the assumption that $\beta_1 \succ \beta_2$.

Inductive step. Now suppose $n > 0$. Then S' is a premise of an inference I with conclusion S''. We'll show that $\gamma_1 = o(S''; \pi_1) \succ o(S''; \pi_2) = \gamma_2$. Then the claim follows by inductive hypothesis (applied to S'' and $\gamma_1 \succ \gamma_2$ instead of S' and $\beta_1 \succ \beta_2$) since the number of inferences between S'' and S is smaller than n.

We show the hardest case, where I is a CUT, and leave the other cases as exercises. So suppose S' is the left premise of a CUT, with right premise S_0'. The level of S', S_0', and S'' in π_1 and π_2 is determined by the degrees of CUT and CJ inferences below I, and since the inferences below I in π_1 are the same as those below I in π_2, the level k of S' and S_0' in π_1 is the same as the level of S' and S_0' in π_2, and similarly for the level ℓ of S'' in π_1 and π_2. The ordinal notation assigned to S_0' in π_1 is the same as that assigned to S_0' in π_2, since it depends only on the sub-proof ending in S_0', and the levels of sequents in that sub-proof are the same in π_1 and π_2. So, we have

$$
\left. \begin{array}{ccc} \vdots\, \pi_1' & & \vdots \\ \beta_1, k \;\; S' & \quad \delta \;\; S_0' \\ \hline \gamma_1, \ell \;\; S'' \end{array} \right\} \pi_1'' \text{CUT}
\qquad
\left. \begin{array}{ccc} \vdots\, \pi_2' & & \vdots \\ \beta_2, k \;\; S' & \quad \delta \;\; S_0' \\ \hline \gamma_2, \ell \;\; S'' \end{array} \right\} \pi_2'' \text{CUT}
$$

$$
\vdots\, \pi_3' \qquad\qquad\qquad \vdots\, \pi_3'
$$

$$
S \qquad\qquad\qquad\qquad S
$$

The ordinal notation assigned to our CUT I in π_1 is $\beta_1 \# \delta$, and in π_2 it is $\beta_2 \# \delta$. Since $\beta_1 \succ \beta_2$, we have $\beta_1 \# \delta \succ \beta_2 \# \delta$ by Proposition 8.33. The ordinal notations assigned to S'' are $\gamma_1 = \omega_{k-\ell}(\beta_1 \# \delta) \succ \omega_{k-\ell}(\beta_2 \# \delta) = \gamma_2$. Since π_3' is π_3 except for the topmost inference (the CUT being considered), it has fewer inferences than π_3, namely $n - 1$. So the inductive hypothesis applies to π_1'', π_2'', S'' and π_3' in the roles of π_1', π_2', S', and π_3. □

Problem 9.10. Complete the proof of Proposition 9.9 by verifying the remaining cases of the inductive step.

9.2   Eliminating inductions from the end-part

We first show that step 1 of the reduction procedure applied to a proof π results in a proof π^* with $o(\pi^*) \prec o(\pi)$. Recall that at the outset we guarantee that the proof we start with is regular, contains only atomic axioms, and contains no free variables which are not eigenvariables. We remove all CJ inferences from the end-part by replacing them with successive CUT inferences as described in Section 7.4. We now show that when this is done, the ordinal notation assigned to the resulting proof is smaller than that assigned to the original proof.

Take a lowermost induction inference I in the end-part of π. It is of the form:

$$\vdots \pi'(a)$$

$$\alpha = \omega^{\alpha_1} + \cdots + \omega^{\alpha_m}, k \; \dfrac{F(a), \Gamma \Rightarrow \Theta, F(a')}{\omega_{k-\ell}(\omega^{\alpha_1 \# 1}), \ell \; F(0), \Gamma \Rightarrow \Theta, F(\overline{n})} \; \text{CJ}$$

$$\vdots \pi_0$$

$$\Pi \Rightarrow \Xi$$

Since I is a lowermost CJ inference in π, the part π_0 of π below it contains no free variables, hence the induction term in the conclusion of I must be a numeral $\overline{n}$ (see Proposition 7.27). If the premise has the ordinal notation $\alpha = \omega^{\alpha_1} + \cdots + \omega^{\alpha_m}$ assigned, and has level k, then the conclusion has the ordinal notation $\omega_{k-\ell}(\omega^{\alpha_1 \# 1})$ assigned (where ℓ is the level of the conclusion in π).

There is a trivial case we can dispense with first: If $n = 0$, then the conclusion of the CJ inference is $F(0), \Gamma \Rightarrow \Theta, F(0)$. By Proposition 5.17, there is a proof of $F(0) \Rightarrow F(0)$ from atomic initial sequents, from which we obtain $F(0), \Gamma \Rightarrow \Theta, F(0)$ by weakenings and interchanges only. This proof uses only operational and structural rules other than CUT, so the ordinal notation assigned to it is $\prec \omega^1$. This is certainly smaller than the ordinal notation assigned to the CJ inference, $\omega_{k-\ell}(\omega^{\alpha_1 \# 1})$.

If $n > 1$, we obtain π^* by replacing the sub-proof ending in I in π as follows:

$$\vdots \pi'(0) \qquad \vdots \pi'(\overline{1})$$

$$\alpha, k \; \dfrac{F(0), \Gamma \Rightarrow \Theta, F(\overline{1}) \qquad \alpha, k \; F(\overline{1}), \Gamma \Rightarrow \Theta, F(\overline{2})}{\alpha \# \alpha, k \; F(0), \Gamma, \Gamma \Rightarrow \Theta, \Theta, F(\overline{2})} \; \text{CUT}$$

$$\dfrac{\alpha \# \alpha, k \; F(0), \Gamma \Rightarrow \Theta, F(\overline{2}) \qquad \vdots \pi'(\overline{2})}{}$$

$$\dfrac{\alpha \# \alpha, k \; F(0), \Gamma \Rightarrow \Theta, F(\overline{2}) \qquad \alpha, k \; F(\overline{2}), \Gamma \Rightarrow \Theta, F(\overline{3})}{\alpha \# \alpha \# \alpha, k \; F(0), \Gamma, \Gamma \Rightarrow \Theta, \Theta, F(\overline{3})} \; \text{CUT}$$

$$\alpha \# \alpha \# \alpha, k \; F(0), \Gamma \Rightarrow \Theta, F(\overline{3})$$

$$\iddots \qquad \vdots \pi'(\overline{n-1})$$

$$\alpha \# \cdots \# \alpha, k \; \dfrac{F(0), \Gamma \Rightarrow \Theta, F(\overline{n-1}) \qquad \alpha, k \; F(\overline{n-1}), \Gamma \Rightarrow \Theta, F(\overline{n})}{\omega_{k-\ell}(\alpha \# \cdots \# \alpha), \ell \; F(0), \Gamma, \Gamma \Rightarrow \Theta, \Theta, F(\overline{n})} \; \text{CUT}$$

$$\omega_{k-\ell}(\alpha \# \cdots \# \alpha), \ell \; F(0), \Gamma \Rightarrow \Theta, F(\overline{n})$$

$$\vdots \pi_0$$

$$\Pi \Rightarrow \Xi$$

Since the ordinal assignment to sequents and inferences in a proof does not depend on the terms in the proof, but only on the type and arrangement of inferences in it, $o(S; \pi'(\overline{n})) = o(S'; \pi'(a))$ for any sequent S in $\pi'(\overline{n})$ and its corresponding sequent S' in $\pi'(a)$. So, if the ordinal notation assigned to the end-sequent of $\pi'(a)$ in π is α, the ordinal notation assigned to the end-sequent of $\pi'(\overline{n})$ in the resulting proof is also α.

Here the ordinal notation assigned to the conclusions $F(0), \Gamma \Rightarrow \Theta, F(\overline{i})$ of the sequence of CUT inferences is $\alpha \# \cdots \# \alpha$ with i α's. This is because the ordinal notation assigned to a CUT inference is the natural sum of the ordinal notations assigned to their premises. Since the degree of the cut-formulas $F(\overline{i})$ in all CUTS is the same as the degree of the induction formula $F(a)$, every CUT except the last has a CUT of the same degree below it, so none of the new CUTS can be a level transition except the last one. Hence, in each CUT except the last one, the level of the premises equals the level of the conclusion (namely k), and so the ordinal notation assigned to the conclusion is the same as the ordinal notation assigned to the CUT itself, i.e., the natural sum of the ordinal notations assigned to the premises. Only the conclusion of the last CUT can be assigned an ω-stack over the natural sum assigned to the CUT itself.

It is also clear that the level of the premises of the CUTS is equal to k, i.e., the level of the premise of I in π, and the level of the conclusion of the last CUT is equal to the level ℓ of the conclusion of I in π. The level of a sequent depends only on the degrees of CUT inferences below it in the proof. The inferences below the conclusion of I in π and those of the conclusion of the last CUT in the new proof are the same: those in π_0. So, the level of the conclusion of the last CUT in the new proof is ℓ. In π, I either is a level transition ($k > \ell$), or it is not ($k = \ell$). If it is, then the degree of I, i.e., of $F(a)$, is larger than the degree of any CUT below I. In the new proof, each CUT other than the last one has a CUT below it of degree k (namely, one of the CUTS on $F(\overline{i})$), and no CUT of larger degree, so the levels of the premises of the new CUT inferences are all $= k$. If it is not a level transition, then $\ell = k$ is the degree of some CUT below I in π_0 and $\ell \geq \deg(F(a))$. In the new proof, each of the new CUT inferences then has a CUT of degree ℓ below it, and so the levels of their premises and conclusions equal $\ell = k$.

What's left to show is that the ordinal notation assigned to π^* is smaller than the ordinal notation assigned to π. By Proposition 9.9, it is enough to show that the conclusion of the last of the new CUT inferences in π^* is assigned a smaller ordinal notation than the conclusion of I in π. In other words, we have to show that $\omega_{k-\ell}(\omega^{\alpha_1 \# 1}) > \omega_{k-\ell}(\alpha \# \cdots \# \alpha)$. By Proposition 8.41, it suffices to show that $\omega^{\alpha_1 \# 1} > \alpha \# \cdots \# \alpha$.

The ordinal notation $\alpha \# \cdots \# \alpha$ is the sum of the terms ω^{α_i} occurring in α, sorted by non-increasing size, and n copies of each, i.e.,

$$\underbrace{\omega^{\alpha_1} + \cdots + \omega^{\alpha_1}}_{n\ \omega^{\alpha_1}\text{'s}} + \cdots + \underbrace{\omega^{\alpha_m} + \cdots + \omega^{\alpha_m}}_{n\ \omega^{\alpha_m}\text{'s}}$$

Now if α_1 is 0, then every α_i is 0, so α is m, and $\alpha \# \cdots \# \alpha$ is the notation $1 + \cdots + 1$ with $j = nm$ copies of 1. On the other hand, $\alpha_1 \# 1$ is 1, so the ordinal notation assigned to the CJ inference I in π is ω^1. But $\omega^1 \succ j$, since ω^1, i.e., ω^{ω^0}, is of height 2 while $\omega^0 + \ldots$ is of height 1.

If α_1 is not $\mathbf{0}$, then $\omega^{\alpha_1 \# \mathbf{1}}$ and $\omega^{\alpha_1} + \ldots$ are of the same height. Of two ordinal notations of the same height, whichever has the lower exponent in the first position where they disagree is smaller. Since α_1 and $\alpha_1 \# \mathbf{1}$ are not identical, $\omega^{\alpha_1} + \ldots$ and $\omega^{\alpha_1 \# \mathbf{1}}$ disagree in the first position. Since $\alpha_1 \prec \alpha_1 \# \mathbf{1}$ (Corollary 8.37), $\omega^{\alpha_1 \# \mathbf{1}} \succ \omega^{\alpha_1} + \ldots .$[5]

This means that the ordinal notation assigned to $F(0), \Gamma \Rightarrow \Theta, F(\overline{n})$ in the new proof is smaller than the ordinal notation assigned to the corresponding sequent in the old proof. This inequality is preserved in the concluding part π_0 (by Proposition 9.9, which we may appeal to here as we are operating on the last CJ inference in the end-part), so the ordinal notation assigned to the end-sequent of the new proof π^* is smaller than the ordinal notation assigned to the end-sequent of π.[6]

Proposition 9.11. *If π is a proof with end-sequent $\Pi \Rightarrow \Xi$, and the end-part of π contains no free variables other than eigenvariables belonging to induction inferences, then there is a proof π^* of $\Pi \Rightarrow \Xi$ with $o(\pi^*) \prec o(\pi)$ and an end-part without induction inferences.*

Proof. Apply the procedure described until no induction inferences remain in the end-part of the proof. As the ordinal notations assigned to new proofs are strictly smaller, the procedure must eventually end. □

Again, as a CJ inference is replaced by CUT inferences, the proof $\pi'(a)$ above the CJ inference is copied multiple times. It may contain additional CJ inferences, which are thereby multiplied. So the number of CJ inferences in the proof—even in the end-part of the proof—may increase. However, since the ordinal notations assigned decrease, eventually all CJ inferences are removed from the end-part. For instance, consider the proof $\eta_5(a, b)$ of $a + b = b + a$ from Example 7.5.

We'll replace a and b by numerals, say by $\overline{1}$ and $\overline{2}$, and abbreviate:

$$a + b = b + a \qquad \text{by} \qquad A(a, b),$$

---

[5] We see here why the assignment of ordinal notations to CJ inferences was chosen as $\omega^{\alpha_1 \# \mathbf{1}}$. The ordinal notation assigned to the last CUT is $\alpha \# \cdots \# \alpha$ (n copies of α). Since we do not have a bound on n, we must assign an ordinal notation to the CJ inference greater than any such sum. Since if $\alpha = \omega^{\alpha_1} + \ldots$ such natural sum also starts with ω^{α_1}, $\omega^{\alpha_1 \# \mathbf{1}}$ fits the bill.

[6] Recall that we have assumed that our proof contains no $+$ or $\cdot$. This allows us to assume that the t in the conclusion $F(0), \Gamma \Rightarrow \Theta, F(t)$ of a CJ inference is a numeral $\overline{n}$, if t contains no free variables. In the general case, t might contain $+$ and $\cdot$ as well. Then we would have to take $n = \text{val}(t)$ and add a CUT with $F(\overline{n}) \Rightarrow F(t)$. If we include identity axioms among the mathematical initial sequents, such sequents have proofs without CJ inferences or complex CUTs, with ordinal notations $\prec \omega^1$ assigned to their end-sequents. Hence, the ordinal notation assigned to the new proof is still smaller than that of the original proof using the CJ inference instead of a sequence of CUTs.

i.e., we consider the proof $\eta_5(\bar{1},\bar{2})$ of $A(\bar{1},\bar{2})$. We'll also display the structure of the sub-proofs that make up $\eta_5(a,b)$: $\eta_4(a)$ is a proof of $\Rightarrow a + 0 = 0 + a$, i.e., $\Rightarrow A(a,0)$. We abbreviate

$$
\begin{array}{lll}
0 + a = a & \text{by} & B(a) \text{ and}\\
a = 0 + a & \text{by} & B'(a).
\end{array}
$$

On the other side, $\eta_1(g,a)$ is a proof of $\Rightarrow g' + a = g + a'$; we abbreviate

$$
\begin{array}{lllll}
g' + a = g + a' & \text{by} & C(g,a), & \bar{1} + 0 = \bar{1} & \text{by} \quad D,\\
g + a' = g' + a & \text{by} & C'(g,a), & \bar{1} + g' = g + \bar{1}' & \text{by} \quad E.
\end{array}
$$

With these abbreviations, $\eta_5(\bar{1},\bar{2})$ has the form:

The sub-proofs indicated by ... all consist only of axioms and atomic CUTS. There are no operational rules, and all CUTS and CJ inferences have degree 0. In particular, there are no essenial CUTS. The entire proof is its own-end-part, and the level of every sequent is 0. The proof is not simple, however, because of the presence of multiple CJ inferences in the end-part. Let's see how these are removed and how the assigned ordinal notations change.

In order to make the structure of the proof clearer, let's abbreviate the sequents corresponding to premises of a CJ inference by $S(a)$ where a represents the eigenvariable of the CJ inference, and $S'(\bar{n})$ is the corresponding conclusion. So we abbreviate $B(f) \Rightarrow B(f')$ by $S_B(f)$ and the conclusion $B(0) \Rightarrow B(\bar{1})$ by $S'_B(\bar{1})$; $C(g,e) \Rightarrow C(g,e')$ and $C(g,0) \Rightarrow C(g,\bar{1})$ by $S_C(g,e)$ and $S'_C(g,\bar{1})$; and $A(\bar{1},g) \Rightarrow A(\bar{1},g')$ and $A(\bar{1},0) \Rightarrow A(\bar{1},\bar{2})$ by $S_A(\bar{1},g)$ and $S'_A(\bar{1},\bar{2})$, respectively. We'll simply indicate the sequents not involved in CJ inferences by S_i. We'll also add the ordinal notations to the sequents; we'll leave the verification that these are correct to the reader.

Then the proof has the form:

$$
\cfrac{
  \cfrac{
    \cfrac{
      1\ S_2 \qquad \cfrac{3\ S_B(f)}{\omega^1\ S'_B(\overline{1})}\ \text{CJ}
    }{\omega^1 + 1\ S_3}\ \text{CUT}
  }{
    1\ S_1 \qquad \cfrac{\omega^1 + 1\ S_3}{\omega^1 + 2\ S_4}\ \text{SYM}
  }\ \text{TR}
  \qquad
  \cfrac{
    \cfrac{
      \cfrac{
        \cfrac{7\ S_7}{6\ S_6}\ \eta'_3
        \qquad
        \cfrac{
          6\ S_C(g,e) \qquad \eta''_1(g,e)
        }{\omega^1\ S'_C(g,\overline{1})}\ \text{CJ}
      }{\omega^1 + 7\ S_8}\ \text{CUT}
    }{\omega^1 + 8\ S_9}\ \text{SYM}
  }{\omega^1 + 15\ S_A(\overline{1},g)}\ \text{TR}
}{\omega^2 + \omega^1 + 4\ S_8}
$$

The example also illustrates how the ordinal notation assigned to CJ inferences increases. Unless it is a level transition, a CJ inference only increases the exponent of the first ω by one. So in a proof like ours with no level transitions, CJ inferences may be assigned ω^n—if there is a thread ending in the premise with $n-1$ CJ inferences. Without level transitions, the ordinal notation assigned to a proof remains below ω^{ω^1}.[7]

There are two lowermost CJ inferences—those involving $S_B(f)$ and $S_A(\overline{1},g)$. The first one is easily removed. Notice that $S_B(0)$ is $B(0) \Rightarrow B(\overline{1})$, i.e., it is identical to $S'_B(\overline{1})$. In this case, the CJ inference is replaced by zero CUT inferences, simply by the sub-proof $\eta'_2(0)$ of the instance $S_B(0)$ of the premise:

$$
\cfrac{
  1\ S_1 \qquad
  \cfrac{
    \cfrac{
      1\ S_2 \qquad
      \cfrac{3\ \begin{matrix}S_B(0)=\\ S'_B(\overline{1})\end{matrix}\ \eta'_2(0)}{4\ S_3}\ \text{CUT}
    }{5\ S_4}\ \text{SYM}
  }{7\ S_5}\ \text{TR}
  \qquad \dots
}{\omega^2 + 7\ S_{10}}
$$

We now replace the other lowermost CJ inference, concluding $S'_A(\overline{1},\overline{2})$. It is replaced by a CUT between $S_A(\overline{1},0)$ and $S_A(\overline{1},\overline{1})$. There are now two copies of the sub-proof

---

[7] The full range of ordinal notations $< \varepsilon_0$ is only used if the CJ inferences are also level transitions. This typically happens if the induction formula is not atomic, and it is eventually removed using a complex CUT. For instance, if the CJ inference has degree d, and the formulas in the conclusion $A(0) \Rightarrow A(\overline{n})$ eventually participate in complex CUTs—the level of the premise is at least d and the level of the conclusion of the lowermost CUT with cut-formulas $A(0)$ and $A(\overline{n})$ is $< d$. Some CUT or CJ in between must be a level transition, raising the ordinal notation assigned by an ω-stack.

ending in the premise of the CJ inference, but the conclusion of the CUT is assigned an ordinal notation $\prec \omega^2$:

$$
\cfrac{
\cfrac{
7\ \dot{S}_5
\qquad
\cfrac{
\cfrac{
\cfrac{
\cfrac{
6\ \dot{S}'_6
\qquad
\cfrac{
\cfrac{
7\ \dot{S}'_7
\qquad
\cfrac{
\vdots\ \eta''_1(0,e) \qquad 6\ S_C(g,e)
}{
\omega^1\ S'_C(0,\overline{1})
}\ \text{CJ}
}{
\omega^1 + 7\ S'_8
}\ \text{CUT}
}{
\omega^1 + 8\ S'_9
}\ \text{SYM}
}{
\omega^1 + 15\ S_A(\overline{1},0)
}\ \text{TR}
\qquad
\vdots\ \eta'_3
}{}
}{}
}{}
}{}
\qquad
\cfrac{
6\ \dot{S}''_6
\qquad
\cfrac{
\cfrac{
\cfrac{
7\ \dot{S}''_7
\qquad
\cfrac{
\vdots\ \eta''_1(\overline{1},e)\qquad 6\ S_C(\overline{1},e)
}{
\omega^1\ S'_C(\overline{1},\overline{1})
}\ \text{CJ}
}{
\omega^1 + 7\ S''_8
}\ \text{CUT}
}{
\omega^1 + 8\ S''_9
}\ \text{SYM}
}{
\omega^1 + 15\ S_A(\overline{1},\overline{1})
}\ \text{TR}
}{
\omega^1 + \omega^1 + 30\ S'_A(\overline{1},\overline{2})
}\ \text{CUT}
}{
\omega^1 + \omega^1 + 37\ S_{10}
}\ \text{CUT}
$$

There are now two CJ inferences that are lowermost, concluding $S'_C(0,\overline{1})$ and $S'_C(\overline{1},\overline{1})$. Since the induction term is $\overline{1}$, we again have the simplest case: there are no new CUTs; the CJ inferences are simply replaced by the premises of the CJ inferences, with the eigenvariable e replaced by 0. We have:

$$
\cfrac{
\cfrac{
7\ \dot{S}_5
\qquad
\cfrac{
\cfrac{
\cfrac{
6\ \dot{S}'_6
\qquad
\cfrac{
\cfrac{
7\ \dot{S}'_7
\qquad
\cfrac{
\vdots\ \eta''_1(0,0)\qquad S_C(0,0)=6\ S'_C(0,\overline{1})
}{
13\ S'_8
}\ \text{CUT}
}{
14\ S'_9
}\ \text{SYM}
}{
21\ S_A(\overline{1},0)
}\ \text{TR}
\qquad
\vdots\ \eta'_3
}{}
}{}
}{}
\qquad
\cfrac{
6\ \dot{S}''_6
\qquad
\cfrac{
\cfrac{
\cfrac{
7\ \dot{S}''_7
\qquad
\cfrac{
\vdots\ \eta''_1(\overline{1},0)\qquad S_C(\overline{1},0)=6\ S'_C(\overline{1},\overline{1})
}{
13\ S''_8
}\ \text{CUT}
}{
14\ S''_9
}\ \text{SYM}
}{
21\ S_A(\overline{1},\overline{1})
}\ \text{TR}
}{
42\ S'_A(\overline{1},\overline{2})
}\ \text{CUT}
}{
49\ S_{10}
}\ \text{CUT}
$$

All CJ inferences have been removed, and the ordinal notation has decreased to below ω^1, i.e., the proof is now simple.

9.3   Removing weakenings

In step 2, we take proof π in which all CJ inferences have been removed from the end-part, and remove all weakenings from the end-part. The procedure produces a new, simpler proof π^*. Its end-sequent may not be the same as the end-sequent of the original proof. But if π proves $\Gamma \Rightarrow \Theta$, then π^* proves $\Gamma^* \Rightarrow \Theta^*$ with $\Gamma^* \subseteq \Gamma$ and $\Theta^* \subseteq \Theta$. The proof of this result (Proposition 7.34) was relatively simple: we proceeded by induction on the number of inferences in the end-part. If there are no inferences in the end-part (i.e., the proof consists only of an axiom), there is nothing to do. Otherwise we distinguish cases according to the last inference in the end-part. If that last inference is a weakening, it is removed. If it is any other inference, we either leave the inferences (if the simplified proof produced by the inductive hypothesis produces sequents in which the auxiliary formulas of the inference are still present), or remove the inference (if not). This works since we're only operating on the end-part, which contains no operational inferences. We'll now show that $o(\pi^*) \preceq o(\pi)$.

The complication here is that we cannot simply do this by induction on the number of inferences in the end-part. For the ordinal notation assigned to an inference in a proof depends not only on the inferences above it, but also on the level of those sequents. And the level of a sequent may change if CUT inferences are subsequently removed. For instance, suppose we are working through the end-part of π, removing weakenings and also any inferences that become superfluous. For example, imagine we have removed a weakening that introduced a formula B which subsequently is a cut-formula. For simplicity, assume that this B is not the ancestor of the principal formula of a contraction, so with the weakening that introduces B, step 2 also eliminates the CUT it is cut-formula of. So, we might have the following situation:

$$
\cfrac{
  \alpha, k\ \Gamma_1 \Rightarrow \Theta_1, A \qquad
  \cfrac{
    \beta, k\ A, \Gamma_2 \Rightarrow \Theta_2
  }{
    \beta, k\ A, \Gamma_2 \Rightarrow \Theta_2, B
  }\ \text{WR}
}{
  \alpha \# \beta, k\ \Gamma_1, \Gamma_2 \Rightarrow \Theta_1, \Theta_2, B
}\ \text{CUT}
$$

$$
\cfrac{
  \gamma, k\ \Gamma \Rightarrow \Theta, B \qquad\qquad \delta, k\ B, \Delta \Rightarrow \Lambda
}{
  \omega_{k-\ell}(\gamma \# \delta), \ell\ \Gamma, \Delta \Rightarrow \Theta, \Lambda
}\ \text{CUT}
$$

Here we've indicated both the ordinal notation assigned to a sequent and its level. By removing some inferences, the sequent $\Gamma \Rightarrow \Theta$ has turned into a sub-sequent $\Gamma' \Rightarrow \Theta'$. Suppose, for simplicity, that $k > \ell$, the degrees of A and B are both k, and there are no other CUTs of level $> \ell$ between the two indicated ones. Now the proof of Proposition 7.34 delivers a simplified proof in which the weakening with B is removed, and the lower CUT on B is redundant. We end up with a proof of this form:

$$
\cfrac{
  \alpha, k\ \Gamma_1 \Rightarrow \Theta_1, A \qquad \beta, k\ A, \Gamma_2 \Rightarrow \Theta_2
}{
  \omega_{k-\ell}(\alpha \# \beta), \ell\ \Gamma_1, \Gamma_2 \Rightarrow \Theta_1, \Theta_2
}\ \text{CUT}
$$

$$
\gamma', \ell\ \Gamma' \Rightarrow \Theta'
$$

Removing the CUT on B has removed the lower CUT of degree k. The higher CUT on A (also of degree k) is now a level transition, and the ordinal notation assigned to its conclusion has changed from $\alpha \# \beta$ to $\omega_{k-\ell}(\alpha \# \beta)$. The ordinal notation assigned to the sequent $\Gamma \Rightarrow \Theta$ has consequently also changed, from some γ to γ', where $\gamma' \succ \gamma$. How can we be sure that $\gamma' \preceq \omega_{k-\ell}(\gamma \# \delta)$, which is the ordinal notation assigned to the conclusion of the CUT that we've removed?

To convince ourselves that the proof resulting from removing weakenings is not just simpler structurally, but also has an overall ordinal notation assigned to it that is no larger than the ordinal notation assigned to the original proof, we use the following strategy. The ordinal notations assigned to inferences and sequents in a proof assigned as per Definition 9.3 depend only on the inferences in the proof and the levels of sequents in it. If we label every sequent in π with its level (as defined by Definition 9.1), and then compute the ordinal notations $o_\ell(S; \pi)$ assigned to each sequent S on the basis of the inferences in π and the labels we've just assigned to each sequent, the resulting ordinal notations are obviously the same. Then, we apply the simplification procedure of Proposition 7.34 to obtain a new proof π^*, but keep the labels intact. Each sequent S^* in the new proof corresponds to a sequent S in the old proof (but not vice versa: some sequents may have been removed). It's easy to show by induction that the ordinal notation $o_\ell(S^*; \pi^*)$ assigned to a sequent in the new proof *on the basis of the labels* is no greater than the ordinal notation $o_\ell(S; \pi)$ assigned to it in the old proof. But the ordinal notation $o_\ell(S^*; \pi^*)$ assigned to a sequent in the new proof on the basis of the labels may not be the same as the ordinal notation $o(S^*; \pi^*)$ assigned to it on the basis of the *actual levels* of sequents in the new proof. As we've seen in the example, the levels of sequents may change. We'll successively correct the labels of the new proof until they match the actual levels, showing that in each step, the ordinal notation assigned to the end-sequent (and thus the entire proof) does not increase. This "correction" is done by decreasing labels on sequents by 1 until the label on a sequent matches its actual level. This relies on the fact, which we'll prove, that the labels assigned to sequents in the old proof are never greater than the actual level of the corresponding sequent in the new proof.

Proposition 9.12. *Suppose π is a proof of $\Gamma \Rightarrow \Theta$ in which the end-part contains no free variables and no induction inferences, and in which each sequent is labelled by its level in π. Then there is a proof π^* such that:*

1. *The end-part of π^* contains no weakenings*

2. *The end-sequent of π^* is $\Gamma^* \Rightarrow \Theta^*$, where $\Gamma^* \subseteq \Gamma$ and $\Theta^* \subseteq \Theta$.*

3. *Each sequent S^* in π^* corresponds to exactly one sequent S in π and is labelled by the same number as its corresponding sequent in π.*

4. *In particular, the end-sequent of π^* corresponds to the end-sequent of π.*

5. *$o_\ell(S^*; \pi^*) \preceq o_\ell(S; \pi)$, where $o_\ell(S; \pi)$ is defined just like $o(S; \pi)$ (i.e., as in Definition 9.3) using the labels of sequents in π instead of levels of sequents.*

Proof. We define π^* by induction on the basis of π as before in the proof of Proposition 7.34.

Induction basis: If π is just an initial sequent S labelled k, then π^* is S labelled k, and S in π^* corresponds to S in π. We have $o_\ell(S; \pi^*) = \mathbf{1} = o_\ell(S; \pi)$.

Inductive step: If π ends in an operational inference, π^* is π with the same labels as in π, and each sequent S in π^* corresponds to S in π. Clearly, $o_\ell(S; \pi^*) = o_\ell(S; \pi)$ for all S.

Now suppose π ends in a weakening, e.g.,

$$\vdots \pi_1$$

$$\frac{\alpha, k\ \Gamma \Rightarrow \Theta}{\alpha, k\ \Gamma \Rightarrow \Theta, A}\ \text{WR}$$

Here k is the label assigned to the indicated sequents and α the ordinal notation assigned to them according to these labels.

By inductive hypothesis, there is a labelled proof π_1^* of end-sequent $\Gamma^* \Rightarrow \Theta^*$ corresponding to $\Gamma \Rightarrow \Theta$ and $o_\ell(\Gamma^* \Rightarrow \Theta^*; \pi_1^*) \preceq o_\ell(\Gamma \Rightarrow \Theta; \pi) = \alpha$. We let $\pi^* = \pi_1^*$. Each sequent in π^* except the last corresponds to whatever sequent it corresponds to in π according to the correspondence between sequents in π_1^* and π_1. The last sequent $\Gamma^* \Rightarrow \Theta^*$ of π^* corresponds to the last sequent of π (in this case, $\Gamma \Rightarrow \Theta, A$). Since $o_\ell(\Gamma^* \Rightarrow \Theta^*; \pi^*) = o_\ell(\Gamma^* \Rightarrow \Theta^*; \pi_1^*)$, we have $o_\ell(\Gamma^* \Rightarrow \Theta^*; \pi^*) \preceq \alpha$.

Now suppose the last inference is a contraction or interchange, say, CL:

$$\vdots \pi_1$$

$$\frac{\alpha, k\ A, A, \Gamma \Rightarrow \Theta}{\alpha, k\ A, \Gamma \Rightarrow \Theta}\ \text{CL}$$

where $\alpha = o_\ell(A, \Gamma \Rightarrow \theta; \pi)$. By induction hypothesis, there is a proof π_1^* of one of

$$\Gamma^* \Rightarrow \Theta^*$$
$$A, \Gamma^* \Rightarrow \Theta^*$$
$$A, A, \Gamma^* \Rightarrow \Theta^*$$

with label k. In the first two cases, we let $\pi^* = \pi_1^*$, in the last, it is

$$\vdots \pi_1^*$$

$$\frac{\alpha^*, k\ A, A, \Gamma^* \Rightarrow \Theta^*}{\alpha^*, k\ A, \Gamma^* \Rightarrow \Theta^*}\ \text{CL}$$

where $\alpha^* = o_\ell(A, A, \Gamma^* \Rightarrow \Theta^*; \pi_1^*)$ is the ordinal notation assigned to the end-sequent of π_1^* on the basis of its labels. In all three cases, the end-sequent S^* of π^*

corresponds to the end-sequent $A, \Gamma \Rightarrow \Theta$ of π. By inductive hypothesis, $\alpha^* \preceq \alpha$. So, the ordinal notation $o_\ell(S^*; \pi^*) = \alpha^* \preceq \alpha$.

Finally, the case of CUT. The proof π ends in

$$
\begin{array}{cc}
\vdots\, \pi_1 & \vdots\, \pi_2 \\[4pt]
\dfrac{\alpha_1, k\ \Gamma \Rightarrow \Theta, A \qquad \alpha_2, k\ A, \Delta \Rightarrow \Lambda}{\omega_{k-\ell}(\alpha_1 \,\#\, \alpha_2), \ell\ \Gamma, \Delta \Rightarrow \Theta, \Lambda} & \text{CUT}
\end{array}
$$

Let $\alpha = \omega_{k-\ell}(\alpha_1 \,\#\, \alpha_2) = o_\ell(\Gamma, \Delta \Rightarrow \Theta, \Lambda; \pi)$. By inductive hypothesis, we have a proof π_1^* of S_1^* which is

$$\Gamma^* \Rightarrow \Theta^*, A \text{ or}$$
$$\Gamma^* \Rightarrow \Theta^*$$

and a proof π_2^* of S_2^* which is

$$A, \Delta^* \Rightarrow \Lambda^* \text{ or}$$
$$\Delta^* \Rightarrow \Lambda^*$$

In each, the end-sequent S_i^* has label k ($i = 1$ or 2). We let $\alpha_i^* = o_\ell(S_i^*; \pi_i^*) \preceq \alpha_i$. We have two cases.

First, either S_1^* or S_2^* no longer contains the occurrence of the cut-formula A. If π_1^* is a proof of $\Gamma^* \Rightarrow \Theta^*$, we let π^* be π_1^*, its end-sequent S^* corresponds to the end-sequent $\Gamma, \Delta \Rightarrow \Theta, \Lambda$ of π. Since by inductive hypothesis $\alpha_1^* \preceq \alpha_1$, we have $\alpha_1^* \preceq \alpha_1 \,\#\, \alpha_2$ by Proposition 8.33. By Proposition 8.40, $\alpha_1^* \preceq \omega_{k-\ell}(\alpha_1 \,\#\, \alpha_2) = \alpha$. Similarly, if π_2^* is a proof of $\Delta^* \Rightarrow \Lambda^*$, we let π^* be π_2^* and have $\alpha_2^* \preceq \omega_{k-\ell}(\alpha_1 \,\#\, \alpha_2)$. In either case, $o_\ell(S^*; \pi^*) \preceq o_\ell(S; \pi)$. (In these cases, removing weakenings has removed a cut-formula, the CUT is made redundant, and the sub-proof ending in the other premise of the CUT is removed.)

In the second case, the end-sequents of both proofs contain the cut-formula A. Then the proof π^* is

$$
\begin{array}{cc}
\vdots\, \pi_1^* & \vdots\, \pi_2^* \\[4pt]
\dfrac{\alpha_1^*, k\ \Gamma^* \Rightarrow \Theta^*, A \qquad \alpha_2^*, k\ A, \Delta^* \Rightarrow \Lambda^*}{\omega_{k-\ell}(\alpha_1^* \,\#\, \alpha_2^*), \ell\ \Gamma^*, \Delta^* \Rightarrow \Theta^*, \Lambda^*} & \text{CUT}
\end{array}
$$

The end-sequent S^* of π^* corresponds to the end-sequent of π. By inductive hypothesis, $\alpha_1^* \preceq \alpha_1$ and $\alpha_2^* \preceq \alpha_2$. By Proposition 8.34, $\alpha_1^* \,\#\, \alpha_2^* \preceq \alpha_1 \,\#\, \alpha_2$. By Proposition 8.41, $o_\ell(S^*; \pi^*) = \omega_{k-\ell}(\alpha_1^* \,\#\, \alpha_2^*) \preceq \omega_{k-\ell}(\alpha_1 \,\#\, \alpha_2) = o_\ell(\Gamma, \Delta \Rightarrow \Theta, \Lambda; \pi)$. $\square$

Proposition 9.13. *If π^* is a labelled proof provided by Proposition 9.12 and S^* a sequent in it labelled k, then the level of S^* in π^* is $\leq k$.*

Proof. The level of a sequent is determined by the maximum degree of a CUT or CJ inference between it and the end-sequent of π^*. Any such inference I^* with degree ℓ in π^* corresponds to an inference I in π which is below the sequent S in π that corresponds to S^*. Since the CUT or induction formula of I^* is the same as that of I, the degrees of I^* and I are the same. So, there cannot be a CUT or CJ inference below S^* in π^* of degree larger than k. □

Proposition 9.14. *Let π^* be a labelled proof and I an inference in it with premises labelled k and conclusion labelled $\ell < k$, and let π^{**} result from π^* by changing the labels of the premises of I to $k - 1$. Then $o_\ell(\pi^{**}) \preceq o_\ell(\pi^*)$.*

Proof. If I is an inference with one premise, we have labelled proofs π^* and π^{**} of the following forms:

$$
\begin{array}{cc}
\vdots\ \pi_1^* & \vdots\ \pi_1^* \\[2pt]
\cfrac{k'}{\alpha_1^*, k\ \ \underline{S_1}}\ I' & \cfrac{k'}{\omega^{\alpha_1}, k-1\ \ \underline{S_1}}\ I' \\[2pt]
\alpha^*, \ell\ \ S \quad I & \alpha^{**}, \ell\ \ S \quad I \\[2pt]
\vdots\ \pi_0^* & \vdots\ \pi_0^*
\end{array}
$$

Here I' is the inference ending in the premise of I, and k' is the label of the premises of I'. The sub-proof π_1^* ending in the premise(s) of I' and the part of the proof π_0^* below I are the same in both proofs; the only difference is the label of the sequent S_1.

Let's verify that the ordinal notation $\alpha_1^{**} = o_\ell(S_1; \pi^{**})$ assigned to the premise S_1 of I in π^{**} is indeed, as claimed, ω^{α_1}. Let $\beta' = o_\ell(I'; \pi^*)$ be the ordinal notation assigned to the inference I' in π^*. Since the inferences and labels above I' in π^* and π^{**} are identical, $o_\ell(I'; \pi^{**}) = \beta'$ as well. In π^*, where S_1 has label k, we have

$$\alpha_1^* = \omega_{k'-k}(\beta')$$

In π^{**}, this ordinal notation changes to

$$\alpha_1^{**} = \omega_{k'-k+1}(\beta'), \text{ i.e.,}$$
$$\alpha_1^{**} = \omega^{\alpha_1^*}$$

Let $\beta^* = o_\ell(I; \pi^*)$ and $\beta^{**} = o_\ell(I; \pi^{**})$. Now we distinguish cases according to the type of inference I: If I is a weakening, contraction, or interchange,

$$\beta^* = \alpha_1^* \quad \text{and} \qquad \beta^{**} = \omega^{\alpha_1^*}, \quad \text{and so}$$
$$\alpha^* = \omega_{k-\ell}(\alpha_1^*) \quad \text{and} \qquad \alpha^{**} = \omega_{k-\ell-1}(\omega^{\alpha_1^*}) = \omega_{k-\ell}(\alpha_1^*).$$

In this case, $\alpha^* = \alpha^{**}$. If I is an operational inference,

$$\beta^* = \alpha_1^* \# 1 \qquad \text{and} \qquad \beta^{**} = \omega^{\alpha_1^*} \# 1, \qquad \text{and so}$$
$$\alpha^* = \omega_{k-\ell}(\alpha_1^* \# 1) \qquad \text{and} \qquad \alpha^{**} = \omega_{k-\ell-1}(\omega^{\alpha_1^*} \# 1).$$

If we let $m = k - \ell - 1$ we can write this as

$$\alpha^* = \omega_{m+1}(\alpha_1^* \# 1) = \omega_m(\omega^{\alpha_1^* \# 1}) \qquad \alpha^{**} = \omega_m(\omega^{\alpha_1^*} \# 1).$$

By Proposition 8.41, to show that $\alpha^* \succeq \alpha^{**}$ it suffices to show that $\omega^{\alpha_1^* \# 1} \succeq \omega^{\alpha_1^*} \# 1$. By Corollary 8.37, $\alpha_1^* \# 1 \succ \alpha_1^*$. We get $\omega^{\alpha_1^* \# 1} \succeq \omega^{\alpha_1^*} \# 1$ by Proposition 8.25 and the fact that $\omega^{\alpha_1^*} \# 1$ is just $\omega^{\alpha_1^*} + 1$.

Finally, suppose I is CJ and $\alpha_1^* = \omega^\gamma + \dots$. Then we have:

$$\beta^* = \omega^{\gamma \# 1} \qquad \text{and} \qquad \beta^{**} = \omega^{\alpha_1^* \# 1}, \qquad \text{and so}$$
$$\alpha^* = \omega_{m+1}(\omega^{\gamma \# 1}) \qquad \text{and} \qquad \alpha^{**} = \omega_m(\omega^{\alpha_1^* \# 1}) = \omega_{m+1}(\alpha_1^* \# 1).$$

Since $\alpha_1^* = \omega^\gamma + \dots$ and $\gamma \prec \gamma \# 1$, we have $\omega^{\gamma \# 1} \succ \alpha_1^* \# 1$. Thus, by Proposition 8.41, $\alpha^* \succ \alpha^{**}$.

Now suppose I is an inference with two premises (i.e., an operational inference or CUT). Then π^* and π^{**} are

We see as before that $\alpha_1^{**} = \omega^{\alpha_1^*}$ and $\alpha_2^{**} = \omega^{\alpha_2^*}$. If I is an operational inference with two premises or CUT, we have

$$\beta^* = \alpha_1^* \# \alpha_2^* \qquad \text{and} \qquad \beta^{**} = \omega^{\alpha_1^*} \# \omega^{\alpha_2^*}, \qquad \text{and so}$$
$$\alpha^* = \omega_m(\omega^{\alpha_1^* \# \alpha_2^*}) \qquad \text{and} \qquad \alpha^{**} = \omega_m(\omega^{\alpha_1^*} \# \omega^{\alpha_2^*}).$$

Since neither α_1^* nor $\alpha_2^* = 0$, $\alpha_1^* \# \alpha_2^* \succ \alpha_1^*$ and $\succ \alpha_2^*$. So, $\omega^{\alpha_1^* \# \alpha_2^*} \succ \omega^{\alpha_1^*} \# \omega^{\alpha_2^*}$.

So far we've assumed that there is an inference I' immediately above the premise(s) of I. If there is not, S_1 (or S_2) is an axiom, in both π^* and π^{**}. Consequently, $\alpha_1^* = \alpha_1^{**} = 1$ (or $\alpha_2^* = \alpha_2^{**} = 1$). We leave the verification that in these cases $\alpha^{**} \preceq \alpha^*$ as an exercise.

The inferences and labels in π_0^* are the same in π^* and π^{**}. We've shown that $o_\ell(S; \pi^{**}) \preceq o_\ell(S; \pi^*)$. By an argument similar to the proof of Proposition 9.9, we have that $o_\ell(\pi^{**}) \preceq o_\ell(\pi^*)$. □

Problem 9.15. Complete the proof of Proposition 9.14 by verifying the case in which there are no inferences above I.

Let's consider an example. Take $A(a)$ to be $a = 0$, then $A(a') \Rightarrow$ and $\Rightarrow A(0)$ are axioms. Suppose B is a formula of degree 3, e.g., $\neg\neg\neg C$ where C is atomic. Let π be the following proof:

```
1,3   A(a') ⇒
-------------------- ¬R
2,3        ⇒ ¬A(a')
-------------------- WL
2,3   ¬A(a) ⇒ ¬A(a')
-------------------- CJ
ω¹,3   ¬A(0) ⇒ ¬A(4̄)                        1,3   ⇒ A(0)
--------------------------- ¬R              ------------------- WR
ω¹+1,3      ⇒ ¬A(4̄), ¬¬A(0)   2,3  ¬¬A(0) ⇒ A(0)   1,3   ⇒ A(0), ¬A(0)
                                                              ¬L
                              ------------------------------------ CUT
                ω¹+3,3   ⇒ ¬A(4̄), A(0)
                ------------------------ WR
                ω¹+3,3   ⇒ ¬A(4̄), A(0), B        3,3  B,C ⇒
                         ---------------------------------- CUT
                ω₂(ω¹+6),1  C ⇒ ¬A(4̄), A(0)    1,1        A(4̄) ⇒ A(4̄)
                --------------------------- IR            ---------------- ¬L
                ω₂(ω¹+6),1  C ⇒ A(0), ¬A(4̄)    2,1  ¬A(4̄), A(4̄) ⇒
                           -------------------------------------------- CUT
                      ω₁(ω₂(ω¹+6)+2),0   C, A(4̄) ⇒ A(0)
```

Now we simplify the end-part by removing the WR together with the CUT on B which is now redundant. If we leave the labels as they are in π, we get:

```
1,3   A(a') ⇒
-------------------- ¬R
2,3        ⇒ ¬A(a')
-------------------- WL
2,3   ¬A(a) ⇒ ¬A(a')
-------------------- CJ
ω¹,3   ¬A(0) ⇒ ¬A(4̄)                        1,3   ⇒ A(0)
--------------------------- ¬R              ------------------- WR
ω¹+1,3      ⇒ ¬A(4̄), ¬¬A(0)   2,3  ¬¬A(0) ⇒ A(0)   1,3   ⇒ A(0), ¬A(0)
                                                              ¬L
                              ------------------------------------ CUT
                ω₂(ω¹+3),1   ⇒ ¬A(4̄), A(0)    1,1        A(4̄) ⇒ A(4̄)
                -------------------------- IR            ---------------- ¬L
                ω₂(ω¹+3),1   ⇒ A(0), ¬A(4̄)    2,1  ¬A(4̄), A(4̄) ⇒
                          -------------------------------------------- CUT
                      ω₁(ω₂(ω¹+3)+2),0   A(4̄) ⇒ A(0)
```

The premise of IR in π^* corresponds to the premise of IR in π, so gets label 1. We've updated the ordinal assignments based on these labels. As shown in Proposition 9.12, the ordinal notations assigned to each sequent on the basis of the labels are no greater than the ordinal notations assigned to the corresponding sequents in π. However, not all labels give the correct level of a sequent. In the example, with the removal of the CUT of degree 3, the CUT on $\neg\neg A(0)$ of degree 2 determines the level of everything above it to be 2. So the actual ordinal notation assigned to the conclusion of this CUT is ω^{ω^1+3}, which is greater than the ordinal notation assigned to its corresponding sequent in π. We proceed to "correct" the levels by successively decreasing levels of lowermost sequents with incorrect labels by one. The first step is to reduce the levels of the premises of the CUT on $\neg\neg A(0)$ to their correct level 2. This has an interesting effect:

$$\cfrac{\cfrac{\cfrac{\cfrac{\cfrac{1,3 \quad A(a') \Rightarrow}{2,3 \qquad \Rightarrow \neg A(a')} \text{\tiny ¬R}}{2,3 \quad \neg A(a) \Rightarrow \neg A(a')} \text{\tiny WL}}{\omega^1,3 \quad \neg A(0) \Rightarrow \neg A(\overline{4})} \text{\tiny CJ}}{\omega^{\omega^1+1},2 \qquad \Rightarrow \neg A(\overline{4}), \neg\neg A(0)} \text{\tiny ¬R} \qquad \cfrac{1,3 \quad \Rightarrow A(0)}{1,3 \quad \Rightarrow A(0), \neg A(0)} \text{\tiny WR}}{\cfrac{\cfrac{\omega_1(\omega^{\omega^1+1}+\omega^2),1 \quad \Rightarrow \neg A(\overline{4}), A(0)}{\omega_1(\omega^{\omega^1+1}+\omega^2),1 \qquad \Rightarrow A(0), \neg A(\overline{4})} \text{\tiny IR}}{}}$$

The ordinal notations assigned to the premises of the cut on $\neg\neg A(0)$ have actually increased. However, the ordinal notation assigned to its conclusion (and therefore also to the end-sequent) has decreased:

$$\omega_1(\omega_1(\omega^{\omega^1+1}+\omega^2)+2) \prec \omega_1(\omega_1(\omega^{\omega^1+3})+2),$$

since $\omega^{\omega^1+1}+\omega^2 \prec \omega^{\omega^1+3}$. Continuing in this way, by readjusting the levels to their correct values, we finally have:

$$\cfrac{\cfrac{\cfrac{\cfrac{\cfrac{1,2 \quad A(a') \Rightarrow}{2,2 \qquad \Rightarrow \neg A(a')} \text{\tiny ¬R}}{2,2 \quad \neg A(a) \Rightarrow \neg A(a')} \text{\tiny WL}}{\omega^1,2 \quad \neg A(0) \Rightarrow \neg A(\overline{4})} \text{\tiny CJ}}{\omega^1+1,2 \qquad \Rightarrow \neg A(\overline{4}), \neg\neg A(0)} \text{\tiny ¬R} \qquad \cfrac{1,2 \quad \Rightarrow A(0)}{1,2 \quad \Rightarrow A(0), \neg A(0)} \text{\tiny WR}}{\cfrac{\cfrac{\omega_1(\omega^1+3),1 \quad \Rightarrow \neg A(\overline{4}), A(0)}{\omega_1(\omega^1+3),1 \quad \Rightarrow A(0), \neg A(\overline{4})} \text{\tiny IR}}{\omega_1(\omega_1(\omega^1+3))+2),0 \quad A(\overline{4}) \Rightarrow A(0)}}}$$

9.4   Reduction of suitable CUTS

Now we verify that the ordinal assignments decrease when we carry out step 3. Recall that a complex CUT I in the end-part of π is suitable if the CUT formulas C in the left and right premise are both descendants of principal formulas of operational inferences on the boundary. We showed in Proposition 7.36 that if the end-part of π contains no CJ or weakening inferences, and the proof is not simple, there must exist a suitable CUT in the end-part of π. In Section 7.4 we outlined a reduction procedure for such suitable CUTS. However, the result of the reduction is generally a more complex, larger proof in which parts of the original proof are duplicated. Hence, just giving the reduction procedure did not establish that repeatedly reducing suitable CUTS (together with steps 1 and 2, of course) eventually results in a simple proof. In order to do this, we have to show that the reduction of suitable CUTS decreases the ordinal notation assigned to our proof.

In Section 7.4 we only described a simple case of the reduction of suitable CUTS. The reason is that the reduction of suitable CUTS in the general case depends on the definitions of levels of sequents: we replace not just the sub-proof ending in the CUT itself but a sub-proof which ends in the topmost level transition below the complex CUT in question. (The simple case described already is the one where the complex CUT is a level transition.) We now finish the consistency proof by showing how a proof containing a suitable CUT in the end-part can be transformed into a proof with a lower ordinal notation assigned to it.

Proposition 9.16. *Let π be a proof in which the end-part contains no CJ or weakening inferences, and let π^* result from reducing an uppermost suitable CUT I in π. Then $o(\pi^*) \prec o(\pi)$.*

Proof. Suppose the cut-formula of I is C. Since I is a complex CUT, C is not atomic. There are different cases to consider, based on the main operator of C.

We carry out the case for $\forall$ and leave the rest as exercises. So suppose $C = \forall x\, F(x)$, and also suppose the complex CUT on C is a level transition. Then π looks like this:

$$
\left.
\begin{array}{cc}
\vdots\ \pi_1(a) & \vdots\ \pi_2 \\[4pt]
\dfrac{\beta_1\ \ \Gamma_1 \Rightarrow \Theta_1, F(a)}{\beta_1 + 1\ \ \Gamma_1 \Rightarrow \Theta_1, \forall x\, F(x)}\ \forall R &
\dfrac{\beta_2\ \ F(\overline{n}), \Delta_1 \Rightarrow \Lambda_1}{\beta_2 + 1\ \ \forall x\, F(x), \Delta_1 \Rightarrow \Lambda_1}\ \forall L \\[6pt]
\ddots\ \ \pi_1' & \pi_2'\ \ \cdots
\end{array}
\right\} \pi'
$$

$$
\dfrac{\alpha_1, r\ \ \Gamma \Rightarrow \Theta, \forall x\, F(x) \qquad \alpha_2, r\ \ \forall x\, F(x), \Delta \Rightarrow \Lambda}{\omega_{r-s}(\alpha_1 \# \alpha_2), s\ \ \Gamma, \Delta \Rightarrow \Theta, \Lambda}\ \text{CUT}\ I
$$

$$
\vdots\ \pi_4
$$

$$
\Pi \Rightarrow \Xi
$$

We've indicated the levels r and s of the premises and conclusion of the complex CUT we're focusing on. Since the CUT is a level transition, $r > s$. If the left and right premises are assigned α_1 and α_2, respectively, the ordinal notation assigned to the conclusion of the CUT is $\omega_{r-s}(\alpha_1 \# \alpha_2)$. Since it's a suitable CUT in the end-part, the cut-formulas on the left and right have ancestors which are principal formulas of $\forall R$ and $\forall L$ inferences on the boundary, respectively. These inferences increase the ordinal notation β_i assigned to the premise by 1, i.e., the ordinal notation assigned to the conclusion is $\beta_i \# 1$.

We'll replace the sub-proof π' of π ending with $\Gamma, \Delta \Rightarrow \Theta, \Lambda$ by the following proof:

$$
\begin{array}{c}
\left.
\begin{array}{cc}
\vdots \pi_L & \vdots \pi_R \\
\dfrac{\gamma_L, r-1\ \Gamma, \Delta \Rightarrow F(\overline{n}), \Theta, \Lambda}{\gamma_L, r-1\ \Gamma, \Delta \Rightarrow \Theta, \Lambda, F(\overline{n})} & \dfrac{\gamma_R, r-1\ \Gamma, \Delta, F(\overline{n}) \Rightarrow \Theta, \Lambda}{\gamma_R, r-1\ F(\overline{n}), \Gamma, \Delta \Rightarrow \Theta, \Lambda}
\end{array}
\right\} \pi''
\end{array}
$$

$$
\dfrac{\omega_{(r-1)-s}(\gamma_L \,\#\, \gamma_R), s\ \ \Gamma, \Delta, \Gamma, \Delta \Rightarrow \Theta, \Lambda, \Theta, \Lambda}{\omega_{(r-1)-s}(\gamma_L \,\#\, \gamma_R), s\ \ \Gamma, \Delta \Rightarrow \Theta, \Lambda} \quad \text{CUT } I'
$$

Here, the sub-proof π_L is:

$$
\vdots \pi_1(\overline{n})
$$

$$
\dfrac{\dfrac{\beta_1\ \Gamma_1 \Rightarrow \Theta_1, F(\overline{n})}{\dfrac{\beta_1\ \Gamma_1 \Rightarrow F(\overline{n}), \Theta_1}{\beta_1\ \Gamma_1 \Rightarrow F(\overline{n}), \Theta_1, \forall x\, F(x)}\ \text{WR}}}{\ }
$$

$$
\cdots \pi_1'' \qquad\qquad\qquad \dfrac{\beta_2\ F(\overline{n}), \Delta_1 \Rightarrow \Lambda_1}{\beta_2 + 1\ \forall x\, F(x), \Delta_1 \Rightarrow \Lambda_1}\ \text{VL} \quad \cdots \pi_2'
$$

$$
\vdots \pi_2
$$

$$
\dfrac{\alpha_1', r\ \ \Gamma \Rightarrow F(\overline{n}), \Theta, \forall x\, F(x) \qquad \alpha_2, r\ \ \forall x\, F(x), \Delta \Rightarrow \Lambda}{\gamma_L = \omega^{\alpha_1' \# \alpha_2}, r-1\ \ \Gamma, \Delta \Rightarrow F(\overline{n}), \Theta, \Lambda} \quad \text{CUT } I_L
$$

and π_R is:

$$
\vdots \pi_2
$$

$$
\vdots \pi_1(a)
$$

$$
\dfrac{\beta_1\ \Gamma_1 \Rightarrow \Theta_1, F(a)}{\beta_1 + 1\ \Gamma_1 \Rightarrow \Theta_1, \forall x\, F(x)}\ \text{VR} \qquad \dfrac{\dfrac{\beta_2\ F(\overline{n}), \Delta_1 \Rightarrow \Lambda_1}{\beta_2\ \Delta_1, F(\overline{n}) \Rightarrow \Lambda_1}}{\beta_2\ \forall x\, F(x), \Delta_1, F(\overline{n}) \Rightarrow \Lambda_1}\ \text{WL}
$$

$$
\cdots \pi_1' \qquad\qquad\qquad \cdots \pi_2''
$$

$$
\dfrac{\alpha_1, r\ \ \Gamma \Rightarrow \Theta, \forall x\, F(x) \qquad \alpha_2', r\ \ \forall x\, F(x), \Delta, F(\overline{n}) \Rightarrow \Lambda}{\gamma_R = \omega^{\alpha_1 \# \alpha_2'}, r-1\ \ \Gamma, \Delta, F(\overline{n}) \Rightarrow \Theta, \Lambda} \quad \text{CUT } I_R
$$

First note that the CUT I' at the end of π'' is now a CUT on $F(\overline{n})$ whereas the CUT I at the end of π' is a CUT on $\forall x\, F(x)$. The CUT I is a level transition, so the degree of $\forall x\, F(x)$ is equal to the level of its premises—all CUT inferences below I are of degree $\leq r-1$, with maximal degree s. Hence, the level of the new CUT is the degree of $F(\overline{n})$, i.e., $r-1$.

Both π_L and π_R end in a CUT (I_L and I_R, respectively) on $\forall x\, F(x)$ of degree r, the conclusion of which is one of the premises of the final CUT I' in π''. As we've just argued, the level of these premises is $r - 1$. Since the final CUTS I_L and I_R in π_L and π_R are of degree r, the level of *their* premises is r. This will now let us verify that the ordinal notations to the sequents in π_L and π_R are as indicated.

First, consider π_L. The sub-proof leading to the right premise $\forall x\, F(x), \Delta \Rightarrow \Lambda$ of the CUT I_L is exactly the same as in π', and so the level of its final sequent is r, i.e., the same as the level of the final sequent $\forall x\, F(x), \Delta \Rightarrow \Lambda$ of the right sub-proof in π'. Consequently, the ordinal notations assigned to sequents in the right sub-proof are the same as in π', i.e., the ordinal notation assigned to the right premise of the CUT I_L is α_2.

The left sub-proof of π_L is obtained by replacing the $\forall$R inference by a weakening of $\forall x\, F(x)$, and moving the $F(\overline{n})$ to the very left of the succedent. The lower part of the sub-proof (labelled π_1'') is just like π_1' (consists of the same inferences in the same order) except that every sequent on a thread through $\Gamma_1 \Rightarrow F(\overline{n}), \Theta_1, \forall x\, F(x)$ contains an additional occurrence of $F(\overline{n})$ immediately to the right of the sequent arrow. Since interchanges and weakenings leave the ordinal notation the same, the ordinal notations assigned to sequents in the left sub-proof differ only in that the contribution $\beta_1 + 1$ of the $\forall$R inference in π' is replaced by just β_1.

We've already seen that the level of the final sequent in π_1'' is r, i.e., the same as the level of the final sequent in π_1', and that the levels of all sequents in the left sub-proof of π_L are the same as the corresponding levels in π'. The ordinal notation α_1' assigned to the left premise $\Gamma \Rightarrow F(\overline{n}), \Theta, \forall x\, F(x)$ of the CUT I_L differs from α_1 only in that in it a sub-term $\beta_1 + 1$ was replaced by β_1. Since $\beta_1 \prec \beta_1 + 1$ (Corollary 8.37), we get $\alpha_1' \prec \alpha_1$ (by induction on the number of inferences in the left sub-proof, using Propositions 8.36 and 8.41 repeatedly).[8]

The verification that the ordinal notations of π_R are as indicated proceeds the same way. Here, the left sub-proof is identical to the left sub-proof of π_1 with all sequents having the same levels, so the ordinal notation assigned to the left premise of the CUT I_R is the same as in π', i.e., α_1. In the right sub-proof we replace the $\forall$L inference with interchanges and a weakening, and so the ordinal notation assigned to the right premise of the CUT I_R is α_2' which differs from α_2 by replacing $\beta_2 + 1$ in it by β_2 appropriately. We again have that $\alpha_2' \prec \alpha_2$.

The ordinal notation assigned to the end-sequent $\Gamma, \Delta \Rightarrow \Theta, \Lambda$ of π^* is

$$\omega_{(r-1)-s}(\gamma_L \# \gamma_R) \quad \text{where}$$
$$\gamma_L = \omega^{\alpha_1' \# \alpha_2} \text{ and}$$
$$\gamma_R = \omega^{\alpha_1 \# \alpha_2'}.$$

---

[8] If you want to give a detailed proof that the reduction of an ordinal notation in the proof reduces all ordinal notations below it, generalize the proof of Proposition 9.9 accordingly.

We want to show that this is smaller than the ordinal notation assigned to the corresponding sequent in π, namely

$$\omega_{r-s}(\alpha_1 \# \alpha_2).$$

In other words, we want to show that

$$\omega_{(r-1)-s}(\omega^{\alpha_1' \# \alpha_2} \# \omega^{\alpha_1 \# \alpha_2'}) \prec \omega_{r-s}(\alpha_1 \# \alpha_2).$$

This follows from Proposition 8.41 if we can show that

$$\omega^{\alpha_1' \# \alpha_2} \# \omega^{\alpha_1 \# \alpha_2'} \prec \omega^{\alpha_1 \# \alpha_2}.$$

In general, $\omega^\mu \# \omega^\nu$ is either $\omega^\mu + \omega^\nu$ if $\mu \geq \nu$, and otherwise it is $\omega^\nu + \omega^\mu$ (Problem 8.32). So either

$$\omega^{\alpha_1' \# \alpha_2} \# \omega^{\alpha_1 \# \alpha_2'} = \omega^{\alpha_1' \# \alpha_2} + \omega^{\alpha_1 \# \alpha_2'}, \text{ or}$$
$$= \omega^{\alpha_1 \# \alpha_2'} + \omega^{\alpha_1' \# \alpha_2}.$$

Since $\alpha_1' \prec \alpha_1$ we have $\alpha_1' \# \alpha_2 \prec \alpha_1 \# \alpha_2$ by Proposition 8.36, and similarly $\alpha_1 \# \alpha_2' \prec \alpha_1 \# \alpha_2$ since $\alpha_2' \prec \alpha_2$. Hence we have both

$$\omega^{\alpha_1' \# \alpha_2} + \omega^{\alpha_1 \# \alpha_2'} \prec \omega^{\alpha_1 \# \alpha_2} \text{ and}$$
$$\omega^{\alpha_1 \# \alpha_2'} + \omega^{\alpha_1' \# \alpha_2} \prec \omega^{\alpha_1 \# \alpha_2},$$

and thus our desired result.

Since the final part π_4 of the proof remains the same, by Proposition 9.9, the ordinal notation assigned to the end-sequent of the new proof π^* is smaller than that of the old proof π.

We'll now consider the case where the topmost suitable CUT I is not a level transition. This case is somewhat trickier. If I is not a level transition, its premises and conclusion have the same level, say, r. There must be other complex CUTS below I in π which are level transitions, since $r > 0$ and the level of the end-sequent is 0. (Note that the level transition cannot be a CJ inference since we have already removed all CJ inferences from the end-part.) If we select a topmost such CUT J, its premises will have level r and the conclusion some level $s < r$. Since it is a level transition, its cut-formula B has degree r. We'll replace the sub-proof ending in J to obtain the reduced proof π^*.

The original proof π has this form:

$$
\begin{array}{c}
\vdots\ \pi_1(a) \\
\cline{1-1}
\beta_1\ \Gamma_1 \Rightarrow \Theta_1, F(a) \\
\hline
\beta_1 + 1\ \Gamma_1 \Rightarrow \Theta_1, \forall x\, F(x)
\end{array}\ \forall\text{R}
$$

$$
\begin{array}{c}
\vdots\ \pi_2 \\
\cline{1-1}
\beta_2\ F(\overline{n}), \Delta_1 \Rightarrow \Lambda_1 \\
\hline
\beta_2 + 1\ \forall x\, F(x), \Delta_1 \Rightarrow \Lambda_1
\end{array}\ \forall\text{L}
$$

$$
\cfrac{\alpha_1, r\ \Gamma \Rightarrow \Theta, \forall x\, F(x) \qquad \alpha_2, r\ \forall x\, F(x), \Delta \Rightarrow \Lambda}{\alpha_1\,\#\,\alpha_2, r\ \Gamma, \Delta \Rightarrow \Theta, \Lambda}\ \text{CUT}\ I
$$

$$
\cfrac{\lambda_1, r\ \Gamma_3 \Rightarrow \Theta_3, B \qquad \lambda_2, r\ B, \Gamma_4 \Rightarrow \Theta_4}{\omega_{r-s}(\lambda_1\,\#\,\lambda_2), s\ \Gamma_3, \Gamma_4 \Rightarrow \Theta_3, \Theta_4}\ \text{CUT}\ J
$$

$$
\vdots\ \pi_5
$$

$$
\Pi \Rightarrow \Xi
$$

Of course, the CUT I may also lie in the sub-proof leading to the right premise of CUT J, but this case is symmetric.

The reduced proof π^* then is

$$
\begin{array}{c}
\vdots\ \pi_L \\
\cline{1-1}
\mu_1, t\ \Gamma_3, \Gamma_4 \Rightarrow F(\overline{n}), \Theta_3, \Theta_4 \\
\hline
\mu_1, t\ \Gamma_3, \Gamma_4 \Rightarrow \Theta_3, \Theta_4, F(\overline{n})
\end{array}
\qquad
\begin{array}{c}
\vdots\ \pi_R \\
\cline{1-1}
\mu_2, t\ \Gamma_3, \Gamma_4, F(\overline{n}) \Rightarrow \Theta_3, \Theta_4 \\
\hline
\mu_2, t\ F(\overline{n}), \Gamma_3, \Gamma_4 \Rightarrow \Theta_3, \Theta_4
\end{array}
$$

$$
\cfrac{\omega_{t-s}(\mu_1\,\#\,\mu_2), s\ \Gamma_3, \Gamma_4, \Gamma_3, \Gamma_4 \Rightarrow \Theta_3, \Theta_4, \Theta_3, \Theta_4}{\omega_{t-s}(\mu_1\,\#\,\mu_2), s\ \Gamma_3, \Gamma_4 \Rightarrow \Theta_3, \Theta_4}\ \text{CUT}\ I'
$$

$$
\vdots\ \pi_5
$$

$$
\Pi \Rightarrow \Xi
$$

We will describe π_L and π_R below; for now, let's focus on the levels of the sequents in this bottom part. The part π_5 of the proof below the new CUT I' is the same in π^* and in π. So, the levels of all sequents below I' are the same as in π. In particular, the level of the conclusion of I' is s.

What is the level t of the premises of I'? We don't know what exactly it is, but we can show that $s \le t < r$. Of course, the level of the conclusion of an inference is always $\le$ the levels of the premises, so obviously $s \le t$. To see that $t < r$, consider the degree d of the cut-formula $F(\overline{n})$. If it is greater than s, then the I' is a level transition, and $t = d$ (because d is the degree of the CUT I'). Recall that the CUT I was not a level transition, so its degree is $\le r$, the level of its conclusion.

Its cut-formula is $\forall x \, F(x)$, which is of degree $d + 1$. So we have $d + 1 \leq r$ and hence $t = d < r$. On the other hand, if I' is not a level transition, i.e., $d \leq s$, then $t = s$, and since $r > s$, we have $r > t$. In either case we have: $t < r$.

We've now verified that the levels in the bottom part of π^* are as indicated, and that the level t of the new CUT I' is such that $s \leq t < r$. Now we'll describe the proofs π_L and π_R and compute the ordinal notations μ_1 and μ_2 assigned to their end-sequents.

The left sub-proof π_L is

$$
\begin{array}{c}
\vdots \ \pi_1(\overline{n}) \\
\beta_1 \ \Gamma_1 \Rightarrow \Theta_1, F(\overline{n}) \\
\hline
\beta_1 \ \Gamma_1 \Rightarrow F(\overline{n}), \Theta_1 \\
\hline
\beta_1 \ \Gamma_1 \Rightarrow F(\overline{n}), \Theta_1, \forall x \, F(x)
\end{array} \ \text{WR}
$$

$$
\begin{array}{c}
\vdots \ \pi_2 \\
\beta_2 \ F(\overline{n}), \Delta_1 \Rightarrow \Lambda_1 \\
\hline
\beta_2 + 1 \ \forall x \, F(x), \Delta_1 \Rightarrow \Lambda_1
\end{array} \ \text{VL}
$$

$$
\begin{array}{c}
\cdots \pi_1'' \qquad\qquad\qquad\qquad \cdots \pi_2' \\
\alpha_1', r \ \Gamma \Rightarrow F(\overline{n}), \Theta, \forall x \, F(x) \qquad \alpha_2, r \ \forall x \, F(x), \Delta \Rightarrow \Lambda \\
\hline
\alpha_1' \, \# \, \alpha_2, r \ \Gamma, \Delta \Rightarrow F(\overline{n}), \Theta, \Lambda
\end{array} \ \text{CUT } I_L
$$

$$
\begin{array}{c}
\vdots \ \pi_3' \qquad\qquad\qquad\qquad\qquad\qquad\qquad \pi_4 \\
\lambda_L, r \ \Gamma_3 \Rightarrow F(\overline{n}), \Theta_3, B \qquad \lambda_2, r \ B, \Gamma_4 \Rightarrow \Theta_4 \\
\hline
\mu_1 = \omega_{r-t}(\lambda_L \, \# \, \lambda_2), t \ \Gamma_3, \Gamma_4 \Rightarrow F(\overline{n}), \Theta_3, \Theta_4
\end{array} \ \text{CUT } J_L
$$

We again replace the $\forall$R inference by interchanges moving $F(\overline{n})$ to the left side of the succedent and weaken by $\forall x \, F(x)$. The proofs π_1'' and π_3' are just like π_1' and π_3 except that every sequent below $\Gamma_1 \Rightarrow F(n), \Theta_1, \forall x \, F(x)$ and $\Gamma, \Delta \Rightarrow F(\overline{n}), \Theta, \Lambda$ likewise contains $F(\overline{n})$ on the far left of the succedent. Otherwise, however, the inferences applied are the same.

Recall that the degree of the cut-formula B of J_L is r, and since $r > t$, J_L is a level transition. Since every sequent between I_L and the left premise of J_L has a CUT of degree r below it (namely, J_L) and since π_3' contains no CUTs of larger degree (since π_3 did not), the levels of sequents in π_3' are the same as the corresponding sequents in π_3. The levels of sequents in $\pi_1(\overline{n})$, π_1'', π_2, and π_2' are the same as the corresponding levels of sequents in π. Hence as before we have $\alpha_1' \prec \alpha_1$, which in turn guarantees $\alpha_1' \, \# \, \alpha_2 \prec \alpha_1 \, \# \, \alpha_2$ and $\lambda_L \prec \lambda_1$.

The proof π_R is:

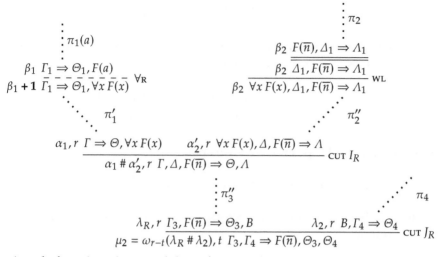

Again as before, $\beta_2 \prec \beta_2 + 1$ and thus $\alpha_2' \prec \alpha_2$. This in turn yields $\alpha_1 \# \alpha_2' \prec \alpha_1 \# \alpha_2$ and $\lambda_R \prec \lambda_1$.

What's left to show is that the ordinal notation assigned to the end-sequent $\Pi \Rightarrow \Xi$ of π^* is smaller than the ordinal notation assigned to the corresponding sequent in π. We do this by showing that the ordinal notation $\omega_{t-s}(\mu_1 \# \mu_2)$ assigned to the conclusion $\Gamma_3, \Gamma_4 \Rightarrow \Theta_3, \Theta_4$ of CUT I' in π^* is smaller than the ordinal notation $\omega_{r-s}(\lambda_1 \# \lambda_2)$ assigned to the conclusion of J in π. (The result then follows from Proposition 9.9, since the parts of π and π^* below these sequents are both π_5, i.e., the same.)

Recall that $s \le t < r$, so we can write the difference of $r - s$ as the sum of differences $t - s$ and $r - t$:

$$r - s = (t - s) + (r - t) \text{ so}$$
$$\omega_{r-s}(\lambda_1 \# \lambda_2) = \omega_{t-s}(\omega_{r-t}(\lambda_1 \# \lambda_2))$$

since in general $\omega_n(\omega_m(\nu)) = \omega_{m+n}(\nu)$, which follows immediately from the way $\omega_n(\nu)$ was defined (Definition 8.39). So to show

$$\omega_{t-s}(\mu_1 \# \mu_2) \prec \omega_{r-s}(\lambda_1 \# \lambda_2)$$

it suffices to show (by Proposition 8.41) that

$$\mu_1 \# \mu_2 \prec \omega_{r-t}(\lambda_1 \# \lambda_2)$$

We had that

$$\mu_1 = \omega_{r-t}(\lambda_L \# \lambda_2) \qquad \text{and}$$
$$\mu_2 = \omega_{r-t}(\lambda_R \# \lambda_2)$$

So we have to show that

$$\omega_{r-t}(\lambda_L \mathbin{\#} \lambda_2) \mathbin{\#} \omega_{r-t}(\lambda_R \mathbin{\#} \lambda_2) \prec \omega_{r-t}(\lambda_1 \mathbin{\#} \lambda_2).$$

We know that $\lambda_L \prec \lambda_1$. So by Proposition 8.33, $\lambda_L \mathbin{\#} \lambda_2 \prec \lambda_1 \mathbin{\#} \lambda_2$. Similarly, since $\lambda_R \prec \lambda_1$, $\lambda_R \mathbin{\#} \lambda_2 \prec \lambda_1 \mathbin{\#} \lambda_2$. By Proposition 8.41,

$$\omega_{r-t}(\lambda_L \mathbin{\#} \lambda_2) \prec \omega_{r-t}(\lambda_1 \mathbin{\#} \lambda_2) \text{ and}$$
$$\omega_{r-t}(\lambda_R \mathbin{\#} \lambda_2) \prec \omega_{r-t}(\lambda_1 \mathbin{\#} \lambda_2).$$

By Problem 8.32 and Proposition 8.25, we obtain

$$\omega_{r-t}(\lambda_L \mathbin{\#} \lambda_2) \mathbin{\#} \omega_{r-t}(\lambda_R \mathbin{\#} \lambda_2) \prec \omega_{r-t}(\lambda_1 \mathbin{\#} \lambda_2).$$

Verifying that the ordinal notations decrease in cases other than CUTs on $\forall x\, F(x)$ is left as an exercise. □

Problem 9.17. Verify that the ordinal notation assigned to the result of reducing a complex CUT on $A \wedge B$ is smaller than that of the original proof.

Problem 9.18. Review your answers for Problem 7.30 and Problem 7.32 to convince yourself that the assignments of ordinal notations after reducing complex CUTs for the case for $\exists$ are the same as for $\forall$, and for $\vee$ are the same as for the case for $\wedge$. Do the same for the reduction of CUTs with main connective $\supset$ (Problem 7.31) and $\neg$ (see Section 7.11).

9.5 A simple example, revisited

Let's revisit our example from Section 7.10. In Section 9.1 we used it to illustrate the assignment of ordinal notations to proofs. Here is the proof again, now with levels of sequents indicated.

The end-part contains no CJ inferences nor weakenings. So there is a suitable CUT in the end-part: the CUT on $\forall x\, F(x)$. It is a level transition, so we only have to consider the simpler case. To find the reduced proof π^* we'll first construct the corresponding π_L and π_R. Our π_L is:

$$
\pi_2(\bar{3})\left\{
\begin{array}{c}
\vdots\ \pi_1(a) \\
\dfrac{1,1\ \Rightarrow F(0) \qquad \dfrac{9,1\ F(a)\Rightarrow F(a')}{\omega^1,1\ F(0)\Rightarrow F(\bar{3})}\ \text{CJ}}{\omega^1+1,1\ \Rightarrow F(\bar{3})}\ \text{CUT}
\end{array}
\right.
$$

$$
\dfrac{\dfrac{\omega^1+1,1\ \Rightarrow F(\bar{3})}{\omega^1+1,1\ \Rightarrow F(\bar{3}),\forall x\,F(x)}\ \text{WR} \qquad \dfrac{1,1\ F(\bar{3})\Rightarrow F(\bar{3})}{2,1\ \forall x\,F(x)\Rightarrow F(\bar{3})}\ \forall_\mathrm{L}}{\dfrac{\omega^{\omega^1+3},0\ \Rightarrow F(\bar{3}),F(\bar{3})}{\omega^{\omega^1+3},0\ \Rightarrow F(\bar{3})}\ \text{CR}}\ \text{CUT}
$$

And π_R is:

$$
\pi_2(b)\left\{
\begin{array}{c}
\vdots\ \pi_1(a) \\
\dfrac{1,1\ \Rightarrow F(0) \qquad \dfrac{9,1\ F(a)\Rightarrow F(a')}{\omega^1,1\ F(0)\Rightarrow F(b)}\ \text{CJ}}{\omega^1+1,1\ \Rightarrow F(b)}\ \text{CUT}
\end{array}
\right.
$$

$$
\dfrac{\dfrac{\omega^1+1,1\ \Rightarrow F(b)}{\omega^1+2,1\ \Rightarrow \forall x\,F(x)}\ \forall_\mathrm{R} \qquad \dfrac{1,1\ F(\bar{3})\Rightarrow F(\bar{3})}{1,1\ \forall x\,F(x),F(\bar{3})\Rightarrow F(\bar{3})}\ \text{WL}}{\omega^{\omega^1+3},0\ F(\bar{3})\Rightarrow F(\bar{3})}\ \text{CUT}
$$

Together we have our proof π^*:

$$
\dfrac{\vdots\ \pi_L \qquad \vdots\ \pi_R}{\dfrac{\omega^{\omega^1+3},0\ \Rightarrow F(\bar{3}) \qquad \omega^{\omega^1+3},0\ F(\bar{3})\Rightarrow F(\bar{3})}{\omega^{\omega^1+3}+\omega^{\omega^1+3},0\ \Rightarrow F(\bar{3})}}\ \text{CUT}
$$

Since $\omega^1+3 < \omega^1+4$, we have

$$
\omega^{\omega^1+3}+\omega^{\omega^1+3} < \omega^{\omega^1+4}.
$$

Note that in the original proof, there was only one CJ inference. Since it occurred above a boundary inference (the $\forall_\mathrm{R}$), it was not in the end-part. The new proof has two CJ inferences. The one on the right still lies above a boundary inference, but on the left the $\forall_\mathrm{R}$ was replaced by a WR—and so now the end-part contains a CJ inference (on the left side). To further reduce the proof, we now replace the sub-proof ending in this cj inference by a series of cuts.

$$
\cfrac{
\cfrac{
\cfrac{
\cfrac{
\begin{array}{cc}
\vdots\,\pi_1(0) & \vdots\,\pi_1(\bar{1}) \\
\mathbf{9},1\ F(0) \Rightarrow F(\bar{1}) & \mathbf{9},1\ F(\bar{1}) \Rightarrow F(\bar{2})
\end{array}
}{\mathbf{18},1\ F(0) \Rightarrow F(\bar{2})}\ \text{CUT}\quad
\cfrac{\vdots\,\pi_1(\bar{2})}{\mathbf{9},1\ F(\bar{2}) \Rightarrow F(\bar{3})}
}{
\cfrac{\mathbf{1},1\ \Rightarrow F(0) \qquad \mathbf{27},1\ F(0) \Rightarrow F(\bar{3})}{\mathbf{28},1\ \Rightarrow F(\bar{3})}\ \text{CUT}
}\ \text{CUT}\ \ \cfrac{\mathbf{28},1\ \Rightarrow F(\bar{3}),\forall x\,F(x)}{}\,\text{WR}
}{
\cfrac{\omega^{30},0\ \Rightarrow F(\bar{3}),F(\bar{3})}{\omega^{30},0\ \Rightarrow F(\bar{3})}\ \text{CR}
}
\qquad
\cfrac{\mathbf{1},1\ F(\bar{3}) \Rightarrow F(\bar{3})}{\mathbf{2},1\ \forall x\,F(x) \Rightarrow F(\bar{3})}\,\forall\text{L}
$$

$$
\cfrac{\omega^{30},0\ \Rightarrow F(\bar{3}) \qquad\qquad \omega^{\omega^1+3},0\ F(\bar{3}) \Rightarrow F(\bar{3})}{\omega^{\omega^1+3}+\omega^{30},0\ \Rightarrow F(\bar{3})}\ \text{CUT}
$$

Note that although there are complex cuts in the end-part of this new proof, in both cases the cut-formulas $\forall x\,F(x)$ are not descendants of boundary inferences. Instead, in each case, one of the cut-formulas is a descendant of a weakening in the end-part. So we apply step 2, removing weakenings from the end-part. First, we remove the wr on the left side:

$$
\pi_L'\left\{
\cfrac{
\cfrac{
\cfrac{
\begin{array}{cc}
\vdots\,\pi_1(0) & \vdots\,\pi_1(\bar{1}) \\
\mathbf{9},0\ F(0) \Rightarrow F(\bar{1}) & \mathbf{9},0\ F(\bar{1}) \Rightarrow F(\bar{2})
\end{array}
}{\mathbf{18},0\ F(0) \Rightarrow F(\bar{2})}\ \text{CUT}\quad
\cfrac{\vdots\,\pi_1(\bar{2})}{\mathbf{9},0\ F(\bar{2}) \Rightarrow F(\bar{3})}
}{
\mathbf{1},0\ \Rightarrow F(0) \qquad \mathbf{27},0\ F(0) \Rightarrow F(\bar{3})
}\ \text{CUT}
}{\mathbf{28},0\ \Rightarrow F(\bar{3})}
\qquad
\cfrac{\vdots\,\pi_R}{\omega^{\omega^1+3},0\ F(\bar{3}) \Rightarrow F(\bar{3})}
\right.
$$

$$
\omega^{\omega^1+3}+\mathbf{28},0\ \Rightarrow F(\bar{3})\ \ \text{CUT}
$$

The left sub-proof now only contains atomic cuts. Let's look at the right sub-proof again:

$$
\cfrac{
\cfrac{\vdots\,\pi_L'}{\mathbf{28},0\ \Rightarrow F(\bar{3})}
\qquad
\cfrac{
\cfrac{
\cfrac{\mathbf{1},1\ \Rightarrow F(0) \quad
\cfrac{\vdots\,\pi_1(a)}{\cfrac{\mathbf{9},1\ F(a) \Rightarrow F(a')}{\omega^1,1\ F(0) \Rightarrow F(b)}\,\text{CJ}}
}{\cfrac{\omega^1+1,1\ \Rightarrow F(b)}{\omega^1+2,1\ \Rightarrow \forall x\,F(x)}\,\forall\text{R}}\ \text{CUT}
\qquad
\cfrac{\mathbf{1},1\ F(\bar{3}) \Rightarrow F(\bar{3})}{\mathbf{1},1\ \forall x\,F(x),F(\bar{3}) \Rightarrow F(\bar{3})}\,\text{WL}
}{\omega^{\omega^1+3},0\ F(\bar{3}) \Rightarrow F(\bar{3})}\ \text{CUT}
}{}
}{\omega^{\omega^1+3}+\mathbf{28},0\ \Rightarrow F(\bar{3})}\ \text{CUT}
$$

The end-part still contains a weakening, so step 2 applies again. Removing the WL reduces the entire right sub-proof to just the axiom $F(\bar{3}) \Rightarrow F(\bar{3})$. We are left with:

$$
\cfrac{
  \cfrac{
    \cfrac{\mathbf{9},0\ F(0) \Rightarrow F(\bar{1}) \qquad \mathbf{9},0\ F(\bar{1}) \Rightarrow F(\bar{2})}{\mathbf{18},0\ F(0) \Rightarrow F(\bar{2})}\ \text{CUT} \qquad \mathbf{9},0\ F(\bar{2}) \Rightarrow F(\bar{3})
  }{
    \cfrac{\mathbf{1},0\ \Rightarrow F(0) \qquad \cfrac{}{\mathbf{27},0\ F(0) \Rightarrow F(\bar{3})}\ \text{CUT}}{\mathbf{28},0\ \Rightarrow F(\bar{3})}
  }\qquad \mathbf{1},1\ F(\bar{3}) \Rightarrow F(\bar{3})
}{
  \mathbf{29},0\ \Rightarrow F(\bar{3})
}\ \text{CUT}
$$

The proof is now its own end-part, and contains no CJ inferences or complex CUTs. We are done: the proof is simple.

Let's recap. We started with a proof of an atomic end-sequent. In our case, that proof was already regular, with all initial sequents atomic, and with the property that all free variables in it were eigenvariables. In general, we would have to first apply the relevant results to ensure this: Proposition 5.17 guarantees that initial sequents can be replaced by atomic initial sequents. Proposition 7.16 ensures that any proof can be turned into a regular proof. Proposition 7.20 allows us to replace free variables that are not eigenvariables in a proof by 0.

We applied the reduction steps 1–3 in the relevant order successively. In the example, the end-part contained neither weakenings nor CJ inferences. Proposition 7.36 guarantees that a proof that is not its own end-part contains a suitable CUT. So we first applied step 3, the reduction of an uppermost suitable CUT in the end-part. Our example contained only one complex CUT in the end-part, the CUT on $\forall x\, F(x)$. As we showed in Proposition 9.16, the ordinal notation assigned to the proof after reduction of the CUT decreased. The new proof still was not its own end-part, but its end-part now did contain a CJ inference. So step 1, the elimination of CJ inferences from the end-part, was required. As we showed in Proposition 9.11, the elimination of CJ inferences decreased the ordinal notation assigned to the proof. Since there was only one CJ inference, step 1 only had to be applied once. The resulting proof contained weakenings, so this time we applied step 2, the removal of weakenings. This step not only removed the weakenings from the end-part, but also the remaining CJ inference and the remaining complex CUTs. This resulted in a simple proof. This was a happy accident: in general, removal of weakenings does not even guarantee that we obtain a new proof with a weakening-free end-part which is assigned a smaller ordinal notation. If the CJ inference or complex CUTs had not just disappeared, steps 1 or 3 would have applied again.

A

The Greek alphabet

Alpha	α	A
Beta	β	B
Gamma	γ	Γ
Delta	δ	Δ
Epsilon	ε	E
Zeta	ζ	Z
Eta	η	H
Theta	θ	Θ
Iota	ι	I
Kappa	κ	K
Lambda	λ	Λ
Mu	μ	M
Nu	ν	N
Xi	ξ	Ξ
Omicron	o	O
Pi	π	Π
Rho	ρ	P
Sigma	σ	Σ
Tau	τ	T
Upsilon	υ	Υ
Phi	φ	Φ
Chi	χ	X
Psi	ψ	Ψ
Omega	ω	Ω

B

Set-theoretic notation

Throughout this book, we use some standard mathematical terminology and symbols, which we collect here. See, e.g., Zach (2019b) for an introduction.

Definition B.1. A *set* is a collection of objects considered irrespective of the way they are specified. If $a_1, \ldots, a_n$ are any objects, the set consisting of all and only those objects is denoted

$$\{a_1, \ldots, a_n\}.$$

If $\phi(x)$ is some property an object x might have, the set of all objects that have the property is written

$$\{x : \phi(x)\}.$$

The objects making up the set are called its *elements*. We write $x \in A$ to indicate that x is an element of the set A. The *empty set* $\emptyset$ is the unique set which has no elements.

Definition B.2. The set of natural numbers is

$$\mathbb{N} = \{0, 1, 2, \ldots\}.$$

Definition B.3. A set A is a *subset* of a set B, $A \subseteq B$, if and only if every element of A is also an element of B. If $A \subseteq B$ but $A \neq B$ we write $A \subsetneq B$ and say that A is a *proper subset* of B.

Definition B.4. If A, B are sets, then their *union* $A \cup B$ and intersection $A \cap B$ are the sets

$$A \cup B = \{x : x \in A \text{ or } x \in B\},$$
$$A \cap B = \{x : x \in A \text{ or } x \in B\}.$$

The *difference* $A \setminus B$ is

$$A \setminus B = \{x : x \in A \text{ and } x \notin B\}.$$

Definition B.5. A *pair* of objects $\langle a, b \rangle$ is a way of considering a and b together and in that order. So $\langle a, b \rangle \neq \langle b, a \rangle$.

If A, B are sets then

$$A \times B = \{\langle x, y \rangle : x \in A \text{ and } y \in B\}.$$

Definition B.6. The set $A \times A$ is the set of pairs $\langle a_1, a_2 \rangle$ of elements of A. It is also denoted A^2. The set of triples $\langle a_1, a_2, a_3 \rangle$ is A^3, and so on. A^* is the set of all finite sequences (including the empty sequence).

Definition B.7. A *relation* R on X is a subset of $X \times X$. We write xRy to mean $\langle x, y \rangle \in R$.

C

Axioms, rules, and theorems of axiomatic calculi

C.1   Axioms and rules of inference

$\mathbf{M}_0$ (Section 2.4.1)   **Heyting**

PL1. $A \supset (A \wedge A)$ [2.1]

PL2. $(A \wedge B) \supset (B \wedge A)$ [2.11]

PL3. $(A \supset B) \supset [(A \wedge C) \supset (B \wedge C)]$ [2.12]

PL4. $[(A \supset B) \wedge (B \supset C)] \supset (A \supset C)$ [2.13]

PL5. $B \supset (A \supset B)$ [2.14]

PL6. $(A \wedge (A \supset B)) \supset B$ [2.15]

PL7. $A \supset (A \vee B)$ [3.1]

PL8. $(A \vee B) \supset (B \vee A)$ [3.11]

PL9. $[(A \supset C) \wedge (B \supset C)] \supset [(A \vee B) \supset C]$ [3.12]

PL10. $[(A \supset B) \wedge (A \supset \neg B)] \supset \neg A$ [4.11]

$\mathbf{J}_0$ (Section 2.4.2)

PL11. $\neg A \supset (A \supset B)$ [4.1]

$\mathbf{K}_0$ (Section 2.4.3)

PL12. $\neg\neg A \supset A$

Rule of Inference: From A and $A \supset B$, infer B (modus ponens, MP).

$\mathbf{M}_1, \mathbf{J}_1, \mathbf{K}_1$ (Section 2.13)

QL1. $\forall x\, A(x) \supset A(t)$

QL2. $A(t) \supset \exists x\, A(x)$

Rules of Inference:

QR₁. If $A \supset B(a)$ is derivable, a does not occur in A, and x is a bound variable not occurring in $B(a)$, then $A \supset \forall x\, B(x)$ is derivable.

QR₂. If $B(a) \supset A$ is derivable, a does not occur in A, and x is a bound variable not occurring in $B(a)$, then $\exists x\, B(x) \supset A$ is also derivable.

C.2 Theorems and derived rules

The following theorems and derived rules of $\mathbf{M}_0$, $\mathbf{J}_0$, $\mathbf{K}_0$ are used in Chapter 2 and Appendix D. When the theorem is provable in $\mathbf{M}_0$ we do not specify anything under the provability sign. The derived rules work in all systems.

Derived rules from Section 2.5:

∧INTRO: If $\vdash_{\mathbf{M}_0} A$ and $\vdash_{\mathbf{M}_0} B$, then $\vdash_{\mathbf{M}_0} (A \wedge B)$.

⊃TRANS: If $\vdash_{\mathbf{M}_0} A \supset B$ and $\vdash_{\mathbf{M}_0} B \supset C$, then $\vdash_{\mathbf{M}_0} A \supset C$.

Theorems and rules from Problem 2.7:

E1. $\vdash (A \wedge B) \supset A$	[2.2]
E2. $\vdash A \supset A$	[2.21]
E3. $\vdash (A \wedge B) \supset B$	[2.22]
Rule E4. If $\vdash B$ then $\vdash A \supset (A \wedge B)$	
Rule E5. If $\vdash B$ then $\vdash A \supset B$	
E6. $\vdash [A \supset (B \wedge C)] \supset [A \supset (C \wedge B)]$	
Rule E7. If $\vdash A \supset (B \supset C)$ and $\vdash A \supset (C \supset D)$, then $\vdash A \supset (B \supset D)$	
E8. $\vdash [(A \supset B) \wedge (C \supset D)] \supset [(A \wedge C) \supset (B \wedge D)]$	[2.23]
E9. $\vdash [(A \supset B) \wedge (A \supset C)] \supset [A \supset (B \wedge C)]$	[2.24a]
E10. $\vdash [A \supset (B \wedge C)] \supset [(A \supset B) \wedge (A \supset C)]$	[2.24b]
E11. $\vdash [B \wedge (A \supset C)] \supset [A \supset (B \wedge C)]$	[2.25]
E12. $\vdash B \supset [A \supset (A \wedge B)]$	[2.26]
E13. $\vdash [A \supset (B \supset C)] \supset [(A \wedge B) \supset C]$	[2.27a]
E14. $\vdash [(A \wedge B) \supset C] \supset [A \supset (B \supset C)]$	[2.27b]

Theorems from Problem 2.17:

E1. $\vdash [A \supset (B \supset C)] \supset [B \supset (A \supset C)]$	[2.271]
E2. $\vdash A \supset \{C \supset [(A \wedge B) \supset C]\}$	[2.28]
E3. $\vdash A \supset \{B \supset [A \supset (C \supset B)]\}$	[2.281]
E4. $\vdash (A \supset B) \supset [(B \supset C) \supset (A \supset C)]$	[2.29]
E5. $\vdash (B \supset C) \supset [(A \supset B) \supset (A \supset C)]$	[2.291]

Theorems from Problem 2.18:

E1. $\vdash \neg(A \wedge \neg A)$	[4.13]
E2. $\vdash (A \supset B) \supset (\neg B \supset \neg A)$	[4.2]
E3. $\vdash (A \supset B) \supset (\neg\neg A \supset \neg\neg B)$	[4.22]
E4. $\vdash (A \supset \neg B) \supset (B \supset \neg A)$	[4.21]
E5. $\vdash A \supset \neg\neg A$	[4.3]
E6. $\vdash \neg A \supset \neg\neg\neg A$	[4.31]
E7. $\vdash \neg\neg\neg A \supset \neg A$	[4.32]
E8. $\vdash \neg A \supset (A \supset \neg B)$	[4.4]
E9. $\vdash \neg\neg(A \wedge B) \supset \neg\neg A \wedge \neg\neg B$	[4.61]
E10. $\vdash \neg\neg(A \vee \neg A)$	[4.8]
E11. $\vdash (A \supset B) \supset \neg(A \wedge \neg B)$	[4.9]
E12. $\vdash (A \wedge B) \supset \neg(\neg A \vee \neg B)$	[4.92]

Derived rules from Lemma 2.19:

If $\vdash \neg\neg A \supset A$ and $\vdash \neg\neg B \supset B$, then:

1. $\vdash \neg\neg(A \wedge B) \supset (A \wedge B)$ and
2. $\vdash \neg\neg(A \supset B) \supset (A \supset B)$.

Theorem from Example 2.34:

$\vdash_{S_1} F(t) \supset \neg \forall x \, \neg F(x)$

Derived rule from Example 2.35:

If $\vdash_{M_1} \neg\neg F(a) \supset F(a)$, then $\vdash_{M_1} \neg\neg \forall x \, F(x) \supset \forall x \, F(x)$.

Theorems and rules from Exercise 2.40:

Rule E1. If $\vdash A(a)$ then $\vdash \forall x \, A(x)$.

E2. $\vdash \forall x (A(x) \supset A(x))$.

E3. $\vdash \forall x \, A(x) \supset \exists x \, A(x)$.

E4. $\vdash \forall x (A \supset B(x)) \supset (A \supset \forall x \, B(x))$, with x not occurring in A.

E5. $\vdash \forall x_1 \forall x_2 C(x_1, x_2) \supset \forall x_2 \forall x_1 C(x_2, x_1)$.

E6. $\vdash \forall x (A(x) \supset B(x)) \supset (\forall x \, A(x) \supset \forall x \, B(x))$.

D

Exercises on axiomatic derivations

D.1 Hints for Problem 2.7

We give hints for working through the exercises in Problem 2.7 to get some practice in finding axiomatic derivations. In each case, you should write out the proof in detail. Throughout we will freely make use of the rules ∧INTRO and ⊃TRANS from Section 2.5.

E1. Prove $\vdash_{M_0} (A \wedge B) \supset A$. [2.2]

Start with axioms PL5 in the form $A \supset (B \supset A)$. In axiom PL3, replace B by $(B \supset A)$ and C by B. This yields:

$$(A \supset (B \supset A)) \supset [(A \wedge B) \supset ((B \supset A) \wedge B)].$$

By modus ponens you obtain

$$(A \wedge B) \supset ((B \supset A) \wedge B) \qquad (*)$$

If you manage to find an implication that uses the consequent of (*) as antecedent and has consequent A you could use ⊃TRANS and would be done. Looking at the consequent of (*) you will see that it has the form $(B \supset A) \wedge B$. There is an axiom that allows you to write $(B \wedge (B \supset A)) \supset A$—which one? The consequent of (*) and the antecedent of this axiom don't quite match up: the order of the conjuncts is reversed. Find an axiom that helps.

E2. Prove $\vdash_{M_0} A \supset A$. [2.21]

Take a look at E1. One special case of it is $(A \wedge A) \supset A$. See if you can find an axiom that will allow you to reach $A \supset A$ using ⊃TRANS.

E3. Prove $\vdash_{M_0} (A \wedge B) \supset B$. [2.22]

Always keep in mind what you have already achieved. Look at E1: $(A \wedge B) \supset A$. A and B are schematic meta-variables. We can replace them uniformly (A by B, and B by A) to obtain $(B \wedge A) \supset B$. So, if you could prove a conditional of the form $(A \wedge B) \supset (B \wedge A)$ you would be done by $\supset$TRANS. Can you see how to prove $(A \wedge B) \supset (B \wedge A)$?

E4. Prove that if $\vdash_{M_0} B$ then $\vdash_{M_0} A \supset (A \wedge B)$.

Since $\vdash_{M_0} B$ is given to us we should look for a formula $B \supset C$ which is derivable and to which we can apply modus ponens in order to obtain a C that is conducive to our goal. The proof can be obtained in two steps. Prove first

$$\vdash_{M_0} B \supset ((A \wedge A) \supset (B \wedge A))$$

using axioms PL5 and PL3. Use modus ponens to obtain $\vdash_{M_0} (A \wedge A) \supset (B \wedge A)$. Then use axioms PL2 and PL1 to prove $\vdash_{M_0} A \supset (A \wedge B)$.

E5. Prove that if $\vdash_{M_0} B$ then $\vdash_{M_0} A \supset B$.

Since you are given $\vdash_{M_0} B$, E4 applies. Thus, $\vdash_{M_0} A \supset (A \wedge B)$. There is a previous exercise which would allow you to infer $\vdash_{M_0} A \supset B$ by $\supset$TRANS.

E6. Prove $\vdash_{M_0} [A \supset (B \wedge C)] \supset [A \supset (C \wedge B)]$.

Looking at the formula, it is clear that the problem consists in being able to turn $B \wedge C$ into $C \wedge B$. This suggests that axiom PL2 might be of use: $\vdash_{M_0} (B \wedge C) \supset (C \wedge B)$. By E4 we can write:

$$\vdash_{M_0} [A \supset (B \wedge C)] \supset \{[A \supset (B \wedge C)] \wedge [(B \wedge C) \supset (C \wedge B)]\}$$

But the consequent of this formula corresponds to the antecedent of an instance of axiom PL4. Go on to infer the result.

E7. Prove that if $\vdash_{M_0} A \supset (B \supset C)$ and $\vdash_{M_0} A \supset (C \supset D)$, then $\vdash_{M_0} A \supset (B \supset D)$.

Assume $\vdash_{M_0} A \supset (B \supset C)$ and $\vdash_{M_0} A \supset (C \supset D)$. We will start by using the former. Axiom PL3 gives us

$$\vdash_{M_0} [A \supset (B \supset C)] \supset [(A \wedge A) \supset ((B \supset C) \wedge A)].$$

Since $\vdash_{M_0} A \supset (B \supset C)$, by modus ponens we obtain

$$\vdash_{M_0} (A \wedge A) \supset ((B \supset C) \wedge A)) \tag{*}$$

By PL2 and $\supset$TRANS we get $\vdash_{M_0} (A \wedge A) \supset (A \wedge (B \supset C))$. Let us now exploit the second assumption: $\vdash_{M_0} A \supset (C \supset D)$. We can try to obtain something related to

(*) by using axiom PL3 to add the conjunct $(B \supset C)$ to the antecedent and obtain, after applying $\supset$TRANS,

$$\vdash_{M_0} [A \wedge (B \supset C)] \supset [(C \supset D) \wedge (B \supset C)].$$

So by $\supset$TRANS,

$$\vdash_{M_0} (A \wedge A) \supset [(C \supset D) \wedge (B \supset C)].$$

Observe that the conclusion you want is $\vdash_{M_0} A \supset (B \supset D)$. We could get that using $\supset$TRANS if we can prove both $A \supset (A \wedge A)$ and $[(C \supset D) \wedge (B \supset C)] \supset (B \supset D)$.

E8. Prove $\vdash_{M_0} [(A \supset B) \wedge (C \supset D)] \supset [(A \wedge C) \supset (B \wedge D)].$ [2.23]

By E3 we have:

$$\vdash_{M_0} [(A \supset B) \wedge (C \supset D)] \supset (C \supset D)$$

Moreover, axiom PL3 yields

$$\vdash_{M_0} (C \supset D) \supset [(C \wedge B) \supset (D \wedge B)].$$

Apply $\supset$TRANS to obtain

$$\vdash_{M_0} [(A \supset B) \wedge (C \supset D)] \supset [(C \wedge B) \supset (D \wedge B)] \qquad (*)$$

Now prove

$$\vdash_{M_0} [(A \supset B) \wedge (C \supset D)] \supset [(A \wedge C) \supset (B \wedge C)]$$

You should now be able to reverse the order of $B \wedge C$ so as to obtain

$$\vdash_{M_0} [(A \supset B) \wedge (C \supset D)] \supset [(A \wedge C) \supset (C \wedge B)] \qquad (**)$$

But now notice that (**) and (*) are in the appropriate form for an application of E7. Infer

$$\vdash_{M_0} [(A \supset B) \wedge (C \supset D)] \supset [(A \wedge C) \supset (D \wedge B)]$$

Now turn $D \wedge B$ into $B \wedge D$ and complete the proof.

E9. Prove $\vdash_{M_0} [(A \supset B) \wedge (A \supset C)] \supset [A \supset (B \wedge C)].$ [2.24a]

By axiom PL1,

$$\vdash_{M_0} A \supset (A \wedge A)$$

From this we get, by E5:

$$\vdash_{M_0} [(A \supset B) \wedge (A \supset C)] \supset [A \supset (A \wedge A)]$$

On the other hand, E8 gives us:

$$\vdash_{M_0} [(A \supset B) \wedge (A \supset C)] \supset [(A \wedge A) \supset (B \wedge C)]$$

One of our previous exercises should now allow you to reach the desired conclusion from the last two lines.

E10. Prove $\vdash_{M_0} [A \supset (B \wedge C)] \supset [(A \supset B) \wedge (A \supset C)]$. [2.24b]

First, use E1, E3 to obtain:

$$\vdash_{M_0} (B \wedge C) \supset B \qquad\qquad (*)$$
$$\vdash_{M_0} (B \wedge C) \supset C \qquad\qquad (**)$$

Now we apply axiom PL4 twice:

$$\vdash_{M_0} [(A \supset (B \wedge C)) \wedge ((B \wedge C) \supset B)] \supset (A \supset B)$$
$$\vdash_{M_0} [(A \supset (B \wedge C)) \wedge ((B \wedge C) \supset C)] \supset (A \supset C)$$

Now given (*) and (**) use E4 to obtain

$$\vdash_{M_0} (A \supset (B \wedge C)) \supset [(A \supset (B \wedge C)) \wedge ((B \wedge C) \supset B)] \text{ and}$$
$$\vdash_{M_0} (A \supset (B \wedge C)) \supset [(A \supset (B \wedge C)) \wedge ((B \wedge C) \supset C)]$$

Two applications of $\supset$TRANS and an appeal to E9 will yield the theorem.

E11. Prove $\vdash_{M_0} [B \wedge (A \supset C)] \supset [A \supset (B \wedge C)]$. [2.25]

Start with the following instance of axiom PL5: $B \supset (A \supset B)$. Then apply axiom PL3 (replace C by the formula $A \supset C$) to obtain

$$\vdash_{M_0} [B \wedge (A \supset C)] \supset [(A \supset B) \wedge (A \supset C)].$$

From here you should see how to reach the conclusion using a previously given theorem.

E12. Prove $\vdash_{M_0} B \supset [A \supset (A \wedge B)]$. [2.26]

Start with an instance of E2, $\vdash_{M_0} A \supset A$. By E5, $\vdash_{M_0} B \supset (A \supset A)$. But axioms PL5 gives $\vdash_{M_0} B \supset (A \supset B)$. Apply $\wedge$INTRO, and conclude the proof using E9 twice.

E13. Prove $\vdash_{M_0} [A \supset (B \supset C)] \supset [(A \wedge B) \supset C]$. [2.27a]

We try to work with a conditional theorem that displays in the antecedent the formula $A \supset (B \supset C)$ and then try to transform it to obtain the result. By axiom PL3,

$$\vdash_{M_0} [A \supset (B \supset C)] \supset [(A \wedge B) \supset ((B \supset C) \wedge B)].$$

By now you should know the routine for reversing $(B \supset C) \wedge B$ into $B \wedge (B \supset C)$ to obtain

$$\vdash_{M_0} [A \supset (B \supset C)] \supset [(A \wedge B) \supset (B \wedge (B \supset C))].$$

Axiom PL6 yields $\vdash_{M_0} [B \wedge (B \supset C)] \supset C$. Now use E5 to get

$$\vdash_{M_0} [A \supset (B \supset C)] \supset [(B \wedge (B \supset C)) \supset C]$$

Use all you have to justify the required theorem by applying one of our previous exercises.

E14. Prove $\vdash_{M_0} [(A \wedge B) \supset C] \supset [A \supset (B \supset C)]$.   [2.27b]

Use E12 to start with $\vdash_{M_0} A \supset [B \supset (B \wedge A)]$. First reverse $B \wedge A$ into $A \wedge B$. This is a bit tricky since $B \wedge A$ occurs nested inside two conditionals. Now you have $A \supset [B \supset (A \wedge B)]$. Apply E4 (where you replace A by $(A \wedge B) \supset C$). Use the resulting formula and E11 to obtain

$$[(A \wedge B) \supset C] \supset \{A \supset [((A \wedge B) \supset C) \wedge (B \supset (A \wedge B))]\}$$

The last step requires reversing the order of the conjuncts in square brackets and finally an application of the Rule E7.

D.2   Hints for Problem 2.18

The deduction theorem (Theorem 2.16) will help us to find proofs of some of the exercises in Problem 2.18. In all these cases, the hints consist of derived formulas (or assumptions when the deduction theorem should be used). We leave out the justifications. You should give these justifications, fill out the missing steps, and reconstruct the proof in detail. The exercises presuppose a good understanding of the theorems proved in Problem 2.7. Feel free to use the deduction theorem whenever it is convenient.

E1. Prove $\vdash_{M_0} \neg(A \wedge \neg A)$.   [4.13]

Use the following:

$$\vdash_{M_0} [((A \wedge \neg A) \supset A) \wedge ((A \wedge \neg A) \supset \neg A)] \supset \neg(A \wedge \neg A)$$
$$\vdash_{M_0} (A \wedge \neg A) \supset A$$
$$\vdash_{M_0} (A \wedge \neg A) \supset \neg A$$

E2. Prove $\vdash_{M_0} (A \supset B) \supset (\neg B \supset \neg A)$.   [4.2]

Use the deduction theorem by assuming $(A \supset B)$ and $\neg B$. Notice that

$$\vdash_{M_0} [(A \supset B) \wedge (A \supset \neg B)] \supset \neg A.$$

E3. Prove $\vdash_{M_0} (A \supset B) \supset (\neg\neg A \supset \neg\neg B)$.   [4.22]

Use E2 in combination with the deduction theorem.

In addition to working out the proof using the deduction theorem, you might also want to try the following proof, which does not appeal to the deduction theorem: start with two instances of E2 and conclude with an application of $\supset$TRANS.

E4. Prove $\vdash_{M_0} (A \supset \neg B) \supset (B \supset \neg A)$.   [4.21]

Note that $\vdash_{M_0} B \supset (A \supset B)$. Assume $A \supset \neg B$ and B. From these assumptions, obtain

$$(A \supset B) \wedge (A \supset \neg B)$$
$$\neg A$$

Then use the deduction theorem twice.

E5. Prove $\vdash_{M_0} A \supset \neg\neg A$.   [4.3]

$$\vdash_{M_0} (\neg A \supset \neg A) \supset (A \supset \neg\neg A)$$

E6. Prove $\vdash_{M_0} \neg A \supset \neg\neg\neg A$.   [4.31]

This is a simple consequence of one of the previous theorems.

E7. Prove $\vdash_{M_0} \neg\neg\neg A \supset \neg A$.   [4.32]

Use

$$\vdash_{M_0} A \supset \neg\neg A$$
$$\vdash_{M_0} (A \supset \neg\neg A) \supset (\neg\neg\neg A \supset \neg A)$$

E8. Prove $\vdash_{M_0} \neg A \supset (A \supset \neg B)$.   [4.4]

Assume $\neg A$ and A. Under those assumptions, you should derive:

$$B \supset \neg A$$
$$B \supset A$$

Then use PL10.

In addition to working out the proof using the deduction theorem, you might also want to try the following proof, which does not appeal to the deduction theorem: start with an instance of Axiom PL6, then an instance of E4, and finally an application of $\supset$TRANS.

E9. Prove $\vdash_{M_0} \neg\neg(A \wedge B) \supset (\neg\neg A \wedge \neg\neg B)$. [4.61]

Assume $\neg\neg(A \wedge B)$. By E1 and E3 of Problem 2.7 we have:

$$\vdash_{M_0} (A \wedge B) \supset A$$
$$\vdash_{M_0} (A \wedge B) \supset B$$

Use E2 and ∧INTRO to get $\neg\neg A \wedge \neg\neg B$. Finally, use the deduction theorem.

E10. Prove $\vdash_{M_0} \neg\neg(A \vee \neg A)$. [4.8]

Use the following:

$$\vdash_{M_0} \neg(A \vee \neg A) \supset \neg A$$
$$\vdash_{M_0} \neg(A \vee \neg A) \supset \neg\neg A$$

E11. Prove $\vdash_{M_0} (A \supset B) \supset \neg(A \wedge \neg B)$. [4.9]

Assume $A \supset B$.

$\vdash_{M_0} (A \supset B) \supset ((A \wedge \neg B) \supset (B \wedge \neg B))$
$\vdash_{M_0} ((A \wedge \neg B) \supset (B \wedge \neg B)) \supset (((A \wedge \neg B) \supset B) \wedge ((A \wedge \neg B) \supset \neg B))$
$\vdash_{M_0} \neg(A \wedge \neg B)$

E12. Prove $\vdash_{M_0} (A \wedge B) \supset \neg(\neg A \vee \neg B)$. [4.92]

$\vdash_{M_0} \neg A \supset \neg(A \wedge B)$
$\vdash_{M_0} \neg B \supset \neg(A \wedge B)$
$\vdash_{M_0} ((\neg A \supset \neg(A \wedge B)) \wedge (\neg B \supset \neg(A \wedge B))) \supset ((\neg A \vee \neg B) \supset \neg(A \wedge B))$
$\vdash_{M_0} \neg\neg(A \wedge B) \supset \neg(\neg A \vee \neg B)$

We now give a hint for the proof of Lemma 2.19(2):

2. Suppose $\vdash_{M_0} \neg\neg A \supset A$ and $\vdash_{M_0} \neg\neg B \supset B$. Prove $\vdash_{M_0} \neg\neg(A \supset B) \supset (A \supset B)$.

Assume $\neg\neg(A \supset B)$ and A.

$\vdash_{M_0} (A \wedge (A \supset B)) \supset B$
$\vdash_{M_0} (A \wedge (A \supset B)) \supset [A \supset ((A \supset B) \supset B)]$
$\vdash_{M_0} (A \supset B) \supset B$
$\vdash_{M_0} ((A \supset B) \supset B) \supset (\neg\neg(A \supset B) \supset \neg\neg B)$

D.3   Exercises with quantifiers

Here are hints for completing some of the exercises in Problem 2.40.

E1. Prove that if $\vdash_{\mathbf{M}_1} A(a)$ then $\vdash_{\mathbf{M}_1} \forall x\, A(x)$.

For B any sentence that is a theorem of $\mathbf{M}_0$, prove $\vdash B \supset A(a)$.

E2. Prove $\vdash_{\mathbf{M}_1} \forall x\, (A(x) \supset A(x))$.

Use E2 of Problem 2.7.

E3. Prove $\vdash_{\mathbf{M}_1} \forall x\, A(x) \supset \exists x\, A(x)$.

Use the fact that $\vdash_{\mathbf{M}_1} \forall x\, A(x) \supset A(a)$.

E

Natural deduction

E.1 Inference rules

$$\frac{A \quad B}{A \wedge B} \wedge \text{I} \qquad\qquad \frac{A \wedge B}{A} \wedge \text{E} \qquad\qquad\qquad \frac{A \wedge B}{B} \wedge \text{E}$$

$$\begin{array}{c} [A] \\ \vdots \\ \dfrac{B}{A \supset B} \end{array} \supset \text{I} \qquad\qquad\qquad\qquad \frac{A \supset B \quad A}{B} \supset \text{E}$$

$$\frac{A}{A \vee B} \vee \text{I} \qquad\qquad \frac{B}{A \vee B} \vee \text{I} \qquad\qquad \frac{A \vee B \quad \overset{\displaystyle[A]}{\underset{\displaystyle C}{\vdots}} \quad \overset{\displaystyle[B]}{\underset{\displaystyle C}{\vdots}}}{C} \vee \text{E}$$

$$\begin{array}{c} [A] \\ \vdots \\ \dfrac{\bot}{\neg A} \end{array} \neg \text{I} \qquad\qquad\qquad\qquad \frac{\neg A \quad A}{\bot} \neg \text{E}$$

$$\frac{\bot}{D} \bot_J \qquad\qquad\qquad\qquad \begin{array}{c} [\neg A] \\ \vdots \\ \dfrac{\bot}{A} \end{array} \bot_K$$

$$\frac{A(c)}{\forall x\, A(x)} \forall \text{I} \qquad\qquad\qquad\qquad \frac{\forall x\, A(x)}{A(t)} \forall \text{E}$$

$$\frac{A(t)}{\exists x\, A(x)} \exists \text{I} \qquad\qquad\qquad\qquad \frac{\exists x\, A(x) \quad \overset{\displaystyle[A(c)]}{\underset{\displaystyle C}{\vdots}}}{C} \exists \text{E}$$

NK consists of all rules; **NJ** of all except $\bot_K$; **NM** of all except $\bot_J$ and $\bot_K$.

E.2   Conversions

E.2.1   *Simplification conversion*

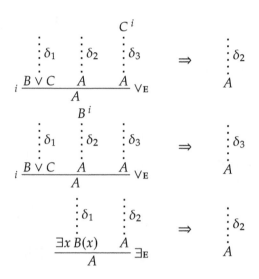

E.2.2   *Permutation conversions*

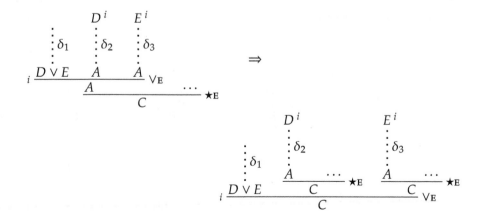

$$
\begin{array}{c}
\begin{array}{cc}
& D(c)^i \\
\vdots\, \delta_1 & \vdots\, \delta_2 \\
\dfrac{\exists x\, D(x) \qquad A}{\dfrac{A}{C}\ \star\mathrm{E}}\ \exists\mathrm{E} \quad \cdots
\end{array}
& \Rightarrow &
\begin{array}{cc}
D(c)^i & \\
\vdots\, \delta_2 & \\
\vdots\, \delta_1 & \\
& \dfrac{A \qquad \cdots}{C}\ \star\mathrm{E} \\
\dfrac{\exists x\, D(x) \qquad \qquad}{C}\ \exists\mathrm{E}
\end{array}
\end{array}
$$

E.2.3   Conversions for $\bot_J$/E-detours

$$
\dfrac{\dfrac{\dfrac{\vdots\,\delta_1}{\bot}}{B \wedge C}\ \bot_J}{B} \quad\Rightarrow\quad \dfrac{\dfrac{\vdots\,\delta_1}{\bot}}{B}\ \bot_J
\qquad\qquad
\dfrac{\dfrac{\dfrac{\vdots\,\delta_1}{\bot}}{B \wedge C}\ \bot_J}{C} \quad\Rightarrow\quad \dfrac{\dfrac{\vdots\,\delta_1}{\bot}}{C}\ \bot_J
$$

$$
\dfrac{\dfrac{\dfrac{\vdots\,\delta_1}{\bot}}{B \supset C}\ \bot_J \qquad \dfrac{\vdots\,\delta_2}{B}}{C}\ \supset\mathrm{E} \quad\Rightarrow\quad \dfrac{\dfrac{\vdots\,\delta_1}{\bot}}{C}\ \bot_J
\qquad\qquad
\dfrac{\dfrac{\dfrac{\vdots\,\delta_1}{\bot}}{\forall x\, B(x)}\ \bot_J}{B(t)}\ \forall\mathrm{E} \quad\Rightarrow\quad \dfrac{\dfrac{\vdots\,\delta_1}{\bot}}{B(t)}\ \bot_J
$$

$$
\dfrac{\dfrac{\dfrac{\vdots\,\delta_1}{\bot}}{B \vee C}\ \bot_J \qquad \dfrac{B^i}{\vdots\,\delta_2}{D} \qquad \dfrac{C^i}{\vdots\,\delta_2}{D}}{D}\ \vee\mathrm{E} \quad\Rightarrow\quad \dfrac{\dfrac{\dfrac{\vdots\,\delta_1}{\bot}}{B}\ \bot_J}{\vdots\,\delta_2}{D}
$$

$$
\dfrac{\dfrac{\vdots\,\delta_1}{\bot}}{\exists x\, B(x)}\ \bot_J \qquad \dfrac{B(a)^i}{\vdots\,\delta_2}{D} \;\;\dfrac{}{C}\ \exists\mathrm{E} \quad\Rightarrow\quad \dfrac{\dfrac{\vdots\,\delta_1}{\bot}}{B(a)}\ \bot_J \; \vdots\,\delta_2 \; D
$$

E.2.4   Conversions for I/E-detours

$$
\dfrac{\dfrac{\dfrac{\vdots\,\delta_1}{B} \quad \dfrac{\vdots\,\delta_2}{C}}{B \wedge C}\ \wedge\mathrm{I}}{B}\ \wedge\mathrm{E} \quad\Rightarrow\quad \dfrac{\vdots\,\delta_1}{B}
\qquad\qquad
\dfrac{\dfrac{\dfrac{\vdots\,\delta_1}{B} \quad \dfrac{\vdots\,\delta_2}{C}}{B \wedge C}\ \wedge\mathrm{I}}{C}\ \wedge\mathrm{E} \quad\Rightarrow\quad \dfrac{\vdots\,\delta_2}{C}
$$

$$
\dfrac{\dfrac{\dfrac{B^1}{\vdots\,\delta_1}{C}}{B \supset C}\ \supset\mathrm{I}^1 \qquad \dfrac{\vdots\,\delta_2}{B}}{C}\ \supset\mathrm{E} \quad\Rightarrow\quad \dfrac{\dfrac{\vdots\,\delta_2}{B}}{\vdots\,\delta_1}{C}
\qquad\qquad
\dfrac{\dfrac{\dfrac{\Gamma}{\vdots\,\delta_1}{B(c)}}{\forall x\, B(x)}\ \forall\mathrm{I}}{B(t)}\ \forall\mathrm{E} \quad\Rightarrow\quad \dfrac{\Gamma}{\vdots\,\delta_1[t/c]}{B(t)}
$$

$$
\cfrac{i\ \cfrac{\cfrac{\vdots\ \delta_1}{B}}{B \vee C}\ {}^{\text{VI}} \quad \cfrac{\overset{B^{\,i}}{\vdots\ \delta_2}}{D} \quad \cfrac{\overset{C^{\,i}}{\vdots\ \delta_3}}{D}}{D}\ {}^{\text{VE}}
\quad\Rightarrow\quad
\cfrac{\cfrac{\cfrac{\vdots\ \delta_1}{B}}{\vdots\ \delta_2}}{D}
$$

$$
\cfrac{i\ \cfrac{\cfrac{\vdots\ \delta_1}{B(t)}}{\exists x\, B(x)}\ {}^{\exists\text{I}} \quad \cfrac{\overset{B(c)^{\,i}}{\vdots\ \delta_2}}{D}}{D}\ {}^{\exists\text{E}}
\quad\Rightarrow\quad
\cfrac{\cfrac{\cfrac{\vdots\ \delta_1}{B(t)}}{\vdots\ \delta_2[t/c]}}{D}
$$

$$
\cfrac{i\ \cfrac{\cfrac{\vdots\ \delta_1}{C}}{B \vee C}\ {}^{\text{VI}} \quad \cfrac{\overset{B^{\,i}}{\vdots\ \delta_2}}{D} \quad \cfrac{\overset{C^{\,i}}{\vdots\ \delta_3}}{D}}{D}\ {}^{\text{VE}}
\quad\Rightarrow\quad
\cfrac{\cfrac{\cfrac{\vdots\ \delta_1}{C}}{\vdots\ \delta_3}}{D}
$$

E.2.5 Andou's classical conversions

Assumes that every $\neg A$ discharged by $\perp_K$ is major premise of $\neg$E.

$$
\cfrac{i\ \cfrac{\cfrac{\cfrac{\neg(B \wedge C)^1 \quad B \wedge C}{\perp}\ {}^{\neg\text{E}}}{\vdots\ \delta_1}}{\cfrac{\perp}{B \wedge C}\ {}^{\perp_K}}\ {}^{\wedge\text{E}}}{B}
\quad\Rightarrow\quad
\cfrac{i\ \cfrac{\cfrac{\neg B^{\,i} \quad \cfrac{B \wedge C}{B}\ {}^{\wedge\text{E}}}{\perp}\ {}^{\neg\text{E}}}{\vdots\ \delta_1}}{\cfrac{\perp}{B}\ {}^{\perp_K}}
$$

$$
\cfrac{\cfrac{i\ \cfrac{\cfrac{\cfrac{\neg(B \supset C)^{\,i} \quad B \supset C}{\perp}\ {}^{\neg\text{E}}}{\vdots\ \delta_1}}{\cfrac{\perp}{B \supset C}\ {}^{\perp_K}} \quad \cfrac{\vdots\ \delta_2}{B}}{C}\ {}^{\supset\text{E}}}{}
\quad\Rightarrow\quad
\cfrac{i\ \cfrac{\cfrac{\neg C^{\,i} \quad \cfrac{B \supset C \quad B}{C}\ {}^{\supset\text{E}}}{\perp}\ {}^{\neg\text{E}}}{\vdots\ \delta_1}}{\cfrac{\perp}{C}\ {}^{\perp_K}}
$$

$$
\cfrac{\cfrac{\vdots\ \delta_1'}{\neg\forall x\, B(x)^{\,i}\quad \forall x\, B(x)}}{\cfrac{\bot}{\cfrac{\vdots\ \delta_1}{i\ \cfrac{\bot}{\forall x\, B(x)}\ \bot_K}{B(t)}\ \forall\text{E}}}\ \neg\text{E}
\qquad\Rightarrow\qquad
\cfrac{\cfrac{\neg B(t)^{\,i}\quad \cfrac{\cfrac{\vdots\ \delta_1'}{\forall x\, B(x)}}{B(t)}\ \forall\text{E}}{\bot}\ \neg\text{E}}{\cfrac{\vdots\ \delta_1}{i\ \cfrac{\bot}{B(t)}\ \bot_K}}
$$

$$
\cfrac{\cfrac{\cfrac{\vdots\ \delta_1'}{\neg(B\vee C)^{\,i}\quad B\vee C}}{\bot}\ \neg\text{E}}{\cfrac{\vdots\ \delta_1}{i\ \cfrac{\bot}{j\ \dfrac{\bot}{B\vee C}}\ \bot_K}}\ \cfrac{B^{\,j}\quad C^{\,j}}{\cfrac{\vdots\ \delta_2\ \vdots\ \delta_3}{D\qquad D}}\quad \cfrac{}{D}\ \vee\text{E}
\qquad\Rightarrow\qquad
\cfrac{\neg D^{\,i}\quad j\ \cfrac{\dfrac{\vdots\,\delta_1'}{B\vee C}\quad \dfrac{B^{\,j}}{\dfrac{\vdots\,\delta_2}{D}}\quad \dfrac{C^{\,j}}{\dfrac{\vdots\,\delta_3}{D}}}{D}\ \vee\text{E}}{\cfrac{\bot}{\cfrac{\vdots\ \delta_1}{i\ \dfrac{\bot}{D}\ \bot_K}}}\ \neg\text{E}
$$

$$
\cfrac{\cfrac{\cfrac{\vdots\ \delta_1'}{\neg\exists x\, B(x)^{\,i}\quad \exists x\, B(x)}}{\bot}\ \neg\text{E}}{\cfrac{\vdots\ \delta_1}{i\ \cfrac{\bot}{j\ \dfrac{\bot}{\exists x\, B(x)}}\ \bot_K}}\ \cfrac{B(c)^{\,j}}{\cfrac{\vdots\ \delta_2}{D}}\quad \cfrac{}{D}\ \exists\text{E}
\qquad\Rightarrow\qquad
\cfrac{\neg D^{\,i}\quad j\ \cfrac{\dfrac{\vdots\,\delta_1'}{\exists x\, B(x)}\quad \dfrac{B(c)^{\,j}}{\dfrac{\vdots\,\delta_2}{D}}}{D}\ \exists\text{E}}{\cfrac{\bot}{\cfrac{\vdots\ \delta_1}{i\ \dfrac{\bot}{D}\ \bot_K}}}\ \neg\text{E}
$$

F

Sequent calculus

Axioms

$$A \Rightarrow A$$

Structural rules

$$\frac{\Gamma \Rightarrow \Theta}{A, \Gamma \Rightarrow \Theta} \text{ WL} \qquad \frac{\Gamma \Rightarrow \Theta}{\Gamma \Rightarrow \Theta, A} \text{ WR}$$

$$\frac{A, A, \Gamma \Rightarrow \Theta}{A, \Gamma \Rightarrow \Theta} \text{ CL} \qquad \frac{\Gamma \Rightarrow \Theta, A, A}{\Gamma \Rightarrow \Theta, A} \text{ CR}$$

$$\frac{\Delta, A, B, \Gamma \Rightarrow \Theta}{\Delta, B, A, \Gamma \Rightarrow \Theta} \text{ IL} \qquad \frac{\Gamma \Rightarrow \Theta, A, B, \Lambda}{\Gamma \Rightarrow \Theta, B, A, \Lambda} \text{ IR}$$

Cut rule

$$\frac{\Gamma \Rightarrow \Theta, A \qquad A, \Delta \Rightarrow \Lambda}{\Gamma, \Delta \Rightarrow \Theta, \Lambda} \text{ CUT}$$

Operational rules

$$\frac{A, \Gamma \Rightarrow \Theta}{A \wedge B, \Gamma \Rightarrow \Theta} \wedge_{L}$$

$$\frac{\Gamma \Rightarrow \Theta, A \quad \Gamma \Rightarrow \Theta, B}{\Gamma \Rightarrow \Theta, A \wedge B} \wedge_{R}$$

$$\frac{B, \Gamma \Rightarrow \Theta}{A \wedge B, \Gamma \Rightarrow \Theta} \wedge_{L}$$

$$\frac{A, \Gamma \Rightarrow \Theta \quad B, \Gamma \Rightarrow \Theta}{A \vee B, \Gamma \Rightarrow \Theta} \vee_{L}$$

$$\frac{\Gamma \Rightarrow \Theta, B}{\Gamma \Rightarrow \Theta, A \vee B} \vee_{R}$$

$$\frac{\Gamma \Rightarrow \Theta, A}{\Gamma \Rightarrow \Theta, A \vee B} \vee_{R}$$

$$\frac{\Gamma \Rightarrow \Theta, A \quad B, \Delta \Rightarrow \Lambda}{A \supset B, \Gamma, \Delta \Rightarrow \Theta, \Lambda} \supset_{L}$$

$$\frac{A, \Gamma \Rightarrow \Theta, B}{\Gamma \Rightarrow \Theta, A \supset B} \supset_{R}$$

$$\frac{\Gamma \Rightarrow \Theta, A}{\neg A, \Gamma \Rightarrow \Theta} \neg_{L}$$

$$\frac{A, \Gamma \Rightarrow \Theta}{\Gamma \Rightarrow \Theta, \neg A} \neg_{R}$$

$$\frac{A(t), \Gamma \Rightarrow \Theta}{\forall x\, A(x), \Gamma \Rightarrow \Theta} \forall_{L}$$

$$\frac{\Gamma \Rightarrow \Theta, A(a)}{\Gamma \Rightarrow \Theta, \forall x\, A(x)} \forall_{R}$$

$$\frac{A(a), \Gamma \Rightarrow \Theta}{\exists x\, A(x), \Gamma \Rightarrow \Theta} \exists_{L}$$

$$\frac{\Gamma \Rightarrow \Theta, A(t)}{\Gamma \Rightarrow \Theta, \exists x\, A(x)} \exists_{R}$$

The rules $\forall_R$ and $\exists_L$ are subject to the eigenvariable condition: the indicated free variable a in the premise does not appear in the lower sequent, i.e., it does not occur in Γ, Θ, $\forall x\, A(x)$, or $\exists x\, A(x)$.

The sequent calculus for classical logic **LK** has all the above rules, and sequents can have any number of formulas in the antecedent or in the succedent. The sequent calculus for intuitionistic logic **LJ** has the same rules, but only sequents with at most one formula in the succedent are allowed. The sequent calculus for minimal logic **LM** also only allows intuitionistic sequents, and excludes the wʀ rule.

G

Outline of the cut-elimination theorem

The proof of the cut-elimination theorem in Chapter 6 is long and involved, and so it is easy to get lost. We need to prove that if a sequent is the conclusion of a derivation which uses CUT then it can also be proved without the use of CUT.

First of all, the proof actually concerns the elimination of the MIX rule rather than the CUT rule. So first we prove that every application of MIX can be obtained using CUT and vice versa (Propositions 6.1 and 6.2). Thus, the theorem to be proved can be obtained from Lemma 6.10 to the effect that if a sequent is the conclusion of a derivation using a single application of MIX in the final step then it is the conclusion of a derivation that does not use MIX.

The proof of the lemma is by double induction and it is "effective." What that means is that we have a mechanical way for transforming a derivation with MIX into one that does not have applications of MIX. Induction is needed to prove that the "simplest" derivations with MIX can be replaced by derivations without MIX. Then we show that the more complex derivations can be reduced to "simpler" derivations.

What do simplicity and complexity mean in this context? There are three sources of complexity at play:

1. The MIX formula to be eliminated can be more or less complex (according to the number of logical symbols occurring in it; this will be captured by the notion of the degree of the MIX and the degree of a proof; Definitions 6.4 and 6.5)

2. The MIX formula to be eliminated has a derivational history that can be longer or shorter (this gives rise to the notion of rank; Definition 6.7)

3. The MIX formula M must figure in the consequent of the left premise $(\Gamma \Rightarrow \Delta, M)$ and in the antecedent $(M, \Theta \Rightarrow \Lambda)$ of the right premise of the MIX. The left rank measures the derivational history of the MIX formula as it occurs in the left premise of the MIX. So the left rank is at least 1. The right rank measures the derivational history of the MIX formula as it occurs in the right premise of the MIX. So the right rank is also at least 1.The rank of a proof is

the sum of the left rank and the right rank and thus it is at least 2. When the rank equals 2, the MIX formula is eliminated as soon as it is introduced, i.e., it is immediately eliminated.

Theorem 6.3 (Cut-elimination theorem, *Hauptsatz*). *The MIX rule is eliminable. That is, given any proof of **LK** − CUT + MIX it is possible to construct another proof with the same end-sequent but that does not contain any MIX.*

First, we introduce the necessary definitions:

1. Introduction of the MIX rule

2. MIX is derivable in **LK** (Proposition 6.1)

3. CUT is derivable in **LK** − CUT + MIX (Proposition 6.2)

4. Degree of a MIX (Definition 6.4)

5. Degree of a proof ending with a MIX (Definition 6.5)

6. Rank of a proof ending with a MIX (Definition 6.7)

As a first step, we modify the proof to ensure it is regular (Proposition 5.23). The Main Lemma (Lemma 6.10) then applies to regular proofs ending in a single MIX inference:

Lemma 6.10 (Main Lemma). *Any regular proof containing a MIX with a single application occurring in its final step can be transformed into a proof with the same end-sequent in which MIX is not applied. A proof that can be transformed in such a way is said to be reducible.*

The proof of the Main Lemma is by double induction on rank and degree of a proof. A proof π_1 is less complex than a proof π_2 iff either $\deg(\pi_1) < \deg(\pi_2)$ or $\deg(\pi_1) = \deg(\pi_2)$ and $\mathrm{rk}(\pi_1) < rk(\pi_2)$

1. **Induction basis.** Every proof π such that $\deg(\pi) = 0$ and $\mathrm{rk}(\pi) = 2$ is reducible.

2. **Inductive step.** If every proof of lower complexity than π is reducible then π is reducible.

The inductive step is divided into two parts: the case when $\mathrm{rk}(\pi) = 2$ and the case when $rk(\pi) > 2$.

A number of cases allow us to remove MIXes directly independently of the degree or the rank. They are covered by three lemmas.

Lemma 6.11. *If π contains a single MIX as its last inference, and one premise of the MIX is an axiom, then there is a proof of the end-sequent without MIX.*

Lemma 6.13. *If π contains a single* MIX *as its last inference, (a) the left premise of the* MIX *is a* WR *or the right premise is a* WL *inference, (b) the* MIX *formula M is the principal formula of the weakening, and (c) M does not occur in the succedent or antecedent, respectively, of the premises of the weakening, then there is a proof of the end-sequent without* MIX.

Lemma 6.15. *If π ends in a* MIX, *and the* MIX *formula occurs in the antecedent of the left premise of the* MIX *or the succedent of the right premise of the* MIX, *then π is reducible.*

Now more in detail.

1. **Induction basis.** The induction basis is covered by

Lemma 6.14. *If $\mathrm{dg}(\pi) = 0$ and $\mathrm{rk}(\pi) = 2$, then there is a proof of the same end-sequent without* MIX.

  A. One of the premises of the MIX is an axiom (Lemma 6.11).
  B. One of the premises is the result of weakening (Lemma 6.13).
  C. One of the premises is the result of contraction or interchange.
  D. One of the premises is the result of an operational rule.

2. **Inductive step.** If every proof of lower complexity than π is reducible then π is reducible.

The inductive step is divided into two parts: the case when $\mathrm{rk}(\pi) = 2$ and the case when $rk(\pi) > 2$.

We begin with $\deg(\pi) > 0$ and $\mathrm{rk}(\pi) = 2$. This is proved in Section 6.4, which consists of a proof of the following:

Lemma 6.16. *Suppose $\mathrm{dg}(\pi) > 0$ and $\mathrm{rk}(\pi) = 2$, and assume that every proof π' with $\mathrm{dg}(\pi') < \mathrm{dg}(\pi)$ is reducible. Then there is a proof of the same end-sequent as π without* MIX.

The proof is divided into cases:

  A. One of the premises of the MIX is an axiom (Lemma 6.11).
  B. One of the premises is the result of weakening (Lemma 6.13).
  C. One of the premises is the result of contraction or interchange.
  D. One of the premises is the result of an operational rule. There are sub-cases depending on the form of the MIX formula: (D1) $A \wedge B$; (D2) $A \vee B$; (D3) $A \supset B$; (D4) $\neg A$; (D5) $\forall x\, F(x)$; (D6) $\exists x\, F(x)$.

The second case is where $rk(\pi) > 2$. This is covered by Section 6.5, which contains the proof of:

Lemma 6.17. *Suppose π is a proof with $\mathrm{rk}(\pi) > 2$, and every proof π' with $\mathrm{dg}(\pi') < \mathrm{dg}(\pi)$ or $\mathrm{dg}(\pi') = \mathrm{dg}(\pi)$ and $\mathrm{rk}(\pi') < \mathrm{rk}(\pi)$ is reducible. Then there is a proof of the same end-sequent as π with equal or lower degree, and lower rank.*

The proof is divided into two general cases:

I The right rank is > 1.
II The right rank equals 1 and the left rank is > 1.

Case I is dealt with by

Lemma 6.18. *Suppose π is a proof with $\mathrm{rk}_r(\pi) > 1$, and every proof π' with $\mathrm{dg}(\pi') < \mathrm{dg}(\pi)$ or $\mathrm{dg}(\pi') = \mathrm{dg}(\pi)$ and $\mathrm{rk}(\pi') < \mathrm{rk}(\pi)$ is reducible. Then there is a proof of the same end-sequent as π with equal or lower degree, and lower rank.*

A. The right premise is the conclusion of wl, il, or cl.
B. The rule that ends with the right premise of the mix is a rule with one premise but not wl, il, or cl. The sub-cases according to which rule ends in the right premise are: (B1) wr; (B2) ir; (B3) cr; (B4) ∧l (two cases); (B5) ∨r; (B6) ¬r (two cases); (B7) ¬l (two cases); (B8) ∀l (two cases); (B9) ∃l (two cases); (B10) ⊃r; (B11) ∀r; (B12) ∃r.
C. The rule that ends with the right sequent of the mix is an operational rule with two premises. The cases are: (C1) ∧r; (C2) ∨l (two cases); (C3) ⊃l (several sub-cases).

Case II is dealt with by:

Lemma 6.19. *Suppose π is a proof with $\mathrm{rk}_r(\pi) = 1$, $\mathrm{rk}_l(\pi) > 1$, and every proof π' with $\mathrm{dg}(\pi') < \mathrm{dg}(\pi)$ or $\mathrm{dg}(\pi') = \mathrm{dg}(\pi)$ and $\mathrm{rk}(\pi') < \mathrm{rk}(\pi)$ is reducible. Then there is a proof of the same end-sequent as π with equal or lower degree, and lower rank.*

The proof is symmetric to that of case I, but special attention needs to be devoted to ⊃l.

Bibliography

Andou, Yuuki (1995), "A normalization-procedure for the first order classical natural deduction with full logical symbols," *Tsukuba Journal of Mathematics*, 19(1), pp. 153–162, DOI: 10.21099/tkbjm/1496162804.

— (2003), "Church-Rosser property of a simple reduction for full first-order classical natural deduction," *Annals of Pure and Applied Logic*, 119(1), pp. 225–237, DOI: 10.1016/S0168-0072(02)00051-9.

Avigad, Jeremy and Richard Zach (2020), "The epsilon calculus," in *Stanford Encyclopedia of Philosophy*, ed. by Edward N. Zalta, Fall 2020, https://plato.stanford.edu/archives/fall2020/entries/epsilon-calculus/.

Bernays, Paul (1918), "Beiträge zur axiomatischen Behandlung des Logik-Kalküls," Habilitationsschrift, Universität Göttingen. Bernays Nachlaß, WHS, ETH Zürich Archive, publ. as "Beiträge zur axiomatischen Behandlung des Logik-Kalküls," in Ewald and Sieg (2013), pp. 222–271.

Bimbó, Katalin (2014), *Proof Theory: Sequent Calculi and Related Formalisms*, Boca Raton: CRC Press.

Boolos, George (1984), "Don't eliminate cut," *Journal of Philosophical Logic*, 13(4), pp. 373–378, DOI: 10.1007/BF00247711.

Boričić, Branislav R. (1985), "On sequence-conclusion natural deduction systems," *Journal of Philosophical Logic*, 14(4), pp. 359–377, DOI: 10.1007/BF00649481.

Borisavljević, Mirjana (2003), "Two measures for proving Gentzen's Hauptsatz without mix," *Archive for Mathematical Logic*, 42(4), pp. 371–387, DOI: 10.1007/s00153-002-0155-x.

Buss, Samuel R., ed. (1998), *Handbook of Proof Theory*, Amsterdam: Elsevier.

Cagnoni, Donatella (1977), "A note on the elimination rules," *Journal of Philosophical Logic*, 6(1), pp. 269–281, DOI: 10.1007/BF00262062.

Cellucci, Carlo (1992), "Existential instantiation and normalization in sequent natural deduction," *Annals of Pure and Applied Logic*, 58(2), pp. 111–148, DOI: 10.1016/0168-0072(92)90002-H.

Curry, Haskell Brooks (1963), *Foundations of Mathematical Logic*, New York, NY: McGraw-Hill.

Curry, Haskell Brooks, J. Roger Hindley, and Jonathan Paul Seldin (1972), *Combinatory Logic*, Studies in Logic and the Foundations of Mathematics, 65, Amsterdam: North-Holland, vol. 2.

David, René and Karim Nour (2003), "A short proof of the strong normalization of classical natural deduction with disjunction," *The Journal of Symbolic Logic*, 68(4), pp. 1277–1288, DOI: 10.2178/jsl/1067620187.

Diller, Justus (2019), *Functional Interpretations: From the Dialectica Interpretation to Functional Interpretations of Analysis and Set Theory*, World Scientific.

Dummett, Michael (2000), *Elements of Intuitionism*, 2nd ed., Oxford Logic Guides, 39, Oxford: Oxford University Press.

Ewald, William Bragg, ed. (1996), *From Kant to Hilbert: A Source Book in the Foundations of Mathematics*, Oxford: Oxford University Press, vol. 2.

Ewald, William Bragg and Wilfried Sieg, eds. (2013), *David Hilbert's Lectures on the Foundations of Arithmetic and Logic, 1917–1933*, Berlin and Heidelberg: Springer.

Feferman, Solomon (1964), "Systems of predicative analysis," *Journal of Symbolic Logic*, 29(1), pp. 1–30, DOI: 10.2307/2269764.

— (1988), "Hilbert's Program relativized: proof-theoretical and foundational reductions," *Journal of Symbolic Logic*, 53(2), pp. 364–384, DOI: 10.2307/2274509.

Frege, Gottlob (1879), *Begriffsschrift, eine der arithmetischen nachgebildete Formelsprache des reinen Denkens*, Halle: Nebert, trans. as "*Begriffschrift*, a formula language, modeled upon that of arithmetic, for pure thought," in van Heijenoort (1967), pp. 1–82.

Gentzen, Gerhard (1933), "Über das Verhältnis zwischen intuitionistischer und klassischer Logik," Galley proofs, pub. as "Über das Verhältnis zwischen intuitionistischer und klassischer Logik," *Archiv für mathematische Logik und Grundlagenforschung*, 16 (1974), pp. 119–132, DOI: 10.1007/BF02015371, trans. as "On the relation between intuitionist and classical arithmetic," in Gentzen (1969), pp. 53–67.

— (1935a), "Die Widerspruchsfreiheit der reinen Zahlentheorie," Galley proofs, pub. as "Der erste Widerspruchsfreiheitsbeweis für die klassische Zahlentheorie," *Archiv für mathematische Logik und Grundlagenforschung*, 16(3–4) (1974), pp. 97–118, DOI: 10.1007/BF02015370, trans. as "The consistency of elementary number theory," in Gentzen (1969), pp. 132–213.

— (1935b), "Untersuchungen über das logische Schließen I," *Mathematische Zeitschrift*, 39(1), pp. 176–210, DOI: 10.1007/BF01201353, trans. as "Investigations into logical deduction," in Gentzen (1969), pp. 68–131.

— (1935c), "Untersuchungen über das logische Schließen II," *Mathematische Zeitschrift*, 39(1), pp. 405–431, DOI: 10.1007/BF01201363.

— (1936), "Die Widerspruchsfreiheit der reinen Zahlentheorie," *Mathematische Annalen*, 112(1), pp. 493–565, DOI: 10.1007/BF01565428, trans. as "The consistency of elementary number theory," in Gentzen (1969), pp. 132–213.

— (1938), "Neue Fassung des Widerspruchsfreiheitsbeweises für die reine Zahlentheorie," *Forschungen zur Logik und zur Grundlegung der exakten Wissenschaften*, neue Folge, 4, pp. 19–44, trans. as "New version of the consistency proof for elementary number theory," in Gentzen (1969), pp. 252–286.

— (1943), "Beweisbarkeit und Unbeweisbarkeit von Anfangsfällen der transfiniten Induktion in der reinen Zahlentheorie," *Mathematische Annalen*, 119(1), pp. 140–161, DOI: 10.1007/BF01564760, trans. as "Provability and nonprovability of restricted transfinite induction in elementary number theory," in Gentzen (1969), pp. 287–308.

— (1969), *The Collected Papers of Gerhard Gentzen*, ed. by Manfred E. Szabo, Amsterdam: North-Holland.

— (2008), "Gentzen's proof of normalization for natural deduction," trans. by Jan von Plato, *Bulletin of Symbolic Logic*, 14(2), pp. 240–257, DOI: 10.2178/bsl/1208442829.

Girard, Jean-Yves (1987), *Proof Theory and Logical Complexity*, Studies in Proof Theory, 1, Naples: Bibliopolis.

Glivenko, Valerii (1929), "Sur quelques points de la logique de M. Brouwer," *Académie Royale de Belgique, Bulletin*, 15, pp. 183–188, trans. as "On some points of the logic of Mr. Brouwer," in Mancosu (1998a), pp. 301–305.

Gödel, Kurt (1931), "Über formal unentscheidbare Sätze der *Principia Mathematica* und verwandter Systeme I," *Monatshefte für Mathematik und Physik*, 38, pp. 173–198, repr. and trans. as "On formally undecidable propositions of *Principia mathematica* and related systems I," in Gödel (1986), pp. 144–195.

— (1933), "Zur intuitionistischen Arithmetik und Zahlentheorie," *Ergebnisse eines mathematisches Kolloquiums*, 4, pp. 34–38, repr. and trans. as "On intuitionistic arithmetic and number theory," in Gödel (1986), pp. 286–295.

— (1958), "Über eine bisher noch nicht benütze Erweiterung des finiten Standpunktes," *Dialectica*, 12(3–4), pp. 280–287, DOI: 10.1111/j.1746-8361.1958.tb01464.x, repr. and trans. as "On a hitherto unutilized extension of the finitary standpoint," in *Collected works*, ed. by Solomon Feferman et al., Collected Works, 2, Oxford: Oxford University Press, 1990, pp. 217–251.

— (1986), *Collected Works: Publications 1929–1936*, ed. by Solomon Feferman, Collected Works, 1, Oxford: Oxford University Press.

Goodstein, Reuben Louis (1944), "On the restricted ordinal theorem," *The Journal of Symbolic Logic*, 9(2), pp. 33–41, DOI: 10.2307/2268019.

Hendricks, Vincent F., Stig Andur Pedersen, and Klaus Frovin Jørgensen, eds. (2000), *Proof Theory: History and Philosophical Significance*, Dordrecht: Springer, DOI: 10.1007/978-94-017-2796-9.

Herbrand, Jaques (1930), *Recherches sur la théorie de la démonstration*, Doctoral Dissertation, University of Paris, trans. as "Investigations in proof theory: the properties of true propositions," in van Heijenoort (1967), pp. 524–581.

— (1931), "Sur la non-contradiction de l'arithmétique," *Journal für die Reine und Angewandte Mathematik*, 166, pp. 1–8, DOI: 10.1515/crll.1932.166.1, trans. as "On the consistency of arithmetic," in van Heijenoort (1967), pp. 618–628.

Hertz, Paul (1929), "Über Axiomensysteme für beliebige Satzsysteme," *Mathematische Annalen*, 101(1), pp. 457–514, DOI: 10.1007/BF01454856.

Heyting, Arend (1930), "Die formalen Regeln der intuitionistischen Logik," *Sitzungsberichte der Preussischen Akademie der Wissenschaften*, pp. 42–56, trans. as "The formal rules of intuitionistic logic," in Mancosu (1998a), pp. 311–327.

— (1956), *Intuitionism: An Introduction*, Amsterdam: North-Holland.

Hilbert, David (1899), "Grundlagen der Geometrie," in *Festschrift zur Feier der Enthüllung des Gauss-Weber-Denkmals in Göttingen*, 1st ed., Leipzig: Teubner, pp. 1–92, trans. as *Foundations of Geometry*, Chicago: Open Court, 1902.

— (1900), "Mathematische Probleme," *Nachrichten von der Königlichen Gesellschaft der Wissenschaften zu Göttingen, Math.-Phys. Klasse*, pp. 253–297, http://reso lver.sub.uni-goettingen.de/purl?GDZPPN002498863, trans. as "Mathematical problems," in Ewald (1996), vol. 2, pp. 1096–1105.

— (1905), "Über die Grundlagen der Logik und der Arithmetik," in *Verhandlungen des dritten Internationalen Mathematiker-Kongresses in Heidelberg vom 8. bis 13. August 1904*, ed. by A. Krazer, Leipzig: Teubner, pp. 174–185, trans. as "On the foundations of logic and arithmetic," in van Heijenoort (1967), pp. 129–138.

— (1917), "Axiomatisches Denken," *Mathematische Annalen*, 78(1–4), pp. 405–415, DOI: 10.1007/BF01457115, trans. as "Axiomatic thought," in Ewald (1996), vol. 2, pp. 1105–1115.

— (1918), "Prinzipien der Mathematik," Lecture notes by Paul Bernays. Winter-Semester 1917–18. Typescript. Bibliothek, Mathematisches Institut, Universität Göttingen. publ. "Prinzipien der Mathematik," in Ewald and Sieg (2013).

— (1922), "Neubegründung der Mathematik: Erste Mitteilung," *Abhandlungen aus dem Seminar der Hamburgischen Universität*, 1, pp. 157–177, trans. as "The new grounding of mathematics: first report," in Mancosu (1998a), pp. 198–214.

— (1923), "Die logischen Grundlagen der Mathematik," *Mathematische Annalen*, 88(1–2), pp. 151–165, DOI: 10.1007/BF01448445, trans. as "The logical foundations of mathematics," in Ewald (1996), vol. 2, pp. 1134–1148.

Hilbert, David and Wilhelm Ackermann (1928), *Grundzüge der theoretischen Logik*, 1st ed., Berlin: Springer, repr. "Grundzüge der theoretischen Logik," in Ewald and Sieg (2013), pp. 806–916.

Iemhoff, Rosalie (2020), "Intuitionism in the philosophy of mathematics," in *The Stanford Encyclopedia of Philosophy*, ed. by Edward N. Zalta, Fall 2020, https://plato.stanford.edu/entries/intuitionism/.

Jaśkowski, Stanisław (1934), *On the Rules of Suppositions in Formal Logic*, Studia Logica, 1, Warsaw: Seminarjum Filozoficzne. Wydz. Matematyczno-Przyrodniczy UW.

Johansson, Ingebrigt (1937), "Der Minimalkalkül, ein reduzierter intuitionistischer Formalismus," *Compositio Mathematica*, 4, pp. 119–136, http://www.numdam.org/item/CM_1937__4__119_0/.

Kirby, Laurie and Jeff Paris (1982), "Accessible independence results for Peano arithmetic," *Bulletin of the London Mathematical Society*, 14(4), pp. 285–293, DOI: 10.1112/blms/14.4.285.

Leivant, Daniel (1979), "Assumption classes in natural deduction," *Zeitschrift für mathematische Logik und Grundlagen der Mathematik*, 25(1–2), pp. 1–4, DOI: 10.1002/malq.19790250102.

Mancosu, Paolo, ed. (1998a), *From Brouwer to Hilbert: The Debate on the Foundations of Mathematics in the 1920s*, New York and Oxford: Oxford University Press.

— (1998b), "Hilbert and Bernays on metamathematics," in Mancosu (1998a), pp. 149–188.

Mancosu, Paolo and Richard Zach (2015), "Heinrich Behmann's 1921 lecture on the decision problem and the algebra of logic," *Bulletin of Symbolic Logic*, 21(2), pp. 164–187, DOI: 10.1017/bsl.2015.10.

Mancosu, Paolo, Richard Zach, and Calixto Badesa (2009), "The development of mathematical logic from Russell to Tarski: 1900–1935," in *The Development of Modern Logic*, ed. by Leila Haaparanta, New York and Oxford: Oxford University Press, pp. 324–478, DOI: 10.1093/acprof:oso/9780195137316.003.0029.

Menzler-Trott, Eckart (2016), *Logic's Lost Genius: The Life of Gerhard Gentzen*, History of Mathematics, 33, American Mathematical Society.

Negri, Sara and Jan von Plato (2001), *Structural Proof Theory*, Cambridge: Cambridge University Press.

Orevkov, Vladimir P. (1982), "Lower bounds for increasing complexity of derivations after cut elimination," *Journal of Soviet Mathematics*, 20(4), pp. 2337–2350, DOI: 10.1007/BF01629444.

— (1993), *Complexity of Proofs and Their Transformations in Axiomatic Theories*, Translations of Mathematical Monographs, 128, Providence, RI: American Mathematical Society.

Parigot, Michel (1992), "$\lambda\mu$-calculus: an algorithmic interpretation of classical natural deduction," in *Logic programming and automated reasoning*, International Conference on Logic for Programming Artificial Intelligence and Reasoning, Lecture Notes in Computer Science, 624, Berlin, Heidelberg: Springer, pp. 190–201, DOI: 10.1007/BFb0013061.

Pierce, Benjamin C. (2002), *Types and Programming Languages*, Cambridge, MA: MIT Press.

Pohlers, Wolfram (2009), *Proof Theory: The First Step into Impredicativity*, Berlin, Heidelberg: Springer, DOI: 10.1007/978-3-540-69319-2.

Prawitz, Dag (1965), *Natural Deduction: A Proof-Theoretical Study*, Stockholm Studies in Philosophy, 3, Stockholm: Almqvist & Wiksell.

— (1971), "Ideas and results in proof theory," in *Proceedings of the Second Scandinavian Logic Symposium*, ed. by Jens E. Fenstad, Studies in Logic and the Foundations of Mathematics, 63, Amsterdam: North-Holland, pp. 235–307, DOI: 10.1016/S0049-237X(08)70849-8.

Raggio, Andrés R. (1965), "Gentzen's Hauptsatz for the systems NI and NK," *Logique et Analyse*, 8(30), pp. 91–100.

Rathjen, Michael and Wilfried Sieg (2020), "Proof theory," in *The Stanford Encyclopedia of Philosophy*, ed. by Edward N. Zalta, Fall 2020, https://plato.stanford.edu/archives/fall2020/entries/proof-theory/.

Schütte, Kurt (1965), "Predicative well-orderings," in *Studies in Logic and the Foundations of Mathematics*, ed. by John N. Crossley and Michael A. E. Dummett, Formal Systems and Recursive Functions, 40, Amsterdam: North-Holland, pp. 280–303, DOI: 10.1016/S0049-237X(08)71694-X.

— (1977), *Proof Theory*, Berlin and New York: Springer.

Seldin, Jonathan Paul (1989), "Normalization and excluded middle. I," *Studia Logica*, 48(2), pp. 193–217, DOI: 10.1007/BF02770512.

Sørensen, Morten Heine and Pawel Urzyczyn (2006), *Lectures on the Curry-Howard Isomorphism*, Studies in Logic and the Foundations of Mathematics, 149, New York: Elsevier.

Stålmarck, Gunnar (1991), "Normalization theorems for full first order classical natural deduction," *The Journal of Symbolic Logic*, 56(1), pp. 129–149, DOI: 10.2307/2274910.

Statman, Richard (1974), *Structural Complexity of Proofs*, PhD Thesis, Stanford, CA: Stanford University.

Takeuti, Gaisi (1967), "Consistency proofs of subsystems of classical analysis," *Annals of Mathematics*, 86(2), pp. 299–348, DOI: 10.2307/1970691.

— (1987), *Proof Theory*, 2nd ed., Studies in Logic, 81, Amsterdam: North-Holland.

Tarski, Alfred, Andrzej Mostowski, and Raphael M. Robinson (1953), *Undecidable Theories*, Studies in Logic and the Foundations of Mathematics, Amsterdam: North-Holland.

Tennant, Neil (1978), *Natural Logic*, Edinburgh: Edinburgh University Press.

Troelstra, Anne Sjerp (1990), "On the early history of intuitionistic logic," in *Mathematical Logic*, ed. by Petio P. Petkov, New York and London: Plenum Press, pp. 3–17.

Troelstra, Anne Sjerp and Helmut Schwichtenberg (2000), *Basic Proof Theory*, 2nd ed., Cambridge: Cambridge University Press.

Van Atten, Mark (2017), "The development of intuitionistic logic," in *The Stanford Encyclopedia of Philosophy*, ed. by Edward N. Zalta, Winter 2017, https://plato.stanford.edu/archives/win2017/entries/intuitionistic-logic-development/.

Van Heijenoort, Jean, ed. (1967), *From Frege to Gödel: A Source Book in Mathematical Logic, 1879–1931*, Cambridge, MA: Harvard University Press.

Von Neumann, Johann (1927), "Zur Hilbertschen Beweistheorie," *Mathematische Zeitschrift*, 26(1), pp. 1–46, DOI: 10.1007/BF01475439.

Von Plato, Jan (2009), "Gentzen's logic," in *Logic from Russell to Church*, ed. by Dov M. Gabbay and John Woods, Handbook of the History of Logic, 5, Amsterdam: North-Holland, pp. 667–721, DOI: 10.1016/S1874-5857(09)70017-2.

— (2018), "The development of proof theory," in *The Stanford Encyclopedia of Philosophy*, ed. by Edward N. Zalta, https://plato.stanford.edu/archives/win2018/entries/proof-theory-development/.

Wadler, Philip (2015), "Propositions as types," *Communications of the ACM*, 58(12), pp. 75–84, DOI: 10.1145/2699407.

Weyl, Hermann (1918), *Das Kontinuum*, Leipzig: Veit, trans. as *The Continuum*, New York: Dover, 1994.

Whitehead, Alfred North and Bertrand Russell (1910–1913), *Principia Mathematica*, 3 vols., Cambridge: Cambridge University Press.

Zach, Richard (1999), "Completeness before Post: Bernays, Hilbert, and the development of propositional logic," *Bulletin of Symbolic Logic*, 5(3), pp. 331–366, DOI: 10.2307/421184.

— (2003), "The practice of finitism: epsilon calculus and consistency proofs in Hilbert's Program," *Synthese*, 137(1/2), pp. 211–259, DOI: 10.1023/A:1026247421 383.

— (2004), "Hilbert's '*Verunglückter Beweis*', the first epsilon theorem, and consistency proofs," *History and Philosophy of Logic*, 25(2), pp. 79–94, DOI: 10.1080/01445340310001606930.

— (2019a), "Hilbert's program," in *Stanford Encyclopedia of Philosophy*, ed. by Edward N. Zalta, Fall 2019, https://plato.stanford.edu/archives/fall2019/entries/hilbert-program/.

— (2019b), *Sets, Logic, Computation: An Open Introduction to Metalogic*, https://slc.openlogicproject.org/.

— (2019c), "The significance of the Curry-Howard isomorphism," in *Philosophy of Logic and Mathematics. Proceedings of the 41st International Ludwig Wittgenstein Symposium*, ed. by Gabriele M. Mras, Paul Weingartner, and Bernhard Ritter, Publications of the Austrian Ludwig Wittgenstein Society, New Series, 26, Berlin: De Gruyter, pp. 313–325, DOI: 10.1515/9783110657883-018.

— (2021), "Cut elimination and normalization for generalized single and multi-conclusion sequent and natural deduction calculi," *Review of Symbolic Logic*, DOI: 10.1017/S1755020320000015.

Zimmermann, Ernst (2002), "Peirce's rule in natural deduction," *Theoretical Computer Science*, 275(1), pp. 561–574, DOI: 10.1016/S0304-3975(01)00296-1.

Index